CHEMICAL REACTION ENGINEERING

Second Edition

OCTAVE LEVENSPIEL

Department of Chemical Engineering
Oregon State University

JOHN WILEY & SONS, INC.

New York London Sydney Toronto

WILEY INTERNATIONAL EDITION

Library of Congress Catalog Card Number: 72-178146

ISBN 0-471-53016-6

Printed in the United States of America.

10 9 8 7 6 5 4

PREFACE

Chemical reaction engineering is that engineering activity concerned with the exploitation of chemical reactions on a commercial scale. Its goal is the successful design and operation of chemical reactors, and probably more than any other activity it sets chemical engineering apart as a distinct branch of the engineering profession.

In a typical situation the engineer is faced with a host of questions: what information is needed to attack a problem, how best to obtain it, and then how to select a reasonable design from the many available alternatives. The purpose of this book is to teach how to answer these questions reliably and wisely. To do this I emphasize qualitative arguments, simple design methods, graphical procedures, and frequent comparison of capabilities of the major reactor types. This approach should help develop a strong intuitive sense for good design which can then guide and reinforce the formal methods.

This is a teaching book; thus, simple ideas are treated first, and are then extended to the more complex. Also, emphasis is placed throughout on the development of a common design strategy for all systems, homogeneous and heterogeneous.

This is an introductory book. The pace is leisurely, and where needed, time is taken to consider why certain assumptions are made, to discuss why an alternative approach is not used, and to indicate the limitations of the treatment when applied to real situations. Although the mathematical level is not particularly difficult (elementary calculus and the linear first-order differential equation is all that is needed), this does not mean that the ideas and concepts being taught are particularly simple. To develop new ways of thinking and new intuitions is not easy.

I feel that problem-solving, that process of applying the concepts to new situations, is essential to learning. Consequently, I include a large number of problems. Some are quite simple and only require qualitative reasoning, others complement the text material, while a few challenge the brave soul to venture beyond. Light relief is provided by a sprinkling of unconventional problems which serve to show that the methods being developed here are rather general and can be used in quite unrelated situations.

This brings me to this new edition. First of all, I should say that in spirit it follows the original, and I try to keep things simple; however, as is inevitable with all second editions, it is longer than the original. (I wonder if this could be another example of that insidious second law at work.) In any case, the first half of this edition covers substantially the same material as the original, although there are all sorts of minor changes sprinkled throughout as well as a few major changes. The last half of the book, the part dealing with nonideal flow and heterogeneous systems, has been completely rewritten and significantly expanded; however, the level of the presentation remains unchanged.

In this new edition I have removed text material which I have regularly bypassed in my teaching. I have also discarded about 80 problems which I found unsuited for one reason or another. These problems have been replaced by about 160 new and mostly simple problems. I strongly favor these mini-problems as an efficient teaching device.

I am often asked how and what to teach at this and that level. Here are my thoughts: I find that Chapters 1–8, 11, and 14 form a reasonable basis for undergraduate instruction; and if only one course is given, rather than thoroughly covering fewer chapters, I'd prefer a briefer treatment of all of them. For example, in my teaching of a one-term course to students with a physical chemistry background I leap over Chapters 1, 2, and 3 in a lecture or two plus a few representative problems mainly as a means of presenting definitions of terms. Then since Chapters 6, 7, and 8 reinforce the ideas of Chapter 5, I only very briefly outline this material and right away plunge into problem-solving. Finally, in Chapter 14 I leave out the sections on product distribution and fluidized bed design.

For the graduate program Chapter 9 onward seems suitable. Although the material is not particularly difficult mathematically, still, there are so many tempting sidelines to explore and new concepts to absorb, that I find it impractical to try to cover this material in less than two terms.

The above plan happens to suit me; however, each teacher should develop his own. Actually, I've read how this material has been successfully used in a programmed learning course. Perhaps more should be done along this line.

With pleasure I come to the acknowledgments. First, I want to express appreciation to Tom Fitzgerald, a helpful colleague and fountain of ideas, both good and bad. I hope I've been able to discriminate between them. Thanks also to Milorad Dudukovic of Belgrade for his extensive review and criticism of the manuscript, to Moto Suzuki of Tokyo for his useful comments on Chapter 10 and to Soon Jai Khang of Seoul for his help with Chapter 15. To the many teachers and fellow engineers who have written me over the years with helpful suggestions, my appreciation. I hope you won't be disappointed with this new edition. Finally, especial thanks to Mary Jo, who has been helpful in so many ways.

Otter Rock, Oregon OCTAVE LEVENSPIEL

CONTENTS

11 INTRODUCTION TO DESIGN FOR HETEROGENEOUS REACTING SYSTEMS 349

12 FLUID-PARTICLE REACTIONS 357

13 FLUID-FLUID REACTIONS 409

NOTATION

Symbols and constants which are defined and used locally are not included here. Cgs units are given to show the dimensions of the symbols.

$a, b, \ldots, r, s, \ldots$ — stoichiometric coefficients for reacting substances A, B, ..., R, S, ...

a — interfacial area per volume of tower, cm^2/cm^3; Ch. 13 only

a_i, a_s — interfacial area per volume of liquid, cm^2/cm^3; Ch. 13 only

a — activity of a catalyst pellet, dimensionless; see Eq. 15.4; Ch. 15 only

A, B, ... — reactants

C_A — concentration of A, mol/cm^3

C_p — molar specific heat, $cal/mol \cdot °C$

C curve — tracer response to an ideal pulse input; see Eq. 9.3

$\mathbf{C_p}$ — specific heat of fluid stream per mole of key reactant, usually A, $cal/mol \cdot A \cdot °C$

$\mathbf{C'_p}$ — specific heat of unreacted feed stream per mole of key reactant, $cal/mol \cdot °C$

$\mathbf{C''_p}$ — specific heat of product stream when key reactant is completely converted, $cal/mol \cdot °C$

d — diameter, cm

d — order of deactivation; see p. 542

d_p — diameter of particle, cm

d_t — diameter of tube, cm

D — dispersion or axial dispersion coefficient, cm^2/sec; see p. 272

$\mathscr{D}$ — molecular diffusion coefficient, cm^2/sec

$\mathscr{D}_e$	effective diffusion coefficient in a porous structure, cm^2/sec
E	activation energy; see Eq. 2.32; for units see footnote, p. 22
E	enhancement factor in mass transfer with reaction; see p. 414
E	enzyme
$\mathbf{E}$	exit age distribution function, sec^{-1}; see p. 255
$\mathscr{E}$	effectiveness factor, dimensionless; see Eq. 14.11
f	fugacity, atm; Ch. 8 only
f	volume fraction of a phase; Ch. 13 only
F	volumetric feed rate of solids, or mass feed rate for negligible density change, cm^3/sec or gm/sec; Ch. 12 only
F_A	molar flow rate of substance A, mol/sec
$F(R_i)$	feed rate of solids of size R_i, cm^3/sec or gm/sec; Ch. 12 only
F curve	tracer response to a unit step input, fraction of tracer in exit stream; see p. 257
G	$= G' p_U/\pi$, upward molar flow rate of inerts in the gas phase per unit cross section of tower, $mol/cm^2 \cdot sec$; Ch. 13 only
G'	upward molar flow rate of all gas per unit cross section of tower, $mol/cm^2 \cdot sec$; Ch. 13 only
$\Delta G°$	standard free energy of a reaction for the stoichiometry as written, cal; see Eq. 1.2 or 8.9
h	height of absorption column, cm
H	enthapy, cal
H	phase distribution coefficient; for gas phase systems $H = p/C$, $atm \cdot cm^3/mol$, Henry's law constant; Ch. 13 only
$\mathbf{H}'$	enthalpy of unreacted feed stream per mole of entering key reactant A, cal/mol A
$\mathbf{H}''$	enthalpy of product stream if key reactant is completely converted, cal/mol
ΔH_r	heat of reaction at temperature T for the stoichiometry as written, cal; see Eq. 1.1 or 8.1
$\Delta \mathbf{H}_r$	heat of reaction per mole of key reactant, cal/mol

I	coalescence parameter; see Eq. 10.14
J	intensity of segregation; see Eq. 10.13
k	reaction rate constant, $(\text{mol/cm}^3)^{1-n}/\text{sec}$; see Eq. 2.6
k_d	rate constant for the deactivation reaction, Ch. 15 only
k_{eff}	effective thermal conductivity, cal/cm·sec·°C; Ch. 14 only
k_g	mass transfer coefficient, cm/sec; see Eq. 12.4. In Ch. 13 k_g refers specifically to the gas phase, $\text{mol/cm}^2\cdot\text{sec}\cdot\text{atm}$; see Eq. 13.2
k_l	mass transfer coefficient in the liquid phase, cm/sec; see Eq. 13.2
k_s	first order reaction rate constant based on unit surface, cm/sec; see Ch. 11 or p. 470
K	equilibrium constant of a reaction for the stoichiometry as written, dimensionless; see Eq. 1.2 or 8.9
K_{bc}	bubble-cloud interchange coefficient in fluidized beds, sec^{-1}; see Eq. 9.76
K_{ce}	cloud-emulsion interchange coefficient in fluidized beds, sec^{-1}; see Eq. 9.77
K_f, K_p, K_y, K_C	see Eq. 8.10
K_g	overall mass transfer coefficient on the gas phase basis, $\text{mol/cm}^2\cdot\text{sec}\cdot\text{atm}$
l	length, cm
L	length of reactor, cm
L	$= L'C_U/C_T$, downward molar flow rate of liquid inerts per unit cross section of tower, $\text{mol/cm}^2\cdot\text{sec}$; Ch. 13 only
L'	downward molar flow rate of all liquid per unit cross section of tower, $\text{mol/cm}^2\cdot\text{sec}$; Ch. 13 only
mL	$= L\sqrt{k/\mathscr{D}}$, Thiele modulus, dimensionless; Chs. 14 and 15
M	film conversion parameter, dimensionless; see Eq. 14.25
n	order of reaction; see Eq. 2.5
N	number of equal-size mixed reactors in series
N_A	moles of component A

P_A	partial pressure of component A, atm
P_A^*	partial pressure of A in gas which would be in equilibrium with C_A in liquid; hence $P_A^* = H_A C_A$, atm
P	poison
Q	heat transfer rate to reacting system, cal/sec
Q	flux of material, mol/cm²·sec; Chs. 11 and 12 only
Q	heat added per mole of key component which enters, cal/mol
r	radial position within a particle
r_A	rate of reaction based on volume of fluid, moles A formed/cm³·sec; see Eq. 1.3
r'_A	rate of reaction based on unit mass of catalyst, moles A formed/gm·sec; see Eq. 1.4 or 14.34
r''_A	rate of reaction based on unit surface, moles A formed/cm²·sec, see Eq. 1.5
r'''_A	rate of reaction based on unit volume of solid, moles A formed/cm³·sec; see Eq. 1.6
r''''_A	rate of reaction based on unit volume of reactor, moles A formed/cm³·sec; see Eq. 1.7
r_c	radius of unreacted core, cm
R	radius of particle
R	ideal gas law constant, = 1.98 cal/gm-mol·°K = 1.98 Btu/lb-mol·°R = 82.06 cm³·atm/gm-mol·°K
R, S, . . .	products of reaction
R	recycle ratio; see Eq. 6.15
s	space velocity, sec⁻¹; see Eq. 5.7
S	surface, cm²
t	time, sec
$\bar{t}$	$= V/v$, reactor holding time or mean residence time of fluid in a flow reactor, sec; see pp. 258 and 259
$\bar{t}(R_i)$	mean residence time of particles of size R_i, sec

T	temperature, °K
v	volumetric flow rate, cm^3/sec
V	volume, cm^3
V_r	$= V/\epsilon$, reactor volume, if different from volume occupied by reacting fluid, cm^3
V_s	volume of solid, cm^3
W	mass, gm
$W(R_i)$	mass of particles of size R_i, gm
X_A	fraction of reactant A converted into product; see Eqs. 3.10, 3.11, and 3.72
$\mathbf{X}_A$	$= C_A/C_U$, moles A/mole inert in liquid; Ch. 13 only
y_A	$= P_A/\pi$, mol fraction of A in gas
$\mathbf{Y}_A$	$= P_A/P_U$, moles A/mole inert in gas, Ch. 13 only
z	$= l/L$, fractional distance through a reactor

GREEK SYMBOLS

α	wake to bubble ratio in fluidized beds, see p. 312
γ	solid fraction in fluidized beds, see Eq. 14.66
$\delta(t)$	Dirac delta function, an ideal pulse occurring at time $t = 0$, sec^{-1}; see p. 262
$\delta(t - t_0)$	Dirac delta function occurring at time t_0
ϵ	porosity or fraction voids in a packed or fluidized bed
ε_A	fractional volume change on complete conversion of A; see Eqs. 3.70 to 3.72
θ	$= t/\bar{t}$, reduced time, dimensionless; see Eq. 9.25
κ	elutriation velocity constant, sec^{-1}; see Eq. 12.67
π	total pressure, atm
ρ	molar density, mol/cm^3
σ^2	variance of a tracer curve or distribution function, sec^2; see p. 261

τ	$= C_{A0}V/F_{A0}$, space-time, sec; see Eq. 5.6
τ'	$= C_{A0}W/F_{A0}$, weight-time, $gm \cdot sec/cm^3$; see Eq. 15.23
$\boldsymbol{\tau}$	time for complete conversion of a single solid particle, sec
$\boldsymbol{\tau}(R_i)$	time for complete conversion of particles of size R_i, sec
φ	instantaneous fractional yield; see Eq. 7.7
$\varphi(M/N)$	instantaneous fractional yield of M with respect to N, moles M formed/mol N formed or reacted away; see p. 170
Φ	over-all fractional yield; see Eq. 7.8

SUBSCRIPTS

b	batch
b	bubble phase of a fluidized bed
c	cloud phase of a fluidized bed
c	at unreacted core
d	deactivation
d	deadwater
e	emulsion phase of a fluidized bed
e	equilibrium conditions
f	leaving or final
g	refers to gas phase or main gas stream
i	entering
l	refers to liquid phase
m	mixed flow
mf	at minimum fluidizing conditions
p	plug flow
s	at surface, or based on surface
T	total moles in liquid phase; Ch. 13 only
U	carrier or inert component in a phase

0	entering or reference
θ	using dimensionless time units; see p. 271

ABBREVIATIONS

ei(α)	$= -\mathrm{Ei}(-\alpha)$, exponential integral; see Eq. 10.9
Re	$= du\rho/\mu$, Reynolds number, dimensionless
RTD	residence time distribution; see p. 255
Sc	$= \mu/\rho\mathscr{D}$, Schmidt number, dimensionless

I

INTRODUCTION

Every industrial chemical process is designed to produce economically a desired product from a variety of starting materials through a succession of treatment steps. Figure 1 shows a typical situation. The raw materials undergo a number of physical treatment steps to put them in the form in which they can be reacted chemically. They then pass through the reactor. The products of the reaction must then undergo further physical treatment—separations, purifications, etc.—for the final desired product to be obtained.

Design of equipment for the physical treatment steps is studied in the unit operations. In this book we are concerned with the chemical treatment step of a process. Economically this may be an inconsequential unit, say a simple mixing tank. More often than not, however, the chemical treatment step is the heart of the process, the thing that makes or breaks the process economically.

Design of the reactor is no routine matter, and many alternatives can be proposed for a process. In searching for the optimum it is not just the cost of reactor that must be minimized. One design may have low reactor cost, but the materials leaving the unit may be such that their treatment requires much higher cost than alternative designs. Hence, the economics of the over-all process must be considered.

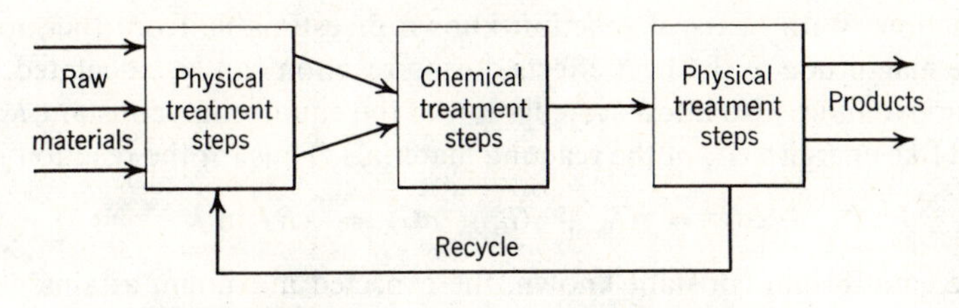

FIGURE 1. Typical chemical process.

Reactor design uses information, knowledge, and experience from a variety of areas—thermodynamics, chemical kinetics, fluid mechanics, heat transfer, mass transfer, and economics. Chemical reaction engineering is the synthesis of all these factors with the aim of properly designing a chemical reactor.

The design of chemical reactors is probably the one activity which is unique to chemical engineering, and it is probably this function more than anything else which justifies the existence of chemical engineering as a distinct branch of engineering.

In chemical reactor design there are two questions which must be answered:

1. What changes can we expect to occur?
2. How fast will they take place?

The first question concerns thermodynamics, the second the various rate processes—chemical kinetics, heat transfer, etc. Tying these all together and trying to determine how these processes are interrelated can be an extremely difficult problem; hence we start with the simplest of situations and build up our analysis by considering additional factors until we are able to handle the more difficult problems.

Let us first of all briefly take an over-all view of the subject, some of it being review, some new. Discussion on thermodynamics and chemical kinetics will set the stage for an outline of the way we cover the subject.

Thermodynamics

Thermodynamics gives two important pieces of information needed in design, the heat liberated or absorbed during reaction and the maximum possible extent of reaction.

Chemical reactions are invariably accompanied by the liberation or absorption of heat, the magnitude of which must be known for proper design. Consider the reaction

$$aA \rightarrow rR + sS, \qquad \Delta H_r \begin{cases} \text{positive, endothermic} \\ \text{negative, exothermic} \end{cases} \tag{1}$$

The heat of reaction at temperature T is the heat transferred *from* surroundings *to* the reacting system when a moles of A disappear to form r moles of R and s moles of S, with the system measured at the same temperature and pressure before and after reaction. With heats of reaction known or estimable from thermochemical data, the magnitude of the heat effects during reaction can be calculated.

Thermodynamics also allows calculation of the equilibrium constant K from the standard free energies $G°$, of the reacting materials. Thus for the reaction just given

$$\Delta G° = rG_R° + sG_S° - aG_A° = -RT \ln K \tag{2}$$

With the equilibrium constant known, the expected maximum attainable yield of the products of reaction can be estimated.

Chemical Kinetics

Under appropriate conditions feed materials may be transformed into new and different materials which constitute different chemical species. If this occurs only by rearrangement or redistribution of the constituent atoms to form new molecules, we say that a chemical reaction has occurred. Chemistry is concerned with the study of such reactions. It studies the mode and mechanism of reactions, the physical and energy changes involved and the rate of formation of products.

It is the last-mentioned area of interest, chemical kinetics, which is of primary concern to us. Chemical kinetics searches for the factors that influence the rate of reaction. It measures this rate and proposes explanations for the values found. Its study is important for a number of reasons:

1. For physical chemists it is the tool for gaining insight into the nature of reacting systems, for understanding how chemical bonds are made and broken, and for estimating their energies and stability.

2. For the organic chemist the value of chemical kinetics is greater still because the mode of reaction of compounds provides clues to their structure. Thus relative strengths of chemical bonds and molecular structure of compounds can be investigated by this tool.

3. In addition, it is the basis for important theories in combustion and dissolution and provides a method to study heat and mass transfer and suggests methods for tackling rate phenomena in other fields of study.

4. For the chemical engineer the kinetics of a reaction must be known if he is to satisfactorily design equipment to effect these reactions on a technical scale. Of course, if the reaction is rapid enough so that the system is essentially at equilibrium, design is very much simplified. Kinetic information is not needed, and thermodynamic information alone is sufficient.

Now our approach to chemical kinetics, the way we express kinetic laws, depends in large part on the type of reaction we are dealing with, and it may be well to consider next the classification of chemical reactions.

Classification of Reactions

There are many ways of classifying chemical reactions. In chemical reaction engineering probably the most useful scheme is the breakdown according to the number and types of phases involved, the big division being between the *homogeneous* and *heterogeneous* systems. A reaction is homogeneous if it takes place in one phase alone. A reaction is heterogeneous if it requires the presence of at least two phases to proceed at the rate that it does. It is immaterial whether the reaction takes place in one, two, or more phases, or at an interface, or whether the reactants and products are distributed among the phases or are all contained within a single phase. All that counts is that at least two phases are necessary for the reaction to proceed as it does.

Sometimes this classification is not clear-cut as with the large class of biological reactions, the enzyme-substrate reactions. Here the enzyme acts as a catalyst in the manufacture of proteins. Since enzymes themselves are highly complicated large-molecular-weight proteins of colloidal size, 10 to 100 mμ, enzyme-containing solutions represent a gray region between homogeneous and heterogeneous systems. Other examples for which the distinction between homogeneous and heterogeneous systems is not sharp are the very rapid chemical reactions, such as the burning gas flame. Here large nonhomogeneity in composition and temperature may exist. Strictly speaking, then, we do not have a single phase, for a phase implies uniform temperature, pressure, and composition throughout. The answer to the question of how to classify these borderline cases is simple. It depends on how we *choose* to treat them, and this in turn depends on which description we think to be more useful. Thus only in the context of a given situation can we decide how best to treat these borderline cases.

Cutting across this classification is the catalytic reaction whose rate is altered by materials that are neither reactants nor products. These foreign materials, called catalysts, need not be present in large amounts. Catalysts act somehow as go-betweens, either hindering or accelerating the reaction process while being modified relatively slowly if at all.

TABLE 1. Classification of chemical reactions useful in reactor design

	Noncatalytic	Catalytic
Homogeneous	Most gas-phase reactions	Most liquid-phase reactions
	Fast reactions such as burning of a flame	Reactions in colloidal systems Enzyme and microbial reactions
Heterogeneous	Burning of coal Roasting of ores Attack of solids by acids Gas-liquid absorption with reaction Reduction of iron ore to iron and steel	Ammonia synthesis Oxidation of ammonia to produce nitric acid Cracking of crude oil Oxidation of SO_2 to SO_3

Table 1 shows the classification of chemical reactions according to our scheme with a few examples of typical reactions of each type.

Variables Affecting the Rate of Reaction

Many variables may affect the rate of a chemical reaction. In homogeneous systems the temperature, pressure, and composition are obvious variables. In heterogeneous systems more than one phase is involved, hence the problem becomes more complex. Material may have to move from phase to phase during reaction; hence the rate of mass transfer can become important. For example, in the burning of a coal briquette the diffusion of oxygen through the gas film surrounding the particle, and through the ash layer at the surface of the particle, can play an important role in limiting the rate of reaction. In addition, the rate of heat transfer may also become a factor. Consider, for example, an exothermic reaction taking place at the interior surfaces of a porous catalyst pellet. If the heat released by reaction is not removed fast enough, a severe nonuniform temperature distribution can occur within the pellet, which in turn will result in differing point rates of reaction. These heat and mass transfer effects become increasingly important the faster the rate of reaction, and in very fast reactions, such as burning flames, they become controlling. Thus heat and mass transfer may play important roles in determining the rates of heterogeneous reactions.

In all cases, if the over-all reaction consists of a number of steps in series, it is the slowest step of the series that exerts the greatest influence and can be said to control. A big problem is to find out which variables affect each of these steps and to what degree. Only when we know the magnitude of each factor do we have a clear picture of the effect of these variables on the rate of reaction, and only then do we have the confidence to extrapolate these rates to new and different conditions.

Definition of Reaction Rate

We next ask how to *define* the rate of reaction in a meaningful and useful way. To answer this, let us adopt a number of definitions of rate of reaction, all interrelated and all intensive rather than extensive measures. But first we must select one reaction component for consideration and define the rate in terms of this component i. If the rate of change in number of moles of this component due to reaction is dN_i/dt, then the rate of reaction in its various forms is defined as follows. Based on unit volume of reacting fluid,

$$r_i = \frac{1}{V}\frac{dN_i}{dt} = \frac{\text{moles } i \text{ formed}}{(\text{volume of fluid})(\text{time})} \tag{3}$$

Based on unit mass of solid in fluid-solid systems,

$$r_i' = \frac{1}{W}\frac{dN_i}{dt} = \frac{\text{moles } i \text{ formed}}{(\text{mass of solid})(\text{time})} \tag{4}$$

Based on unit interfacial surface in two-fluid systems or based on unit surface of solid in gas-solid systems,

$$r_i'' = \frac{1}{S}\frac{dN_i}{dt} = \frac{\text{moles } i \text{ formed}}{\text{(surface)(time)}} \tag{5}$$

Based on unit volume of solid in gas-solid systems

$$r_i''' = \frac{1}{V_s}\frac{dN_i}{dt} = \frac{\text{moles } i \text{ formed}}{\text{(volume of solid)(time)}} \tag{6}$$

Based on unit volume of reactor, if different from the rate based on unit volume of fluid,

$$r_i'''' = \frac{1}{V_r}\frac{dN_i}{dt} = \frac{\text{moles } i \text{ formed}}{\text{(volume of reactor)(time)}} \tag{7}$$

In homogeneous systems the volume of fluid in the reactor is often identical to the volume of reactor. In such a case V and V_r are identical and Eqs. 3 and 7 are used interchangeably. In heterogeneous systems all the above definitions of reaction rate are encountered, the definition used in any particular situation often being a matter of convenience.

The rate of reaction is a function of the state of the system

$$r_i = f(\text{state of the system})$$

The form of this functional relationship remains the same, no matter how we choose to define the rate of reaction. It is only the constants of proportionality and their dimensions that change when we switch from one definition to another.

From Eqs. 3 to 7 these intensive definitions of reaction rate are related by

$$\begin{pmatrix}\text{volume}\\ \text{of fluid}\end{pmatrix} r_i = \begin{pmatrix}\text{mass of}\\ \text{solid}\end{pmatrix} r_i' = \begin{pmatrix}\text{surface}\\ \text{of solid}\end{pmatrix} r_i'' = \begin{pmatrix}\text{volume}\\ \text{of solid}\end{pmatrix} r_i''' = \begin{pmatrix}\text{volume}\\ \text{of reactor}\end{pmatrix} r_i''''$$

or

$$Vr_i = Wr_i' = Sr_i'' = V_s r_i''' = V_r r_i'''' \tag{8}$$

Over-all Plan

Our over-all plan is to start with homogeneous systems (Chapters 2 to 10) to see how rate expressions are suggested from theory (Chapter 2), how they are determined experimentally (Chapter 3), and how they are applied to the design of batch and flow chemical reactors involving ideal flow (Chapters 4 to 8) and the nonideal flow of fluids in real reactors (Chapters 9 and 10). The additional complications of design for heterogeneous systems are then introduced (Chapter 11), and a brief introduction to the specific problems of noncatalytic fluid-solid systems, two-fluid systems, and solid-catalyzed fluid systems are then considered in turn (Chapters 12 to 15).

2

KINETICS OF HOMOGENEOUS REACTIONS

In homogeneous reactions all reacting materials are found within a single phase, be it gas, liquid, or solid. In addition, if the reaction is catalytic, the catalyst must also be present within this phase. Though there are a number of ways of defining the rate of reaction, the intensive measure based on unit volume of reacting fluid is used practically exclusively for homogeneous systems. Thus the rate of reaction of any reaction component A is defined as

$$r_A = \frac{1}{V}\left(\frac{dN_A}{dt}\right)_{\text{by reaction}} = \frac{\text{(moles of A which appear by reaction)}}{\text{(unit volume)(unit time)}} \tag{1}$$

By this definition, if A is a reaction product, the rate is positive; if it is a reactant which is being consumed, the rate is negative; thus $-r_A$ is the rate of disappearance of reactant.

Now we may expect the progress of this class of reactions to depend on the composition of the materials within the phase as well as the temperature and pressure of the system. Shape of container, surface properties of solid materials in contact with the phase, and the diffusional characteristic of the fluid should not affect the rate of homogeneous reaction. Thus we may write for the rate of reaction of component A

$$\begin{aligned} r_A &= f(\text{state of the system}) \\ &= f(\text{temperature, pressure, composition}) \end{aligned}$$

These variables of pressure, temperature, and composition are interdependent in that the pressure is determined given the temperature and composition of the phase.* Thus we may write without loss of generality

$$r_A = f(\text{temperature, composition})$$

In this chapter we are concerned with the forms of this functional relationship. In turn we consider explanations from chemical theory for the composition dependency and temperature dependency of the rate expression, and also the question of predictability of rates of reaction.

CONCENTRATION-DEPENDENT TERM OF A RATE EQUATION

Before we can find the form of the concentration term in a rate expression, we must distinguish between different types of reactions. This distinction is based on the form and number of kinetic equations used to describe the progress of reaction. Also, since we are concerned with the concentration-dependent term of the rate equation, we assume that the temperature of the system is kept constant.

Single and Multiple Reactions

First of all, when materials react to form products it is usually easy to decide after examining the stoichiometry, preferably at more than one temperature, whether we should consider a single reaction or a number of reactions to be occurring.

When a single stoichiometric equation and single rate equation are chosen to represent the progress of the reaction, we have a *single reaction.* When more than one stoichiometric equation is used to represent the observed changes, then more than one kinetic expression is needed to follow the changing composition of all the reaction components, and we have *multiple reactions.*

Multiple reactions may be classified as:

series reactions,

$$A \rightarrow R \rightarrow S$$

parallel reactions, which are of two types

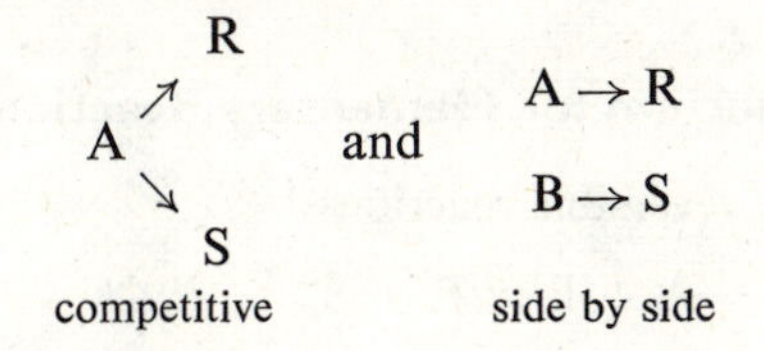

* Strictly speaking, this interdependency only applies at equilibrium; however, for lack of any better supposition, we assume that it also is true for nonequilibrium systems which are not changing too rapidly.

and more complicated schemes, an example of which is

$$A + B \rightarrow R$$
$$R + B \rightarrow S$$

Here reaction proceeds in parallel with respect to B, but in series with respect to A, R, and S.

Elementary and Nonelementary Reactions

Consider a single reaction with stoichiometric equation

$$A + B \rightarrow R$$

If we postulate that the rate-controlling mechanism involves the collision or interaction of a single molecule of A with a single molecule of B, then the number of collisions of molecules A with B is proportional to the rate of reaction. But at a given temperature the number of collisions is proportional to the concentration of reactants in the mixture, hence the rate of disappearance of A is given by

$$-r_A = kC_A C_B$$

Such reactions in which the rate equation corresponds to a stoichiometric equation are called *elementary reactions.*

When there is no correspondence between stoichiometry and rate then we have a *nonelementary reaction*. The classical example of a nonelementary reaction is that between hydrogen and bromine,

$$H_2 + Br_2 \rightarrow 2HBr$$

which has a rate expression*

$$r_{HBr} = \frac{k_1[H_2][Br_2]^{1/2}}{k_2 + [HBr]/[Br_2]}$$

Nonelementary reactions are explained by assuming that what we observe as a single reaction is in reality the over-all effect of a sequence of elementary reactions. The reason for observing only a single reaction rather than two or more elementary reactions is that the amount of intermediates formed is negligibly small and therefore escapes detection. We take up these explanations later.

Kinetic View of Equilibrium for Elementary Reactions

Consider the elementary reversible reactions

$$A + B \rightleftarrows R + S, \qquad K_C, K$$

* To eliminate much writing, at various places in this chapter we use square brackets to indicate concentrations. Thus,

$$C_{HBr} = [HBr]$$

The rate of formation of R by the forward reaction is

$$r_{R,\text{forward}} = k_1 C_A C_B$$

and its rate of disappearance by the reverse reaction is

$$-r_{R,\text{reverse}} = k_2 C_R C_S$$

At equilibrium there is no net formation of R, hence

$$r_{R,\text{forward}} + r_{R,\text{reverse}} = 0$$

or

$$\frac{k_1}{k_2} = \frac{C_R C_S}{C_A C_B} \tag{2}$$

In addition K_C is defined as*

$$K_C = \frac{C_R C_S}{C_A C_B} \tag{3}$$

So at equilibrium these two conditions can be combined to give

$$K_C = \frac{k_1}{k_2} = \frac{C_R C_S}{C_A C_B}$$

On the other hand, when not at equilibrium Eqs. 2 and 3 do not hold. In addition, since K_C and k_1/k_2 are constants independent of concentration and are equal to each other at one concentration, the equilibrium concentration, they must be equal to each other at all concentrations.

To summarize: for the elementary reaction considered,

$$K_C = \frac{k_1}{k_2} \left[= \frac{C_R C_S}{C_A C_B} \right]_{\text{only at equilibrium}} \tag{4}$$

For nonelementary reactions we cannot relate equilibrium, reaction rates, and concentrations in this simple way; however, Denbigh (1955) does treat this situation and shows what are the restrictions imposed by thermodynamics on the possible forms of the kinetic equation.

So, kinetics views equilibrium as a dynamic steady state involving a constant interchange of reactant and product molecules, rather than a static situation with everything at rest.

We can now view equilibrium in one of three ways.

1. From thermodynamics, we say that a system is in equilibrium with its surroundings of given temperature and pressure if the free energy of the system is at its lowest possible value. Thus for any movement away from equilibrium,

$$(\Delta G)_{p,T} > 0$$

* See any chemical engineering thermodynamics textbook or the brief thermodynamics review in Chapter 8.

2. From statistical mechanics, equilibrium is the state of the system consisting of the greatest number of equally likely molecular configurations which are macroscopically indistinguishable and can be considered to be identical. Thus from the gross point of view the state of the system that has the overwhelmingly great probability of occurring is called the equilibrium state.

3. From chemical kinetics, the system is at equilibrium if the rates of change of all the forward and reverse elementary reactions are equal.

These three criteria depend in turn on energy, probability, and rate considerations. Actually the thermodynamic and probabilistic views are enunciations of the same statement in different languages. The kinetic point of view, however, has further implications, for it requires knowledge of the mechanism of reaction for systems not at equilibrium. Thus in terms of understanding what is occurring, the kinetic point of view is more illuminating.

Molecularity and Order of Reaction

The *molecularity* of an elementary reaction is the number of molecules involved in the reaction, and this has been found to have the values of one, two, and occasionally three. Note that the molecularity refers only to an elementary reaction.

Often we find that the rate of progress of a reaction, involving say materials A, B, . . ., D, can be approximated by an expression of the following type:

$$r_A = kC_A{}^a C_B{}^b \cdots C_D{}^d, \qquad a + b + \cdots + d = n \tag{5}$$

where $a, b, \ldots, d$ are not necessarily related to the stoichiometric coefficients. We call the powers to which the concentrations are raised the *order of the reaction.* Thus the reaction is

*a*th order with respect to A
*b*th order with respect to B
*n*th order overall

Since the order refers to the empirically found rate expression, it can have a fractional value and need not be an integer. However, the molecularity of a reaction must be an integer since it refers to the mechanism of reaction, and can only apply to an elementary reaction.

For rate expressions not of the form of Eq. 5, for example Eq. 11 or the HBr reaction, it makes no sense to use the term reaction order.

Rate Constant *k*

When the rate expression for a homogeneous chemical reaction is written in the form of Eq. 5, the dimensions of the rate constant k for the nth-order reaction are

$$(\text{time})^{-1}(\text{concentration})^{1-n} \tag{6a}$$

which for a first-order reaction become simply

$$(\text{time})^{-1} \tag{6b}$$

Representation of a Reaction Rate

In expressing a rate we may use any measure equivalent to concentration, for example partial pressure, in which case

$$r_A = kp_A{}^a p_B{}^b \cdots p_D{}^d$$

Whatever measure we use leaves the order unchanged; however, it will affect the rate constant k.

For brevity, elementary reactions are often represented by an equation showing both the molecularity and the rate constant. For example,

$$2A \xrightarrow{k_1} 2R \tag{7}$$

represents a bimolecular irreversible reaction with second-order rate constant k_1, implying that the rate of reaction is

$$-r_A = r_R = k_1 C_A{}^2$$

It would not be proper to write Eq. 7 as

$$A \xrightarrow{k_1} R$$

for this would imply that the rate expression is

$$-r_A = r_R = k_1 C_A$$

Thus we must be careful to distinguish between the one equation which represents the elementary reaction and the many possible representations of the stoichiometry.

We should note that writing the elementary reaction with the rate constant, as shown by Eq. 7, may not be sufficient to avoid ambiguity. At times it may be necessary to specify the component in the reaction to which the rate constant is referred. For example, consider the reaction

$$B + 2D \xrightarrow{k_2} 3T \tag{8}$$

If the rate is measured in terms of B, the rate equation is

$$-r_B = k_2' C_B C_D{}^2$$

If it refers to D, the rate equation is

$$-r_D = k_2'' C_B C_D{}^2$$

Or if it refers to the product T, then

$$r_T = k_2''' C_B C_D{}^2$$

But from the stoichiometry

$$-r_B = -\tfrac{1}{2}r_D = \tfrac{1}{3}r_T$$

hence

$$k_2' = \tfrac{1}{2}k_2'' = \tfrac{1}{3}k_2'''$$

In Eq. 8, which of these three primed k_2 values are we referring to? We cannot tell. Hence, to avoid ambiguity when the stoichiometry involves different numbers of molecules of the various components, we must specify the component being considered.

To sum up, the condensed form of expressing the rate can be ambiguous. To eliminate any possible confusion, write the stoichiometric equation followed by the complete rate expression, and give the units of the rate constant.

Kinetic Models for Nonelementary Reactions

To explain the kinetics of nonelementary reactions we assume that a sequence of elementary reactions is actually occurring but that we cannot measure or observe the intermediates formed because they are only present in very minute quantities. Thus we observe only the initial reactants and final products, or what appears to be a single reaction. For example, if the kinetics of the reaction

$$A_2 + B_2 \rightarrow 2AB$$

indicates that the reaction is nonelementary, we may postulate a series of elementary steps to explain the kinetics, such as

$$A_2 \rightleftarrows 2A^*$$

$$A^* + B_2 \rightleftarrows AB + B^*$$

$$A^* + B^* \rightleftarrows AB$$

where the asterisks refer to the unobserved intermediates. To test our postulational scheme, we must see whether its predicted kinetic expression corresponds to experiment.

The types of intermediates we may postulate are suggested by the chemistry of the materials. These may be grouped as follows.

Free Radicals. Free atoms or larger fragments of stable molecules which contain one or more unpaired electrons are called free radicals. The unpaired electron is designated by a "dot" in the chemical symbol for the substance. Some free radicals are relatively stable, such as triphenylmethyl,

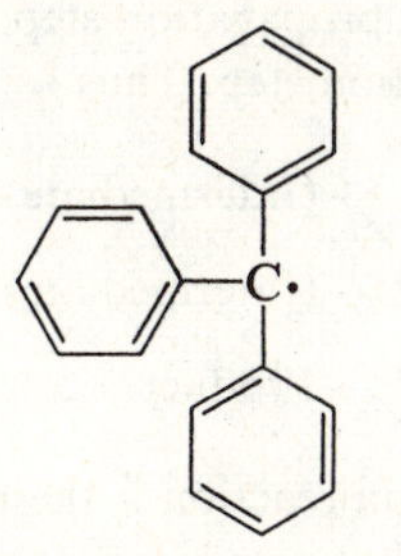

but as a rule they are unstable and highly reactive, such as

$$CH_3\cdot,\quad C_2H_5\cdot,\quad I\cdot,\quad H\cdot,\quad CCl_3\cdot$$

Ions and Polar Substances. Electrically charged atoms, molecules, or fragments of molecules, such as

$$N_3^-,\quad Na^+,\quad OH^-,\quad H_3O^+,\quad NH_4^+,\quad CH_3OH_2^+,\quad I^-$$

are called ions. These may act as active intermediates in reactions.

Molecules. Consider the consecutive reactions

$$A \rightarrow R \rightarrow S$$

Ordinarily these are treated as multiple reactions. If the product material R is highly reactive, however, its mean lifetime will be very small and its concentration in the reacting mixture can become too small to measure. In such a situation R is not observed and can be considered to be a reactive intermediate.

Transition Complexes. The numerous collisions between reactant molecules result in a wide distribution of energies among the individual molecules. This can result in strained bonds, unstable forms of molecules, or unstable association of molecules which can then either decompose to give products, or by further collisions return to molecules in the normal state. Such unstable forms are called transition complexes.

Postulated reaction schemes involving these four kinds of intermediates can be of two types.

Nonchain Reactions. In the nonchain reaction the intermediate is formed in the first reaction and then disappears as it reacts further to give the product. Thus

$$\text{Reactants} \rightarrow (\text{Intermediates})^*$$
$$(\text{Intermediates})^* \rightarrow \text{Products}$$

Chain Reactions. In chain reactions the intermediate is formed in a first reaction, called the chain initiation step. It then combines with reactant to form product and more intermediate in the chain propagation step. Occasionally the intermediate is destroyed in the chain termination step. Thus

$$\text{Reactant} \rightarrow (\text{Intermediate})^* \qquad \text{Initiation}$$

$$(\text{Intermediate})^* + \text{Reactant} \rightarrow (\text{Intermediate})^* + \text{Product} \qquad \text{Propagation}$$

$$(\text{Intermediate})^* \rightarrow \text{Product} \qquad \text{Termination}$$

The essential feature of the chain reaction is the propagation step. In this step the intermediate is not consumed but acts simply as a catalyst for the conversion of material. Thus each molecule of intermediate can catalyze a long chain of reactions before being finally destroyed.

The following are examples of mechanisms of various kinds.

1. *Free radicals, chain reaction mechanism.* The reaction

$$H_2 + Br_2 \rightarrow 2HBr$$

with experimental rate

$$r_{HBr} = \frac{k_1[H_2][Br_2]^{1/2}}{k_2 + [HBr]/[Br_2]}$$

can be explained by the following scheme:

$$Br_2 \rightleftarrows 2Br\cdot \qquad \text{Initiation and termination}$$

$$Br\cdot + H_2 \rightleftarrows HBr + H\cdot \qquad \text{Propagation}$$

$$H\cdot + Br_2 \rightarrow HBr + Br\cdot \qquad \text{Propagation}$$

2. *Molecular intermediates, nonchain mechanism.* The general class of enzyme-catalyzed fermentation reactions

$$A \xrightarrow[\text{enzyme}]{\text{with}} R$$

is viewed to proceed as follows:

$$A + \text{enzyme} \rightleftarrows (A\cdot\,\text{enzyme})^*$$

$$(A\cdot\,\text{enzyme})^* \rightarrow R + \text{enzyme}$$

In such reactions the concentration of intermediate may become more than negligible, in which case a special analysis, first proposed by Michaelis and Menten (1913), is required.

3. *Ionic intermediates, catalyzed nonchain mechanism.* The kinetics of the acid-catalyzed hydration of the unsaturated hydrocarbon isobutene

$$\underset{\text{isobutene}}{CH_3{-}\overset{\overset{\displaystyle CH_3}{|}}{C}{=}CH_2} + H_2O \underset{}{\overset{\text{dilute } HNO_3}{\rightleftarrows}} \underset{tert\text{-butyl alcohol}}{CH_3{-}\overset{\overset{\displaystyle CH_3}{|}}{C}OH{-}CH_3}$$

is consistent with a multistep mechanism involving formation of a number of intermediates, all polar. Thus in general

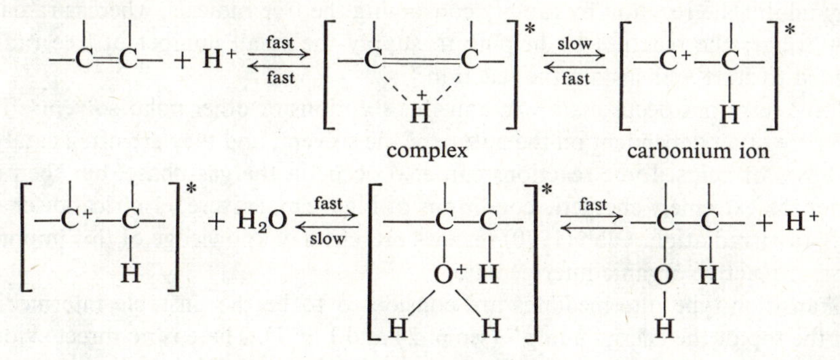

4. *Transition complex, nonchain mechanism.* The spontaneous decomposition of azomethane

$$(CH_3)_2N_2 \rightarrow C_2H_6 + N_2 \quad \text{or} \quad A \rightarrow R + S$$

exhibits under various conditions first-order, second-order, or intermediate kinetics. This type of behavior can be explained by postulating the existence of an energized and unstable form for the reactant. Thus

$A + A \rightarrow A^* + A$ Formation of energized molecule

$A^* + A \rightarrow A + A$ Return to stable form by collision

$A^* \rightarrow R + S$ Spontaneous decomposition into products

Lindemann (1922) first suggested this type of intermediate.

5. *Transition complex, nonchain reaction.* The intermediate in the reaction

$$H_2 + I_2 \rightleftarrows 2HI$$

with elementary second-order kinetics is an example of another type of transition complex, this one consisting of an association of molecules. Thus

$$\begin{matrix} H \\ | \\ H \end{matrix} + \begin{matrix} I \\ | \\ I \end{matrix} \rightleftarrows \begin{bmatrix} H\text{---}I \\ \vdots \quad \vdots \\ H\text{---}I \end{bmatrix}^* \rightleftarrows \begin{matrix} H{-}I \\ + \\ H{-}I \end{matrix}$$

This reaction is called a four-center-type reaction.

At first, free radicals were hypothesized to explain observed kinetics without any direct evidence of their actual existence. In recent years, however, with the development of more sensitive experimental techniques such as high-resolution spectroscopic analyses and reaction freezing at very low temperature, the existence of many free radicals has been directly verified. Today it is thought that such substances play a role in many types of reactions. In general, free-radical reactions occur in the gas phase at high temperature. More often than not they occur by a chain mechanism and may be greatly affected by radiation and traces of impurities. Such impurities may inhibit the reaction by rapidly consuming the free radicals, whereas radiation may trigger the reaction by helping to supply the small number of free radicals needed to start and sustain the reaction.

Ionic reactions occur mainly in aqueous solutions or other polar solvents. Their rates are often dependent on the nature of the solvent, and they are often catalyzed by bases or acids. Ionic reactions can also occur in the gas phase, but then only under the extremely energetic conditions of high temperature, electrical discharge, or X-ray irradiation. Olah (1970) reviews present-day knowledge of this important class of reactive organic intermediate.

Transition-type intermediates are considered to be the unstable intermediates "at the top of the energy hump" (see p. 24 and Fig. 1). There is no direct evidence for their existence; however, their use does explain observed data.

Intermediates consisting of rapidly decomposing molecules have real existence and have been observed in a variety of reactions, both gas and liquid.

Testing Kinetic Models

Two problems make the search for the correct mechanism of reaction difficult. First, the reaction may proceed by more than one mechanism, say free radical and ionic, with relative rates which change with conditions. Second, more than one mechanism can be consistent with kinetic data. Resolving these problems is difficult and requires an extensive knowledge of the chemistry of the substances involved. Leaving these aside, let us see how to test the correspondence between experiment and a proposed mechanism which involves a sequence of elementary reactions.

In matching the predicted rate expression with experiment we rely on the following two rules. (1) If component i takes part in more than one reaction, its net rate of change is the sum total of the rates of change of that component in each of the elementary reactions, or

$$r_{i,\text{net}} = \sum_{\substack{\text{all elementary} \\ \text{reactions}}} r_i \tag{9}$$

(2) Because intermediates are present in such small quantities, their rates of change in the system after a very short time can never be great; hence with negligible error these rates are taken to be zero. This is called the steady-state approximation. It is

needed if we are to solve the attendant mathematics, and our justification is that the predicted results based on this assumption very often agree with experiment.

The trial-and-error procedure involved in searching for a mechanism is illustrated in the following example.

EXAMPLE 1. Search for the reaction mechanism

The irreversible reaction

$$2A + B = A_2B \tag{10}$$

has been studied kinetically, and the rate of formation of product has been found to be well correlated by the following rate equation:

$$r_{A_2B} = \frac{0.72C_A{}^2C_B}{1 + 2C_A} = \frac{0.72[A]^2[B]}{1 + 2[A]} \tag{11}$$

What reaction mechanism is suggested by this rate expression if the chemistry of the reaction suggests that the intermediate consists of an association of reactant molecules and that a chain reaction does not occur?

SOLUTION

If this were an elementary reaction, the rate would be given by

$$r_{A_2B} = kC_A{}^2C_B = k[A]^2[B] \tag{12}$$

Since Eqs. 11 and 12 are not of the same type, the reaction evidently is nonelementary. Consequently, let us try various mechanisms and see which gives a rate expression similar in form to the experimentally found expression. We start with simple two-step models, and if these are unsuccessful we will try more complicated three-, four-, or five-step models.

Model 1. Hypothesize a two-step reversible scheme involving the formation of an intermediate substance A_2^*, not actually seen and hence thought to be present only in small amounts. Thus

$$\begin{aligned} 2A &\underset{k_2}{\overset{k_1}{\rightleftarrows}} A_2^* \\ A_2^* + B &\underset{k_4}{\overset{k_3}{\rightleftarrows}} A_2B \end{aligned} \tag{13}$$

which really involves four elementary reactions

$$2A \xrightarrow{k_1} A_2^* \tag{14}$$

$$A_2^* \xrightarrow{k_2} 2A \tag{15}$$

$$A_2^* + B \xrightarrow{k_3} A_2B \tag{16}$$

$$A_2B \xrightarrow{k_4} A_2^* + B \tag{17}$$

Let the k values refer to the components disappearing; thus k_1 refers to A, k_2 refers to A_2^*, etc.

Now write the expression for the rate of formation of A_2B. Since this component is involved in Eqs. 16 and 17, its over-all rate of change is the sum of the individual rates. Thus

$$r_{A_2B} = k_3[A_2^*][B] - k_4[A_2B] \tag{18}$$

Since the concentration of intermediate A_2^* is not measurable, the above rate expression cannot be tested in its present form. So, replace $[A_2^*]$ by concentrations that can be measured, such as [A], [B], or [AB]. This is done in the following manner. From the four elementary reactions which all involve A_2^* we find

$$r_{A_2^*} = \tfrac{1}{2}k_1[A]^2 - k_2[A_2^*] - k_3[A_2^*][B] + k_4[A_2B] \tag{19}$$

Because the concentration of A_2^* is always extremely small we may assume that its rate of change is zero or

$$r_{A_2^*} = 0 \tag{20}$$

This is the steady-state approximation. Combining Eqs. 19 and 20 we then find

$$[A_2^*] = \frac{\tfrac{1}{2}k_1[A]^2 + k_4[A_2B]}{k_2 + k_3[B]} \tag{21}$$

which when replaced in Eq. 18 gives the rate of formation of A_2B in terms of measurable quantities. Thus

$$r_{A_2B} = \frac{\tfrac{1}{2}k_1k_3[A]^2[B] - k_2k_4[A_2B]}{k_2 + k_3[B]} \tag{22}$$

In searching for a model consistent with observed kinetics we may, if we wish, restrict a more general model by arbitrarily selecting the magnitude of the various rate constants. Since Eq. 22 does not match Eq. 11, let us see if any of its simplified forms will. Thus, if k_2 is very small, this expression reduces to

$$r_{A_2B} = \tfrac{1}{2}k_1[A]^2 \tag{23}$$

If k_4 is very small, r_{A_2B} reduces to

$$r_{A_2B} = \frac{(k_1k_3/2k_2)[A]^2[B]}{1 + (k_3/k_2)[B]} \tag{24}$$

Neither of these special forms, Eqs. 23 and 24, matches the experimentally found rate, Eq. 11; thus the hypothesized mechanism, Eq. 13, is incorrect.

Model 2. As our first guess gave a rate expression, Eq. 24, somewhat like Eq. 11, let us try, for our second guess, a mechanism somewhat similar to Model 1. Let us try the mechanism

$$\begin{aligned} A + B &\underset{k_2}{\overset{k_1}{\rightleftharpoons}} AB^* \\ AB^* + A &\underset{k_4}{\overset{k_3}{\rightleftharpoons}} A_2B \end{aligned} \tag{25}$$

Following the procedure used for Model 1 the desired rate is

$$r_{A_2B} = k_3[AB^*][A] - k_4[A_2B] \tag{26}$$

Next eliminate [AB*] in this expression. With the steady-state approximation we obtain

$$r_{AB^*} = k_1[A][B] - k_2[AB^*] - k_3[AB^*][A] + k_4[A_2B] = 0$$

from which

$$[AB^*] = \frac{k_1[A][B] + k_4[A_2B]}{k_2 + k_3[A]} \tag{27}$$

Replacing Eq. 27 in Eq. 26 to eliminate the concentration of intermediate, we obtain

$$r_{A_2B} = \frac{k_1k_3[A]^2[B] - k_2k_4[A_2B]}{k_2 + k_3[A]} \tag{28}$$

Let us restrict this general model. With k_4 very small we obtain

$$r_{A_2B} = \frac{(k_1k_3/k_2)[A]^2[B]}{1 + (k_3/k_2)[A]} \tag{29}$$

Comparing Eqs. 11 and 29, we see that they are of the same form. Thus the reaction may be represented by the mechanism

$$\begin{aligned} A + B &\underset{k_2}{\overset{k_1}{\rightleftarrows}} AB^* \\ AB^* + A &\xrightarrow{k_3} A_2B \end{aligned} \tag{30}$$

We were fortunate in this example to have represented our data by a form of equation which happened to match exactly that obtained from the theoretical mechanism. Often a number of equation types will fit a set of experimental data equally well, especially for somewhat scattered data. Hence to avoid rejecting the correct mechanism, it is advisable to test the fit of the various theoretically derived equations to the raw data using statistical criteria whenever possible, rather than just matching equation forms.

TEMPERATURE-DEPENDENT TERM OF A RATE EQUATION

Temperature Dependency from Arrhenius' Law

For many reactions and particularly elementary reactions the rate expression can be written as a product of a temperature-dependent term and a composition-dependent term, or

$$\begin{aligned} r_i &= f_1(\text{temperature}) \cdot f_2(\text{composition}) \\ &= k \cdot f_2(\text{composition}) \end{aligned} \tag{31}$$

For such reactions the temperature-dependent term, the reaction rate constant, has been found in practically all cases to be well represented by Arrhenius' law:

$$k = k_0 e^{-E/RT} \tag{32}$$

where k_0 is called the frequency factor and E is called the activation energy of the reaction.* This expression fits experiment well over wide temperature ranges and is strongly suggested from various standpoints as being a very good approximation to the true temperature dependency.

Temperature Dependency from Thermodynamics

The temperature dependency of the equilibrium constant of the elementary reversible reactions such as

$$A \underset{k_2}{\overset{k_1}{\rightleftarrows}} R, \qquad \Delta H_r \tag{33}$$

is given by the van't Hoff equation, Eq. 8.15,

$$\frac{d(\ln K)}{dT} = \frac{\Delta H_r}{RT^2} \tag{34}$$

Because $K = K_C = [R]/[A] = k_1/k_2$ for this reaction, we can then rewrite the van't Hoff relationship as

$$\frac{d(\ln k_1)}{dT} - \frac{d(\ln k_2)}{dT} = \frac{\Delta H_r}{RT^2}$$

Though it does not necessarily follow, the fact that the difference in derivatives is equal to $\Delta H_r/RT^2$ suggests that each derivative alone is equal to a term of that form, or

$$\frac{d(\ln k_1)}{dT} = \frac{E_1}{RT^2} \quad \text{and} \quad \frac{d(\ln k_2)}{dT} = \frac{E_2}{RT^2} \tag{35}$$

where

$$E_1 - E_2 = \Delta H_r \tag{36}$$

* There seems to be a disagreement in the dimensions used to report the activation energy; some authors use calories, and others use calories per mole. On the one hand, calories per mole are clearly indicated by Eq. 32. In contrast to the dimensionally identical thermodynamic quantities $\Delta G°$ and ΔH_r, however, the numerical value of E does not depend on how we represent the reaction stoichiometry (number of moles used). Thus reporting calories per mole may be misinterpreted. To avoid this E is reported here simply as calories.

What moles are we referring to in the units of E? These are always the quantities associated with the molar representation of the rate-controlling step of the reaction. Numerically E can be found without knowing what this is; however, if E is to be compared with analogous quantities from thermodynamics, collision theory, or transition-state theory, this mechanism must be known and its stoichiometric representation must be used throughout.

This whole question can be avoided by using the ratio E/R throughout, since E and R always refer to the same number of moles.

In addition, if the energy terms are assumed to be temperature-independent, Eq. 35 can be integrated to give Arrhenius' equation, Eq. 32.

Temperature Dependency from Collision Theory

The collision rate of molecules in a gas can be found from the kinetic theory of gases. For the bimolecular collisions of like molecules A we have

$$Z_{AA} = \sigma_A{}^2 \mathbf{n}_A{}^2 \sqrt{\frac{4\pi \mathbf{k} T}{\mathbf{M}_A}} = \sigma_A{}^2 \frac{\mathbf{N}^2}{10^6} \sqrt{\frac{4\pi \mathbf{k} T}{\mathbf{M}_A}}\, C_A{}^2$$

$$= \frac{\text{number of collisions of A with A}}{\text{sec} \cdot \text{cm}^3} \tag{37}$$

where σ = diameter of a molecule, cm

$\mathbf{M}$ = (molecular weight)/$\mathbf{N}$, mass of a molecule, gm

$\mathbf{N} = 6.023 \times 10^{23}$ molecules/mol, Avogadro's number

C_A = concentration of A, mol/liter

$\mathbf{n}_A = \mathbf{N}C_A/10^3$, number of molecules of A/cm^3

$\mathbf{k} = R/\mathbf{N} = 1.30 \times 10^{-16}$ erg/°K, Boltzmann constant

For bimolecular collisions of unlike molecules in a mixture of A and B kinetic theory gives

$$Z_{AB} = \left(\frac{\sigma_A + \sigma_B}{2}\right)^2 \mathbf{n}_A \mathbf{n}_B \sqrt{8\pi \mathbf{k} T \left(\frac{1}{\mathbf{M}_A} + \frac{1}{\mathbf{M}_B}\right)}$$

$$= \left(\frac{\sigma_A + \sigma_B}{2}\right)^2 \frac{\mathbf{N}^2}{10^6} \sqrt{8\pi \mathbf{k} T \left(\frac{1}{\mathbf{M}_A} + \frac{1}{\mathbf{M}_B}\right)}\, C_A C_B \tag{38}$$

If every collision between reactant molecules results in the transformation of reactants into product, these expressions give the rate of bimolecular reaction. The actual rate is usually much lower than that predicted, and this indicates that only a small fraction of all collisions result in reaction. This suggests that only the more energetic and violent collisions, or more specifically, only those collisions that involve energies in excess of a given minimum energy E lead to reaction. From the Maxwell distribution law of molecular energies the fraction of all bimolecular collisions that involve energies in excess of this minimum energy is given approximately by

$$e^{-E/RT}$$

when $E \gg RT$. Since we are only considering energetic collisions, this assumption is reasonable. Thus the rate of reaction is given by

$$-r_A = -\frac{1}{V}\frac{dN_A}{dt} = kC_AC_B = \begin{pmatrix}\text{collision rate,}\\ \text{mol/liter}\cdot\text{sec}\end{pmatrix}\begin{pmatrix}\text{fraction of collisions involving}\\ \text{energies in excess of } E\end{pmatrix}$$

$$= \mathbf{Z}_{AB}\frac{10^3}{\mathbf{N}}e^{-E/RT}$$

$$= \left(\frac{\sigma_A + \sigma_B}{2}\right)^2 \frac{\mathbf{N}}{10^3}\sqrt{8\pi\mathbf{k}T\left(\frac{1}{\mathbf{M}_A} + \frac{1}{\mathbf{M}_B}\right)}\, e^{-E/RT}C_AC_B \tag{39}$$

A similar expression can be found for the bimolecular collisions between like molecules. For both, in fact for all bimolecular reactions, Eq. 39 shows that the temperature dependency of the rate constant is given by

$$k \propto T^{1/2}e^{-E/RT} \tag{40}$$

Temperature Dependency from Transition-state Theory

A more detailed explanation for the transformation of reactants into products is given by the transition-state theory. This pictures reactants combining to form unstable intermediates called activated complexes which then decompose spontaneously into products. It assumes in addition that an equilibrium exists between the concentration of reactants and activated complex at all times and that the rate of decomposition of complex is the same for all reactions and is given by $\mathbf{k}T/\mathbf{h}$ where $\mathbf{k}$ is the Boltzmann constant and $\mathbf{h} = 6.63 \times 10^{-27}$ erg·sec is the Planck constant. Thus for the forward elementary reaction of a reversible reaction,

$$A + B \underset{k_2}{\overset{k_1}{\rightleftarrows}} AB, \qquad \Delta H_r \tag{41}$$

we have the following conceptual scheme:

$$A + B \underset{k_4}{\overset{k_3}{\rightleftarrows}} AB^* \xrightarrow{k_5} AB \tag{42}$$

with

$$K_C^* = \frac{k_3}{k_4} = \frac{[AB^*]}{[A][B]}$$

and

$$k_5 = \frac{\mathbf{k}T}{\mathbf{h}}$$

The observed rate of the forward reaction is then

$$r_{\text{AB,forward}} = \begin{pmatrix}\text{concentration of} \\ \text{activated complex}\end{pmatrix}\begin{pmatrix}\text{rate of decomposition} \\ \text{of activated complex}\end{pmatrix}$$

$$= \frac{\mathbf{k}T}{\mathbf{h}}[\text{AB}^*]$$

$$= \frac{\mathbf{k}T}{\mathbf{h}} K_C^* C_A C_B \qquad (43)$$

By expressing the equilibrium constant of the activated complex in terms of the standard free energy,

$$\Delta G^* = \Delta H^* - T\Delta S^* = -RT \ln K_C^*$$

or (44)

$$K_C^* = e^{-\Delta G^*/RT} = e^{-\Delta H^*/RT + \Delta S^*/R}$$

the rate becomes

$$r_{\text{AB,forward}} = \frac{\mathbf{k}T}{\mathbf{h}} e^{\Delta S^*/R} e^{-\Delta H^*/RT} C_A C_B \qquad (45)$$

Theoretically both ΔS^* and ΔH^* vary very slowly with temperature. Hence, of the three terms that make up the rate constant in Eq. 45, the middle one, $e^{\Delta S^*/R}$, is so much less temperature-sensitive than the other two terms that we may take it to be constant. So for the forward reaction, and similarly for the reverse reaction of Eq. 41, we have approximately

$$k_1 \propto Te^{-\Delta H_1^*/RT}$$
$$k_2 \propto Te^{-\Delta H_2^*/RT} \qquad (46)$$

where

$$\Delta H_1^* - \Delta H_2^* = \Delta H_r$$

We next look for a relationship between ΔH^* and the Arrhenius activation energy E. Though none can be logically derived, still we can arbitrarily define one. This is generally done by using analogy arguments from thermodynamics. Thus for liquids and solids

$$E = \Delta H^* - RT \qquad (47)$$

and for gases

$$E = \Delta H^* - (\text{molecularity} - 1)RT \qquad (48)$$

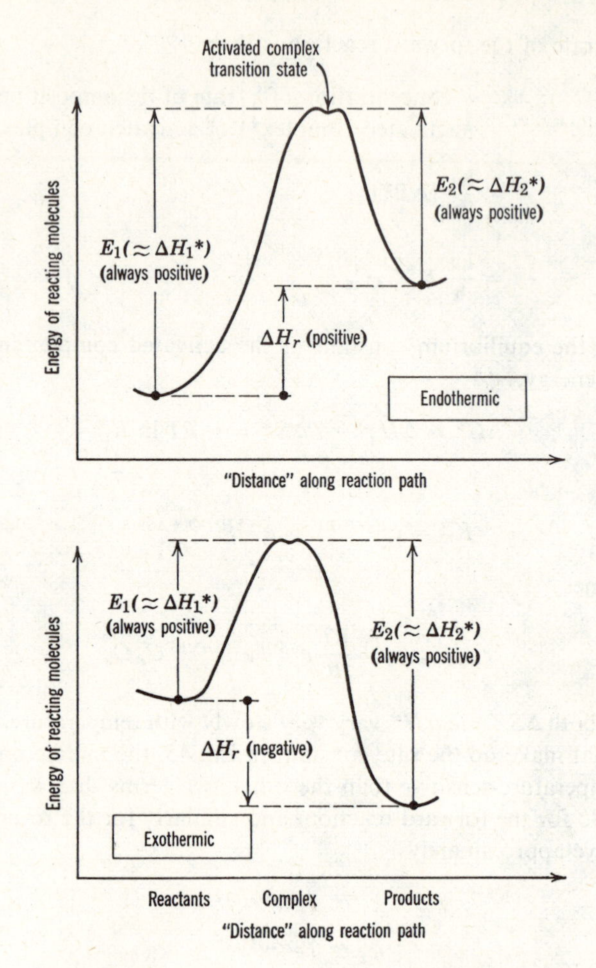

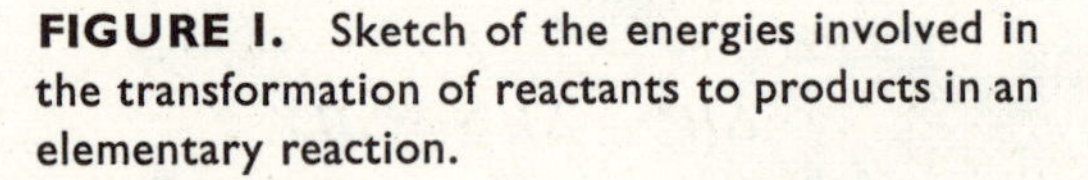

FIGURE 1. Sketch of the energies involved in the transformation of reactants to products in an elementary reaction.

With this definition the difference between E and ΔH^* is small and of the order RT; hence transition-state theory predicts approximately that

$$k \propto Te^{-E/RT} \tag{49}$$

Figure 1 illustrates the energies involved in reactants and complexes in such a scheme.

Comparison of Theories

It is interesting to note the difference in approach between the collision and transition-state theories. Consider A and B colliding and forming an unstable intermediate which then decomposes into product, or

$$A + B \rightarrow AB^* \rightarrow AB \tag{50}$$

Collision theory views the rate to be governed by the number of energetic collisions between reactants. What happens to the unstable intermediate is of no concern. The theory simply assumes that this intermediate breaks down rapidly enough into products so as not to influence the rate of the over-all process. Transition-state theory, on the other hand, views the reaction rate to be governed by the rate of decomposition of intermediate. The rate of formation of intermediate is assumed to be so rapid that it is present in equilibrium concentrations at all times. How it is formed is of no concern. Thus collision theory views the first step of Eq. 50 to be slow and rate-controlling, whereas transition-state theory views the second step of Eq. 50 combined with the determination of complex concentration to be the rate-controlling factors. In a sense, then, these two theories complement each other.

Comparison of Theories with Arrhenius' Law

The expression

$$\begin{aligned} k &\propto T^m e^{-E/RT} \\ &= k_0' T^m e^{-E/RT}, \qquad 0 \leqslant m \leqslant 1 \end{aligned} \tag{51}$$

summarizes the predictions of the simpler versions of these theories for the temperature dependency of the rate constant. For more complicated versions m can be as great as 3 or 4. Now because the exponential term is so much more temperature-sensitive than the T^m term, the variation of k caused by the latter is effectively masked, and we have in effect

$$\begin{aligned} k &\propto e^{-E/RT} \\ &= k_0 e^{-E/RT} \end{aligned} \tag{32}$$

We can show this in another way. Taking logarithms of Eq. 51 and differentiating with respect to T, we find how k varies with temperature. This gives

$$\frac{d(\ln k)}{dT} = \frac{m}{T} + \frac{E}{RT^2} = \frac{mRT + E}{RT^2}$$

As $mRT \ll E$ for most reactions studied, we may ignore the mRT term and may write

$$\frac{d(\ln k)}{dT} = \frac{E}{RT^2} \tag{32}$$

or

$$k \propto e^{-E/RT}$$

This discussion shows that Arrhenius' law is a good approximation to the temperature dependency of both collision and transition-state theories.

Activation Energy and Temperature Dependency

TABLE 1. Temperature rise needed to double the rate of reaction for activation energies and average temperatures shown; hence shows temperature sensitivity of reactions

	Activation Energy E		
Temperature	10,000 cal	40,000 cal	70,000 cal
0°C	11°C	3°C	2°C
400°C	70	17	9
1000°C	273	62	37
2000°C	1037	197	107

TABLE 2. Relative rates of reaction as a function of activation energy and temperature

	Activation Energy E		
Temperature	10,000 cal	40,000 cal	70,000 cal
0°C	10^{48}	10^{24}	1
400°C	7×10^{52}	10^{43}	2×10^{33}
1000°C	2×10^{54}	10^{49}	10^{44}
2000°C	10^{55}	10^{52}	2×10^{49}

The temperature dependency of reactions is determined by the activation energy and temperature level, as illustrated in Fig. 2 and Tables 1 and 2. These findings are summarized as follows:

1. From Arrhenius' law a plot of $\ln k$ vs $1/T$ gives a straight line, with large slope for large E and small slope for small E.

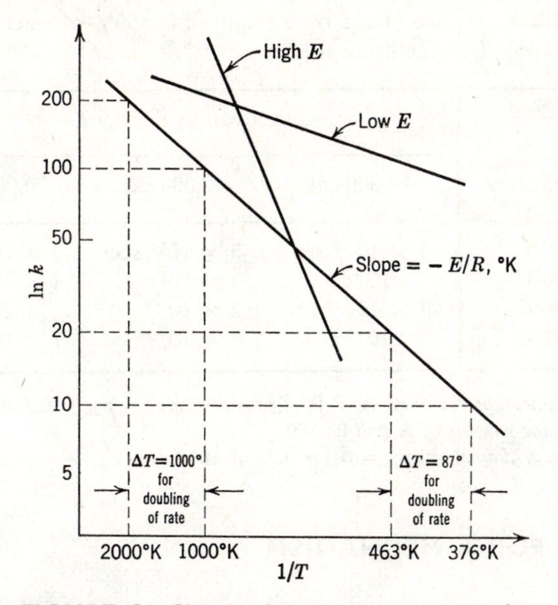

FIGURE 2. Sketch showing temperature dependency of the reaction rate.

2. Reactions with high activation energies are very temperature-sensitive; reactions with low activation energies are relatively temperature-insensitive.

3. A given reaction is much more temperature-sensitive at low temperature than at high temperature.

4. From the Arrhenius law the frequency factor k_0 does not affect the temperature sensitivity of a reaction. In an actual reaction there may be a slight temperature dependency of this term as predicted by Eq. 51; however, this is rather minor and can be ignored.

Rate of Reaction Predicted by the Theories

Experimental values for rates of reaction are generally either in the order of magnitude of, or are below, those predicted by collision theory, and an indication of these predictions is shown in Table 3. Thus collision theory may be used to estimate the upper bound to the expected rate of reaction. Once in a while, however, a reaction is encountered with much higher rates than predicted. This suggests a complex reaction, frequently catalytic.

Occasionally, for the elementary reaction between simpler molecules enough information is available to allow prediction of the rates from transition-state theory. When available, these predictions usually agree more closely with experiment than do the predictions of collision theory.

TABLE 3. Approximate time required for 50% of reactants to react away; from collision theory

	Activation Energy E		
Temperature	10,000 cal	40,000 cal	70,000 cal
0°C	3×10^{-5} sec	3×10^{19} sec	3×10^{43} sec
400°C	10^{-9}	8×10^{2}	4×10^{9}
1000°C	2×10^{-11}	2×10^{-6}	30
2000°C	10^{-13}	9×10^{-11}	7×10^{-8}

Elementary gas reaction: A + B → products, $p_{A0} = p_{B0} = 0.5$ atm.
Molecular weights of A *and* B: 100.
Molecular diameters: $\sigma_A = \sigma_B = 3.35 \times 10^{-8}$ cm.

SEARCHING FOR A MECHANISM

The more we know about what is occurring during reaction, what the reacting materials are, and how they react, the more assurance we have for proper design. This is the incentive to find out as much as we can about the factors influencing a reaction within the limitations of time and effort set by the economic optimization of the many factors involved in the industrial exploitation of a process.

There are three areas of investigation of a reaction, the *stoichiometry*, the *kinetics*, and the *mechanism*. In general, the stoichiometry is studied first, and when this is far enough along the kinetics is then investigated. With empirical rate expressions available, the mechanism is then looked into. In any investigative program considerable feedback of information occurs from area to area. For example, our ideas about the stoichiometry of the reaction may change on the basis of kinetic data obtained, and the form of the kinetic equations themselves may be suggested by mechanism studies. With this kind of interrelationship of the many factors, no straightforward experimental program can be formulated for the study of reactions. Thus it becomes a matter of shrewd scientific detective work, with carefully planned experimental programs especially designed to discriminate between rival hypotheses, which in turn have been suggested and formulated on the basis of all pertinent information available at that time.

Although we cannot delve into the many aspects of this problem, a number of clues which are often used in such experimentation can be mentioned.

1. Stoichiometry can tell whether we have a single reaction or not. Thus a complicated stoichiometry

$$A \rightarrow 1.45R + 0.85S$$

or one which changes with reaction conditions or extent of reaction is clear evidence of multiple reactions.

2. Stoichiometry can suggest whether a single reaction is elementary or not because no elementary reactions with molecularity greater than three have been observed to date. As an example the reaction

$$N_2 + 3H_2 \rightarrow 2NH_3$$

is not elementary.

3. A comparison of the stoichiometric equation with the experimental kinetic expression can show whether we are dealing with an elementary reaction or not.

4. A large difference in the order of magnitude between the experimentally found frequency factor of a reaction and that calculated from collision theory or transition-state theory may suggest a nonelementary reaction; however, this is not necessarily true. For example, certain isomerizations have very low frequency factors and are still elementary.

5. Consider two alternate paths for a simple reversible reaction. If one of these paths is preferred for the forward reaction, the same path must also be preferred for the reverse reaction. This is called the principle of microscopic reversibility. Consider, for example, the forward reaction of

$$2NH_3 \rightleftarrows N_2 + 3H_2$$

At first sight this could very well be an elementary bimolecular reaction with two molecules of ammonia combining to yield directly the four product molecules. From this principle, however, the reverse reaction would then also have to be an elementary reaction involving the direct combination of three molecules of hydrogen with one of nitrogen. Since such a process is rejected as improbable, the bimolecular forward mechanism must also be rejected.

6. The principle of microreversibility also indicates that changes involving bond rupture, molecular syntheses, or splitting are likely to occur one at a time, each then being an elementary step in the mechanism. From this point of view the simultaneous splitting of the complex into the four product molecules in the reaction

$$2NH_3 \rightarrow (NH_3)_2^* \rightarrow N_2 + 3H_2$$

is very unlikely. This rule does not apply to changes which involve a shift in electron density along a molecule, which may take place in a cascade-like manner. For example, the transformation

$$\underset{\text{vinyl allyl ether}}{CH_2{=}CH{-}CH_2{-}O{-}CH{=}CH_2} \rightarrow \underset{n\text{-pentaldehyde-ene 4}}{CH_2{=}CH{-}CH_2{-}CH_2{-}CHO}$$

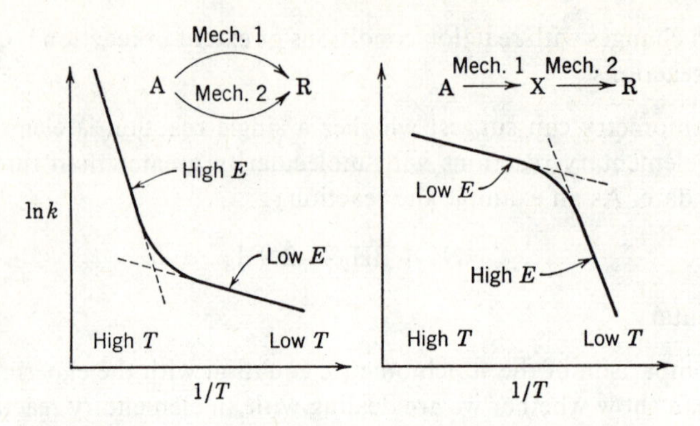

FIGURE 3. A change in activation energy indicates a shift in controlling mechanism of reaction.

can be explained in terms of the following shifts in electron density:

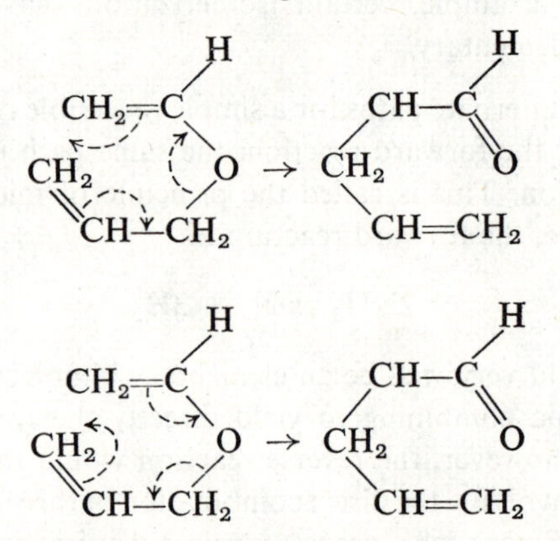

7. A change in activation energy with temperature indicates a shift in controlling mechanism of reaction. Recalling that a higher E represents a more temperature sensitive reaction, a rise in E value with temperature indicates that the controlling mechanism has shifted to an alternate or parallel path; on the other hand, a drop in E value indicates that the controlling mechanism has shifted from one of a succession of elementary steps to another in the series. These conclusions are illustrated in Fig. 3.

PREDICTABILITY OF REACTION RATE FROM THEORY

The rate expression in general involves two factors, the temperature-dependent and the concentration-dependent factors. Consider the prediction of these factors in turn.

Concentration-dependent Term

If a reaction has available a number of competing paths (e.g., noncatalytic and catalytic) it will in fact proceed by all of these paths, although primarily by the one of least resistance. This path usually dominates. Only a knowledge of the energies of all possible intermediates will allow prediction of the dominant path and its corresponding rate expression. As such information cannot be found in the present state of knowledge, *a priori* prediction of the form of the concentration term is not possible. Actually, the form of the experimentally found rate expression is often the clue used to investigate the energies of the intermediates of a reaction.

Temperature-dependent Term

Assuming that we already know the mechanism of reaction and whether it is elementary or not, we may then proceed to the prediction of the frequency factor and activation energy terms of the rate constant.

Frequency factor predictions from either collision or transition-state theory may come within a factor of 100 of the correct value; however, in specific cases predictions may be much further off.

Though activation energies can be estimated from transition-state theory, reliability is poor, and it is probably best to estimate them from the experimental findings for reactions of similar compounds. For example, the activation energies of the following homologous series of reactions

$$RI + C_6H_5ONa \xrightarrow{\text{ethanol}} C_6H_5OR + NaI$$

where R is

CH_3	C_7H_{15}	iso-C_3H_7	*sec*-C_4H_9
C_2H_5	C_8H_{17}	iso-C_4H_9	*sec*-C_6H_{13}
C_3H_7	$C_{16}H_{33}$	iso-C_5H_{11}	*sec*-C_8H_{17}
C_4H_9			*tert*-C_4H_9

all lie between 21.5 and 23.5 kcal.

Use of Predicted Values in Design

The frequent order-of-magnitude predictions of the theories tend to confirm the correctness of their representations, help find the form and the energies of various intermediates, and give us a better understanding of chemical structure. However, theoretical predictions rarely match experiment by a factor of two. In addition, we can never tell beforehand whether the predicted rate will be in the order of magnitude of experiment or will be off by a factor of 10^6. Therefore for engineering design this kind of information should not be relied on, and experimentally found rates should be used in all cases. Thus theoretical studies may be used as a supplementary aid to suggest the temperature sensitivity of a given reaction from a similar type of reaction, to suggest the upper limits of reaction rate, etc. Design invariably relies on experimentally determined rates.

RELATED READINGS

Friess, S. L., and Weissberger, A., Editors, *Techniques of Organic Chemistry, Vol. 8, Investigation of Rates and Mechanisms of Reaction*, Interscience Press, New York, 1953.

Jungers, J. C., *et al.*, *Cinétique chimique appliquée*, Technip, Paris, 1958.

Laidler, K. J., *Chemical Kinetics*, 2nd ed., McGraw-Hill, New York, 1965.

Moore, W. J., *Physical Chemistry*, Prentice-Hall, New York, 1950, Ch. 17.

REFERENCES

Denbigh, K. G., *The Principles of Chemical Equilibrium*, Cambridge University Press, Cambridge, England, 1955, p. 442.

Dolbear, A. E., *Am. Naturalist*, **31**, 970 (1897).

Lindemann, F. A., *Trans. Faraday Soc.*, **17**, 598 (1922).

Michaelis, L., and Menten, M. L., *Biochem. Z.*, **49**, 333 (1913). This treatment is discussed by Laidler (1965) and by Freiss and Weissberger (1953), see Related Readings.

Ogg, R., *J. Chem. Phys.*, **15**, 337 (1947).

Olah, G. A., *Science*, **168**, 1298 (1970).

Rice, F. O., and Herzfeld, K. F., *J. Am. Chem. Soc.*, **56**, 284 (1934).

PROBLEMS

1. A reaction has the stoichiometric equation $A + B = 2R$. What is the order of reaction?

2. Given the reaction $2NO_2 + \frac{1}{2}O_2 = N_2O_5$, what is the relation between the rates of formation and disappearance of the three components of the reaction?

3. A reaction with stoichiometric equation $\frac{1}{2}A + B = R + \frac{1}{2}S$ has the following rate expression

$$-r_A = 2C_A^{0.5}C_B$$

What is the rate expression for this reaction if the stoichiometric equation is written as $A + 2B = 2R + S$?

4. A certain reaction has a rate given by

$$-r_A = 0.005C_A^2, \quad \text{mol/cm}^3\cdot\text{min}$$

If the concentration is to be expressed in mol/liter and time in hours, what would be the value and units of the rate constant?

5. For a gas reaction at 400°K the rate is reported as

$$-\frac{dp_A}{dt} = 3.66p_A^2, \quad \text{atm/hr}$$

(*a*) What are the units of the rate constant?
(*b*) What is the value of the rate constant for this reaction if the rate equation is expressed as

$$-r_A = -\frac{1}{V}\frac{dN_A}{dt} = kC_A^2, \quad \text{mol/liter}\cdot\text{hr}$$

6. Show that the following scheme

$$N_2O_5 \underset{k_2}{\overset{k_1}{\rightleftarrows}} NO_2 + NO_3^*$$

$$NO_3^* \xrightarrow{k_3} NO^* + O_2$$

$$NO^* + NO_3^* \xrightarrow{k_4} 2NO_2$$

proposed by Ogg (1947) is consistent with, and can explain, the observed first-order decomposition of N_2O_5.

7. The decomposition of reactant A at 400°C for pressures between 1 and 10 atm follows a first-order rate law.

(*a*) Show that a mechanism similar to azomethane decomposition, p. 17,

$$A + A \rightleftarrows A^* + A$$

$$A^* \rightarrow R + S$$

is consistent with the observed kinetics.

Different mechanisms can be proposed to explain first-order kinetics. To claim that this mechanism is correct in the face of the other alternatives requires additional evidence.

(*b*) For this purpose, what further experiments would you suggest we run and what results would you expect to find?

8. Experiment shows that the homogeneous decomposition of ozone proceeds with a rate

$$-r_{O_3} = k[O_3]^2[O_2]^{-1}$$

(*a*) What is the over-all order of reaction?
(*b*) Suggest a two-step mechanism to explain this rate and state how you would further test this mechanism.

9. Under the influence of oxidizing agents, hypophosphorous acid is transformed into phosphorous acid:

$$H_3PO_2 \xrightarrow{\text{oxidizing agent}} H_3PO_3$$

The kinetics of this transformation present the following features. At low concentration of oxidizing agent

$$r_{H_3PO_3} = k[\text{oxidizing agent}][H_3PO_2]$$

At high concentration of oxidizing agent

$$r_{H_3PO_3} = k'[H^+][H_3PO_2]$$

To explain the observed kinetics, it has been postulated that with hydrogen ion as catalyst normal unreactive H_3PO_2 is transformed into an active form, the nature of which is unknown. This intermediate then reacts with the oxidizing agent to give H_3PO_3. Show that this scheme does explain the observed kinetics.

10. Chemicals A, B, and D combine to give R and S with stoichiometry $A + B + D = R + S$, and after the reaction has proceeded to a significant extent, the observed rate is

$$r_R = kC_A C_B C_D / C_R$$

(*a*) What is the order of the reaction?

The following two mechanisms involving formation of active intermediate have been proposed to explain the observed kinetics.

Mechanism I:

$$A + B \rightleftarrows X^* + R$$

$$D + X^* \rightarrow S$$

Mechanism II:

$$A + D \rightleftarrows Y^* + R$$

$$B + Y^* \rightarrow S$$

(*b*) Are these mechanisms consistent with the kinetic data?
(*c*) If neither is consistent, devise a scheme that is consistent with the kinetics. If only one is consistent, what line of investigation may strengthen the conviction that the mechanism selected is correct? If both are consistent, how would you be able to choose between them?

11. A_2B decomposes with stoichiometry $A_2B = AB + \frac{1}{2}A_2$. Much effort has been expended to discover the kinetics of this reaction, but the results are discouraging, and no concise rate equation can be made to fit the data. The following observations can be made from the data, however.

1. At the start of any experimental run the reaction seems to be of first order with respect to reactant.
2. When the reactant is just about gone, the data are well correlated by an equation which is second order with respect to reactant.
3. Introducing product AB into the feed leaves the rate unaffected.
4. Introducing product A_2 into the feed slows down the rate of reaction; however, no proportionality can be found between the amount of A_2 added and the slowing effect.

With the hope that a theoretical treatment may suggest a satisfactory form of rate expression, the following mechanisms are being considered.

Mechanism I:

$$2A_2B \rightleftarrows (A_4B_2)^*$$

$$(A_4B_2)^* \rightleftarrows A_2 + 2AB$$

Mechanism II:

$$A_2B \rightleftarrows A^* + AB$$

$$A_2B + A^* \rightleftarrows A_2 + AB$$

(*a*) Are either of these mechanisms consistent with the experimental findings? If a mechanism is rejected, state on what basis you reject it.
(*b*) If neither of these mechanisms is satisfactory, can you devise one that is consistent with the experimental findings?

12. The primary reaction occurring in the homogeneous decomposition of nitrous oxide is found to be

$$N_2O \rightarrow N_2 + \tfrac{1}{2}O_2$$

with rate

$$-r_{N_2O} = \frac{k_1[N_2O]^2}{1 + k_2[N_2O]}$$

Devise a mechanism to explain this observed rate.

13. The pyrolysis of ethane proceeds with an activation energy of about 75,000 cal. How much faster is the decomposition at 650°C than at 500°C.

14. On typical summer days field crickets nibble, jump, and chirp now and then. But at night when great numbers congregate chirping seems to become a serious business and the chirp rate becomes quite regular. In 1897 Dolbear reported that not only was it regular but that the rate was determined by the temperature as given by

$$(\text{number of chirps in 15 secs}) + 40 = (\text{temperature, °F})$$

Assuming that the chirping rate is a direct measure of the metabolic rate, find the activation energy, in calories, of these crickets in the temperature range of 60 to 80°F.

15. Experiment shows that the primary reaction in the homogeneous decomposition of nitrous oxide proceeds with stoichiometry

$$N_2O \rightarrow N_2 + \tfrac{1}{2}O_2$$

and rate

$$-r_{N_2O} = \frac{k_1[N_2O]^2}{1 + k_2[N_2O]}, \quad \text{mol/liter}\cdot\text{min}$$

where

$$k_1 = 10^{19.39}e^{-81,800/RT}$$

$$k_2 = 10^{8.69}e^{-28,400/RT}$$

(*a*) What is the activation energy for this reaction?
(*b*) Sketch the curve you expect to obtain in a graph such as Fig. 3.

16. At 500°K the rate of a bimolecular reaction is ten times the rate at 400°K. Find the activation energy of this reaction:

(*a*) From Arrhenius' law.
(*b*) From collision theory.
(*c*) What is the percentage difference in rate of reaction at 600°K predicted by these two methods?

17. The formation and decomposition of phosgene has been found to proceed as follows:

$$CO + Cl_2 \underset{k_2}{\overset{k_1}{\rightleftarrows}} COCl_2$$

Forward reaction: $r_{COCl_2} = k_1[Cl_2]^{3/2}[CO]$

Reverse reaction: $-r_{COCl_2} = k_2[Cl_2]^{1/2}[COCl_2]$

(*a*) Are these rate expressions thermodynamically consistent?
(*b*) Determine which of the following mechanisms is consistent with these experimentally found rates.

Mechanism I:

$Cl_2 \rightleftarrows 2Cl^*$ fast, at equilibrium
$Cl^* + CO \rightleftarrows COCl^*$ fast, at equilibrium
$COCl^* + Cl_2 \rightleftarrows COCl_2 + Cl^*$ slow and rate controlling

Mechanism II:

$Cl_2 \rightleftarrows 2Cl^*$ fast, at equilibrium
$Cl^* + Cl_2 \rightleftarrows Cl_3^*$ fast, at equilibrium
$Cl_3^* + CO \rightleftarrows COCl_2 + Cl^*$ slow and rate controlling

18. *Free radical chain reactions.* The thermal decomposition of hydrocarbons frequently exhibit $n = 0.5, 1.0, 1.5, \ldots$ order kinetics, and this behavior can be explained in terms of a free radical chain reaction mechanism, first proposed by Rice and Herzfeld (1934). As an example, suppose that hydrocarbon A decomposes as follows to form product molecules $R_1, R_2, \ldots$, while $X_1\cdot$ and $X_2\cdot$ are free radical intermediates.

Formation of free radical:

$$A \xrightarrow{\text{slow}} R_1 + X_1\cdot \tag{i}$$

Chain propagation steps, usually repeated many times:

$$A + X_1\cdot \rightarrow R_2 + X_2\cdot \tag{ii}$$

$$X_2\cdot \rightarrow R_3 + X_1\cdot \tag{iii}$$

Possible termination steps, representing the destruction of the free radical:

$$2X_1\cdot \rightarrow R_4 \tag{iv}$$

$$X_1\cdot + X_2\cdot \rightarrow R_5 \tag{v}$$

$$2X_2\cdot \rightarrow R_6 \tag{vi}$$

Using initiation and propagation steps (i), (ii), and (iii) determine, in turn, the reaction order of the decomposition of A when termination proceeds according to

(*a*) Step (iv)
(*b*) Step (v)
(*c*) Step (vi)

19. *Enzyme-substrate reactions.* Here a reactant, called the substrate, is converted into product by the action of an enzyme, a high molecular weight (m.w. > 10,000) protein-like substance. An enzyme is highly specific, catalyzing only one particular reaction, or one group of reactions. Thus we have

$$A \xrightarrow{\text{enzyme}} R$$

Many of these reactions exhibit the following rate characteristics:

1. The rate is proportional to the concentration of enzyme introduced into the mixture $[E_0]$.
2. At low reactant concentration the rate is proportional to the reactant concentration.
3. At high reactant concentration the rate levels off to become independent of reactant concentration.

Michaelis and Menten (1913) first explained this general behavior with the following mechanism:

$$A + E \underset{k_2}{\overset{k_1}{\rightleftarrows}} (A.E)^*$$

$$(A.E)^* \xrightarrow{k_3} R + E$$

The particular feature of this model is the assumption that the concentration of intermediate can be appreciable, in which case the total enzyme is distributed as follows:

$$[E_0] = [E] + [(A.E)^*]$$

Since the enzyme concentration cannot easily be followed, develop the rate equation for this reaction in terms of $[E_0]$ and $[A]$ and show that it explains the observed behavior. Use the steady-state approximation in your development.

20. *Photochemical reactions.* These reactions proceed only in the presence of light with a rate dependent on the intensity of incident radiation, but are otherwise homogeneous in nature. This type of reaction does not fit into our homogeneous-heterogeneous classification of reaction types. Examine the definitions in Chapter 1 and beginning of Chapter 2 and then decide how best to modify or interpret these classifications to include photochemical reactions.

If the reaction rate of an otherwise homogeneous reaction is influenced by the magnetic field strength, the electrical field strength (see *Chem. Eng. News*, **44**, 37 (Feb. 28, 1966)), intensity of psychical waves, shear stress on the fluid or what have you, how would you include these factors?

3

INTERPRETATION OF BATCH REACTOR DATA

A rate equation characterizes the rate of reaction and its form may either be suggested by theoretical considerations or simply be the result of an empirical curve-fitting procedure. In any case the value of the constants of the equation can only be found by experiment; predictive methods are inadequate at present.

The determination of the rate equation is usually a two-step procedure; first the concentration dependency is found at fixed temperature and then the temperature dependence of the rate constants is found, yielding the complete rate equation.

Equipment by which empirical information is obtained can be divided into two types, the *batch* and *flow* reactors. The batch reactor is simply a container to hold the contents while they react. All that has to be determined is the extent of reaction at various times, and this can be followed in a number of ways, for example:

1. By following the concentration of a given component.
2. By following the change in some physical property of the fluid such as the electrical conductivity or refractive index.
3. By following the change in total pressure of a constant-volume system.
4. By following the change in volume of a constant-pressure system.

The experimental batch reactor is usually operated isothermally and at constant volume because it is easy to interpret the results of such runs. This reactor is a relatively simple device adaptable to small-scale laboratory setups, and it needs but little auxiliary equipment or instrumentation. Thus it is used whenever possible for obtaining homogeneous kinetic data. This chapter deals with the batch reactor.

The flow reactor is used primarily in the study of the kinetics of heterogeneous reactions, though in a number of instances it is used to complement and offers advantages over the batch reactor in the study of homogeneous reactions. Reactions which are difficult to follow, reactions which yield a variety of products, very rapid reactions, and gas-phase reactions are examples of situations which may be more easily followed in flow reactors. Planning of experiments and interpretation of data obtained in flow reactors are considered in later chapters.

There are two procedures for analyzing kinetic data, the *integral* and the *differential* methods. In the integral method of analysis we guess a particular form of rate equation and, after appropriate integrations and mathematical manipulations, predict that the plot of a certain concentration function versus time should yield a straight line. The data are plotted, and if a reasonably good straight line is obtained then the rate equation is said to satisfactorily fit the data.

In the differential method of analysis we test the fit of the rate expression to the data directly, and without any integration. However, since the rate expression is a differential equation, we must first find $(1/V)(dN/dt)$ from the data before attempting the fitting procedure.

There are advantages and disadvantages to each method. The integral method is easy to use and is recommended when testing specific mechanisms, or relatively simple rate expressions, or when the data are so scattered that we cannot reliably find the derivatives needed in the differential method. The differential method is useful in more complicated situations but requires more accurate or larger amounts of data. The integral method can only test this or that particular mechanism or rate form; the differential method can be used to develop or build up a rate equation to fit the data.

In general, it is suggested that integral analysis be attempted first, and, if not successful, that the differential method be tried. For complicated cases we may need to use special experimental methods which give a partial solution of the problem, or else flow reactors coupled with differential analysis.

CONSTANT-VOLUME BATCH REACTOR

When we mention the constant-volume batch reactor we are really referring to the volume of reaction mixture, and not the volume of reactor. Thus this term actually means a constant-volume, or a *constant-density reaction system.* Most liquid-phase reactions as well as all gas-phase reactions occurring in a constant-volume bomb fall in this class.

In a constant-volume system the measure of reaction rate of component i becomes

$$r_i = \frac{1}{V}\frac{dN_i}{dt} = \frac{d(N_i/V)}{dt} = \frac{dC_i}{dt} \tag{1}$$

or for ideal gases

$$r_i = \frac{1}{RT}\frac{dp_i}{dt} \tag{2}$$

Thus the rate of reaction of any component is given by the rate of change of its concentration or partial pressure; so no matter how we choose to follow the progress of the reaction, we must eventually relate this measure to the concentration or partial pressure if we are to follow the rate of reaction.

For gas reactions with changing number of moles, a simple way of finding the reaction rate is to follow the change in total pressure π of the system. Let us see how this can be done.

Analysis of Total Pressure Data Obtained in a Constant-volume System. For isothermal gas reactions where the number of moles of material change during reaction, let us develop the general expression which relates the changing total pressure of the system π to the changing concentration or partial pressure of any of the reaction components, for given initial conditions and any reaction stoichiometry.

Write the general stoichiometric equation, and under each term indicate the number of moles of that component present:

	aA	$+$	bB	$+\cdots=$	rR	$+$	sS	$+\cdots$
At time 0:	N_{A0}		N_{B0}		N_{R0}		N_{S0}	N_{inert}
At time t:	$N_{A0} - ax$		$N_{B0} - bx$		$N_{R0} + rx$		$N_{S0} + sx$	N_{inert}

Initially the total number of moles present in the system is

$$N_0 = N_{A0} + N_{B0} + \cdots + N_{R0} + N_{S0} + \cdots + N_{inert}$$

but at time t it is

$$N = N_0 + x(r + s + \cdots - a - b - \cdots) = N_0 + x\,\Delta n$$

where

$$\Delta n = r + s + \cdots - a - b - \cdots$$

Assuming that the ideal gas law holds, we may write for any reactant, say A, in the system of volume V

$$C_A = \frac{p_A}{RT} = \frac{N_A}{V} = \frac{N_{A0} - ax}{V}$$

Combining these two expressions we obtain

$$C_A = \frac{N_{A0}}{V} - \frac{a}{\Delta n}\frac{N - N_0}{V}$$

or

$$p_A = C_A RT = p_{A0} - \frac{a}{\Delta n}(\pi - \pi_0) \tag{3}$$

Equation 3 gives the concentration or partial pressure of reactant A as a function of the total pressure π at time t, initial partial pressure of A, p_{A0}, and initial total pressure of the system π_0.

Similarly, for any product R we can find

$$p_R = C_R RT = p_{R0} + \frac{r}{\Delta n}(\pi - \pi_0) \tag{4}$$

Equations 3 and 4 are the desired relationships between total pressure of the system and the partial pressure of reacting materials.

It should be emphasized that if the precise stoichiometry is not known, or if more than one stoichiometric equation is needed to represent the reaction, then the "total pressure" procedure cannot be used.

Integral Method of Analysis of Data

General Procedure. The integral method of analysis always puts a particular rate equation to the test by integrating and comparing the predicted C versus t curve with the experimental C versus t data. If the fit is unsatisfactory, another rate equation is suggested and tested. The procedure may be summarized as follows.

1. In a constant-volume system the rate expression for the disappearance of reactant A will be of the following form:

$$-r_A = -\frac{dC_A}{dt} = f(k, C) \tag{5}$$

or in the more restricted case in which the concentration-dependent terms may be separated from the concentration-independent terms, we have

$$-r_A = -\frac{dC_A}{dt} = kf(C) \tag{6}$$

With either form we proceed as follows; however, it is easier to illustrate the procedure with Eq. 6.

2. Equation 6 is rearranged to give

$$-\frac{dC_A}{f(C)} = k\,dt$$

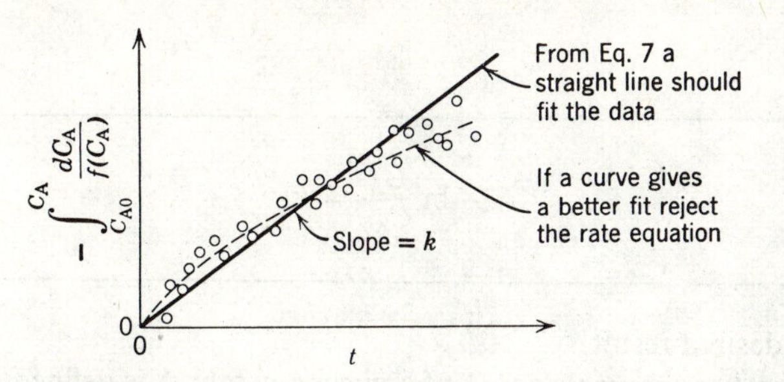

FIGURE 1. Test of a rate equation by the integral method of analysis.

Now $f(C)$ only involves concentrations of materials, which may be expressed in terms of C_A. Thus Eq. 6 may be integrated either analytically or graphically to give

$$-\int_{C_{A0}}^{C_A} \frac{dC_A}{f(C_A)} = k \int_0^t dt = kt \tag{7}$$

3. This concentration function is proportional to time, so a plot such as that of Fig. 1 yields a straight line of slope k for this particular rate equation.

4. From experiment determine numerical values for the integral of Eq. 7, and plot these at the corresponding times on Fig. 1.

5. See whether these data fall on a reasonably straight line passing through the origin. If there is no reason to doubt this, then it may be said that the particular rate equation being tested satisfactorily fits the data. If the data are better fitted by a curved line then this rate equation and its mechanism are rejected, and another rate equation is tried.

The integral method is especially useful for fitting simple reaction types corresponding to elementary reactions. Let us examine a number of these forms.

Irreversible Unimolecular-type First-order Reactions. Consider the reaction

$$A \rightarrow \text{products} \tag{8a}$$

Suppose we wish to test the first-order rate equation of the following type,

$$-r_A = -\frac{dC_A}{dt} = kC_A \tag{8b}$$

for this reaction. Separating and integrating we obtain

$$-\int_{C_{A0}}^{C_A} \frac{dC_A}{C_A} = k \int_0^t dt$$

or

$$-\ln \frac{C_A}{C_{A0}} = kt \tag{9}$$

which is the desired result.

Now the *fractional conversion* X_A of a given reactant A is defined as the fraction of reactant converted into product or

$$X_A = \frac{N_{A0} - N_A}{N_{A0}} \quad (*) \tag{10}$$

Fractional conversion (or simply conversion) is a convenient variable often used in place of concentration in engineering work; therefore most of the results which follow will be presented in terms of both C_A and X_A.

Let us now see how Eq. 9 can be derived using conversions. First of all, for a constant density system V does not change so

$$C_A = \frac{N_A}{V} = \frac{N_{A0}(1 - X_A)}{V} = C_{A0}(1 - X_A) \tag{11*}$$

and

$$-dC_A = C_{A0}\, dX_A$$

and hence Eq. 8 becomes

$$\frac{dX_A}{dt} = k(1 - X_A)$$

rearranging and integrating gives

$$\int_0^{X_A} \frac{dX_A}{1 - X_A} = k \int_0^t dt$$

or

$$-\ln (1 - X_A) = kt \tag{12}$$

which is equivalent to Eq. 9. A plot of $\ln (1 - X_A)$ or $\ln (C_A/C_{A0})$ versus t, as shown in Fig. 2, gives a straight line through the origin for an equation of this type.

* Equation 72 presents a more general relationship between concentration and conversion for variable volume (or variable density) systems.

$(*)\cdot\ N_A = N_{A0}(1 - X_A) = N_{A0} + \alpha x\ ,\ a < 0$

$\therefore\ x = -\frac{X_A N_{A0}}{\alpha_A}$

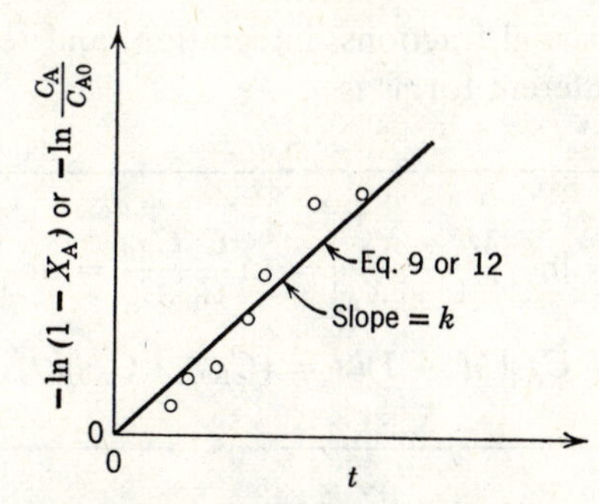

FIGURE 2. Test for the first-order reaction, Eq. 8.

Caution: We should point out that equations such as

$$-\frac{dC_A}{dt} = kC_A^{0.6}C_B^{0.4}$$

are first order but are not amenable to this kind of analysis; hence not all first-order reactions can be treated this way.

Irreversible Bimolecular-type Second-order Reactions. Consider the reaction

$$A + B \rightarrow \text{products} \tag{13a}$$

with corresponding rate equation

$$-r_A = -\frac{dC_A}{dt} = -\frac{dC_B}{dt} = kC_AC_B \tag{13b}$$

Noting that the amounts of A and B which have reacted at any time t are equal and given by $C_{A0}X_A = C_{B0}X_B$ we may write Eq. 13 in terms of X_A as

$$-r_A = C_{A0}\frac{dX_A}{dt} = k(C_{A0} - C_{A0}X_A)(C_{B0} - C_{A0}X_A)$$

Letting $M = C_{B0}/C_{A0}$ be the initial molar ratio of reactants, we obtain

$$-r_A = C_{A0}\frac{dX_A}{dt} = kC_{A0}^2(1 - X_A)(M - X_A)$$

which on separation and formal integration becomes

$$\int_0^{X_A} \frac{dX_A}{(1 - X_A)(M - X_A)} = C_{A0}k\int_0^t dt$$

After breakdown into partial fractions, integration, and rearrangement, the final result in a number of different forms is

$$\ln \frac{1 - X_B}{1 - X_A} = \ln \frac{M - X_A}{M(1 - X_A)} = \ln \frac{C_B C_{A0}}{C_{B0} C_A} = \ln \frac{C_B}{M C_A}$$
$$= C_{A0}(M - 1)kt = (C_{B0} - C_{A0})kt, \qquad M \neq 1 \tag{14}$$

Figure 3 shows two equivalent ways of obtaining a linear plot between the concentration function and time for this second-order rate law.

If C_{B0} is much larger than C_{A0}, C_B remains approximately constant at all times, and Eq. 14 approaches Eq. 9 or 12 for the first-order reaction. Thus the second-order reaction becomes a pseudo first-order reaction.

Caution 1: In the special case where reactants are introduced in the stoichiometric ratio the integrated rate expression becomes indeterminate and this requires taking limits of quotients for evaluation. This difficulty is avoided if we go back to the original differential rate expression and solve it for this particular reactant ratio. Thus for the second-order reaction with equal initial concentrations of A and B, or for the reaction

$$2A \rightarrow \text{products} \tag{15a}$$

the defining second-order differential equation becomes

$$-r_A = -\frac{dC_A}{dt} = kC_A^2 = kC_{A0}^2(1 - X_A)^2 \tag{15b}$$

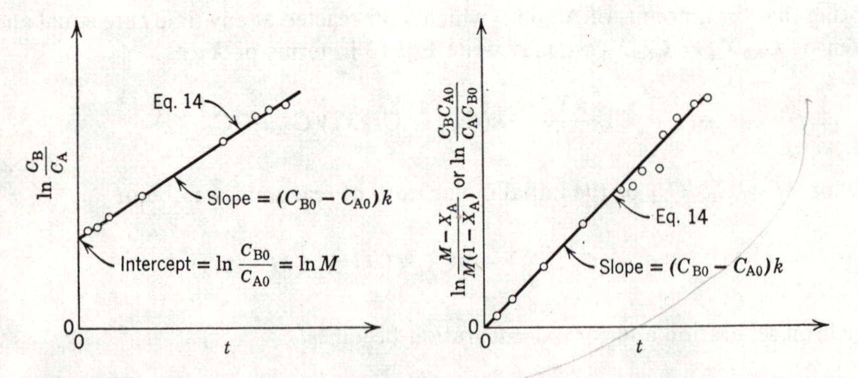

FIGURE 3. Test for the bimolecular mechanism $A + B \rightarrow R$ with $C_{A0} \neq C_{B0}$, or for the second-order reaction, Eq. 13.

which on integration yields

$$\frac{1}{C_A} - \frac{1}{C_{A0}} = \frac{1}{C_{A0}}\frac{X_A}{1 - X_A} = kt \tag{16}$$

Plotting the variables as shown in Fig. 4 provides a test for this rate expression.

In practice we should choose reactant ratios either equal to or widely different from the stoichiometric ratio.

Caution 2: The integrated rate expression depends on the stoichiometry as well as the kinetics. To illustrate, if the reaction

$$A + 2B \rightarrow \text{products} \tag{17a}$$

is first order with respect to both A and B, hence second order overall, or

$$-r_A = -\frac{dC_A}{dt} = kC_AC_B = kC_{A0}^2(1 - X_A)(M - 2X_A) \tag{17b}$$

the integrated form is

$$\ln\frac{C_BC_{A0}}{C_{B0}C_A} = \ln\frac{M - 2X_A}{M(1 - X_A)} = C_{A0}(M - 2)kt, \qquad M \neq 2 \tag{18}$$

With a stoichiometric reactant ratio the integrated form is

$$\frac{1}{C_A} - \frac{1}{C_{A0}} = \frac{1}{C_{A0}}\frac{X_A}{1 - X_A} = 2kt, \qquad M = 2 \tag{19}$$

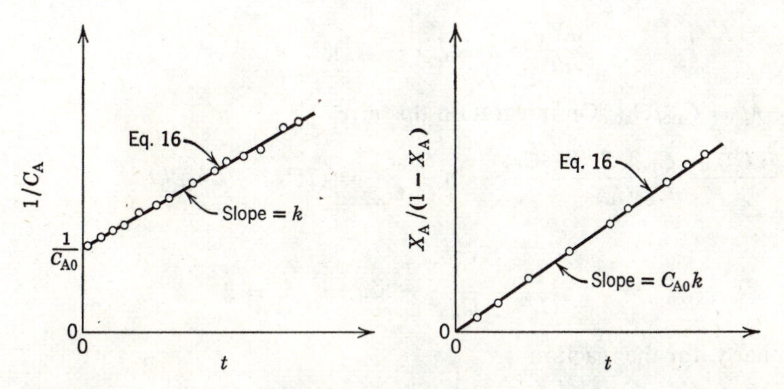

FIGURE 4. Test for the bimolecular mechanisms, $A + B \rightarrow R$ with $C_{A0} = C_{B0}$, or for the second-order reaction of Eq. 15.

These two cautions apply to all reaction types. Thus special forms for the integrated expressions appear when the reactants are used in stoichiometric ratios, or when the reaction is not elementary.

Irreversible Trimolecular-type Third-order Reactions. For the reaction

$$A + B + D \rightarrow \text{products} \tag{20a}$$

let the rate equation be

$$-r_A = -\frac{dC_A}{dt} = kC_AC_BC_D \tag{20b}$$

or in terms of X_A

$$C_{A0}\frac{dX_A}{dt} = kC_{A0}^3(1 - X_A)\left(\frac{C_{B0}}{C_{A0}} - X_A\right)\left(\frac{C_{D0}}{C_{A0}} - X_A\right)$$

On separation of variables, breakdown into partial fractions, and integration, we obtain after manipulation

$$\frac{1}{(C_{A0} - C_{B0})(C_{A0} - C_{D0})}\ln\frac{C_{A0}}{C_A} + \frac{1}{(C_{B0} - C_{D0})(C_{B0} - C_{A0})}\ln\frac{C_{B0}}{C_B}$$
$$+ \frac{1}{(C_{D0} - C_{A0})(C_{D0} - C_{B0})}\ln\frac{C_{D0}}{C_D} = kt \tag{21}$$

Now if C_{D0} is much larger than both C_{A0} and C_{B0}, the reaction becomes second order and Eq. 21 reduces to Eq. 14.

All trimolecular reactions found so far are of the form of Eq. 22 or 25. Thus

$$A + 2B \rightarrow R \qquad \text{with } -r_A = -\frac{dC_A}{dt} = kC_AC_B^2 \tag{22}$$

In terms of conversions the rate of reaction becomes

$$\frac{dX_A}{dt} = kC_{A0}^2(1 - X_A)(M - 2X_A)^2$$

where $M = C_{B0}/C_{A0}$. On integration this gives

$$\frac{(2C_{A0} - C_{B0})(C_{B0} - C_B)}{C_{B0}C_B} + \ln\frac{C_{A0}C_B}{C_AC_{B0}} = (2C_{A0} - C_{B0})^2kt, \qquad M \neq 2 \tag{23}$$

or

$$\frac{1}{C_A^2} - \frac{1}{C_{A0}^2} = 8kt, \qquad M = 2 \tag{24}$$

Similarly, for the reaction

$$A + B \rightarrow R \qquad \text{with } -r_A = -\frac{dC_A}{dt} = kC_AC_B^2 \tag{25}$$

integration gives

$$\frac{(C_{A0} - C_{B0})(C_{B0} - C_B)}{C_{B0}C_B} + \ln \frac{C_{A0}C_B}{C_{B0}C_A} = (C_{A0} - C_{B0})^2 kt, \qquad M \neq 1 \tag{26}$$

or

$$\frac{1}{C_A{}^2} - \frac{1}{C_{A0}{}^2} = 2kt, \qquad M = 1 \tag{27}$$

Empirical Rate Equations of nth Order. When the mechanism of reaction is not known, we often attempt to fit the data with an nth-order rate equation of the form

$$-r_A = -\frac{dC_A}{dt} = kC_A{}^n \tag{28}$$

which on separation and integration yields

$$C_A^{1-n} - C_{A0}^{1-n} = (n - 1)kt, \qquad n \neq 1 \tag{29}$$

The order n cannot be found explicitly from Eq. 29, so a trial-and-error solution must be made. This is not too difficult, however; we select a value for n and calculate k. The value of n which minimizes the variation in k is the desired value of n.

One curious feature of this rate form is that reactions with order $n > 1$ can never go to completion in finite time. On the other hand, for orders $n < 1$ this rate form predicts that the reactant concentration will fall to zero and then become negative at some finite time, found from Eq. 29 as

$$t = \frac{C_{A0}^{1-n}}{(1 - n)k}$$

Since the real concentration cannot fall below zero we should not carry out the integration beyond this time for $n < 1$. Also, as a consequence of this feature, in real systems the observed fractional order will shift upward to unity as a reactant is depleted.

Zero-order Reactions. A reaction is of zero order when the rate of reaction is independent of the concentration of materials; thus

$$-r_A = -\frac{dC_A}{dt} = k \tag{30}$$

Integrating and noting that C_A can never become negative we obtain directly

$$C_{A0} - C_A = C_{A0}X_A = kt \qquad \text{for } t < \frac{C_{A0}}{k} \tag{31}$$

which means that the conversion is proportional to time, as shown in Fig. 5.

As a rule reactions are of zero order only in certain concentration ranges—the higher concentrations. If the concentration is lowered far enough we usually find that the reaction becomes concentration-dependent, in which case the order rises from zero.

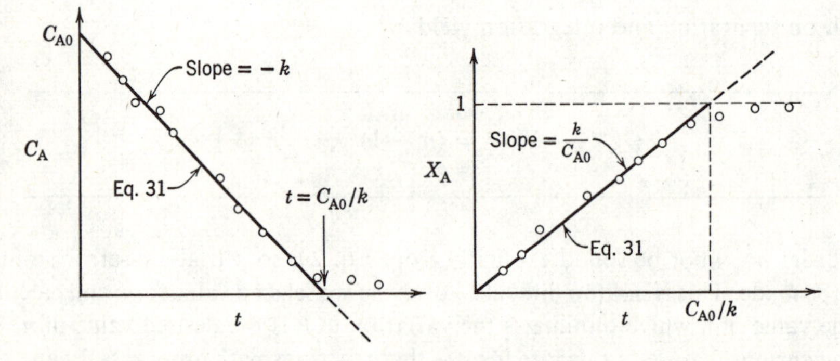

FIGURE 5. Test for a zero-order reaction, or rate equation, Eq. 30.

In general, zero-order reactions are those whose rates are determined by some factor other than the concentration of the reacting materials, e.g., the intensity of radiation within the vat for photochemical reactions, or the surface available in certain solid catalyzed gas reactions. It is important, then, to define the rate of zero order reactions so that this factor is included and properly accounted for. Consider the following illustrations.

In the photochemical reactor completely bathed by radiation the intensity of radiation within a volumetric element of fluid governs the rate, and doubling the volume of fluid doubles the number of moles reacted per unit time, hence the proper definition of reaction rate for this situation is

$$-\frac{1}{\text{volume of fluid}}\frac{dN_A}{dt} = f(\text{intensity of radiation})$$

which is Eq. 30, or the definition of Eq. 1.3.

On the other hand, in a solid catalyzed reaction the number of moles reacted per unit time is independent of the volume of fluid but is directly proportional to the surface of solid exposed. Thus the rate of reaction should be based on unit surface available, or

$$-\frac{1}{\text{surface available}}\frac{dN_A}{dt} = k$$

A knowledge of the physical situation will tell which rate form from among Eqs. 1.3, 1.4, 1.5, or 1.6 should be used to describe the rate. If Eq. 1.3 is used this leads to Eq. 30 giving a *homogeneous* zero-order reacting system. If some other definition of rate is the proper one to use then we have a *heterogeneous* zero-order system. These two systems should not be confused.

Over-all Order of Irreversible Reactions from the Half-life $t_{1/2}$. Sometimes for the irreversible reaction

$$\alpha A + \beta B + \cdots \rightarrow \text{products}$$

we may write

$$-r_A = -\frac{dC_A}{dt} = kC_A{}^aC_B{}^b\cdots$$

If the reactants are present in their stoichiometric ratios, they will remain at that ratio throughout the reaction. Thus for reactants A and B at any time $C_B/C_A = \beta/\alpha$, and we may write

$$-r_A = -\frac{dC_A}{dt} = kC_A{}^a\left(\frac{\beta}{\alpha}C_A\right)^b\cdots = k\left(\frac{\beta}{\alpha}\right)^b\cdots C_A{}^{a+b+\cdots}$$

or

$$-\frac{dC_A}{dt} = k'C_A{}^n \tag{32}$$

integrating for $n \neq 1$ gives

$$C_A^{1-n} - C_{A0}^{1-n} = k'(n-1)t$$

Defining the half-life of the reaction, $t_{1/2}$, as the time needed for the concentration of reactants to drop to one-half the original value, we obtain

$$t_{1/2} = \frac{2^{n-1}-1}{k'(n-1)}C_{A0}^{1-n} \tag{33}$$

This expression shows that a plot of log $t_{1/2}$ versus log C_{A0} gives a straight line of slope $1-n$ as shown in Fig. 6.

The half-life method requires making a series of runs, each at a different initial concentration, and shows that the fractional conversion in a given time rises with

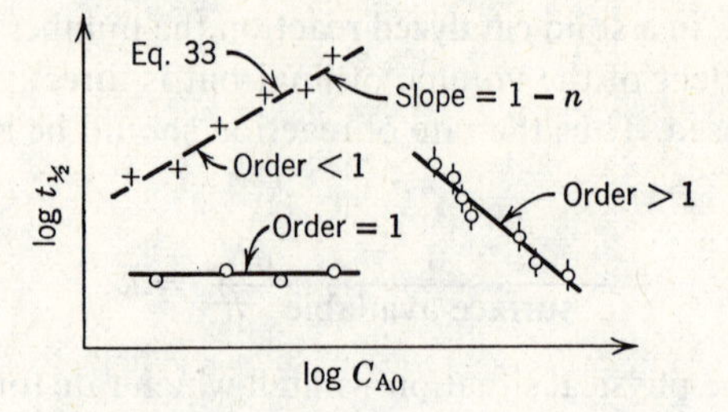

FIGURE 6. Over-all order of reaction from a series of half-life experiments, each at a different initial concentration of reactant.

increased concentration for orders greater than one, drops with increased concentration for orders less than one, and is independent of initial concentration for reactions of first order.

Numerous variations of this procedure are possible. For instance, by having all but one component, say A, in large excess, we can find the order with respect to that one component. For this situation the general expression reduces to

$$-\frac{dC_A}{dt} = k''C_A{}^a$$

where

$$k'' = k(C_{B0}{}^b \cdots) \qquad \text{and} \qquad C_B = C_{B0}$$

This method can be extended to any fractional life data $t_{1/m}$ but cannot be used for reactions in which it is impossible to maintain stoichiometric ratios, such as autocatalytic reactions.

Irreversible Reactions in Parallel. Consider the simplest case, A decomposing by two competing paths, both elementary reactions:

$$A \xrightarrow{k_1} R$$

$$A \xrightarrow{k_2} S$$

The rates of change of the three components are given by

$$-r_A = -\frac{dC_A}{dt} = k_1C_A + k_2C_A = (k_1 + k_2)C_A \tag{34}$$

$$r_R = \frac{dC_R}{dt} = k_1C_A \tag{35}$$

$$r_S = \frac{dC_S}{dt} = k_2C_A \tag{36}$$

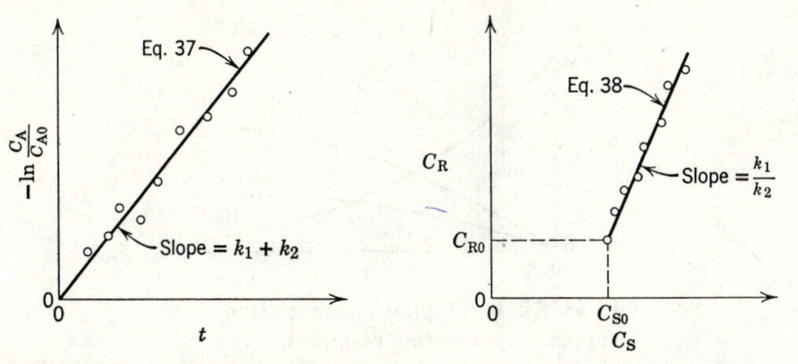

FIGURE 7. Evaluation of the rate constants for two competing elementary first-order reactions of the type

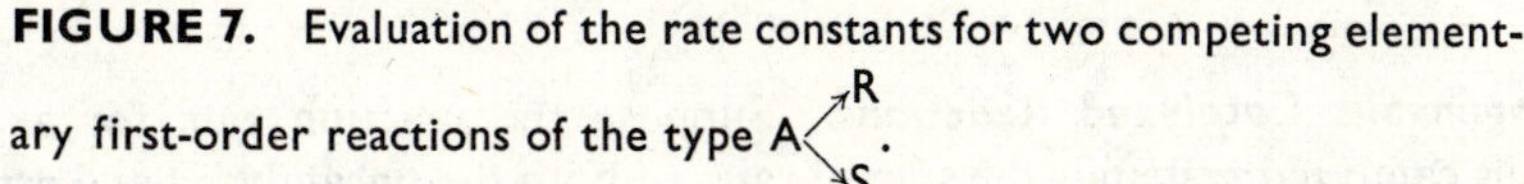

This is the first time we have encountered multiple reactions. For these in general, if it is necessary to write N stoichiometric equations to describe what is happening, then it is necessary to follow the composition of N reaction components to describe the kinetics. Thus in this system following C_A, C_R, or C_S alone will not give both k_1 and k_2. At least two components must be followed. Then from the stoichiometry, noting that $C_A + C_R + C_S$ is constant, we can find the concentration of the third component.

The k values are found using all three differential rate equations. First of all, Eq. 34, which is of simple first order, is integrated to give

$$-\ln \frac{C_A}{C_{A0}} = (k_1 + k_2)t \tag{37}$$

When plotted as in Fig. 7, the slope is $k_1 + k_2$. Then dividing Eq. 35 by Eq. 36 we obtain

$$\frac{r_R}{r_S} = \frac{dC_R}{dC_S} = \frac{k_1}{k_2}$$

which when integrated gives simply

$$\frac{C_R - C_{R0}}{C_S - C_{S0}} = \frac{k_1}{k_2} \tag{38}$$

This result is shown in Fig. 7. Thus the slope of a plot of C_R versus C_S gives the ratio k_1/k_2. Knowing k_1/k_2 as well as $k_1 + k_2$ gives k_1 and k_2. Typical concentration-time curves of the three components in a batch reactor for the case where $C_{R0} = C_{S0} = 0$ and $k_1 > k_2$ are shown in Fig. 8.

Reactions in parallel are considered in more detail in Chapter 7.

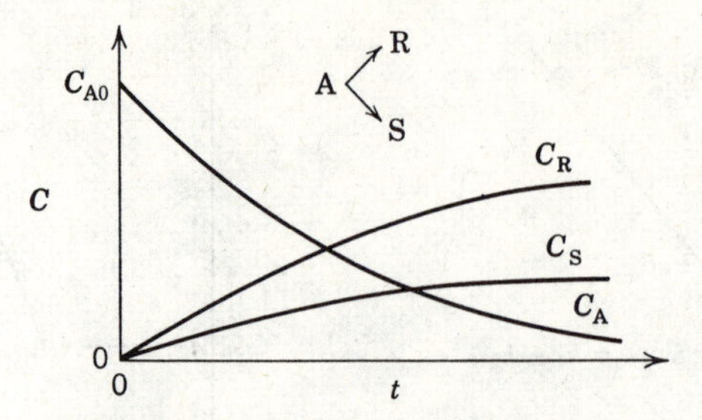

FIGURE 8. Typical concentration-time curves for competing reactions.

Homogeneous Catalyzed Reactions. Suppose the reaction rate for a homogeneous catalyzed system is the sum of rates of both the uncatalyzed and catalyzed reactions,

$$A \xrightarrow{k_1} R$$

$$A + C \xrightarrow{k_2} R + C$$

with corresponding reaction rates

$$-\left(\frac{dC_A}{dt}\right)_1 = k_1 C_A$$

$$-\left(\frac{dC_A}{dt}\right)_2 = k_2 C_A C_C$$

This means that the reaction would proceed even without a catalyst present and that the rate of catalyzed reaction is directly proportional to the catalyst concentration. The over-all rate of disappearance of reactant A is then

$$-\frac{dC_A}{dt} = k_1 C_A + k_2 C_A C_C = (k_1 + k_2 C_C) C_A \tag{39}$$

On integration, noting that the catalyst concentration remains unchanged, we have

$$-\ln \frac{C_A}{C_{A0}} = -\ln(1 - X_A) = (k_1 + k_2 C_C)t = k_{\text{observed}}\, t \tag{40}$$

Making a series of runs with varying catalyst concentration allows us to find k_1 and k_2. This is done by plotting the observed k value against the catalyst concentration as shown in Fig. 9. The slope of such a plot is k_2 and the intercept k_1.

Autocatalytic Reactions. A reaction in which one of the products of reaction acts as a catalyst is called an autocatalytic reaction. The simplest such reaction is

$$A + R \rightarrow R + R \tag{41a}$$

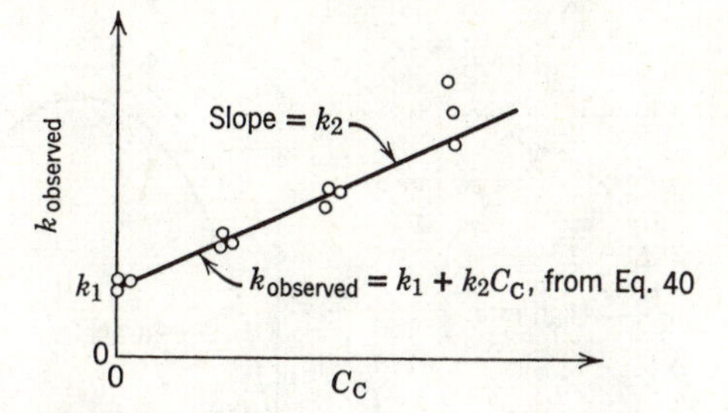

FIGURE 9. Rate constants for a homogeneous catalyzed reaction from a series of runs with different catalyst concentrations.

for which the rate equation is

$$-r_A = -\frac{dC_A}{dt} = kC_AC_R \tag{41b}$$

Because the total number of moles of A and R remain unchanged as A is consumed we may write at any time

$$C_0 = C_A + C_R = C_{A0} + C_{R0} = \text{constant}$$

Thus the rate equation becomes

$$-r_A = -\frac{dC_A}{dt} = kC_A(C_0 - C_A)$$

Rearranging and breaking into partial fractions, we obtain

$$-\frac{dC_A}{C_A(C_0 - C_A)} = -\frac{1}{C_0}\left(\frac{dC_A}{C_A} + \frac{dC_A}{C_0 - C_A}\right) = k\,dt$$

which on integration gives

$$\ln\frac{C_{A0}(C_0 - C_A)}{C_A(C_0 - C_{A0})} = \ln\frac{C_R/C_{R0}}{C_A/C_{A0}} = C_0kt = (C_{A0} + C_{R0})kt \tag{42}$$

In terms of the initial reactant ratio $M = C_{R0}/C_{A0}$ and fractional conversion of A this can be written as

$$\ln\frac{M + X_A}{M(1 - X_A)} = C_{A0}(M + 1)kt = (C_{A0} + C_{R0})kt \tag{43}$$

In an autocatalytic reaction some product R must be present if the reaction is to proceed at all. Starting with a very small concentration of R, we see qualitatively that the rate will rise as R is formed. At the other extreme, when A is just about used up the rate must drop to zero. This result is given in Fig. 10, which shows that

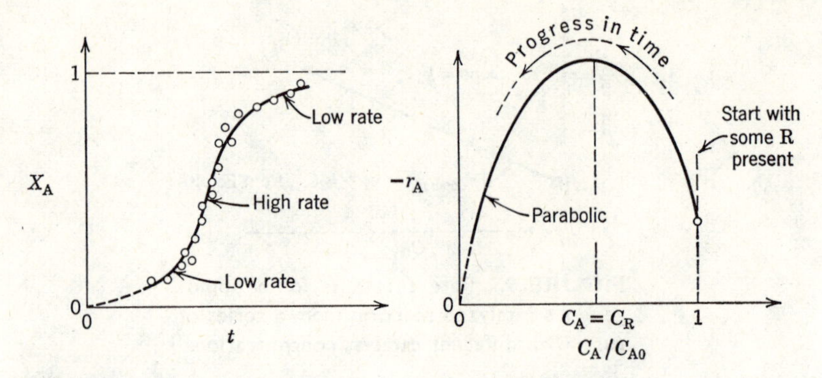

FIGURE 10. Conversion-time and rate-concentration curves for autocatalytic reaction of Eq. 41. This shape is typical for this type of reaction.

the rate follows a parabola, with a maximum where the concentrations of A and R are equal.

To test for an autocatalytic reaction, plot the time and concentration coordinates of Eq. 42 or 43 as shown in Fig. 11 and see whether a straight line passing through zero is obtained.

Autocatalytic reactions are considered in more detail in Chapter 6.

Irreversible Reactions in Series. We first consider consecutive unimolecular-type first-order reactions such as

$$A \xrightarrow{k_1} R \xrightarrow{k_2} S$$

whose rate equations for the three components are

$$r_A = \frac{dC_A}{dt} = -k_1 C_A \tag{44}$$

$$r_R = \frac{dC_R}{dt} = k_1 C_A - k_2 C_R \tag{45}$$

$$r_S = \frac{dC_S}{dt} = k_2 C_R \tag{46}$$

Let us start with a concentration C_{A0} of A, no R or S present, and see how the concentrations of the components change with time. By integration of Eq. 44 we find the concentration of A to be

$$-\ln \frac{C_A}{C_{A0}} = k_1 t \qquad \text{or} \qquad C_A = C_{A0} e^{-k_1 t} \tag{47}$$

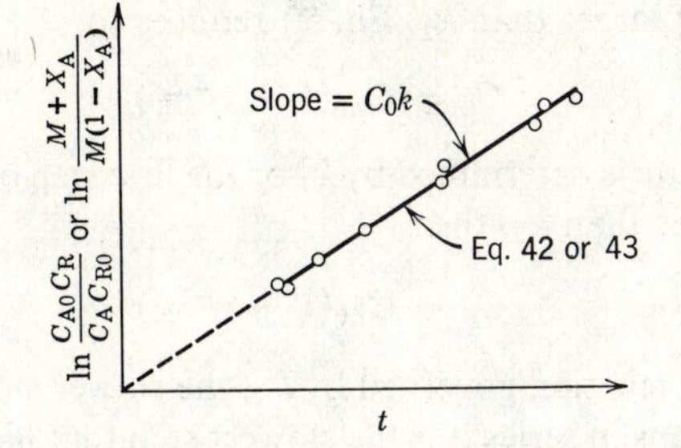

FIGURE 11. Test for the autocatalytic reaction of Eq. 41.

To find the variation in concentration of R, substitute the concentration of A from Eq. 47 into the differential equation governing the rate of change of R, Eq. 45; thus

$$\frac{dC_R}{dt} + k_2C_R = k_1C_{A0}e^{-k_1t} \tag{48}$$

which is a first-order linear differential equation of the form

$$\frac{dy}{dx} + Py = Q$$

By multiplying through with the integrating factor $e^{\int P\,dx}$ the solution is

$$ye^{\int P\,dx} = \int Qe^{\int P\,dx}\,dx + \text{constant}$$

Applying this general procedure to the integration of Eq. 48, we find that the integrating factor is e^{k_2t}. The constant of integration is found to be $-k_1C_{A0}/(k_2 - k_1)$ from the initial condition $C_{R0} = 0$ at $t = 0$, and the final expression for the changing concentration of R is

$$C_R = C_{A0}k_1\left(\frac{e^{-k_1t}}{k_2 - k_1} + \frac{e^{-k_2t}}{k_1 - k_2}\right) \tag{49}$$

Noting that there is no change in total number of moles, the stoichiometry relates the concentrations of reacting components by

$$C_{A0} = C_A + C_R + C_S$$

which with Eqs. 47 and 48 gives

$$C_S = C_{A0}\left(1 + \frac{k_2}{k_1 - k_2}\,e^{-k_1t} + \frac{k_1}{k_2 - k_1}\,e^{-k_2t}\right) \tag{50}$$

Thus we have found how the concentrations of components A, R, and S vary with time.

Now if k_2 is much larger than k_1, Eq. 50 reduces to

$$C_S = C_{A0}(1 - e^{-k_1 t})$$

In other words, the rate is determined by k_1 or the first step of the two-step reaction.
If k_1 is much larger than k_2, then

$$C_S = C_{A0}(1 - e^{-k_2 t})$$

which is a first-order reaction governed by k_2, the slower step in the two-step reaction. Thus for reactions in series it is the slowest step that has the greatest influence on the over-all reaction rate.

As may be expected, the values of k_1 and k_2 also govern the location and maximum concentration of R. This may be found by differentiating Eq. 49 and setting $dC_R/dt = 0$. The time at which the maximum concentration of R occurs is thus

$$t_{\max} = \frac{1}{k_{\text{log mean}}} = \frac{\ln (k_2/k_1)}{k_2 - k_1} \tag{51}$$

The maximum concentration of R is found by combining Eqs. 49 and 51 to give

$$\frac{C_{R,\max}}{C_{A0}} = \left(\frac{k_1}{k_2}\right)^{k_2/(k_2 - k_1)} \tag{52}$$

Figure 12 shows the general characteristics of the concentration-time curves for the three components; A decreases exponentially, R rises to a maximum and then falls, and S rises continuously, the greatest rate of increase of S occurring where R is a maximum.

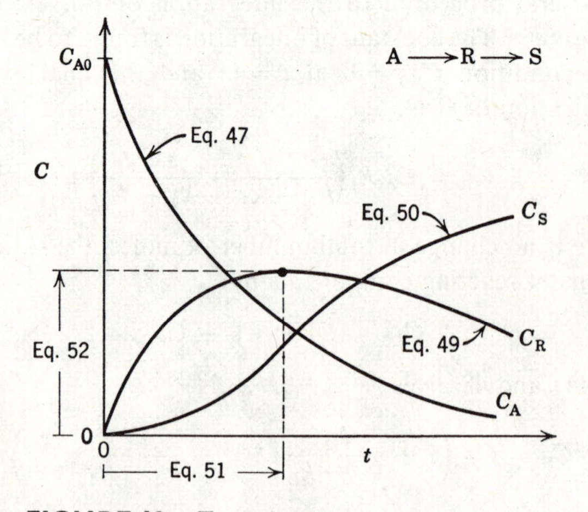

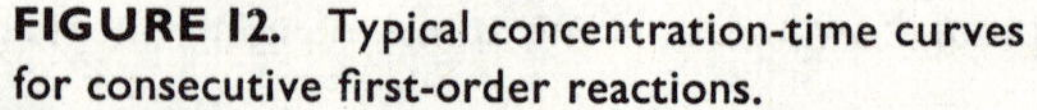
FIGURE 12. Typical concentration-time curves for consecutive first-order reactions.

Chapter 7 discusses series reactions in more detail. Specifically, Fig. 7.7 shows how the location and value of $C_{R,max}$ are related to the k_1/k_2 ratio. These may be used to find k_1 and k_2.

The following suggestions will help in exploring the kinetics of reactions in series for which the orders of the individual steps are unknown.

1. First determine whether the reactions can be treated as irreversible by seeing whether any reactants or intermediates are still present in the mixture after a long time.

2. Then, if the reactions are irreversible, examine the concentration-time curve for the reactants. This will give the reaction order and rate constant for the first step.

3. Find how the maximum concentration of intermediate varies as the reactant concentration is changed. For example, if the first step is of first order and $C_{R,max}/C_{A0}$ is independent of C_{A0}, the second step of the series is first order as well. If, however, $C_{R,max}/C_{A0}$ drops as C_{A0} rises, the disappearance of R becomes more rapid than its formation. Therefore the disappearance, or second step, is more concentration-sensitive and consequently is of higher order than the first step. Similarly, if $C_{R,max}/C_{A0}$ rises as C_{A0} rises, the second step is of lower order than the first step.

For reversible reactions of order other than one the analysis becomes more difficult.

For a longer chain of reactions, say

$$A \rightarrow R \rightarrow S \rightarrow T \rightarrow U$$

the treatment is similar, though more cumbersome than the two-step reaction just considered. Figure 13 illustrates typical concentration-time curves for this situation.

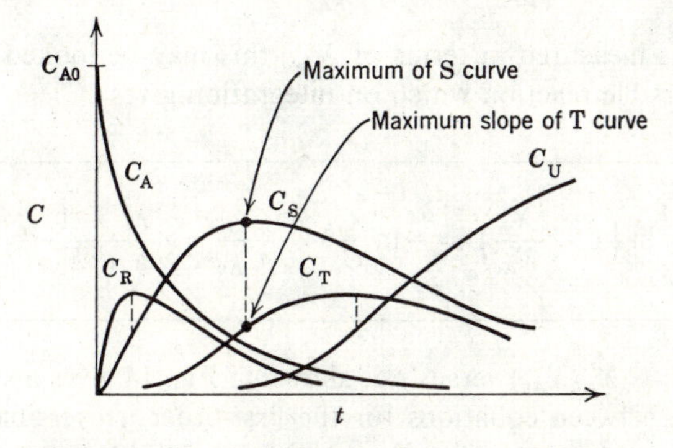

FIGURE 13. Concentration-time curves for a chain of successive first-order reactions.

Again, as with reactions in parallel, flow reactors may be more useful than batch reactors in the study of these multiple reactions.

First-order Reversible Reactions. Though no reaction ever goes to completion, we can consider many reactions to be essentially irreversible because of the large value of the equilibrium constant. These are the situations we have examined up to this point. Let us now consider reactions for which complete conversion cannot be assumed. The simplest case is the opposed unimolecular-type reaction

$$A \underset{k_2}{\overset{k_1}{\rightleftarrows}} R, \qquad K_C = K = \text{equilibrium constant} \tag{53a}$$

Starting with a concentration ratio $M = C_{R0}/C_{A0}$ the rate equation is

$$\frac{dC_R}{dt} = -\frac{dC_A}{dt} = C_{A0}\frac{dX_A}{dt} = k_1 C_A - k_2 C_R$$

$$= k_1(C_{A0} - C_{A0}X_A) - k_2(C_{A0}M + C_{A0}X_A) \tag{53b}$$

Now at equilibrium $dC_A/dt = 0$. Hence from Eq. 53 we find the fractional conversion of A at equilibrium conditions to be

$$K_C = \frac{C_{Re}}{C_{Ae}} = \frac{M + X_{Ae}}{1 - X_{Ae}}$$

and the equilibrium constant to be

$$K_C = \frac{k_1}{k_2}$$

Combining the above three equations we obtain, in terms of the equilibrium conversion,

$$\frac{dX_A}{dt} = \frac{k_1(M+1)}{M + X_{Ae}}(X_{Ae} - X_A)$$

With conversions measured in terms of X_{Ae}, this may be looked on as a pseudo first-order irreversible reaction which on integration gives

$$-\ln\left(1 - \frac{X_A}{X_{Ae}}\right) = -\ln\frac{C_A - C_{Ae}}{C_{A0} - C_{Ae}} = \frac{M+1}{M + X_{Ae}}k_1 t \tag{54}$$

A plot of $-\ln(1 - X_A/X_{Ae})$ versus t as shown in Fig. 14 gives a straight line.

The similarity between equations for the first-order irreversible and reversible reactions can be seen by comparing Eq. 12 with Eq. 54 or by comparing Fig. 2 with Fig. 14. The reversible reaction may be considered to be irreversible if the concen-

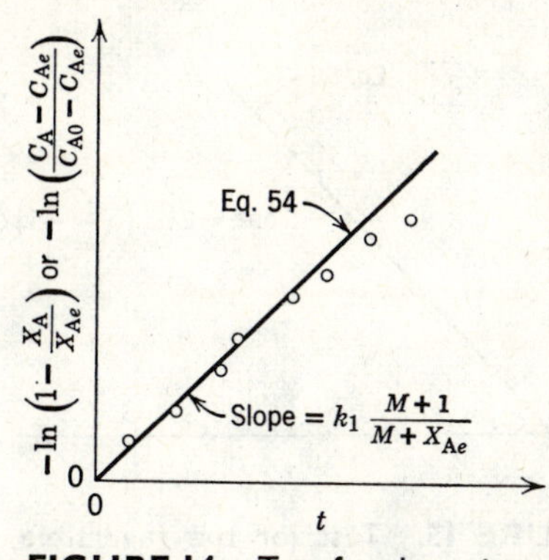

FIGURE 14. Test for the unimolecular type reversible reactions of Eq. 53.

tration is measured by $C_A - C_{Ae}$, or the concentration in excess of the equilibrium value. The conversion is then measured as the fraction of the maximum attainable or equilibrium conversion. On the other hand, we see that the irreversible reaction is simply the special case of the reversible reaction in which $C_{Ae} = 0$, or $X_{Ae} = 1$, or $K_C = \infty$.

Second-order Reversible Reactions. For the bimolecular-type second-order reactions

$$A + B \underset{k_2}{\overset{k_1}{\rightleftarrows}} R + S \tag{55a}$$

$$2A \underset{k_2}{\overset{k_1}{\rightleftarrows}} R + S \tag{55b}$$

$$2A \underset{k_2}{\overset{k_1}{\rightleftarrows}} 2R \tag{55c}$$

$$A + B \underset{k_2}{\overset{k_1}{\rightleftarrows}} 2R \tag{55d}$$

with the restrictions that $C_{A0} = C_{B0}$ and $C_{R0} = C_{S0} = 0$, the integrated rate equations are all identical, as follows

$$\ln \frac{X_{Ae} - (2X_{Ae} - 1)X_A}{X_{Ae} - X_A} = 2k_1\left(\frac{1}{X_{Ae}} - 1\right)C_{A0}t \tag{56}$$

A plot as shown in Fig. 15 can then be used to test the adequacy of these kinetics.

Reversible Reactions in General. For orders other than one and two integration of the rate equation is difficult. So if Eq. 54 or 56 is not able to fit the data then the search for an adequate rate equation is best done by the differential method.

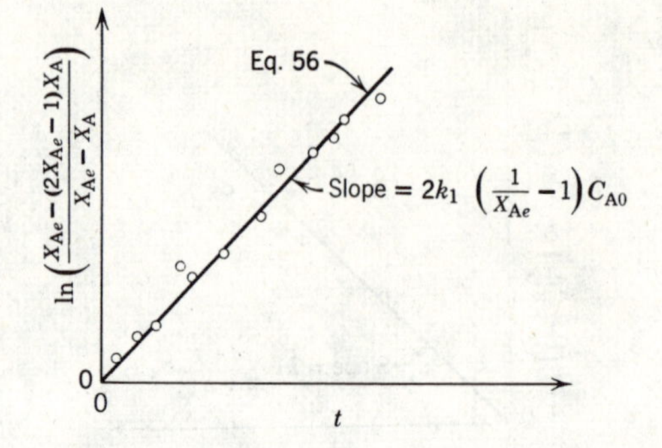

FIGURE 15. Test for the reversible bimolecular reactions of Eq. 55.

Reactions of Shifting Order. In searching for a kinetic equation it may be found that the data are well fitted by one reaction order at high concentrations, but by another order at low concentrations. Let us present some forms of rate equations which are able to fit this type of data.

Shift from Low to High Order as the Reactant Concentration Drops. Consider the reaction

$$A \to R \qquad \text{with } -r_A = -\frac{dC_A}{dt} = \frac{k_1 C_A}{1 + k_2 C_A} \tag{57}$$

From this rate equation we see:

At high C_A (or $k_2C_A \gg 1$)—the reaction is of zero order with rate constant k_1/k_2

At low C_A (or $k_2C_A \ll 1$)—the reaction is of first order with rate constant k_1

This behavior is shown in Fig. 16.

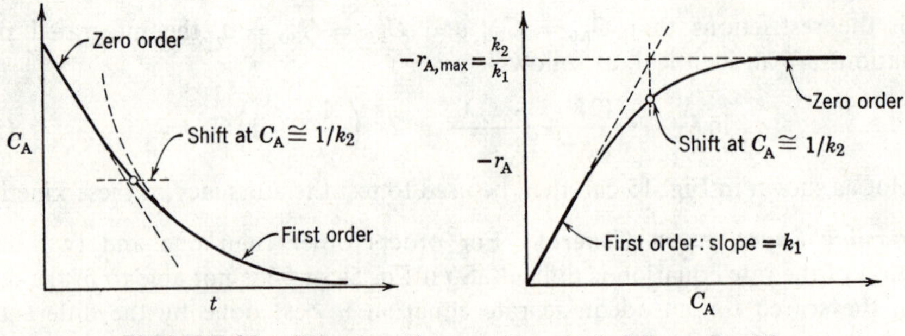

FIGURE 16. Behavior of a reaction which follows Eq. 57.

To apply the integral method, separate variables and integrate Eq. 57. This gives

$$\ln \frac{C_{A0}}{C_A} + k_2(C_{A0} - C_A) = k_1 t \tag{58a}$$

or

$$\frac{\ln (C_{A0}/C_A)}{C_{A0} - C_A} = -k_2 + \frac{k_1 t}{C_{A0} - C_A} \tag{58b}$$

A test of this rate form is then shown in Fig. 17.

By similar reasoning to the above we can show that the general rate form

$$-r_A = -\frac{dC_A}{dt} = \frac{k_1 C_A{}^m}{1 + k_2 C_A{}^n} \tag{59}$$

shifts from order $m - n$ at high concentration to order m at low concentration, the transition taking place where $k_2 C_A{}^n \cong 1$. This type of equation can then be used to fit data of any two orders. Another form which can account for this shift is

$$-r_A = -\frac{dC_A}{dt} = \frac{k_1 C_A{}^m}{(1 + k_2 C_A)^n} \tag{60}$$

Mechanism studies may suggest which form to use.

The rate form of Eq. 57 and some of its generalizations are used to represent a number of widely different kinds of reactions. For example, in homogeneous systems this form is used for enzyme-catalyzed reactions where it is suggested by mechanistic studies (see the Michaelis-Menten mechanism in Chapter 2 and in Problem 2.17). It is also used to represent the kinetics of surface-catalyzed reactions.

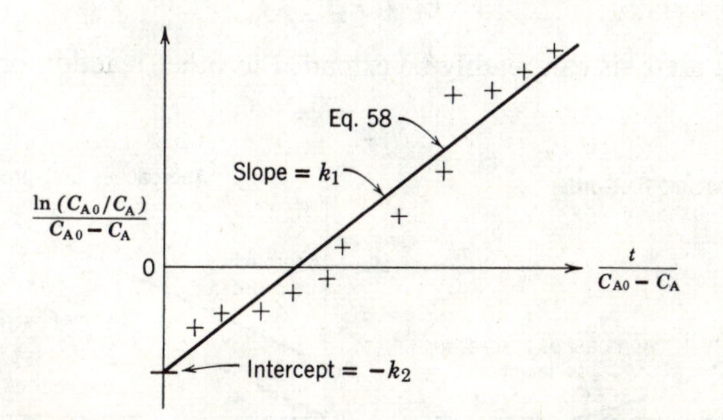

FIGURE 17. Test of the rate equation, Eq. 57, by integral analysis.

In mechanistic studies this form of equation appears whenever the rate-controlling step of a reaction is viewed to involve the association of reactant with some quantity which is present in limited but fixed amounts, for example, the association of reactant with enzyme to form a complex, or the association of gaseous reactant with an active site on the catalyst surface.

Shift from High to Low Order as the Concentration Drops. This behavior can be accounted for by considering two competing reaction paths which are of different orders. As an example consider the zero- and first-order decompositions

$$A \rightarrow R \tag{61a}$$

$$\text{Path 1:} \quad (-r_A)_1 = -\left(\frac{dC_A}{dt}\right)_1 = k_1$$

$$\text{Path 2:} \quad (-r_A)_2 = -\left(\frac{dC_A}{dt}\right)_2 = k_2 C_A$$

The overall disappearance of A is the sum of the individual rates, or

$$(-r_A)_{\text{overall}} = -\left(\frac{dC_A}{dt}\right)_{\text{overall}} = k_1 + k_2 C_A \tag{61b}$$

At low concentrations the lower order term dominates the rate, the opposite is true at high concentrations; hence the reaction approaches zero-order kinetics at low concentrations and first-order at high concentrations. This behavior is shown in Fig. 18.

The easiest way of testing this form of rate equation is to fit by the integral method for first- and then zero-orders so as to obtain the rate constants k_2 and k_1. Then the complete equation is verified by testing graphically the integrated form of Eq. 61*b*, which is

$$-\ln\left(\frac{k_1 + k_2 C_{A0}}{k_1 + k_2 C_A}\right) = k_2 t \tag{62}$$

This type of analysis can readily be extended to other reaction orders.

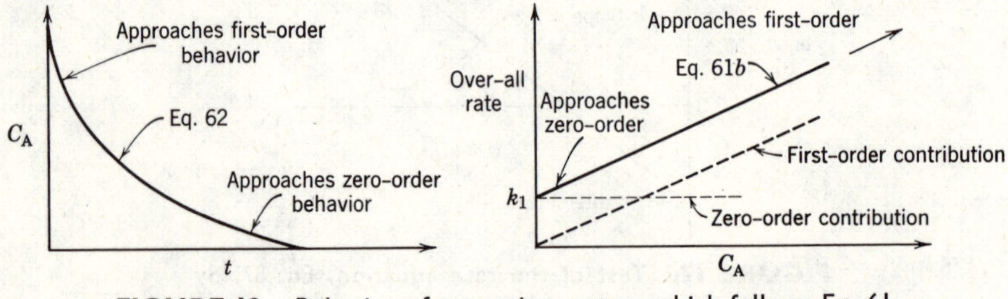

FIGURE 18. Behavior of a reacting system which follows Eq. 61.

Differential Method of Analysis of Data

The differential method of analysis deals directly with the differential rate equation to be tested, evaluating all terms in the equation including the derivative dC_i/dt, and testing the goodness of fit of the equation with experiment.

We may plan the experimental program to test the complete rate equation in question, or to test and fit separately the various parts of the rate equation which are then combined to give the complete rate equation. These two procedures are outlined in turn.

Analysis of the Complete Rate Equation. The analysis of the complete rate equation by the differential method may be summarized as follows.

1. Hypothesize a mechanism and from it obtain a rate equation. As with the integral analysis, it will be of the form

$$-r_A = -\frac{dC_A}{dt} = f(k, C) \tag{5}$$

or

$$-r_A = -\frac{dC_A}{dt} = kf(C) \tag{6}$$

If it is of the latter form, proceed with step 2; if the former, see the remarks following step 6.

2. From experiment obtain concentration-time data and plot them.
3. Draw a smooth curve through these data.
4. Determine the slope of this curve at suitably selected concentration values. These slopes, dC_A/dt are the rates of reaction at these compositions.
5. Evaluate $f(C)$ for each composition.
6. Plot $-(dC_A/dt)$ versus $f(C)$. If we obtain a straight line through the origin the rate equation is consistent with the data; if not then another equation should be tested. Figure 19 illustrates the procedure.

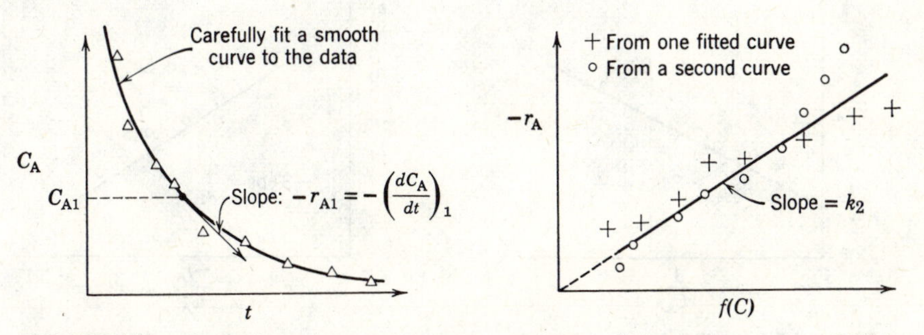

FIGURE 19. Procedure for testing a rate equation of the type $-r_A = kf(C)$ by the differential method of analysis.

The crucial operations in this procedure are steps 3 and 4. Even a slight uncertainty in the slope of the fitted curve will result in a large uncertainty in evaluated slopes. It is therefore recommended that the C_A versus t data be plotted on a large piece of paper, that smooth curves be carefully fitted *freehand* to the data, independently by a number of people, and the resultant slopes be averaged.

If the rate equation to be tested or fitted is of the form of Eq. 5 then the analysis usually becomes much more involved, requiring either a trial-and-error adjustment of the constants or a nonlinear least-squares analysis. In such cases it is usually preferable to turn to the partial analysis of the rate equation, outlined in the following section.

With certain simpler rate equations, however, mathematical manipulation may be able to yield an expression suitable for graphical testing. As an example, consider the rate equation

$$-r_A = -\frac{dC_A}{dt} = \frac{k_1 C_A}{1 + k_2 C_A} \tag{57}$$

which has already been treated by the integral method of analysis. By the differential method we can obtain $-r_A$ versus C_A by steps 1 to 4, above; however, we cannot proceed to steps 5 and 6. So, as suggested, let us manipulate Eq. 57 to obtain a more useful expression. Thus, taking reciprocals we obtain

$$\frac{1}{(-r_A)} = \frac{1}{k_1 C_A} + \frac{k_2}{k_1} \tag{63}$$

and a plot of $1/(-r_A)$ versus $1/C_A$ is linear, as shown in Fig. 20.

Alternatively, a different manipulation (multiply Eq. 63 by $k_1(-r_A)/k_2$) yields another form, also suitable for testing, thus

$$(-r_A) = \frac{k_1}{k_2} - \frac{1}{k_2}\left[\frac{(-r_A)}{C_A}\right] \tag{64}$$

A plot of $-r_A$ versus $(-r_A)/C_A$ is linear, as shown in Fig. 20.

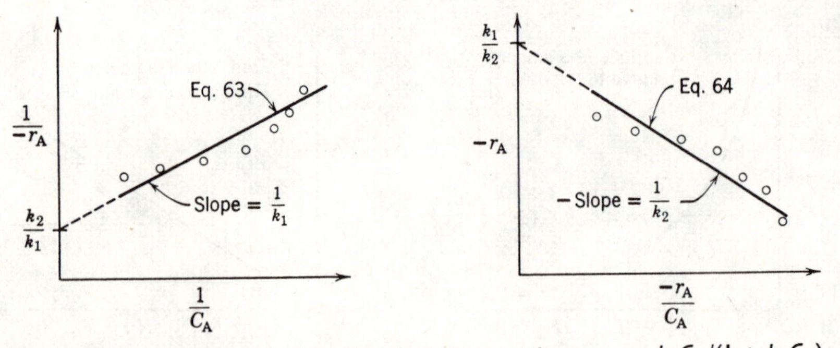

FIGURE 20. Two ways of testing the rate equation $-r_A = k_1C_A/(1 + k_2C_A)$ by differential analysis.

Whenever a rate equation can be manipulated to give a linear plot this becomes a simple way for testing the equation.

Partial Analysis of the Rate Equation. We can avoid the difficulties in testing a complete rate equation of the form of Eq. 5 by planning an experimental program so as to find various parts of the rate equation with different experiments. When we have no particular rate equation in mind, the partial solution approach is especially appealing.

As an illustration consider the reaction

$$A \rightleftarrows R + S$$

which is not elementary. Suppose, on starting with either A or R and S, that we end up with a mixture of all three components. We may expect that one of the number of rate equations such as

$$-\frac{dC_A}{dt} = k_1 C_A{}^a - k_2 C_R{}^r C_S{}^s \tag{65a}$$

or

$$-\frac{dC_A}{dt} = \frac{k_1' C_A - k_2' C_R{}^2 C_S}{1 + k_3' C_S} \tag{66a}$$

will fit the data. How do we check this?

We could use the *method of isolation* in which we make kinetic runs in the absence of certain reaction components. Thus, starting with only reactant A, ending the run before the concentrations of R and S become appreciable, the above rate equations reduce to

$$-\frac{dC_A}{dt} = k_1 C_A{}^a \tag{65b}$$

and

$$-\frac{dC_A}{dt} = k_1' C_A \tag{66b}$$

which are much easier to fit than the complete rate equation. Similarly, starting with pure R and S and ending before the concentration of A becomes appreciable, we have to deal only with

$$\frac{dC_A}{dt} = k_2 C_R{}^r C_S{}^s \tag{65c}$$

and

$$\frac{dC_A}{dt} = \frac{k_2' C_R{}^2 C_S}{1 + k_3' C_S} \tag{66c}$$

A somewhat similar procedure is the *method of initial rates.* Here a series of C_A versus t runs is made for different feed compositions, and each run is extrapolated back to the initial conditions to find the initial rate of reaction.

As an example, to test Eq. 65*b*, different C_{A0} are used, and the procedure is illustrated in Fig. 21.

Sometimes the reaction order obtained by a time run differs from that found by initial rates. This is a sure indication of complex kinetics where products influence the rate in such a way that the net effect is an observed order for the time run which differs from the true order. As an insurance against an incorrect interpretation of kinetic data it is recommended that both concentration-time and initial rate data be taken and the orders compared.

Another technique, the *method of least squares*, is especially useful for fitting equations of the type

$$-\frac{dC_A}{dt} = kC_A{}^aC_B{}^b\cdots \tag{67}$$

where $k, a, b, \ldots$ are to be determined. The technique is described as follows. Take logarithms of Eq. 67. Thus

$$\log\left(-\frac{dC_A}{dt}\right) = \log k + a\log C_A + b\log C_B + \cdots$$

which is of the form

$$y = a_0 + a_1x_1 + a_2x_2 + \cdots$$

This may be solved [see Levenspiel *et al.* (1956)] to yield the values of best fit for $a_0 = \log k$, $a_1 = a$, $a_2 = b$, etc.

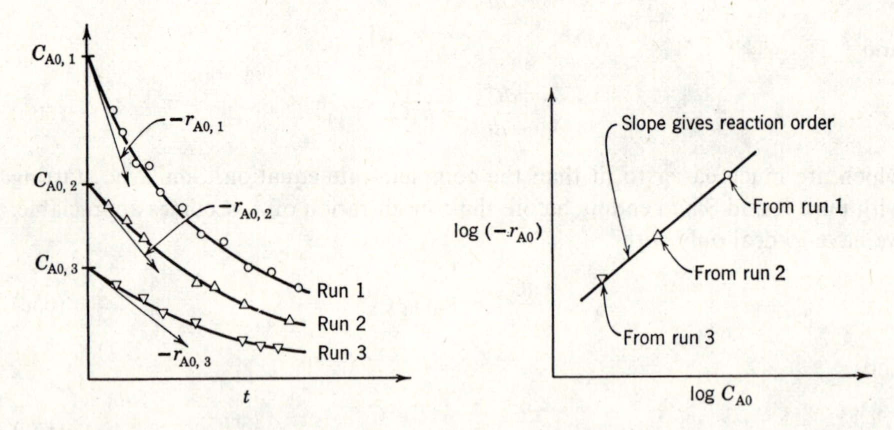

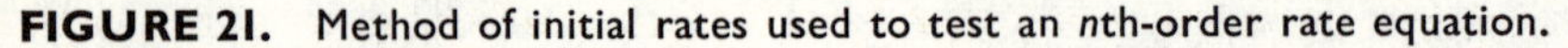
FIGURE 21. Method of initial rates used to test an *n*th-order rate equation.

Alternatively, using the *method of excess*, we may find the orders $a, b, \ldots$ one at a time in separate experiments by keeping in great excess all components other than the one to be examined. For example, if all but A are in great excess, their concentrations will be unchanged. Thus Eq. 67 reduces to

$$-\frac{dC_A}{dt} = k(C_{B0}{}^b \cdots)C_A{}^a$$

which may be solved simply by plotting $-(dC_A/dt)$ versus C_A on log-log paper.

With any given problem we must use good judgment in planning an experimental program. Usually the clues and partial information obtained in any set of runs will guide and suggest the line of further experimentation. Needless to say, after the various parts of the rate equation are found, the resulting complete equation should be checked to see whether all interactions have been accounted for by making an integral kinetic run in which all materials are present and are made to vary over wide concentration ranges. Finally, the rate equation should be of a form, and its constants related in such a way, so that the rate approaches zero as the composition approaches equilibrium.

VARIABLE-VOLUME BATCH REACTOR

The general form for the rate of change of component i in either the constant or variable-volume system is

$$r_i = \frac{1}{V}\frac{dN_i}{dt} = \frac{1}{V}\frac{d(C_iV)}{dt} = \frac{1}{V}\frac{V\,dC_i + C_i\,dV}{dt}$$

or

$$r_i = \frac{dC_i}{dt} + \frac{C_i}{V}\frac{dV}{dt} \tag{68}$$

Hence two terms must be evaluated from experiment to find r_i. Luckily, for the constant-volume system the second term drops out, leaving the simple expression

$$r_i = \frac{dC_i}{dt} \tag{1}$$

In the variable-volume reactor we can avoid the cumbersome two-term expression of Eq. 68 if we use fractional conversion rather than concentration as the primary variable. This simplification is effected if we make the restriction that the volume of reacting system varies linearly with conversion, or

$$V = V_0(1 + \varepsilon_A X_A) \tag{69}$$

where ε_A is the fractional change in volume of the system between no conversion and complete conversion of reactant A. Thus

$$\varepsilon_A = \frac{V_{X_A=1} - V_{X_A=0}}{V_{X_A=0}} \tag{70}$$

As an example of the use of ε_A consider the isothermal gas-phase reaction

$$A \rightarrow 4R$$

By starting with pure reactant A,

$$\varepsilon_A = \frac{4-1}{1} = 3$$

but with 50% inerts present at the start, two volumes of reactant mixture yield, on complete conversion, five volumes of product mixture. In this case

$$\varepsilon_A = \frac{5-2}{2} = 1.5$$

We see, then, that ε_A accounts for both the reaction stoichiometry and the presence of inerts. Noting that

$$N_A = N_{A0}(1 - X_A) \tag{71}$$

we have, on combining with Eq. 69,

$$C_A = \frac{N_A}{V} = \frac{N_{A0}(1 - X_A)}{V_0(1 + \varepsilon_A X_A)} = C_{A0}\frac{1 - X_A}{1 + \varepsilon_A X_A}$$

Thus

$$\frac{C_A}{C_{A0}} = \frac{1 - X_A}{1 + \varepsilon_A X_A} \qquad \text{or} \qquad X_A = \frac{1 - C_A/C_{A0}}{1 + \varepsilon_A C_A/C_{A0}} \tag{72}$$

which is the relationship between conversion and concentration for variable-volume (or variable-density) systems satisfying the linearity assumption of Eq. 69. With these relationships Eq. 68, written for reactant A, becomes

$$-r_A = -\frac{1}{V}\frac{dN_A}{dt} = -\frac{1}{V_0(1 + \varepsilon_A X_A)}\frac{N_{A0}\,d(1 - X_A)}{dt}$$

or

$$-r_A = \frac{C_{A0}}{1 + \varepsilon_A X_A} \frac{dX_A}{dt} \tag{73}$$

which certainly is easier to handle than Eq. 68. Note that when we have a constant-density system ε_A equals zero and Eq. 73 reduces to Eq. 1.

The development to follow is based on the assumption made in Eq. 69 that the volume change of the system varies linearly with conversion. This is a reasonable restrictive assumption which holds for all practical purposes for reactions represented by a single stoichiometric expression operated either isothermally at constant pressure, or sometimes adiabatically at constant pressure. When this assumption does not hold then the actual relationship between V and X_A, or p and X_A, must be used in Eq. 73 and must be used to relate C_A to X_A in Eq. 72. However, analysis becomes more difficult and these conditions are avoided whenever possible in studying the kinetics of reactions.

Differential Method of Analysis

The procedure for differential analysis of isothermal variable-volume data is the same as for the constant-volume situation except that we replace

$$\frac{dC_A}{dt} \quad \text{by} \quad \frac{dC_A}{dt} + C_A \frac{d \ln V}{dt} \quad \text{or preferably by} \quad \frac{-C_{A0}}{1 + \varepsilon_A X_A} \frac{dX_A}{dt}.$$

Integral Method of Analysis

Integral analysis of data requires integration of the rate expression to be tested. The resulting C function versus t is then compared with experiment. Thus for reactant A

$$-r_A = -\frac{1}{V} \frac{dN_A}{dt} = \frac{C_{A0}}{1 + \varepsilon_A X_A} \frac{dX_A}{dt}$$

Integrating formally we obtain

$$C_{A0} \int_0^{X_A} \frac{dX_A}{(1 + \varepsilon_A X_A)(-r_A)} = t \tag{74}$$

which is the expression for batch reactors in which the volume is a linear function of the conversion of material. We now consider rate forms which can be integrated simply. These are few in number when compared to those for the constant-volume reactor.

Since the progress of the reaction is directly related to the volume change by Eq. 69 the integrated forms will be presented in terms of the volume change of the system, when convenient.

Zero-order Reactions. For a homogeneous zero-order reaction the rate of change of any reactant A is independent of the concentration of materials, or

$$-r_A = -\frac{1}{V}\frac{dN_A}{dt} = \frac{C_{A0}}{1 + \varepsilon_A X_A}\frac{dX_A}{dt} = k \tag{75}$$

With Eqs. 74 and 69 we obtain on integration

$$C_{A0}\int_0^{X_A}\frac{dX_A}{1 + \varepsilon_A X_A} = \frac{C_{A0}}{\varepsilon_A}\ln(1 + \varepsilon_A X_A) = \frac{C_{A0}}{\varepsilon_A}\ln\frac{V}{V_0} = kt \tag{76}$$

As shown in Fig. 22, the logarithm of the fractional change in volume versus time yields a straight line of slope $k\varepsilon_A/C_{A0}$.

First-order Reactions. For a unimolecular-type first-order reaction the rate of change of reactant A is

$$-r_A = -\frac{1}{V}\frac{dN_A}{dt} = kC_A \tag{77}$$

This equation is transformed into conversion units by combining it with Eqs. 72 and 73. Thus

$$-r_A = \frac{C_{A0}}{1 + \varepsilon_A X_A}\frac{dX_A}{dt} = \frac{kC_{A0}(1 - X_A)}{1 + \varepsilon_A X_A} \tag{78}$$

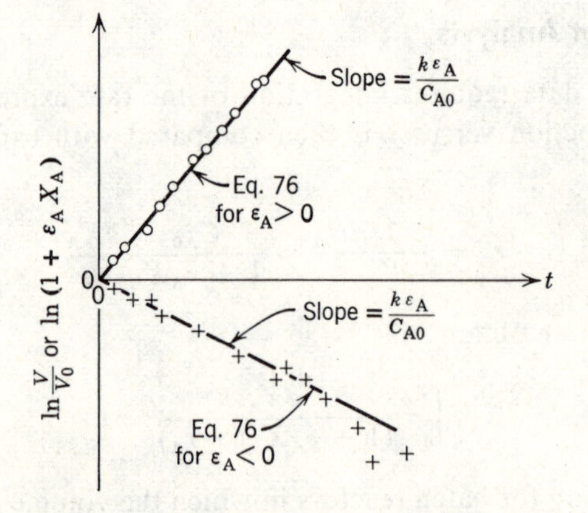

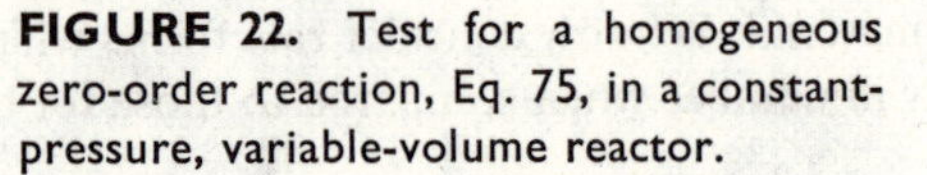

FIGURE 22. Test for a homogeneous zero-order reaction, Eq. 75, in a constant-pressure, variable-volume reactor.

Separating and integrating, we obtain, with Eq. 69,

$$\int_0^{X_A} \frac{dX_A}{1 - X_A} = -\ln(1 - X_A) = -\ln\left(1 - \frac{\Delta V}{\varepsilon_A V_0}\right) = kt \tag{79}$$

A semilogarithmic plot of Eq. 79 as shown in Fig. 23 yields a straight line of slope k.

Comparing this result with that for constant-volume systems, we see that the fractional conversion at any time is the same in both cases; however, the concentration of materials is not the same.

Second-order Reactions. For a bimolecular-type second-order reaction,

$$2A \rightarrow \text{products}$$

or

$$A + B \rightarrow \text{products}, \qquad C_{A0} = C_{B0}$$

the rate of reaction of A is given by

$$-r_A = kC_A^2 \tag{80}$$

With Eqs. 72 and 73 this becomes, in terms of conversion,

$$-r_A = \frac{C_{A0}}{1 + \varepsilon_A X_A}\frac{dX_A}{dt} = kC_{A0}^2\left(\frac{1 - X_A}{1 + \varepsilon_A X_A}\right)^2 \tag{81}$$

Separating variables, breaking into partial fractions, and integrating, we obtain

$$\int_0^{X_A} \frac{1 + \varepsilon_A X_A}{(1 - X_A)^2}\,dX_A = \frac{(1 + \varepsilon_A)X_A}{1 - X_A} + \varepsilon_A \ln(1 - X_A) = kC_{A0}t \tag{82}$$

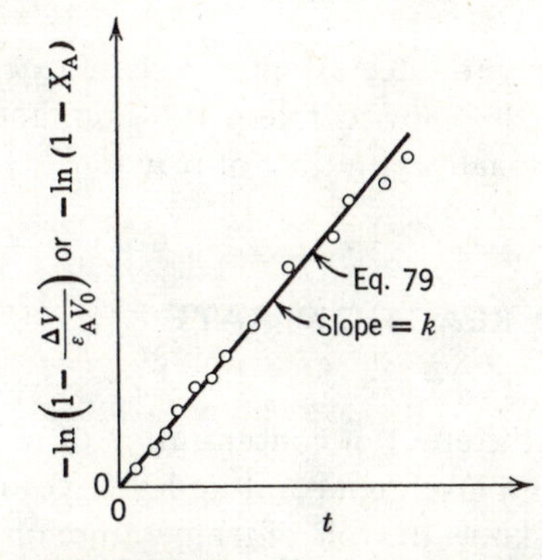

FIGURE 23. Test for a first-order reaction, Eq. 77, in a constant-pressure, variable-volume reactor.

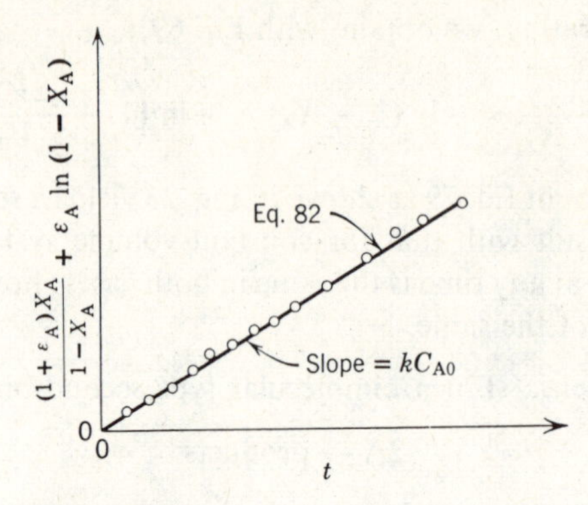

FIGURE 24. Test for the second-order reaction, Eq. 80, in a constant-pressure, variable-volume reactor.

With Eq. 69 this can be transformed into a V versus t relationship. Figure 24 shows how to test for these kinetics.

nth-order and Other Reactions. For an nth-order reaction of the type

$$-r_A = kC_A{}^n = kC_{A0}{}^n\left(\frac{1 - X_A}{1 + \varepsilon_A X_A}\right)^n \tag{83}$$

Eq. 74 becomes

$$\int_0^{X_A} \frac{(1 + \varepsilon_A X_A)^{n-1}}{(1 - X_A)^n}\, dX_A = C_{A0}^{n-1}kt \tag{84}$$

which does not yield simply to integration. For these reactions we must integrate Eq. 74 graphically, using the guessed rate expression, and then test whether the integral increases proportionally with time of reaction.

TEMPERATURE AND REACTION RATE

So far we have examined the effect of concentration of reactants and products on the rate of reaction, all at a given temperature level. To obtain the complete rate equation, we also need to know the role of temperature on reaction rate. Now in a typical rate equation we have

$$-r_A = -\frac{1}{V}\frac{dN_A}{dt} = kf(C)$$

and it is the reaction rate constant, the concentration-independent term, which is affected by the temperature, whereas the concentration-dependent terms $f(C)$ usually remain unchanged at different temperatures.

For elementary reactions, theory predicts that the rate constant should be temperature-dependent in the following manner:

1. From Arrhenius' law,

$$k \propto e^{-E/RT}$$

2. From collision or transition-state theory,

$$k \propto T^m e^{-E/RT}$$

In Chapter 2 we showed that the latter expression very often reduces to the former because the exponential term is so temperature-sensitive that any variation caused by the T^m term is completely masked.

Even for nonelementary reactions where the rate constants may be products of k values for the elementary reactions, these composite rate constants have been found to vary as $e^{-E/RT}$.

Thus, after finding the concentration dependency of the reaction rate, we can then examine for the variation of the rate constant with temperature by an Arrhenius-type relationship

$$k = k_0 e^{-E/RT} \tag{85}$$

This is conveniently determined by plotting $\ln k$ versus $1/T$, as shown in Fig. 25.

Finally, as mentioned in Chapter 2, a shift in E with temperature reflects a change in controlling mechanism of reaction. Since this is likely to be accompanied by a change in concentration dependency this possibility should also be examined.

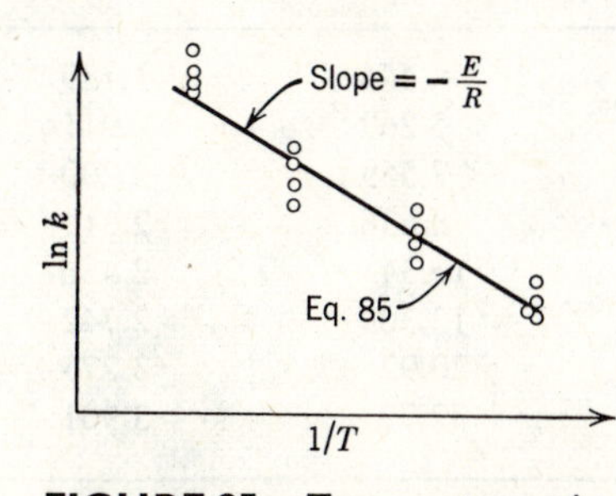

FIGURE 25. Temperature dependency of a reaction according to Arrhenius' law.

EXAMPLE 1. Kinetics from a batch reactor

Bodenstein and Lind (1906) studied the reaction

$$H_2 + Br_2 = 2HBr$$

and found on the basis of a careful analysis of good experimental data that the kinetics was well represented by the expression

$$-r_{H_2} = -\frac{d[H_2]}{dt} = \frac{k_1[H_2][Br_2]^{1/2}}{k_2 + [HBr]/[Br_2]} \tag{86}$$

suggesting a nonelementary reaction. Interestingly enough, this result was explained independently and almost simultaneously about 13 years later by Christiansen (1919), Herzfeld (1919), and Polanyi (1920) in terms of a chain reaction mechanism.

Using some of the data of Bodenstein and Lind, let us illustrate the variety of methods which may be used to explore the kinetics of a reaction.

Data. Choose eight time-concentration runs, four of which use equal concentrations of H_2 and Br_2. At the start of all runs no HBr is present. Columns 1 and 2 of Table 1 show the data of one of the four runs which use equal concentrations of H_2 and Br_2. Table 2 shows initial rate data for these eight runs. This information will be used later.

Search for a Simple Rate Equation by Integral Analysis. First see whether the kinetics of this reaction can be described by a simple rate equation. For this use integral analysis. With Column 3 of Table 1, Fig. 26 tests for a second-order rate equation. The data do not lie on a straight line; hence we do not have an elementary bimolecular-type reaction. Figure 27 then tests for a first-order rate equation. Again the data of Table 1 do not lie on a straight line; hence the reaction is not of first order. Similar linearity tests with other simple forms of rate equations do not give any straight line fit.

TABLE 1. Time versus concentration run

(1)[a] Time, min	(2)[a] $[H_2] = [Br_2]$ mol/liter	(3)[b] $1/[H_2]$	(4)[c] $[H_2]^{-0.39}$	(5)[c] $[H_2]^{-0.39} - [H_2]_0^{-0.39}$
0	0.2250	4.444	1.789	0
20	0.1898	5.269	1.911	0.122
60	0.1323	7.559	2.200	0.411
90	0.1158	8.636	2.319	0.530
128	0.0967	10.34	2.486	0.697
180	0.0752	13.30	2.742	0.953
300	0.0478	20.92	3.275	1.486
420	0.0305	32.79	3.901	2.112

[a] Original data.
[b] Used in Fig. 26.
[c] Used in Fig. 31.

TABLE 2. Initial rate data

(1) $[H_2]_0$	(2) $[Br_2]_0$	(3) $(-r_{H_2})10^3$
0.2250	0.2250	1.76
0.9000	0.9000	10.9
0.6750	0.6750	8.19
0.4500	0.4500	4.465
0.5637	0.2947	4.82
0.2881	0.1517	1.65
0.3103	0.5064	3.28
0.1552	0.2554	1.267

Findings of Integral Analysis. Kinetically the reaction is not elementary; neither can it be satisfactorily represented by any of the simpler rate expressions.

Possible Lines of Investigation. Let us now see whether we can fit the rate by an expression of the form

$$-r_{H_2} = k[H_2]^a[Br_2]^b \qquad \text{with } a + b = n \tag{87}$$

To do this we may proceed in one of two ways: we may use the C versus t data of a single run or we may use initial rate data for the eight runs. If HBr influences the rate of reaction, the first procedure will lead to complications because HBr is formed during reaction. These complications are avoided if we analyze the initial rate data. Let us follow the latter procedure since it has less chance of getting us into trouble.

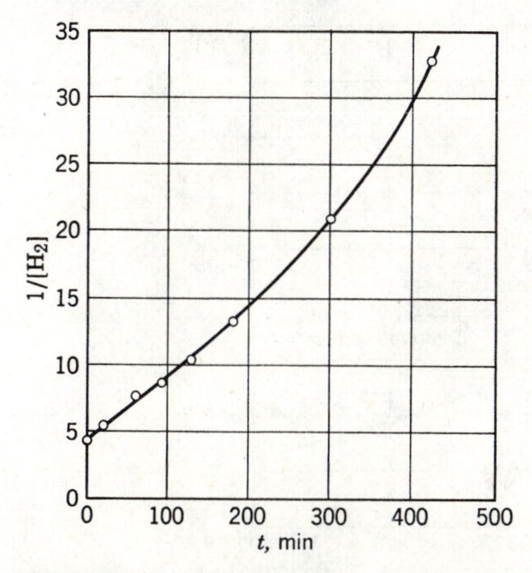

FIGURE 26. Test for second-order kinetics by the integral method, according to Fig. 4.

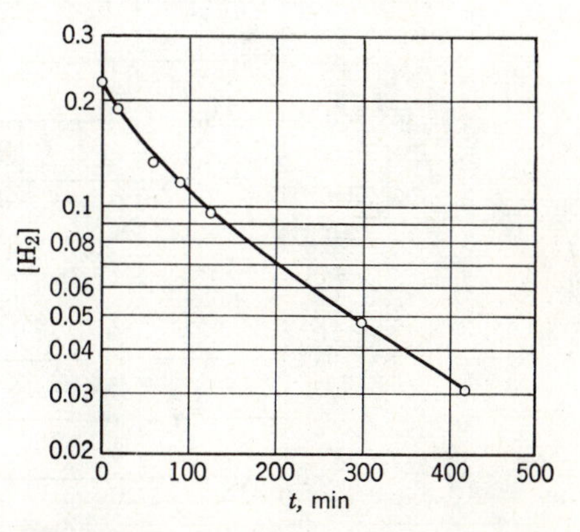

FIGURE 27. Test for first-order kinetics by the integral method, according to Fig. 2.

Find Initial Rate Data. Initial rates of reaction are found either by plotting the C versus t curve and finding the slopes at $t = 0$ as shown in Fig. 28 for the run of Table 1, or by evaluating $\Delta C/\Delta t$ for the first two data points in a kinetic run. The latter procedure was used by Bodenstein and Lind to obtain the data of Table 2.

Over-all Order of Reaction. For equal concentrations of H_2 and Br_2 Eq. 87 becomes

$$-r_{H_2} = k[H_2]^a[Br_2]^b = k[H_2]^{a+b} = k[H_2]^n$$

and taking logs this gives

$$\log(-r_{H_2}) = \log k + n \log [H_2] \tag{88}$$

Thus from the first four runs of Table 2 we can find the over-all order of reaction. This is shown in Fig. 29. Fitting by eye we find the slope to be

$$n = a + b = 1.35$$

Reaction Order with Respect to Individual Reactants. Knowing the over-all order, we can find the order with respect to each component with the following manipulation:

$$-r_{H_2} = k[H_2]^a[Br_2]^b = k[H_2]^a[Br_2]^{n-a} = k[Br_2]^n\left(\frac{[H_2]}{[Br_2]}\right)^a$$

Taking logs and using the value of n which was found, we obtain

$$\log \frac{-r_{H_2}}{[Br_2]^{1.35}} = \log k + a \log \frac{[H_2]}{[Br_2]} \tag{89}$$

Plotted as in Fig. 30 the slope, determined by eye, is found to be $a = 0.90$. Hence from the initial rate data we find

$$-r_{H_2} = k[H_2]^{0.90}[Br_2]^{0.45} \tag{90}$$

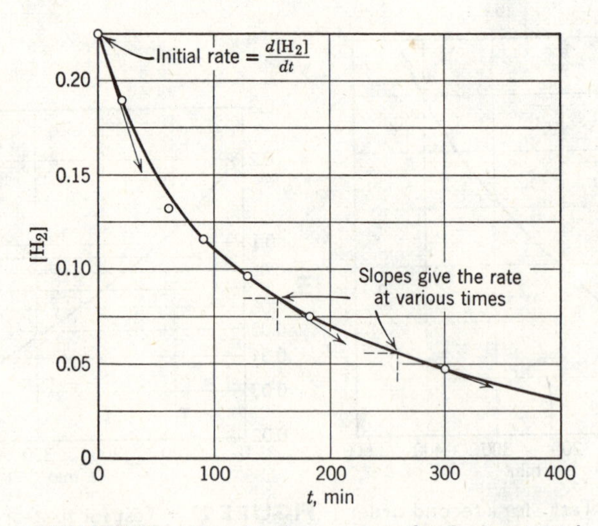

FIGURE 28. Graphical procedure for finding initial rate of reaction.

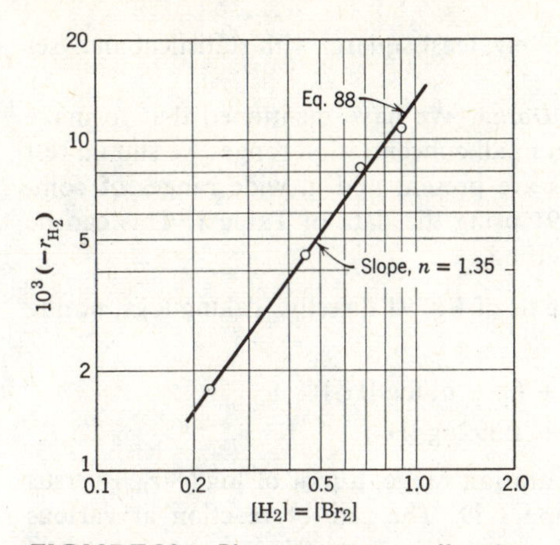

FIGURE 29. Plot to give over-all order of reaction.

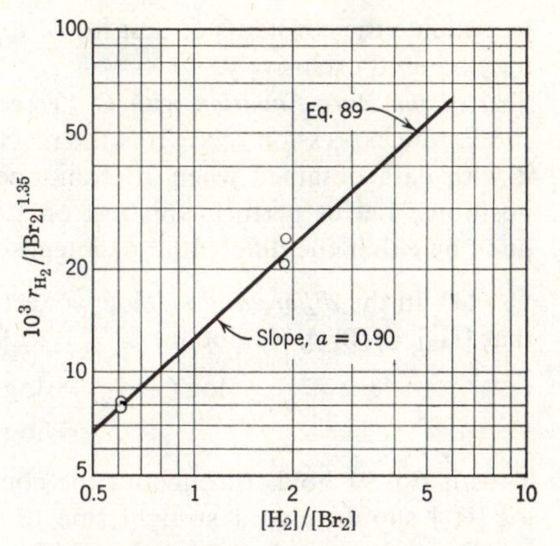

FIGURE 30. Determination of order with respect to hydrogen once the over-all order is known.

Order of Reaction by the Method of Least Squares. The method of least squares can give the order of reaction with respect to all the components all at one time. In effect, this procedure replaces the two preceding steps.

Taking logs of the rate equation to be fitted, Eq. 87, we obtain

$$\log(-r_{H_2}) = \log k + a \log [H_2] + b \log [Br_2]$$

which is of the form

$$y = a_0 + ax_1 + bx_2$$

The best estimate of the coefficients by the least-squares criterion gives

$$a = 0.93$$

$$b = 0.46$$

Thus the rate equation is

$$-r_{H_2} = k[H_2]^{0.93}[Br_2]^{0.46} \tag{91}$$

which agrees closely with the results obtained when fitting graphically by eye.

Evaluation of Least-squares Versus Graphical Procedures. The advantage of the method of least squares is that it allows us to find the individual orders of reaction all at one time by an objective method which is not influenced by the experimenter's biases in fitting lines to experimental points. Though it does give the constants of best fit, yet, without additional statistical analyses this method gives no indication of how good the fit is. The graphical method allows us to estimate at each step the goodness of fit of the rate equation to the data.

Probably the best procedure is to use the graphical method to find whether the selected equation form is satisfactory, and if it is, then to use the method of least squares to evaluate the constants of best fit. Alternatively, least squares with statistical analyses may be used exclusively.

Test of the Rate Equation with C Versus t Data. We have mentioned that to make sure a rate expression represents the reaction in all concentration ranges we should test it with data obtained when all components are present and in wide ranges of compositions. Let us perform this test on Eq. 91 using the data of Table 1. This can be done by either the differential or integral method.

(1) In the *differential method* we test the fit of Eq. 91 directly. Taking logs, noting that $[H_2] = [Br_2]$, we obtain

$$\log(-r_{H_2}) = \log k + (a + b)\log[H_2]$$
$$= \log k + 1.39 \log[H_2]$$

Thus if Eq. 91 holds throughout the concentration range, a plot of $\log(-r_{H_2})$ versus $\log[H_2]$ should give a straight line of slope 1.39. The rate of reaction at various conditions is found by taking slopes of the C versus t curve at various points, as shown in Fig. 28.

(2) In the *integral method* we must first integrate Eq. 91 and then test the fit. Thus for $[H_2] = [Br_2]$

$$-r_{H_2} = -\frac{d[H_2]}{dt} = k[H_2]^{1.39}$$

Separating and integrating we obtain

$$-\int_{[H_2]_0}^{[H_2]} \frac{d[H_2]}{[H_2]^{1.39}} = k\int_0^t dt$$

or

$$[H_2]^{-0.39} - [H_2]_0^{-0.39} = 0.39kt \qquad (92)$$

Thus if Eq. 91 fits at all concentrations, a plot of

$$([H_2]^{-0.39} - [H_2]_0^{-0.39}) \text{ versus } t$$

should yield a straight line.

Now which procedure should we use? Differential analysis requires taking slopes of C versus t curves and the errors and uncertainty which thereby result may well mask any trend from linearity which may exist. In addition the differential analysis is more time-consuming than integral analysis which is quite straightforward. Thus we shall use integral analysis.

Figure 31 is a plot of Eq. 92 based on values tabulated in Columns 4 and 5 of Table 2. It seems to suggest that the data do not fall on a straight line. Actually, some of the other runs show this more clearly. So, although the initial rates (with no product present) fit Eq. 91, the time runs (where product is present) most likely do not.

Concluding Remarks. We can say at this point that the rate is well represented by the expression

$$-r_{H_2} = k[H_2]^{0.93}[Br_2]^{0.46} \qquad (91)$$

when no HBr is present. The fact that the order changes during a run suggests that the product of reaction, HBr, influences the rate and enters into the rate equation, and that

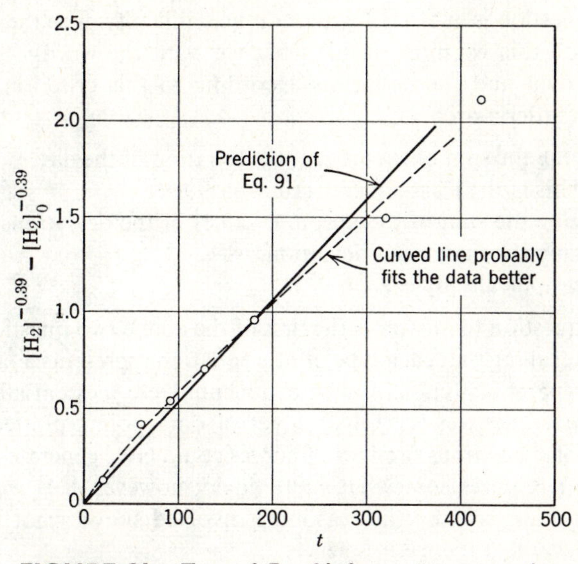

FIGURE 31. Test of Eq. 91 for a time run where the products of reaction are present in significant quantities.

Eq. 91 only partially represents the facts, just as Eqs. 65*b* and 66*b* are only partial representations of Eqs. 65*a* and 66*a*.

Since reaction orders predicted by theory are usually integers or simple fractions, as the problems in Chapter 2 show, it could be suspected that the rate expression should really be

$$-r_{H_2} = k'[H_2][Br_2]^{0.5} \tag{93}$$

The difference in goodness of fit of Eqs. 91 and 93 is quite minor and can be explained in terms of experimental scatter.

Actually, Eq. 93 is the correct partial representation of the over-all reaction rate, Eq. 86, as found by Bodenstein and Lind; hence we are well on our way to a solution of the problem.

THE SEARCH FOR A RATE EQUATION

In searching for a rate equation and mechanism to fit a set of experimental data, we would like answers to two questions.

1. Have we the correct mechanism and corresponding type of rate equation?
2. Once we have the right form of rate equation, do we have the best values for the rate constants in the equation?

The second question is answered without much difficulty once the equation form is decided. To do this we must decide what we want the words "best value" to mean, and we then find the constants according to this criterion. Some of the commonly used criteria are:

1. Minimizing the sum of squares of the deviation of the data points about the rate equation. This is the least-squares criterion.
2. Minimizing the sum of the absolute values of the deviations.
3. Minimizing the maximum deviation.
4. Fitting graphically by eye.

The difficult question to answer is the first of the above two questions. Let us see why this is so. Recalling that each type of rate equation represents a family of curves on a given plot (parabola, cubic, simple exponential, etc.), essentially what we are asking in the first question is "Given a set of data points plotted in a certain manner, which of the various families of curves could have generated these data?" To answer this requires somewhat subtle logic; however, it is worthwhile considering this question because the reasoning involved shows what we mean by a true, correct, or verified theory in science.

Suppose we have a set of data and we wish to find out whether any one of the families of curves—parabolas, cubics, hyperbolas, exponentials, etc.—really fits these data better than any other. This question cannot be answered simply; neither can high-powered mathematical or statistical methods help in deciding for us. The one exception to this conclusion occurs when one of the families being compared is a straight line. For this situation we can simply, consistently, and fairly reliably tell whether the straight line does not fit better than any other family of curves. Thus for the family of straight lines we have what is essentially a negative test, one which allows us to reject that family when there is sufficient evidence against it, but which can never tell whether it fits better than all others. Since the family of straight lines can represent a proposed mechanism with its rate, or more generally, any scientific explanation, hypothesis, or theory to be tested, the above discussion shows that all we are able to do is reject hypotheses and theories, but never prove them to be true or correct. So, when we say that a theory or hypothesis is correct, what we mean is that the evidence to date is insufficient to reject it. It is clearly understood that our acceptance of any theory or hypothesis is only provisional and temporary.

All the rate equations in this chapter were manipulated mathematically into a linearized form because of this particular property for the family of straight lines which allows it to be tested and rejected.

Three methods are commonly used to test for the linearity of a set of points. These are as follows.

Calculation of k from Individual Data Points. With a mechanism at hand, the rate constant can be found for each experimental point by either the integral

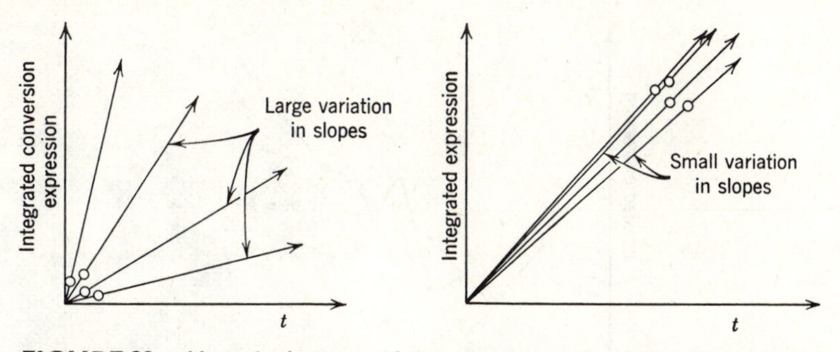

FIGURE 32. How the location of the experimental points influences the scatter in calculated k values.

or differential method. If no trend in k values is discernible, the rate equation is considered to be satisfactory and the k values are averaged.

Now the k values calculated this way are the slopes of lines joining the individual points to the origin. So for the same magnitude of scatter on the graph the k values calculated for points near the origin (low conversion) will vary widely, whereas those calculated for points far from the origin will show little variation (Fig. 32). This fact can make it difficult to decide whether k is constant and, if so, what is its best mean value.

Calculation of k from Pairs of Data Points. k values can be calculated from successive pairs of experimental points. For large data scatter, however, or for points close together, this procedure will give widely different k values from which k_{mean} will be difficult to determine. In fact, finding k_{mean} by this procedure for points located at equal intervals on the x axis is equivalent to considering only the two extreme data points while ignoring all the data points in between. This fact can easily be verified. Figure 33 illustrates this procedure.

This is a poor method in all respects and is not recommended for testing the linearity of data or for finding mean values of rate constants.

Graphical Method. Actually, the above methods do not require making a plot of the data to obtain k values. With the graphical method the data are plotted and then examined for deviations from linearity. The decision whether a straight line gives a satisfactory fit is usually made intuitively by using good judgment when looking at the data. When in doubt we should take more data, and occasionally (comparing second-degree polynomials with straight lines or nth degree and $n - 1$ degree polynomials) we may be able to use statistics to help us arrive at a decision.

The graphical procedure is probably the safest, soundest, and most reliable method for evaluating the fit of rate equations to the data, and should be used whenever possible. For this reason we stress this method here.

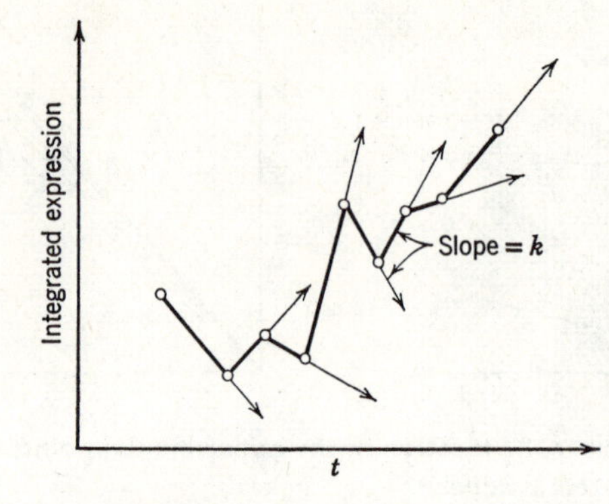

FIGURE 33. Calculated k values from successive experimental points are likely to fluctuate widely.

RELATED READINGS

Frost, A. A., and Pearson, R. G., *Kinetics and Mechanism*, 2nd ed., John Wiley & Sons, New York, 1961.

S.M. Walas. Reaction Kinetics for Chemical Engineers. Mc. Graw Hill N.Y. 1959

REFERENCES

Bodenstein, M., and Lind, S. C., *Z. physik. Chem.* (Leipzig), **57**, 168 (1906).
Christiansen, J. A., *Kgl. Dansk Videnskab. Selskab Mat.-Fys. Medd.*, **1**, 14 (1919).
Hellin, M., and Jungers, J. C., *Bull. soc. chim. France*, 386 (1957).
Herzfeld, K. F., *Z. Elektrochem.*, **25**, 301 (1919); *Ann. Physik.*, **59**, 635 (1919).
Levenspiel, O., Weinstein, N. J., and Li, J. C. R., *Ind. Eng. Chem.*, **48**, 324 (1956).
Polanyi, M., *Z. Elektrochem.*, **26**, 50 (1920).

PROBLEMS

1. If $-r_A = -(dC_A/dt) = 0.2$ mol/liter·sec when $C_A = 1$ mol/liter, what is the rate of reaction when $C_A = 10$ mol/liter.
Note: The order of reaction is not known.

2. Liquid A decomposes by first-order kinetics, and in a batch reactor 50% of A is converted in a 5-minute run. How much longer would it take to reach 75% conversion?

3. Repeat the previous problem for second-order kinetics.

4. A 10-minute experimental run shows that 75% of liquid reactant is converted to product by a $\frac{1}{2}$ order rate. What would be the amount converted in a half-hour run?

5. In a homogeneous isothermal liquid polymerization, 20% of the monomer disappears in 34 min for initial monomer concentration of 0.04 and also for 0.8 mol/liter. What is the rate of disappearance of the monomer?

6. After 8 minutes in a batch reactor, reactant ($C_{A0} = 1$ mol/liter) is 80% converted; after 18 minutes, conversion is 90%. Find a rate equation to represent this reaction.

7. Snake-Eyes Magoo is a man of habit. For instance, his Friday evenings are all alike—into the joint with his week's salary of $180, steady gambling at "2-up" for two hours, then home to his family leaving $45 behind. Snake-Eyes' betting pattern is predictable. He always bets in amounts proportional to his cash at hand, and his losses are also predictable—at a rate proportional to his cash at hand. This week Snake-Eyes received a raise, so he played for three hours, but as usual went home with $135. How much was his raise?

8. Find the over-all order of the irreversible reaction

$$2H_2 + 2NO \rightarrow N_2 + 2H_2O$$

from the following constant-volume data using equimolar amounts of hydrogen and nitric oxide:

Total pressure, mm Hg	200	240	280	320	360
Half-life, sec	265	186	115	104	67

9. The first-order reversible liquid reaction

$$A = R, \qquad C_{A0} = 0.5 \text{ mol/liter}, \qquad C_{R0} = 0$$

takes place in a batch reactor. After 8 minutes, conversion of A is 33.3% while equilibrium conversion is 66.7%. Find the rate equation for this reaction.

10. In units of moles, liters, and seconds, find the rate expression for the decomposition of ethane at 620°C from the following information obtained at atmospheric pressure. The decomposition rate of pure ethane is 7.7%/sec, but with 85.26% inerts present the decomposition rate drops to 2.9%/sec.

11. The aqueous reaction $A \rightarrow R + S$ proceeds as follows,

Time, min	0	36	65	100	160	∞
C_A, mol/liter	0.1823	0.1453	0.1216	0.1025	0.0795	0.0494

with

$$C_{A0} = 0.1823 \text{ mol/liter}$$
$$C_{R0} = 0$$
$$C_{S0} \simeq 55 \text{ mol/liter}$$

Find the rate equation for this reaction.

12. On p. 54 it was stated that the half-life method for finding reaction orders can be extended to any fractional-life data. Do this, defining $t_{1/m}$ as the time required for the reactant concentration to drop to $1/m$th of its original value.

13. Hellin and Jungers (1957) present the data in Table P13 on the reaction of sulfuric acid with diethylsulfate in aqueous solution at 22.9°C:

$$H_2SO_4 + (C_2H_5)_2SO_4 = 2C_2H_5SO_4H$$

Initial concentrations of H_2SO_4 and $(C_2H_5)_2SO_4$ are each 5.5 mol/liter. Find a rate equation for this reaction.

TABLE P13

Time, min	$C_2H_5SO_4H$, mol/liter	Time, min	$C_2H_5SO_4H$, mol/liter
0	0	180	4.11
41	1.18	194	4.31
48	1.38	212	4.45
55	1.63	267	4.86
75	2.24	318	5.15
96	2.75	368	5.32
127	3.31	379	5.35
146	3.76	410	5.42
162	3.81	∞	(5.80)

14. A small reaction bomb fitted with a sensitive pressure-measuring device is flushed out and then filled with pure reactant A at 1-atm pressure. The operation is carried out at 25°C, a temperature low enough that the reaction does not proceed to any appreciable extent. The temperature is then raised as rapidly as possible to 100°C by plunging the bomb into boiling water, and the readings in Table P14 are obtained. The stoichiometry of the reaction is $2A \rightarrow B$, and after leaving the bomb in the bath over the weekend the contents are analyzed for A; none can be found. Find a rate equation in units of moles, liters, and minutes which will satisfactorily fit the data.

TABLE P14

t, min	π, atm	t, min	π, atm
1	1.14	7	0.850
2	1.04	8	0.832
3	0.982	9	0.815
4	0.940	10	0.800
5	0.905	15	0.754
6	0.870	20	0.728

15. Betahundert Bashby likes to play the gaming tables for relaxation. He does not expect to win and he doesn't, so he picks games in which losses are a given small fraction of the money bet. He plays steadily without a break, and the size of his bets are proportional to the money he has. If at "galloping dominoes" it takes him four hours to lose half of his money and if it takes him two hours to lose half of his money at "chuk-a-luck," how long can he play both games simultaneously if he starts with \$100.00 and quits when he has \$1.00 left, which is just enough for a quick nip and carfare home?

16. Nitrogen pentoxide decomposes as follows:

$$N_2O_5 \rightarrow \tfrac{1}{2}O_2 + N_2O_4, \qquad -r_{N_2O_5} = (2.2 \times 10^{-3}\ \text{min}^{-1})C_{N_2O_5}$$

$$N_2O_4 \rightleftarrows 2NO_2, \qquad \text{instantaneous, } K_p = 45\ \text{mm Hg}$$

Find the partial pressures of the contents of a constant-volume bomb after 6.5 hr if we start with pure N_2O_5 at atmospheric pressure.

17. At room temperature sucrose is hydrolyzed by the catalytic action of the enzyme sucrase as follows:

$$\text{sucrose} \xrightarrow{\text{sucrase}} \text{products}$$

Starting with a sucrose concentration $C_{A0} = 1.0$ millimol/liter and an enzyme concentration $C_{E0} = 0.01$ millimol/liter, the following kinetic data are obtained in a batch reactor (concentrations calculated from optical rotation measurements):

C_A, millimol/liter	0.84	0.68	0.53	0.38	0.27	0.16	0.09	0.04	0.018	0.006	0.0025
t, hr	1	2	3	4	5	6	7	8	9	10	11

Determine whether these data can be reasonably fitted by a kinetic equation of the Michaelis-Menten type, or

$$-r_A = \frac{k_3 C_A C_{E0}}{C_A + M} \qquad \text{where } M = \text{Michaelis constant}$$

If the fit is reasonable evaluate the constants k_3 and M.

Note: See Problem 2.19 for the mechanism leading to the Michaelis-Menten equation.

18. For the reactions in series

$$A \xrightarrow{k_1} R \xrightarrow{k_2} S, \qquad k_1 = k_2$$

find the maximum concentration of R and when it is reached.

19. A small reaction bomb fitted with a sensitive pressure-measuring device is flushed out and filled with a mixture at 76.94% reactant A and 23.06% inert at 1-atm pressure and 14°C, a temperature low enough that the reaction does not proceed to any appreciable extent.

TABLE P19

t, min	π, atm	t, min	π, atm
0.5	1.5	3.5	1.99
1	1.65	4	2.025
1.5	1.76	5	2.08
2	1.84	6	2.12
2.5	1.90	7	2.15
3	1.95	8	2.175

The temperature is raised rapidly to 100°C, and the readings in Table P19 are obtained. The stoichiometry of the reaction is $A \rightarrow 2R$ and after sufficient time the reaction proceeds to completion. Find a rate equation in units of moles, liters, and minutes which will satisfactorily fit the data.

20. The following data are obtained at 0°C in a constant-volume batch reactor using pure gaseous A:

Time, min	0	2	4	6	8	10	12	14	∞
Partial pressure of A, mm	760	600	475	390	320	275	240	215	150

The stoichiometry of the decomposition is $A \rightarrow 2.5R$. Find a rate equation which satisfactorily represents this decomposition.

21. Find the first-order rate constant for the disappearance of A in the gas reaction $2A \rightarrow R$ if, on holding the pressure constant, the volume of the reaction mixture, starting with 80% A, decreases by 20% in 3 min.

22. Find the first-order rate constant for the disappearance of A in the gas reaction $A \rightarrow 1.6R$ if the volume of the reaction mixture, starting with pure A, increases by 50% in 4 min. The total pressure within the system stays constant at 1.2 atm, and the temperature is 25°C.

23. A zero-order homogeneous gas reaction $A \rightarrow rR$ proceeds in a constant-volume bomb, 20% inerts, and the pressure rises from 1 to 1.3 atm in 2 min. If the same reaction takes place in a constant-pressure batch reactor, what is the fractional volume change in 4 min if the feed is at 3 atm and consists of 40% inerts?

24. A zero-order homogeneous gas reaction with stoichiometry $A \rightarrow rR$ proceeds in a constant-volume bomb, $\pi = 1$ when $t = 0$ and $\pi = 1.5$ when $t = 1$. If the same reaction, same feed composition, and initial pressure proceeds in a constant-pressure setup, find V at $t = 1$ if $V = 1$ at $t = 0$.

25. The first-order homogeneous gaseous decomposition $A \rightarrow 2.5R$ is carried out in an isothermal batch reactor at 2 atm with 20% inerts present, and the volume increases by 60% in 20 min. In a constant-volume reactor, find the time required for the pressure to reach 8 atm if the initial pressure is 5 atm, 2 atm of which consist of inerts.

26. The gas reaction $2A \rightarrow R + 2S$ is approximately second order with respect to A. When pure A is introduced at 1 atm into a constant-volume batch reactor, the pressure rises 40% in 3 min. For a constant-pressure batch reactor find (*a*) the time required for the same conversion, and (*b*) the fractional increase in volume at that time.

27. Pure gaseous A is prepared under refrigeration and is introduced into a thin-walled capillary which acts as reaction vessel as shown in Fig. P27. No appreci-

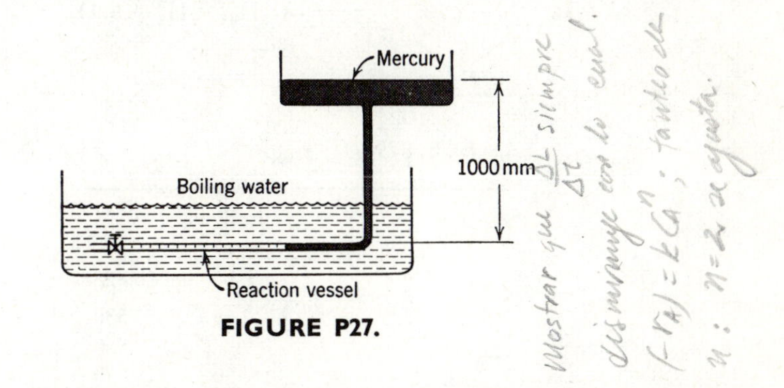

FIGURE P27.

able reaction occurs during handling. The reaction vessel is rapidly plunged into a bath of boiling water, reactant A decomposes to completion according to the reaction $A \rightarrow R + S$, and the following data are obtained:

Time, min	0.5	1	1.5	2	3	4	6	10	∞
Length of capillary occupied by reaction mixture, cm	6.1	6.8	7.2	7.5	7.85	8.1	8.4	8.7	9.4

Find a rate equation in units of moles, liters, and minutes for this decomposition.

28. The presence of substance C seems to increase the rate of reaction of A with B, $A + B \rightarrow AB$. We suspect that C acts catalytically by combining with one of the reactants to form an intermediate, which then reacts further. From the rate data in Table P28 suggest a mechanism and rate equation for this reaction.

TABLE P28

[A]	[B]	[C]	r_{AB}
1	3	0.02	9
3	1	0.02	5
4	4	0.04	32
2	2	0.01	6
2	4	0.03	20
1	2	0.05	12

29. Determine the complete rate equation in units of moles, liters, and seconds for the thermal decomposition of tetrahydrofuran from the half-life data in Table P29.

$$\text{(tetrahydrofuran ring: } H_2C{-}CH_2,\ H_2C\ \ CH_2,\ O) \rightarrow \begin{cases} C_2H_4 + CH_4 + CO \\ C_3H_6 + H_2 + CO \\ \text{etc.} \end{cases}$$

TABLE P29

π_0, mm Hg	$t_{1/2}$, min	T, °C
214	14.5	569
204	67	530
280	17.3	560
130	39	550
206	47	539

4

INTRODUCTION TO REACTOR DESIGN

So far we have considered the mathematical expression called the *rate equation* which describes the progress of a homogeneous reaction. The rate equation for a reacting component i is an intensive measure and it tells how rapidly component i forms or disappears in a given environment as a function of the conditions there, or

$$r_i = \frac{1}{V}\left(\frac{dN_i}{dt}\right)_{\text{by reaction}} = f(\text{conditions within the region of volume } V)$$

This is a differential expression.

In reactor design we want to know what size and type of reactor and method of operation are best for a given job. Since this may require that the conditions in the reactor vary with position as well as time, this question can only be answered by a proper integration of the rate equation for the operation. This may pose difficulties because the temperature and composition of the reacting fluid may vary from point to point within the reactor, depending on the endothermic or exothermic character of the reaction, and depending on the rate of heat addition or removal from the system. In addition, the actual geometry of the reactor will determine the path of the fluid through the vessel and fix the gross mixing patterns which help to dilute rich feed and redistribute material and heat. In effect, then, many factors must be accounted for in predicting the performance of a reactor. How best to treat these factors is the main problem of reactor design.

Equipment in which homogeneous reactions are effected can be one of three general types; the *batch*, the *steady-state flow*, and the *unsteady-state flow* or *semi-*

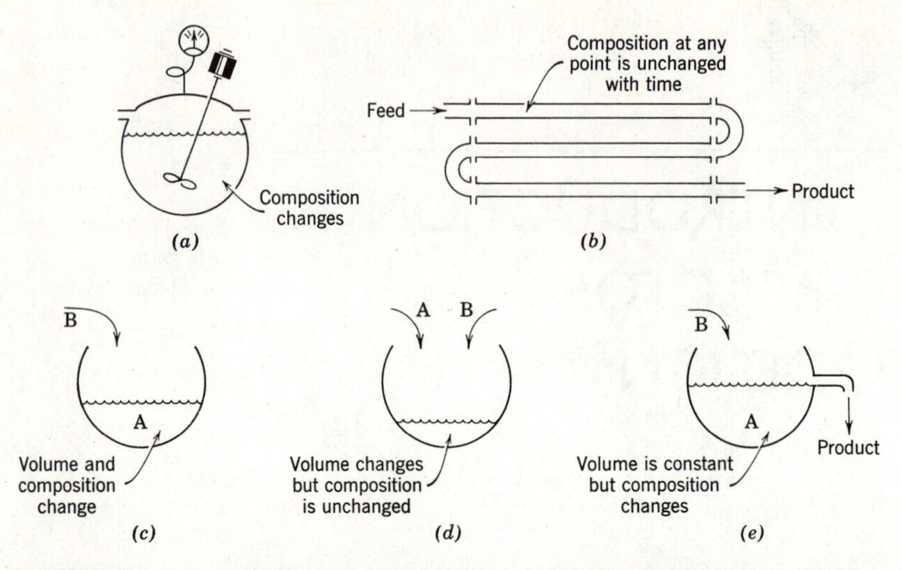

FIGURE 1. Broad classification of reactor types. (*a*) The batch reactor. (*b*) The steady-state flow reactor. (*c*), (*d*), and (*e*) Various forms of the semibatch reactor.

batch reactor. The latter classification includes all reactors that do not fall into the first two categories. These types are shown in Fig. 1.

Let us briefly indicate the particular features and the main areas of application of these reactor types. Naturally these remarks will be amplified further along in the text. The batch reactor is simple, needs little supporting equipment, and is therefore ideal for small-scale experimental studies on reaction kinetics. Industrially it is used when relatively small amounts of material are to be treated. The steady-state

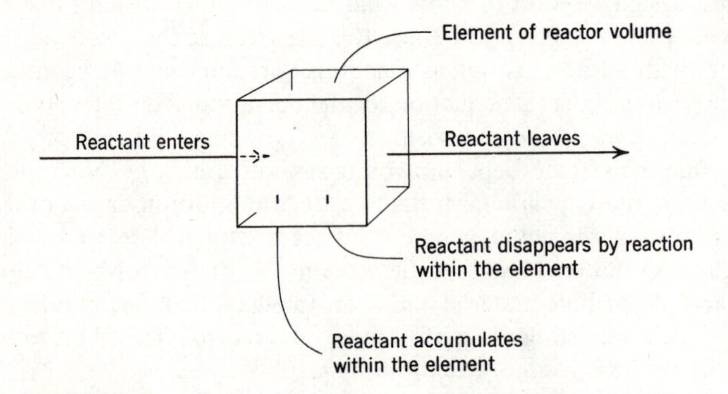

FIGURE 2. Material balance for an element of volume of the reactor.

flow reactor is ideal for industrial purposes when large quantities of material are to be processed and when the rate of reaction is fairly high to extremely high. Supporting-equipment needs are great; however, extremely good product quality control can be obtained. As may be expected, this is the reactor which is widely used in the oil industry. The semibatch reactor is a flexible system but is more difficult to analyze than the other reactor types. It offers good control of reaction speed because the reaction proceeds as reactants are added. Such reactors are used in a variety of applications from the colorimetric titrations in the laboratory to the large open hearth furnaces for steel production.

The starting point for all design is the material balance expressed for any reactant (or product). Thus, as illustrated in Fig. 2, we have

$$\begin{pmatrix}\text{rate of}\\ \text{reactant}\\ \text{flow into}\\ \text{element}\\ \text{of volume}\end{pmatrix} = \begin{pmatrix}\text{rate of}\\ \text{reactant}\\ \text{flow out}\\ \text{of element}\\ \text{of volume}\end{pmatrix} + \begin{pmatrix}\text{rate of reactant}\\ \text{loss due to}\\ \text{chemical reaction}\\ \text{within the element}\\ \text{of volume}\end{pmatrix} + \begin{pmatrix}\text{rate of}\\ \text{accumulation}\\ \text{of reactant}\\ \text{in element}\\ \text{of volume}\end{pmatrix} \tag{1}$$

Where the composition within the reactor is uniform (independent of position), the accounting may be made over the whole reactor. Where the composition is not uniform, it must be made over a differential element of volume and then integrated across the whole reactor for the appropriate flow and concentration conditions. For the various reactor types this equation simplifies one way or another, and the resultant expression when integrated gives the basic performance equation for that type of unit. Thus in the batch reactor the first two terms are zero; in the steady-state flow reactor the fourth term disappears; for the semibatch reactor all four terms may have to be considered.

In nonisothermal operations energy balances must be used in conjunction with material balances. Thus, as illustrated in Fig. 3, we have

$$\begin{pmatrix}\text{rate of heat}\\ \text{flow into}\\ \text{element of}\\ \text{volume}\end{pmatrix} = \begin{pmatrix}\text{rate of heat}\\ \text{flow out of}\\ \text{element of}\\ \text{volume}\end{pmatrix} + \begin{pmatrix}\text{rate of}\\ \text{disappearance}\\ \text{of heat by}\\ \text{reaction within}\\ \text{element of}\\ \text{volume}\end{pmatrix} + \begin{pmatrix}\text{rate of}\\ \text{accumulation}\\ \text{of heat within}\\ \text{element of}\\ \text{volume}\end{pmatrix} \tag{2}$$

Again, depending on circumstances, this accounting may be made either about a differential element of reactor or about the reactor as a whole.

The material balance of Eq. 1 and the energy balance of Eq. 2 are tied together by their third terms because the heat effect is produced by the reaction itself.

Since Eqs. 1 and 2 are the starting points for all design, we consider their integration for a variety of situations of increasing complexity in the chapters to follow.

When we can predict the response of the reacting system to changes in operating conditions (how rates and equilibrium conversion change with temperature and

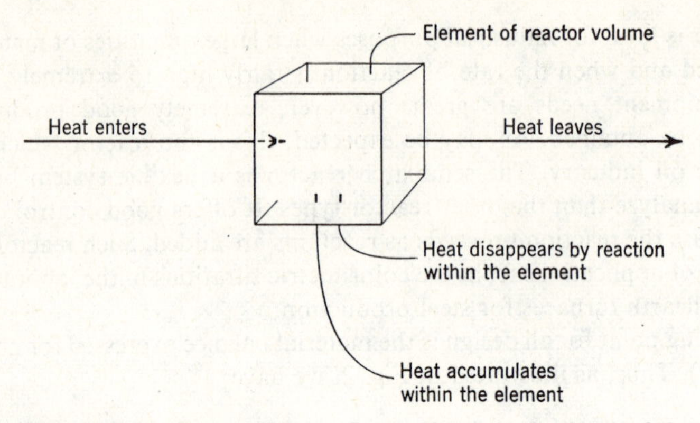

FIGURE 3. Energy balance for an element of volume of the reactor.

pressure), when we are able to compare yields for alternative designs (adiabatic versus isothermal operations, single versus multiple reactor units, flow versus batch system), and when we can estimate the economics of these various alternatives, then and only then will we feel sure that we can arrive at the design well fitted for the purpose at hand. Unfortunately, real situations are rarely simple.

Should we explore all reasonable design alternatives? How sophisticated should our analysis be? What simplifying assumptions should we make? What short cuts should we take? Which factors should we ignore and which should we consider? And how should the reliability and completeness of the data at hand influence our decisions? Good engineering judgment, which only comes with experience, will suggest which course of action to take.

5

SINGLE IDEAL REACTORS

In this chapter we develop the performance equations for a single fluid reacting in the three ideal reactors shown in Fig. 1. We call these homogeneous reactions. Applications and extensions of these equations to various isothermal and non-isothermal operations are considered in the following three chapters.

In the *batch reactor* of Fig. 1*a* the reactants are initially charged into a container, are well mixed, and are left to react for a certain period. The resultant mixture is then discharged. This is an unsteady-state operation where composition changes with time; however, at any instant the composition throughout the reactor is uniform.

The first of the two ideal steady-state flow reactors is variously known as the plug flow, slug flow, piston flow, ideal tubular, and unmixed flow reactor, and it is shown in Fig. 1*b*. We refer to it as the *plug flow reactor* and to this pattern of flow as *plug flow*. It is characterized by the fact that the flow of fluid through the reactor is orderly with no element of fluid overtaking or mixing with any other element ahead or behind. Actually, there may be lateral mixing of fluid in a plug flow reactor; however, there must be no mixing or diffusion along the flow path. The necessary and sufficient condition for plug flow is for the residence time in the reactor to be the same for all elements of fluid.*

The other ideal steady-state flow reactor is called the mixed reactor, the backmix reactor, the ideal stirred tank reactor, or the CFSTR (constant flow stirred tank reactor) and, as its names suggest, it is a reactor in which the contents are well

* The necessary condition follows directly from the definition of plug flow. However, the sufficient condition—that the same residence times implies plug flow—can be established only by using the second law of thermodynamics.

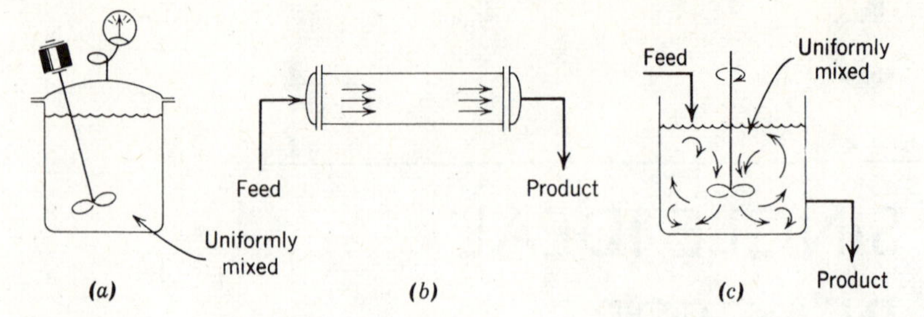

FIGURE 1. The three types of ideal reactors: (*a*) batch reactor, (*b*) plug flow reactor, and (*c*) mixed flow reactor.

stirred and uniform throughout. Thus the exit stream from this reactor has the same composition as the fluid within the reactor. We refer to this type of flow as *mixed flow*, and the corresponding reactor, the *mixed reactor*, or the *mixed flow reactor*.

These three ideals are relatively easy to treat. In addition, one or other usually represents the best way of contacting the reactants—no matter what the operation. For these reasons we try to design real reactors so that their flows approach these ideals, and much of the development in this book centers about them.

In the treatment to follow it should be understood that the term V, called the reactor volume, really refers to the volume of fluid in the reactor. When this differs from the internal volume of reactor, then V_r designates the internal volume while V designates the volume of fluid. For example, in solid catalyzed reactors with voidage ϵ we have

$$V = \epsilon V_r$$

For homogeneous systems, however, we use the term V alone.

Ideal Batch Reactor

Make a material balance for any component A. For such an accounting we usually select the limiting component. In a batch reactor, since the composition is uniform throughout at any instant of time, we may make the accounting about the whole reactor. Noting that no fluid enters or leaves the reaction mixture during reaction Eq. 4.1, written for component A, becomes

$$\overset{0}{\cancel{\text{input}}} = \overset{0}{\cancel{\text{output}}} + \text{disappearance} + \text{accumulation}$$

or

$$+\begin{pmatrix}\text{rate of loss of reactant A}\\ \text{within reactor due to}\\ \text{chemical reaction}\end{pmatrix} = -\begin{pmatrix}\text{rate of accumulation}\\ \text{of reactant A}\\ \text{within the reactor}\end{pmatrix} \tag{1}$$

Evaluating the terms of Eq. 1 we find

$$\begin{array}{l}\text{disappearance of A} \\ \quad\text{by reaction,} \\ \quad\text{moles/time}\end{array} = (-r_A)V = \left(\frac{\text{moles A reacting}}{(\text{time})(\text{volume of fluid})}\right)(\text{volume of fluid})$$

$$\begin{array}{l}\text{accumulation of A,} \\ \quad\text{moles/time}\end{array} = \frac{dN_A}{dt} = \frac{d[N_{A0}(1 - X_A)]}{dt} = -N_{A0}\frac{dX_A}{dt}$$

By replacing in Eq. 1,

$$(-r_A)V = N_{A0}\frac{dX_A}{dt} \tag{2}$$

then rearranging and integrating, we obtain

$$t = N_{A0}\int_0^{X_A} \frac{dX_A}{(-r_A)V} \tag{3}$$

This is the general equation showing the time required to achieve a conversion X_A for either isothermal or nonisothermal operation. The volume of reacting fluid and the reaction rate remain under the integral sign, for in general they both change as reaction proceeds.

This equation may be simplified for a number of situations. If the density of the fluid remains constant we obtain

$$t = C_{A0}\int_0^{X_A} \frac{dX_A}{-r_A} = -\int_{C_{A0}}^{C_A} \frac{dC_A}{-r_A} \tag{4}$$

For all reactions in which the volume of reacting mixture changes proportionately with conversion, such as in single gas-phase reactions with significant density changes, Eq. 3 becomes

$$t = N_{A0}\int_0^{X_A} \frac{dX_A}{(-r_A)V_0(1 + \varepsilon_A X_A)} = C_{A0}\int_0^{X_A} \frac{dX_A}{(-r_A)(1 + \varepsilon_A X_A)} \tag{5}$$

In one form or another, Eqs. 2 to 5 have all been encountered in Chapter 3. They are applicable to both isothermal and nonisothermal operations. For the latter the variation of rate with temperature, and the variation of temperature with

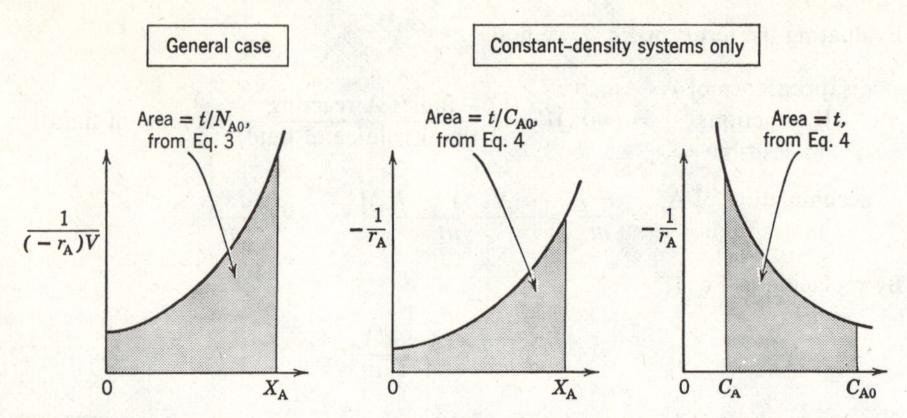

FIGURE 2. Graphical representation of the performance equations for batch reactors, isothermal or nonisothermal.

conversion, must be known before solution is possible. Figure 2 is a graphical representation of two of these equations.

Space-time and Space-velocity

Just as the reaction time t is the natural measure of the processing rate in a batch reactor, so are the space-time and space-velocity the proper performance measures of flow reactors. These terms are defined as follows:

Space-time:

$$\tau = \frac{1}{s} = \begin{pmatrix} \text{time required to process one} \\ \text{reactor volume of feed measured} \\ \text{at specified conditions} \end{pmatrix} = [\text{time}] \tag{6}$$

Space-velocity:

$$s = \frac{1}{\tau} = \begin{pmatrix} \text{number of reactor volumes of} \\ \text{feed at specified conditions which} \\ \text{can be treated in unit time} \end{pmatrix} = [\text{time}^{-1}] \tag{7}$$

Thus a space-velocity of 5 hr^{-1} means that five reactor volumes of feed at specified conditions are being fed into the reactor per hour. A space-time of 2 min means that every two minutes one reactor volume of feed at specified conditions is being treated by the reactor.

Now we may arbitrarily select the temperature, pressure, and state (gas, liquid, or solid) at which we choose to measure the volume of material being fed to the reactor. Certainly, then, the value for space-velocity or space-time depends on the conditions selected. If they are of the stream entering the reactor, the relation between s or τ and the other pertinent variables is

$$\tau = \frac{1}{s} = \frac{C_{A0}V}{F_{A0}} = \frac{\left(\dfrac{\text{moles A entering}}{\text{volume of feed}}\right)(\text{volume of reactor})}{\left(\dfrac{\text{moles A entering}}{\text{time}}\right)}$$

$$= \frac{V}{v_0} = \frac{(\text{reactor volume})}{(\text{volumetric feed rate})} \tag{8}$$

It may be more convenient to measure the volumetric feed rate at some standard state, especially when the reactor is to operate at a number of temperatures. If, for example, the material is gaseous when fed to the reactor at high temperature but is liquid at the standard state, care must be taken to specify precisely what standard state has been chosen. The relation between the space-velocity and space-time for actual entering conditions and at standard conditions (designated by primes) is given by

$$\tau' = \frac{1}{s'} = \frac{C'_{A0}V}{F_{A0}} = \tau\frac{C'_{A0}}{C_{A0}} = \frac{1}{s}\frac{C'_{A0}}{C_{A0}} \tag{9}$$

In most of what follows, we deal with the space-velocity and space-time based on feed at actual entering conditions; however, the change to any other basis is easily made.

Steady-state Mixed Flow Reactor

The performance equation for the mixed reactor is obtained from Eq. 4.1, which makes an accounting of a given component within an element of volume of the system. But since the composition is uniform throughout, the accounting may be made about the reactor as a whole. By selecting reactant A for consideration, Eq. 4.1 becomes

$$\text{input} = \text{output} + \text{disappearance by reaction} + \overset{0}{\text{accumulation}} \tag{10}$$

As shown in Fig. 3, if $F_{A0} = v_0C_{A0}$ is the molar feed rate of component A to the reactor, then considering the reactor as a whole we have

$$\text{input of A, moles/time} = F_{A0}(1 - X_{A0}) = F_{A0}$$

$$\text{output of A, moles/time} = F_A = F_{A0}(1 - X_A)$$

$$\begin{array}{l}\text{disappearance of A}\\ \text{by reaction,}\\ \text{moles/time}\end{array} = (-r_A)V = \left(\frac{\text{moles A reacting}}{(\text{time})(\text{volume of fluid})}\right)\left(\begin{array}{c}\text{volume of}\\ \text{reactor}\end{array}\right)$$

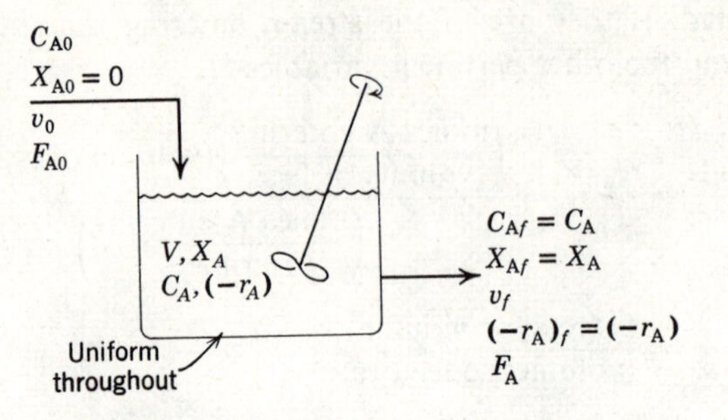

FIGURE 3. Notation for a mixed reactor.

Replacing in Eq. 10, we obtain

$$F_{A0}X_A = (-r_A)V$$

which on rearranging becomes

$$\frac{V}{F_{A0}} = \frac{\tau}{C_{A0}} = \frac{\Delta X_A}{-r_A} = \frac{X_A}{-r_A}$$

or (11)

$$\tau = \frac{1}{s} = \frac{V}{v_0} = \frac{VC_{A0}}{F_{A0}} = \frac{C_{A0}X_A}{-r_A}$$

where X_A and r_A are evaluated at exit stream conditions, which are the same as the conditions within the reactor.

More generally, if the feed on which conversion is based, subscript 0, enters the reactor partially converted, subscript i, and leaves at conditions given by subscript f, we have

$$\frac{V}{F_{A0}} = \frac{\Delta X_A}{(-r_A)_f} = \frac{X_{Af} - X_{Ai}}{(-r_A)_f}$$

or (12)

$$\tau = \frac{VC_{A0}}{F_{A0}} = \frac{C_{A0}(X_{Af} - X_{Ai})}{(-r_A)_f}$$

A situation where this form of equation is used is given on p. 135.

For the special case of constant-density systems $X_A = 1 - C_A/C_{A0}$, in which case the performance equation for mixed reactors can also be written in terms of concentrations, or

$$\frac{V}{F_{A0}} = \frac{X_A}{-r_A} = \frac{C_{A0} - C_A}{C_{A0}(-r_A)}$$

or (13)

$$\tau = \frac{V}{v} = \frac{C_{A0}X_A}{-r_A} = \frac{C_{A0} - C_A}{-r_A}$$

These expressions relate in a simple way the four terms X_A, $-r_A$, V, F_{A0}; thus, knowing any three allows the fourth to be found directly. In design, then, the size of reactor needed for a given duty or the extent of conversion in a reactor of given size is found directly. In kinetic studies each steady-state run gives, without integration, the reaction rate for the conditions within the reactor. The ease of interpretation of data from a mixed reactor makes its use very attractive in kinetic studies, in particular with messy reactions (e.g., multiple reactions and solid catalyzed reactions).

Figure 4 is a graphical representation of these performance equations. For any specific kinetic form the equations can be written out directly. As an example, for

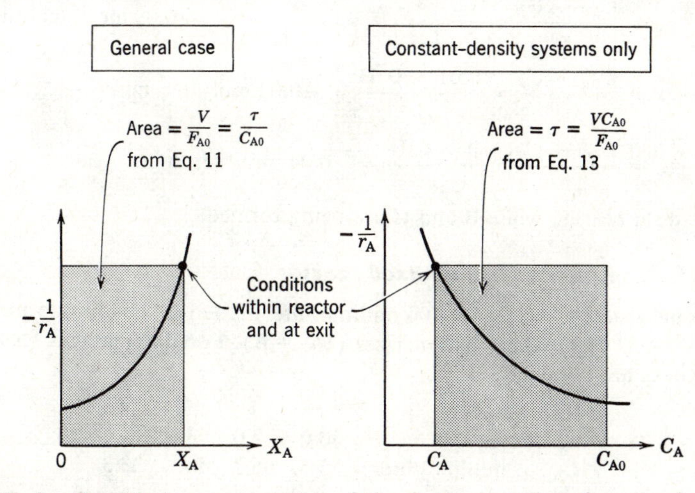

FIGURE 4. Graphical representation of the design equations for mixed reactor.

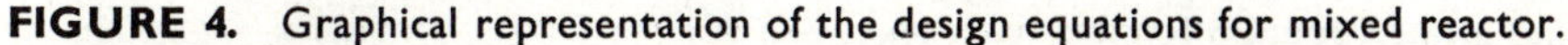

constant density systems $C_A/C_{A0} = 1 - X_A$, thus the performance expression *for first-order reaction* becomes

$$k\tau = \frac{X_A}{1 - X_A} = \frac{C_{A0} - C_A}{C_A} \qquad \text{for } \varepsilon_A = 0 \tag{14}$$

On the other hand, for linear expansion

$$V = V_0(1 + \varepsilon_A X_A) \quad \text{and} \quad \frac{C_A}{C_{A0}} = \frac{1 - X_A}{1 + \varepsilon_A X_A}$$

thus *for first-order reaction* the performance expression becomes

$$k\tau = \frac{X_A(1 + \varepsilon_A X_A)}{1 - X_A} \tag{15}$$

Similar expressions can be written for any other form of rate equation. These expressions can be written either in terms of concentrations or conversions. The latter form is simpler for systems of changing density, while either form can be used for systems of constant density.

EXAMPLE 1. *Reaction rate in a mixed reactor*

One liter/min of liquid containing A and B (C_{A0} = 0.10 mol/liter, C_{B0} = 0.01 mol/liter) flow into a mixed reactor of volume V = 1 liter. The materials react in a complex manner for which the stoichiometry is unknown. The outlet stream from the reactor contains A, B, and C (C_{Af} = 0.02 mol/liter, C_{Bf} = 0.03 mol/liter, C_{Cf} = 0.04 mol/liter). Find the rate of reaction of A, B, and C for the conditions within the reactor.

SOLUTION

For a liquid in a mixed reactor $\varepsilon_A = 0$ and Eq. 13 applied to each of the reacting components, giving for the rate of disappearance:

$$-r_A = \frac{C_{A0} - C_A}{\tau} = \frac{C_{A0} - C_A}{V/v} = \frac{0.10 - 0.02}{1/1} = 0.08 \text{ mol/liter}\cdot\text{min}$$

$$-r_B = \frac{C_{B0} - C_B}{\tau} = \frac{0.01 - 0.03}{1} = -0.02 \text{ mol/liter}\cdot\text{min}$$

$$-r_C = \frac{C_{C0} - C_C}{\tau} = \frac{0 - 0.04}{1} = -0.04 \text{ mol/liter}\cdot\text{min}$$

Thus A is disappearing while B and C are being formed.

EXAMPLE 2. *Kinetics from a mixed reactor*

Pure gaseous reactant A (C_{A0} = 100 millimol/liter) is fed at steady rate into a mixed reactor (V = 0.1 liter) where it dimerizes ($2A \rightarrow R$). For different gas feed rates the following data are obtained:

Run number	1	2	3	4
v_0, liter/hr	30.0	9.0	3.6	1.5
$C_{A,out}$, millimol/liter	85.7	66.7	50	33.3

Find a rate equation for this reaction.

SOLUTION

For this stoichiometry, $2A \rightarrow R$, the expansion factor is

$$\varepsilon_A = \frac{1-2}{2} = -\frac{1}{2}$$

and the corresponding relation between concentration and conversion is

$$\frac{C_A}{C_{A0}} = \frac{1 - X_A}{1 + \varepsilon_A X_A} = \frac{1 - X_A}{1 - \frac{1}{2}X_A}$$

or

$$X_A = \frac{1 - C_A/C_{A0}}{1 + \varepsilon_A C_A/C_{A0}} = \frac{1 - C_A/C_{A0}}{1 - C_A/2C_{A0}}$$

The conversion for each run is then calculated and tabulated in Column 4 of Table E2.

TABLE E2

	Given		Calculated	
Run	v_0	$C_{A,\text{out}}$	X_A	$(-r_A)_{\text{out}} = \dfrac{v_0 C_{A0} X_A}{V}$
1	30.0	85.7	0.25	$\dfrac{(30)(100)(0.25)}{0.1} = 7500$
2	9.0	66.7	0.50	4500
3	3.6	50	0.667	2400
4	1.5	33.3	0.80	1200

From the performance equation, Eq. 11, the rate of reaction for each run is given by

$$(-r_A) = \frac{v_0 C_{A0} X_A}{V}, \qquad \left[\frac{\text{millimol}}{\text{liter} \cdot \text{hr}}\right]$$

These values are tabulated in Column 5 of Table E2.

Having paired values of r_A and C_A (see Table E2) we are ready to test various kinetic expressions. Instead of separately testing for first order (plot r_A versus C_A), second order (plot r_A versus C_A^2), etc., let us test directly for nth-order kinetics. For this take logarithms of $-r_A = kC_A^n$, giving

$$\log(-r_A) = \log k + n \log C_A$$

This shows that nth-order kinetics will give a straight line on a $\log(-r_A)$ versus $\log C_A$ plot. As shown in Fig. E2, the four actual data are reasonably represented by a straight line of slope 2, so the rate equation for this dimerization is

$$-r_A = \left(1.0 \frac{\text{liter}}{\text{hr} \cdot \text{millimol}}\right) C_A^2, \qquad \left[\frac{\text{millimol}}{\text{liter} \cdot \text{hr}}\right]$$

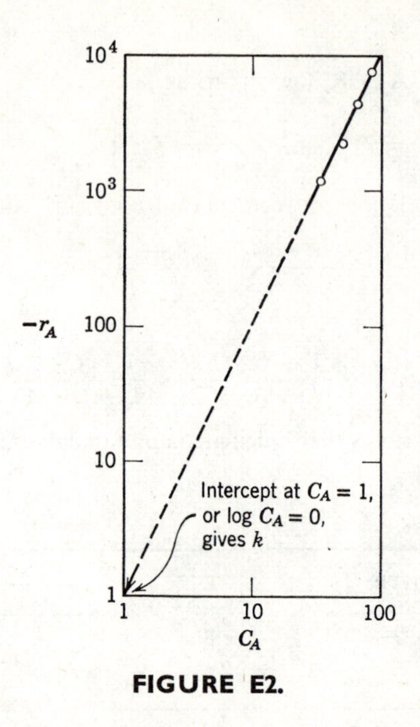

FIGURE E2.

Comment. If we ignore the density change in our analysis (or put $\varepsilon_A = 0$ and use $C_A/C_{A0} = 1 - X_A$) we end up with an incorrect rate equation (reaction order $n \cong 1.6$) which when used in design would give wrong performance predictions.

EXAMPLE 3. Mixed reactor performance

The liquid-phase reaction

$$A + B \underset{k_2}{\overset{k_1}{\rightleftharpoons}} R + S \qquad \begin{aligned} k_1 &= 7 \text{ liter/mol·min} \\ k_2 &= 3 \text{ liter/mol·min} \end{aligned}$$

is to take place in a 120-liter steady-state mixed reactor. Two feed streams, one containing 2.8 mol A/liter and the other containing 1.6 mol B/liter, are to be introduced in equal volumes into the reactor, and 75% conversion of limiting component is desired (see Fig. E3). What should be the flow rate of each stream? Assume a constant density throughout.

SOLUTION

The concentration of components in the mixed feed stream is

$$C_{A0} = 1.4 \text{ mol/liter}$$

$$C_{B0} = 0.8 \text{ mol/liter}$$

$$C_{R0} = C_{S0} = 0$$

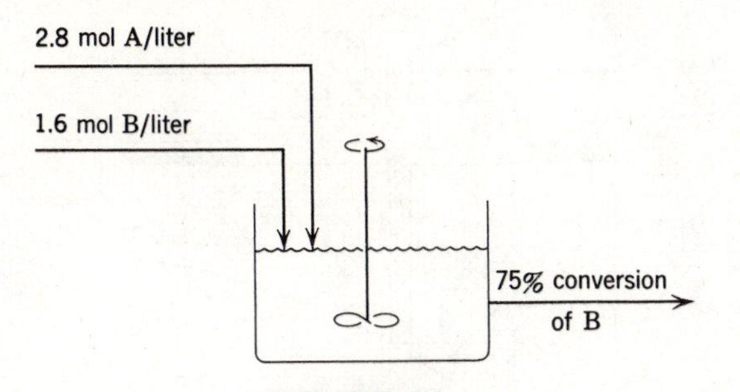

FIGURE E3.

For 75% conversion of B and $\varepsilon = 0$ the composition within the reactor and in the exit stream is

$$C_A = 1.4 - 0.6 = 0.8 \text{ mol/liter}$$
$$C_B = 0.8 - 0.6 = 0.2 \text{ mol/liter}$$
$$C_R = 0.6 \text{ mol/liter}$$
$$C_S = 0.6 \text{ mol/liter}$$

Now the rate of reaction at the conditions within the reactor is

$$-r_A = -r_B = k_1 C_A C_B - k_2 C_R C_S$$
$$= \left(7 \frac{\text{liter}}{\text{mol}\cdot\text{min}}\right)\left(0.8 \frac{\text{mol}}{\text{liter}}\right)\left(0.2 \frac{\text{mol}}{\text{liter}}\right) - \left(3 \frac{\text{liter}}{\text{mol}\cdot\text{min}}\right)\left(0.6 \frac{\text{mol}}{\text{liter}}\right)^2$$
$$= (1.12 - 1.08) \frac{\text{mol}}{\text{liter}\cdot\text{min}} = 0.04 \frac{\text{mol}}{\text{liter}\cdot\text{min}}$$

For no change in density, hence $\varepsilon = 0$, Eq. 13 gives

$$\tau = \frac{V}{v} = \frac{C_{A0} - C_A}{-r_A} = \frac{C_{B0} - C_B}{-r_B}$$

Hence the volumetric flow rate into and out of the reactor is

$$v = \frac{V(-r_A)}{C_{A0} - C_A} = \frac{V(-r_B)}{C_{B0} - C_B}$$
$$= \frac{(120 \text{ liter})(0.04 \text{ mol/liter}\cdot\text{min})}{0.6 \text{ mol/liter}} = 8 \frac{\text{liter}}{\text{min}}$$

or 4 liter/min of each of the two feed streams.

Steady-state Plug Flow Reactor

In a plug flow reactor the composition of the fluid varies from point to point along a flow path; consequently, the material balance for a reaction component must be made for a differential element of volume dV. Thus for reactant A Eq. 4.1 becomes

$$\text{input} = \text{output} + \text{disappearance by reaction} + \underset{\nearrow 0}{\text{accumulation}} \qquad (10)$$

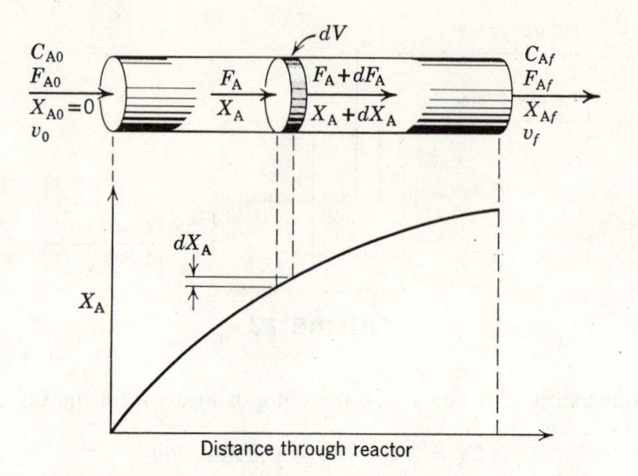

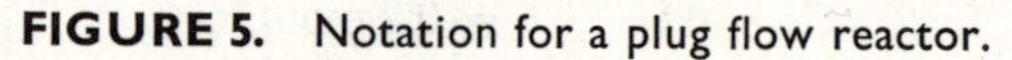

FIGURE 5. Notation for a plug flow reactor.

Referring to Fig. 5, we see for volume dV that

$$\text{input of A, moles/time} = F_A$$

$$\text{output of A, moles/time} = F_A + dF_A$$

$$\begin{aligned}\text{disappearance of A by reaction, moles/time} &= (-r_A)\,dV \\ &= \left(\frac{\text{moles A reacting}}{\text{(time)(volume of fluid)}}\right) \\ &\quad \times \text{(volume of differential element)}\end{aligned}$$

Introducing these three terms in Eq. 10, we obtain

$$F_A = (F_A + dF_A) + (-r_A)\,dV$$

Noting that

$$dF_A = d[F_{A0}(1 - X_A)] = -F_{A0}\,dX_A$$

We obtain on replacement

$$F_{A0}\,dX_A = (-r_A)\,dV \tag{16}$$

This, then, is the equation accounting for A in the differential section of reactor of volume dV. For the reactor as a whole the expression must be integrated. Now F_{A0}, the feed rate, is constant, but r_A is certainly dependent on the concentration of materials or conversion. Grouping the terms accordingly, we obtain

$$\int_0^V \frac{dV}{F_{A0}} = \int_0^{X_{Af}} \frac{dX_A}{-r_A}$$

Thus

$$\frac{V}{F_{A0}} = \frac{\tau}{C_{A0}} = \int_0^{X_{Af}} \frac{dX_A}{-r_A}$$

or (17)

$$\tau = \frac{V}{v_0} = C_{A0} \int_0^{X_{Af}} \frac{dX_A}{-r_A}$$

Equation 17 allows the determination of reactor size for a given feed rate and required conversion. Compare Eqs. 11 and 17. The difference is that in plug flow r_A varies, whereas in mixed flow r_A is constant.

As a more general expression for plug flow reactors, if the feed on which conversion is based, subscript 0, enters the reactor partially converted, subscript *i*, and leaves at a conversion designated by subscript *f*, we have

$$\frac{V}{F_{A0}} = \frac{V}{C_{A0} v_0} = \int_{X_{Ai}}^{X_{Af}} \frac{dX_A}{-r_A}$$

or (18)

$$\tau = \frac{V}{v_0} = C_{A0} \int_{X_{Ai}}^{X_{Af}} \frac{dX_A}{-r_A}$$

A case where this form of equation is used is given on p. 133.

For the special case of *constant-density systems*

$$X_A = 1 - \frac{C_A}{C_{A0}} \qquad \text{and} \qquad dX_A = -\frac{dC_A}{C_{A0}}$$

in which case the performance equation can be expressed in terms of concentrations, or

$$\frac{V}{F_{A0}} = \frac{\tau}{C_{A0}} = \int_0^{X_{Af}} \frac{dX_A}{-r_A} = -\frac{1}{C_{A0}} \int_{C_{A0}}^{C_{Af}} \frac{dC_A}{-r_A}$$

or (19)

$$\tau = \frac{V}{v_0} = C_{A0} \int_0^{X_{Af}} \frac{dX_A}{-r_A} = -\int_{C_{A0}}^{C_{Af}} \frac{dC_A}{-r_A}$$

These performance equations, Eqs. 17 to 19, can be written either in terms of concentrations or conversions. For systems of changing density it is more convenient to use conversions; however, there is no particular preference for constant density systems. Whatever its form, the performance equations interrelate the *rate of reaction*, the *extent of reaction*, the *reactor volume*, and the *feed rate*, and if any one of these quantities is unknown it can be found from the other three.

Figure 6 displays these performance equations and shows that the space-time needed for any particular duty can always be found by numerical or graphical integration. However, for certain simple kinetic forms analytic integration is possible—and convenient. To do this, insert the kinetic expression for r_A in Eq. 17 and integrate. Some of the simpler integrated forms for plug flow are as follows:

Zero-order homogeneous reaction, any constant ε_A

$$k\tau = \frac{kC_{A0}V}{F_{A0}} = C_{A0}X_A \tag{20}$$

First-order irreversible reaction, A → products, any constant ε_A,

$$k\tau = -(1 + \varepsilon_A)\ln(1 - X_A) - \varepsilon_A X_A \tag{21}$$

First-order reversible reaction, $A \rightleftarrows rR$, $C_{R0}/C_{A0} = M$, kinetics approximated or fitted by $-r_A = k_1C_A - k_2C_R$ with an observed equilibrium conversion X_{Ae}, any constant ε_A,

$$k_1\tau = \frac{M + rX_{Ae}}{M + r}\left[-(1 + \varepsilon_A X_{Ae})\ln\left(1 - \frac{X_A}{X_{Ae}}\right) - \varepsilon_A X_A\right] \tag{22}$$

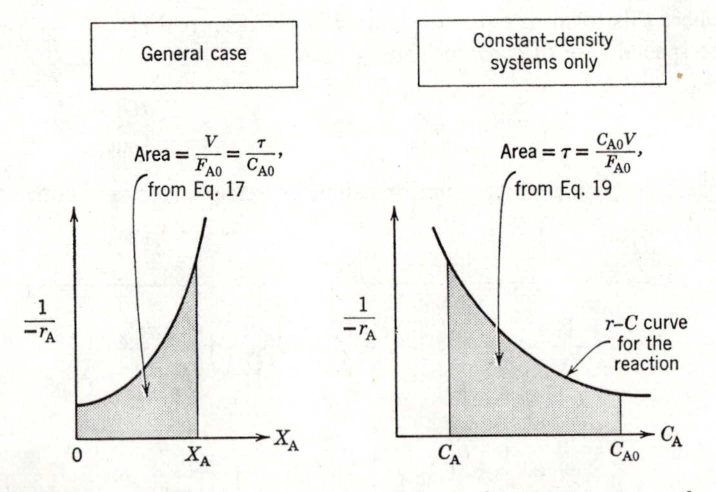

FIGURE 6. Graphical representation of the performance equations for plug flow reactors.

Second-order irreversible reaction, $A + B \rightarrow$ products with equimolar feed or $2A \rightarrow$ products, any constant ε_A,

$$C_{A0}k\tau = 2\varepsilon_A(1+\varepsilon_A)\ln(1-X_A) + \varepsilon_A^2 X_A + (\varepsilon_A+1)^2\frac{X_A}{1-X_A} \tag{23}$$

Where the density is constant, put $\varepsilon_A = 0$ to obtain the simplified performance equation.

By comparing the batch expressions of Chapter 3 with these plug flow expressions we find:

(1) For *systems of constant density* (constant-volume batch and constant-density plug flow) the performance equations are identical, τ for plug flow is equivalent to t for the batch reactor, and the equations can be used interchangeably.

(2) For *systems of changing density* there is no direct correspondence between the batch and the plug flow equations and the correct equation must be used for each particular situation. In this case the performance equations cannot be used interchangeably.

The following illustrative examples show how these expressions are used.

EXAMPLE 4. *Plug flow reactor performance*

A homogeneous gas reaction $A \rightarrow 3R$ has a reported rate at 215°C

$$-r_A = 10^{-2}C_A^{1/2}, \quad [\text{mol/liter}\cdot\text{sec}]$$

Find the space-time needed for 80% conversion of a 50% A–50% inert feed to a plug flow reactor operating at 215°C and 5 atm ($C_{A0} = 0.0625$ mol/liter).

SOLUTION

For this stoichiometry and with 50% inerts two volumes of feed gas would give four volumes of completely converted product gas, thus

$$\varepsilon_A = \frac{4-2}{2} = 1$$

in which case the plug flow performance equation, Eq. 17, becomes

$$\tau = C_{A0}\int_0^{X_{Af}}\frac{dX_A}{-r_A} = C_{A0}\int_0^{X_{Af}}\frac{dX_A}{kC_{A0}^{1/2}\left(\dfrac{1-X_A}{1+\varepsilon_A X_A}\right)^{1/2}} = \frac{C_{A0}^{1/2}}{k}\int_0^{0.8}\left(\frac{1+X_A}{1-X_A}\right)^{1/2} dX_A \tag{i}$$

The integral can be evaluated in any one of three ways: graphically, numerically, or analytically. Let us illustrate these methods.

Graphical Integration. First evaluate the function to be integrated at selected values (see Table E4), and plot this function (see Fig. E4).

Counting squares or estimating by eye we find

$$\text{Area} = \int_0^{0.8}\left(\frac{1+X_A}{1-X_A}\right)^{1/2} dX_A = (1.70)(0.8) = 1.36$$

TABLE E4

X_A	$\frac{1+X_A}{1-X_A}$	$\left(\frac{1+X_A}{1-X_A}\right)^{1/2}$
0	1	1
0.2	$\frac{1.2}{0.8} = 1.5$	1.227
0.4	2.3	1.528
0.6	4	2
0.8	9	3

Numerical Integration. Using Simpson's rule, applicable to an even number of uniformly spaced intervals on the X_A axis, we find for the data of Table E4

$$\int_0^{0.8} \left(\frac{1+X_A}{1-X_A}\right)^{1/2} dX_A = \text{(average height)(total width)}$$

$$= \left[\frac{1(1) + 4(1.227) + 2(1.528) + 4(2) + 1(3)}{12}\right](0.8)$$

$$= 1.331$$

Analytical integration. From a table of integrals

$$\int_0^{0.8} \left(\frac{1+X_A}{1-X_A}\right)^{1/2} dX_A = \int_0^{0.8} \frac{1+X_A}{\sqrt{1-X_A^2}} dX_A = \left(\arcsin X_A - \sqrt{1-X_A^2}\right)\Big|_0^{0.8} = 1.328$$

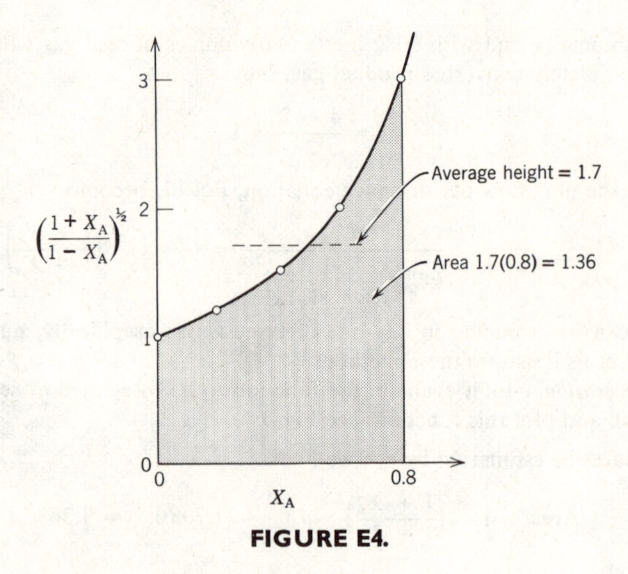

FIGURE E4.

The method of integration recommended depends on the situation. In this problem probably the numerical method is the quickest and simplest and gives a good enough answer for most purposes.

So with the integral evaluated Eq. (i) becomes

$$\tau = \frac{(0.0625 \text{ mol/liter})^{1/2}}{(10^{-2} \text{ mol}^{1/2}/\text{liter}^{1/2}\cdot\text{sec})}(1.33) = 33.2 \text{ sec}$$

EXAMPLE 5. Plug flow reactor volume

The homogeneous gas decomposition of phosphine

$$4PH_3(g) \rightarrow P_4(g) + 6H_2$$

proceeds at 1200°F with first-order rate

$$-r_{PH_3} = (10/\text{hr})C_{PH_3}$$

What size of plug flow reactor operating at 1200°F and 4.6 atm can produce 80% conversion of a feed consisting of 4 lb-mol of pure phosphine per hour.

SOLUTION

Let $A = PH_3$, $R = P_4$, $S = H_2$. Then the reaction becomes

$$4A \rightarrow R + 6S$$

with

$$-r_A = (10/\text{hr})C_A$$

The volume of plug flow reactor is given by Eq. 17; thus

$$V = F_{A0}\int_0^{X_A}\frac{dX_A}{-r_A} = F_{A0}\int_0^{X_A}\frac{dX_A}{kC_A}$$

At constant pressure

$$C_A = C_{A0}\left(\frac{1 - X_A}{1 + \varepsilon_A X_A}\right)$$

hence

$$V = \frac{F_{A0}}{kC_{A0}}\int_0^{X_A}\frac{1 + \varepsilon_A X_A}{1 - X_A}\,dX_A$$

Integrating we obtain Eq. 21 or

$$V = \frac{F_{A0}}{kC_{A0}}\left[(1 + \varepsilon_A)\ln\frac{1}{1 - X_A} - \varepsilon_A X_A\right]$$

Evaluating the individual terms in this expression have:

$$F_{A0} = 4 \text{ lb-mol/hr}$$

$$k = 10/\text{hr}$$

$$C_{A0} = \frac{p_{A0}}{RT} = \frac{4.6 \text{ atm}}{(0.729 \text{ ft}^3\cdot\text{atm/lb-mol}\cdot{}^\circ\text{R})(1660^\circ\text{R})} = 0.0038 \text{ lb-mol/ft}^3$$

$$\varepsilon_A = \frac{7-4}{4} = 0.75$$

$$X_A = 0.8$$

hence the volume of reactor

$$V = \frac{4 \text{ lb-mol/hr}}{(10/\text{hr})(0.0038 \text{ lb-mol/ft}^3)}\left[(1+0.75)\ln\frac{1}{0.2} - 0.75(0.8)\right] = 234 \text{ ft}^3$$

EXAMPLE 6. *Test of a kinetic equation in a plug flow reactor*

We suspect that the gas reaction between A, B, and R is an elementary reversible reaction

$$A + B \underset{k_2}{\overset{k_1}{\rightleftarrows}} R$$

and we plan to test this with experiments in an isothermal plug flow reactor.

(*a*) Develop the isothermal performance equation for these kinetics for a feed of A, B, R, and inerts.

(*b*) Show how to test this equation for an equimolar feed of A and B.

SOLUTION

(*a*) Feed of A, B, R, and inerts. For this elementary reaction the rate is

$$-r_A = k_1 C_A C_B - k_2 C_R = k_1 \frac{N_A}{V}\frac{N_B}{V} - k_2 \frac{N_R}{V}$$

At constant pressure, basing expansion and conversion on substance A,

$$-r_A = k_1 \frac{N_{A0} - N_{A0}X_A}{V_0(1+\varepsilon_A X_A)}\frac{N_{B0} - N_{A0}X_A}{V_0(1+\varepsilon_A X_A)} - k_2\frac{N_{R0} + N_{A0}X_A}{V_0(1+\varepsilon_A X_A)}$$

Letting $M = C_{B0}/C_{A0}$, $M' = C_{R0}/C_{A0}$, we obtain

$$-r_A = k_1 C_{A0}^2 \frac{(1-X_A)(M-X_A)}{(1+\varepsilon_A X_A)^2} - k_2 C_{A0}\frac{M' + X_A}{1+\varepsilon_A X_A}$$

Hence, the design equation for plug flow, Eq. 17, becomes

$$\tau = C_{A0}\int_0^{X_{Af}} \frac{dX_A}{-r_A} = \int_0^{X_{Af}} \frac{(1+\varepsilon_A X_A)^2\, dX_A}{k_1 C_{A0}(1-X_A)(M-X_A) - k_2(M'+X_A)(1+\varepsilon_A X_A)}$$

In this expression ε_A accounts for inerts present in the feed.

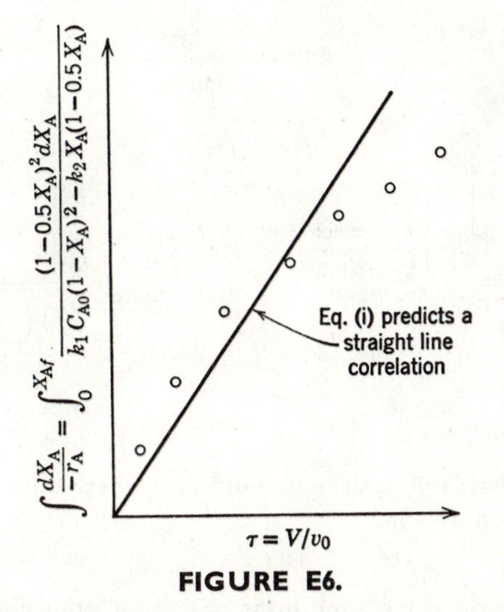

FIGURE E6.

(*b*) Equimolar feed of A and B. For $C_{A0} = C_{B0}$, $C_{R0} = 0$, and no inerts, we have $M = 1$, $M' = 0$, $\varepsilon_A = -0.5$; hence the expression for part *a* reduces to

$$\tau = \int_0^{X_{Af}} \frac{(1 - 0.5X_A)^2\, dX_A}{k_1 C_{A0}(1 - X_A)^2 - k_2 X_A(1 - 0.5X_A)} \tag{i}$$

Having V, v, and X_{Af} for a series of experiments, separately evaluate the left side and the right side of Eq. (i) and make the plot of Fig. E6. If the data fall on a reasonably straight line then the suggested kinetic scheme can be said to be satisfactory in that it fits the data.

Holding Time and Space-time for Flow Systems

To illustrate the distinction between the holding time (or mean residence time) and the space-time for a flow reactor consider the following simple situations illustrated in Fig. 7.

Case 1. Suppose 1 liter/sec of gaseous reactant A is introduced into a mixed reactor. The stoichiometry is $A \rightarrow 3R$, the conversion is 50%, and under these conditions the leaving flow rate is 2 liters/sec. Then by definition the space-time for this operation is

$$\tau_{\text{mixed}} = \frac{V}{v_0} = \frac{1 \text{ liter}}{1 \text{ liter/sec}} = 1 \text{ sec}$$

However, since each element of fluid expands to twice its volume immediately on entering the reactor the holding or mean residence time is

$$\bar{t}_{\text{mixed}} = \frac{V}{v_0(1 + \varepsilon_A X_A)} = \frac{V}{v_f} = \frac{1 \text{ liter}}{2 \text{ liters/sec}} = \tfrac{1}{2} \text{ sec}$$

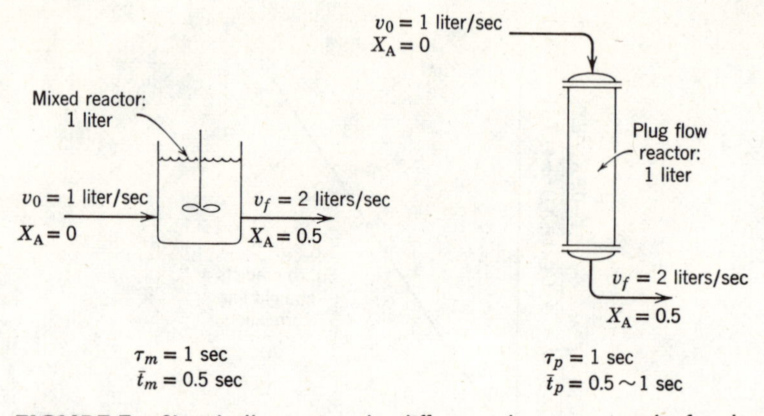

FIGURE 7. Sketch illustrating the difference between τ and t for the gas-phase reaction $A \rightarrow 3R$.

Case 2. Suppose the previous conditions are for a plug flow reactor. Then by definition the space-time is still

$$\tau_{\text{plug}} = \frac{V}{v_0} = \frac{1 \text{ liter}}{1 \text{ liter/sec}} = 1 \text{ sec}$$

However, since the gas reacts progressively as it passes through the plug flow reactor, it expands correspondingly—not immediately on entering, and not all at one time as it leaves—in which case

$$\bar{t}_{\text{plug}} = \tfrac{1}{2} \sim 1 \text{ sec}$$

The precise value of the holding time is determined by the particular kinetics of the system.

Case 3. If this were a liquid-phase system—not gas-phase—expansion would be negligible, one liter would leave for each liter entering; thus, the holding time and space-time would be identical, or

$$\bar{t} = \tau = 1 \text{ sec}$$

These examples show that $\bar{t}$ and τ are not, in general, identical. Now which is the natural performance measure for steady-state flow reactors? For batch systems it is the time of reaction; hence for flow systems we may suspect, by analogy, that the holding time is the appropriate measure. However, holding time does not appear anywhere in the performance equations for flow systems developed in this chapter, Eqs. 11–19, while it is seen that space-time or V/F_{A0} do naturally appear. Hence τ or V/F_{A0} are the proper performance measures for flow systems.

The above simple examples show that in the special case of constant fluid density the space-time is equivalent to the holding time, hence these terms can be used interchangeably. This special case includes practically all liquid phase reactions. However, for fluids of changing density, e.g., nonisothermal gas reactions or gas reaction with changing number of moles, a distinction should be made between τ and $\bar{t}$ and the correct quantity should be used in each situation.

REFERENCE

Pease, R. N., *J. Am. Chem. Soc.*, **51**, 3470 (1929).

PROBLEMS

1. Consider a gas-phase reaction $2A = R + 2S$ with unknown kinetics. If a space velocity of 1/min is needed for 90% conversion of A in a plug flow reactor, find the corresponding space-time and mean residence time or holding time of fluid in the reactor.

2. In an isothermal batch reactor 70% of a liquid reactant is converted in 13 min. What space-time and space-velocity are needed to effect this conversion in a plug flow reactor and in a mixed flow reactor.

3. We are planning to operate a batch reactor to convert A into R. This is a liquid reaction, the stoichiometry is $A \rightarrow R$, and the rate of reaction is given in Table P3. How long must we react each batch for the concentration to drop from $C_{A0} = 1.3$ mol/liter to $C_{Af} = 0.3$ mol/liter.

TABLE P3

C_A, mol/liter	$-r_A$, mol/liter·min
0.1	0.1
0.2	0.3
0.3	0.5
0.4	0.6
0.5	0.5
0.6	0.25
0.7	0.10
0.8	0.06
1.0	0.05
1.3	0.045
2.0	0.042

4. For the reaction of Problem 3, what size of plug flow reactor would be needed for 80% conversion of a feed stream of 1000 mol A/hr at $C_{A0} = 1.5$ mol/liter?

5. (*a*) For the reaction of Problem 3, what size of mixed flow reactor is needed for 75% conversion of a feed stream of 1000 mol A/hr at $C_{A0} = 1.2$ mol/liter?
(*b*) Repeat part (*a*) with the modification that the feed rate is doubled, thus 2000 mol A/hr at $C_{A0} = 1.2$ mol/liter are to be treated.
(*c*) Repeat part (*a*) with the modification that $C_{A0} = 2.4$ mol/liter; however, 1000 mol A/hr are still to be treated down to $C_{Af} = 0.3$ mol/liter.

6. One gaseous feed stream (1 liter/min) containing A ($C_{A0} = 0.01$ mol/liter) and a second gaseous feed stream (3 liter/min) containing B ($C_{B0} = 0.02$ mol/liter) enter a mixed reactor 1 liter in volume and react to form a whole host of products R, S, T, ... Analysis of the outgoing stream of 6 liters/min shows that $C_{Af} =$ 0.0005 mol/liter and $C_{Rf} = 0.001$ mol/liter. All flow rates and concentrations are measured at the uniform temperature and pressure of the reactor. Find the rate of reaction of A and the rate of formation of R in the reactor.

7. A homogeneous liquid phase reaction

$$A \rightarrow R, \qquad -r_A = kC_A^2$$

takes place with 50% conversion in a mixed reactor.

(*a*) What will be the conversion if this reactor is replaced by one 6 times as large—all else remaining unchanged?

(*b*) What will be the conversion if the original reactor is replaced by a plug flow reactor of equal size—all else remaining unchanged?

8. Assuming a stoichiometry $A \rightarrow R$ for a first-order gas reaction we calculate the size of plug flow reactor needed for a given duty (99% conversion of a pure A feed) to be $V = 32$ liters. In fact, however, the reaction stoichiometry is $A \rightarrow 3R$. With this corrected stoichiometry, what is the required reactor volume?

9. A high molecular weight hydrocarbon stream A is fed continuously to a heated high temperature mixed reactor where it thermally cracks (homogeneous gas reaction) into lower molecular weight materials, collectively called R, by a stoichiometry approximated by $A \rightarrow 5R$. By changing the feed rate different extents of cracking are obtained as follows:

F_{A0}, millimol/hr	300	1000	3000	5000
$C_{A,out}$, millimol/liter	16	30	50	60

The internal void volume of the reactor is $V = 0.1$ liter, and at the temperature of the reactor the feed concentration is $C_{A0} = 100$ millimol/liter. Find a rate equation to represent the cracking reaction.

10. In the previous problem suppose the density change is ignored; in other words, suppose we assumed $\varepsilon_A = 0$. What reaction order and kinetic equation would we find for the reaction?

Note: A comparison of reaction orders for these two problems will suggest how serious is the error in ignoring the ε_A factor.

11. The aqueous decomposition of A is studied in an experimental mixed reactor. The results in Table P11 are obtained in steady-state runs. To obtain 75% conversion of reactant in a feed, $C_{A0} = 0.8$ mol/liter, what holding time is needed (*a*) in a plug flow reactor, (*b*) in a mixed flow reactor.

TABLE P11

Concentration of A, mol/liter		Holding Time,
In Feed	In Exit Stream	sec
2.00	0.65	300
2.00	0.92	240
2.00	1.00	250
1.00	0.56	110
1.00	0.37	360
0.48	0.42	24
0.48	0.28	200
0.48	0.20	560

12. From the following data find a satisfactory rate equation for the gas-phase decomposition $A \rightarrow R + S$ taking place isothermally in a mixed reactor.

Run Number	1	2	3	4	5
τ based on inlet feed conditions, sec	0.423	5.10	13.5	44.0	192
X_A (for $C_{A0} = 0.002$ mol/liter)	0.22	0.63	0.75	0.88	0.96

13. The homogeneous gas reaction $A \rightarrow 3R$ follows second-order kinetics. For a feed rate of 4 m^3/hr of pure A at 5 atm and 350°C, an experimental reactor consisting of a 2.5 cm ID pipe 2 m long gives 60% conversion of feed. A commercial plant is to treat 320 m^3/hr of feed consisting of 50% A, 50% inerts at 25 atm and 350°C to obtain 80% conversion.

(*a*) How many 2-m lengths of 2.5 cm ID pipe are required?

(*b*) Should they be placed in parallel or in series?

Assume plug flow in the pipe, negligible pressure drop, and ideal gas behavior.

14. HOLMES: You say he was last seen tending this vat. . . .
SIR BOSS: You mean "overflow stirred tank reactor," Mr. Holmes.

HOLMES: You must excuse my ignorance of your particular technical jargon, Sir Boss.
SIR BOSS: That's all right; however, you must find him, Mr. Holmes. Imbibit was a queer chap; always staring into the reactor, taking deep breaths, and licking his lips, but he was our very best operator. Why, since he left, our conversion of googliox has dropped from 80% to 75%.
HOLMES (*tapping the side of the vat idly*): By the way, what goes on in the vat?
SIR BOSS: Just an elementary second-order reaction, between ethanol and googliox, if you know what I mean. Of course, we maintain a large excess of alcohol, about 100 to 1 and . . . .
HOLMES (*interrupting*): Intriguing, we checked every possible lead in town and found not a single clue.
SIR BOSS (*wiping away the tears*): We'll give the old chap a raise—about twopence per week—if only he'll come back.
DR. WATSON: Pardon me, but may I ask a question?
HOLMES: Why certainly, Watson.
WATSON: What is the capacity of this vat, Sir Boss?
SIR BOSS: A hundred Imperial gallons, and we always keep it filled to the brim. That is why we call it an overflow reactor. You see we are running at full capacity—profitable operation you know.
HOLMES: Well, my dear Watson, we must admit that we're stumped, for without clues deductive powers are of no avail.
WATSON: Ahh, but that is where you are wrong, Holmes. (*Then turning to the manager*) Imbibit was a largish fellow—say about 18 stone—was he not?
SIR BOSS, Why yes, how did you know?
HOLMES (*with awe*): Amazing, my dear Watson!
WATSON (*modestly*): Why it's quite elementary, Holmes. We have all the clues necessary to deduce what happened to the happy fellow. But first of all, would someone fetch me some dill.

With Sherlock Holmes and Sir Boss impatiently waiting, Dr. Watson casually leaned against the vat, slowly and carefully filled his pipe, and—with the keen sense of the dramatic—lit it. There our story ends.

(*a*) What momentous revelation was Dr. Watson planning to make and how did he arrive at this conclusion?
(*b*) Why did he never make it?

15. The data in Table P15 have been obtained on the decomposition of gaseous reactant A in a constant volume batch reactor at 100°C. The stoichiometry of the reaction is $2A \rightarrow R + S$. What size plug flow reactor (in liters) operating at 100°C and 1 atm can treat 100 mol A/hr in a feed consisting of 20% inerts to obtain 95% conversion of A?

TABLE P15

t, sec	p_A, atm	t, sec	p_A, atm
0	1.00	140	0.25
20	0.80	200	0.14
40	0.68	260	0.08
60	0.56	330	0.04
80	0.45	420	0.02
100	0.37		

16. A 55-gal tank (208 liters) is to be used as a mixed reactor to effect the reaction of the previous problem. For identical feed and identical operating conditions, what conversion of A may be expected from this reactor?

17. An aqueous reaction is being studied in a laboratory-size steady-state flow system. The reactor is a flask whose contents (5 liters of fluid) are well stirred and uniform in composition. The stoichiometry of the reaction is $A \rightarrow 2R$, and reactant A is introduced at a concentration of 1 mol/liter. Results of the experimental investigation are summarized in Table P17. Find a rate expression for this reaction.

TABLE P17

Run	Feed Rate, cm³/sec	Temperature of Run, °C	Concentration of R in Effluent, mol/liter
1	2	13	1.8
2	15	13	1.5
3	15	84	1.8

18. The homogeneous gas reaction $A \rightarrow 2B$ is run at 100°C at a constant pressure of 1 atm in an experimental batch reactor. The data in Table P18 were obtained starting with pure A. What size plug flow reactor operated at 100°C and 10 atm would yield 90% conversion of A for a total feed rate of 10 mol/sec, the feed containing 40% inerts?

TABLE P18

Time, min	V/V_0	Time, min	V/V_0
0	1.00	8	1.82
1	1.20	9	1.86
2	1.35	10	1.88
3	1.48	11	1.91
4	1.58	12	1.92
5	1.66	13	1.94
6	1.72	14	1.95
7	1.78		

19. Find the rate equation for the decomposition of liquid A from the results of the kinetic runs made in a mixed reactor at steady state (Table P19).

TABLE P19

Concentration of A Liquid in Feed, mol/liter	Concentration of A in Reactor, mol/liter	Mean Residence Time of Fluid in Reactor, sec	Temperature, °C
1.0	0.4	220	44
1.0	0.4	100	57
1.0	0.4	30	77
1.0	0.1	400	52
1.0	0.1	120	72
1.0	0.1	60	84

20.* At 600°K the gas-phase reaction

$$C_2H_4 + Br_2 \underset{k_2}{\overset{k_1}{\rightleftarrows}} C_2H_4Br_2$$

has rate constants

$$k_1 = 500 \text{ liter/mol} \cdot \text{hr}$$
$$k_2 = 0.032 \text{ hr}^{-1}$$

* Problems 5.20, 5.22, and 14.10 either are taken directly from or are modified versions of problems given in *Chemical Engineering Problems*, 1946, edited by the Chemical Engineering Education Projects Committee, American Institute of Chemical Engineers, 1955.

If a plug flow reactor is to be fed 600 m^3/hr of gas containing 60% Br_2 and 30% C_2H_4 and 10% inerts by volume at 600°K and 1.5 atm compute

(*a*) the maximum possible fractional conversion of C_2H_4 into $C_2H_4Br_2$,

(*b*) the volume of reaction vessel required to obtain 60% of this maximum conversion.

21. It has been reported that the reaction

$$\underset{\text{ethylene chlorhydrin}}{\begin{matrix} CH_2OH \\ | \\ CH_2Cl \end{matrix}} + NaHCO_3 \rightarrow \underset{\text{ethylene glycol}}{\begin{matrix} CH_2OH \\ | \\ CH_2OH \end{matrix}} + NaCl + CO_2$$

is elementary with rate constant $k = 5.2$ liter/mol·hr at 82°C. On the basis of this information we wish to construct a pilot plant to determine the economic feasibility of producing ethylene glycol from two available feeds, a 15 wt % aqueous solution of sodium bicarbonate and a 30 wt % aqueous solution of ethylene chlorhydrin.

(*a*) What volume of tubular (plug flow) reactor will produce 20 kg/hr ethylene glycol at 95% conversion of an equimolar feed produced by intimately mixing appropriate quantities of the two feed streams.

(*b*) What size mixed reactor is needed for the same feed, conversion, and production rate as in part (*a*)?

Assume all operations at 82°C, at which temperature the specific gravity of the mixed reacting fluid is 1.02.

22. Experimental work by Pease (1929) indicates that the pyrolitic polymerization of acetylene in the gas phase follows second-order kinetics and occurs so that 0.009 of the acetylene is converted into an average tetramer complex, $4C_2H_2 \rightarrow (C_2H_2)_4$, in 1 sec at 550°C and 1 atm. An estimation of plant performance is to be made from this information. The units available consist of five identical tubular furnaces, each comprising 47 tubes, each 3.5 m long and 5 cm ID. The five units are to be operated continuously in parallel, are to be fed gas at 20 atm, and are to be maintained at such a temperature that the reaction may be considered to occur isothermally at 550°C. The total gas feed rate to the plant is to be 1000 m^3/hr measured at 20 atm and 550°C. The analysis of the feed gas is 80 vol % acetylene and 20 vol % inerts. For estimation purposes the pressure drop through the system may be neglected, and the ideal gas law may be assumed to apply.

(*a*) Compute the production rate of the tetramer complex for the entire plant in kg/hr.

(*b*) If the five units are operated in series, compute the gas feed rate that may be used if the fraction of conversion of acetylene is to be the same as it is in part (*a*).

6

DESIGN FOR SINGLE REACTIONS

There are many ways of processing a fluid: in a single batch or flow reactor, in a chain of reactors possibly with interstage feed injection or heating, in a reactor with recycle of the product stream using various feed ratios and conditions, and so on. Which scheme should we use? Unfortunately numerous factors may have to be considered in answering this question; for example, the reaction type, planned scale of production, cost of equipment and operations, safety, stability and flexibility of operation, equipment life expectancy, length of time that the product is expected to be manufactured, ease of convertibility of the equipment to modified operating conditions or to new and different processes. With the wide choice of systems available and with the many factors to be considered, no neat formula can be expected to give the optimum setup. Experience, engineering judgment, and a sound knowledge of the characteristics of the various reactor systems are all needed in selecting a reasonably good design and, needless to say, the choice in the last analysis will be dictated by the economics of the over-all process.

The reactor system selected will influence the economics of the process by dictating the size of units needed and by fixing the ratio of products formed. The first factor, reactor size, may well vary a hundredfold among competing designs while the second factor, product distribution, is usually of prime consideration where it can be varied and controlled.

In this chapter we deal with *single reactions*. These are reactions whose progress can be described and followed adequately by using one and only one rate expression coupled with the necessary stoichiometric and equilibrium expressions. For such reactions product distribution is fixed; hence, the important factor in comparing designs is the reactor size. We consider in turn the size comparison of various single and multiple ideal reactor systems. Then we introduce the recycle reactor, and develop its performance equations. Finally, we treat a rather unique type of reaction, the autocatalytic reaction, and show how to apply our findings to it.

Design for multiple reactions, for which the primary consideration is product distribution, is treated in the next chapter.

SIZE COMPARISON OF SINGLE REACTORS

Batch Reactor (*)

First of all, before we compare flow reactors, let us mention the batch reactor briefly. The batch reactor has the advantage of small instrumentation cost and flexibility of operation (may be shut down easily and quickly). It has the disadvantage of high labor and handling cost, often considerable shutdown time to empty, clean out, and refill, and poorer quality control of the product. Hence we may generalize to state that the batch reactor is well suited to produce small amounts of material or to produce many different products from one piece of equipment. On the other hand, for the chemical treatment of materials in large amounts the continuous process is nearly always found to be more economical.

Regarding reactor sizes, a comparison of Eqs. 5.4 and 5.19 for a given duty and for $\varepsilon = 0$ shows that an element of fluid reacts for the same length of time in the batch and in the plug flow reactor. Thus the same volume of these reactors is needed to do a given job. Of course, on a long-term production basis we must correct the size requirement estimate to account for the shutdown time between batches. Thus it is easy to relate the performance capabilities of the batch reactor with the plug flow reactor.

Mixed Versus Plug Flow Reactors, First- and Second-order Reactions

For a given duty the ratio of sizes of mixed and plug flow reactors will depend on the extent of reaction, the stoichiometry, and the form of the rate equation. For the general case a comparison of Eqs. 5.11 and 5.17 will give this size ratio. Let us make this comparison for the large class of reactions approximating the simple nth-order rate law

$$-r_A = -\frac{1}{V}\frac{dN_A}{dt} = kC_A{}^n$$

where n varies anywhere from zero to three. For mixed flow Eq. 5.11 gives

$$\tau_m = \left(\frac{C_{A0}V}{F_{A0}}\right)_m = \frac{C_{A0}X_A}{-r_A} = \frac{1}{kC_{A0}^{n-1}}\frac{X_A(1+\varepsilon_A X_A)^n}{(1-X_A)^n}$$

(*) VER Rutherford Aris
Analisis de Reactores (sec. 10.2)

whereas for plug flow Eq. 5.17 gives

$$\tau_p = \left(\frac{C_{A0}V}{F_{A0}}\right)_p = C_{A0}\int_0^{X_A}\frac{dX_A}{-r_A} = \frac{1}{kC_{A0}^{n-1}}\int_0^{X_A}\frac{(1+\varepsilon_A X_A)^n\,dX_A}{(1-X_A)^n}$$

Dividing we find that

$$\frac{(\tau C_{A0}^{n-1})_m}{(\tau C_{A0}^{n-1})_p} = \frac{\left(\dfrac{C_{A0}{}^n V}{F_{A0}}\right)_m}{\left(\dfrac{C_{A0}{}^n V}{F_{A0}}\right)_p} = \frac{\left[X_A\left(\dfrac{1+\varepsilon_A X_A}{1-X_A}\right)^n\right]_m}{\left[\displaystyle\int_0^{X_A}\frac{(1+\varepsilon_A X_A)^n}{(1-X_A)^n}\,dX_A\right]_p} \tag{1}$$

With density constant, or $\varepsilon = 0$, this expression integrates to

$$\frac{(\tau C_{A0}^{n-1})_m}{(\tau C_{A0}^{n-1})_p} = \frac{\left[\dfrac{X_A}{(1-X_A)^n}\right]_m}{\left[\dfrac{(1-X_A)^{1-n}-1}{n-1}\right]_p}, \qquad n \neq 1$$

or (2)

$$\frac{(\tau C_{A0}^{n-1})_m}{(\tau C_{A0}^{n-1})_p} = \frac{\left(\dfrac{X_A}{1-X_A}\right)_m}{-\ln(1-X_A)_p}, \qquad n = 1$$

Equations 1 and 2 are displayed in graphical form in Fig. 1 to provide a quick comparison of the performance of plug flow with mixed flow reactors. For identical feed composition C_{A0} and flow rate F_{A0} the ordinate of this figure gives directly the volume ratio required for any specified conversion. Figure 1 shows the following.

1. For any particular duty and for all positive reaction orders the mixed reactor is always larger than the plug flow reactor. The ratio of volumes increases with order. For zero-order reactions reactor size is independent of the type of flow.

2. When conversion is small, the reactor performance is only slightly affected by flow type, the volume ratio approaching unity as conversion approaches zero. The ratio increases very rapidly at high conversion; consequently a proper representation of the flow becomes very important in this range of conversion.

3. Density variation during reaction affects design; however, it is normally of secondary importance compared to the difference in flow type. Expansion (or density decrease) during reaction increases the volume ratio, in other words, further decreases the effectiveness of the mixed reactor with respect to the plug flow reactor; density increase during reaction has the opposite effect.

Figures 5 and 6 show the same first- and second-order curves for $\varepsilon = 0$ but also include dashed lines which represent fixed values of the dimensionless reaction rate group, defined as

$k\tau$ for first-order reaction

$kC_{A0}\tau$ for second-order reaction

$kC_{A0}^{n-1}\tau$ for nth-order reaction

NOTA: Para rxns. de orden cero $\dfrac{(\tau C_{A0}^{n-1})_m}{(\tau C_{A0}^{n-1})_p} = \dfrac{(\tau/C_{A0})_m}{(\tau/C_{A0})_p} = 1$

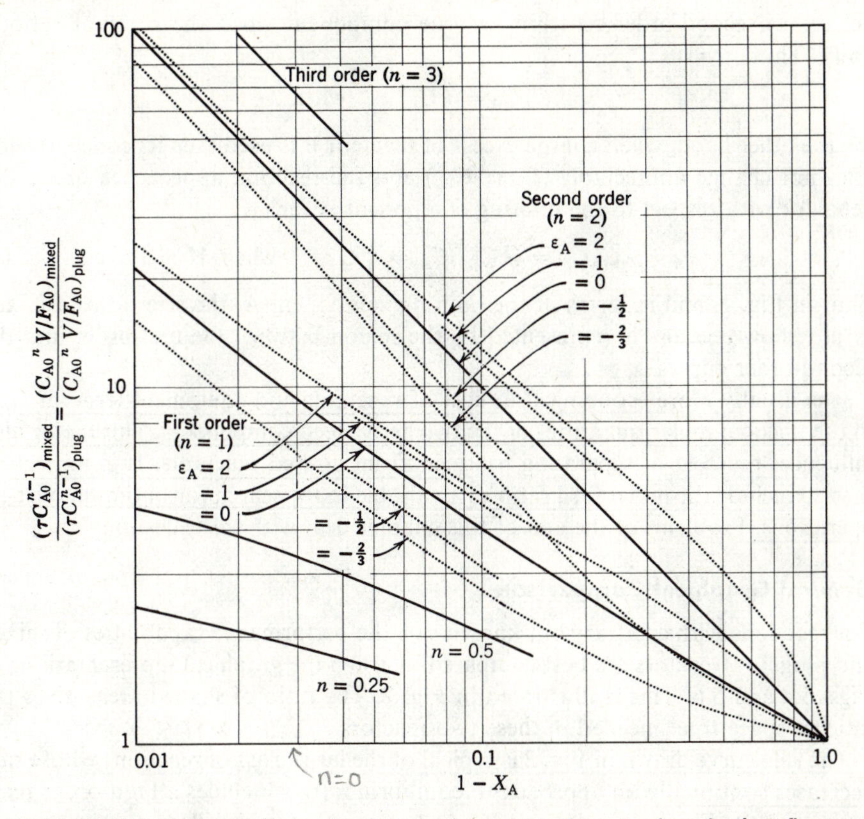

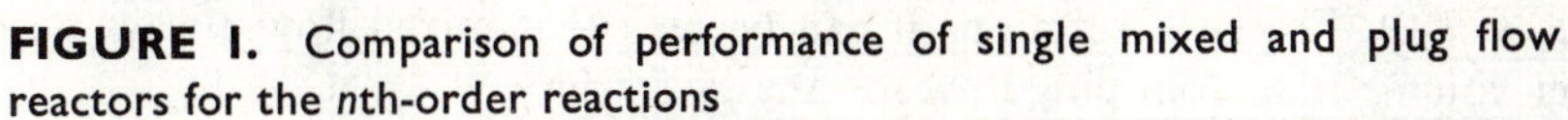

FIGURE 1. Comparison of performance of single mixed and plug flow reactors for the nth-order reactions

$$A \rightarrow \text{products}, \quad -r_A = kC_A^n$$

The ordinate becomes the volume ratio V_m/V_p or space-time ratio τ_m/τ_p if the same quantities of identical feed are used.

With these lines we can compare different reactor types, reactor sizes, and conversion levels.

Example 1 illustrates the use of these charts.

Variation of Reactant Ratio for Second-order Reactions

Second-order reactions of two components and of the type

$$A + B \rightarrow \text{products}, \qquad M = C_{B0}/C_{A0}$$
$$-r_A = -r_B = kC_AC_B \tag{3.13}$$

behave as second-order reactions of one component when the reactant ratio is unity. Thus

$$-r_A = kC_AC_B = kC_A^2 \qquad \text{when } M = 1 \tag{3}$$

On the other hand, when a large excess of reactant B is used then its concentration does not change appreciably ($C_B \cong C_{B0}$) and the reaction approaches first-order behavior with respect to the limiting component A, or

$$-r_A = kC_AC_B = (kC_{B0})C_A = k'C_A \qquad \text{when } M \gg 1 \tag{4}$$

Thus in Fig. 1, and in terms of the limiting component A, the size ratio of mixed to plug flow reactors is represented by the region between the first-order and the second-order curves.

It is usually more economical in cost of reactants and equipment (reactor size) to use unequal molar quantities of the two active feed components. This factor also influences the cost of separating products from unused reactants. It is important then to include the molar feed ratio as a variable in the search for optimum over-all operations. Problems at the end of this chapter deal with this question.

General Graphical Comparison

For reactions with arbitrary but known rate the performance capabilities of mixed and plug flow reactors are best compared by using the graphical representations of Figs. 5.4 and 5.6. This is illustrated in Fig. 2. The ratio of shaded areas gives the ratio of space-times needed in these two reactors.

The rate curve drawn in Fig. 2 is typical of the large class of reactions whose rate decreases continually on approach to equilibrium (this includes all nth-order reactions, $n > 0$). For such reactions it can be seen that mixed flow always needs a larger volume than does plug flow for any given duty.

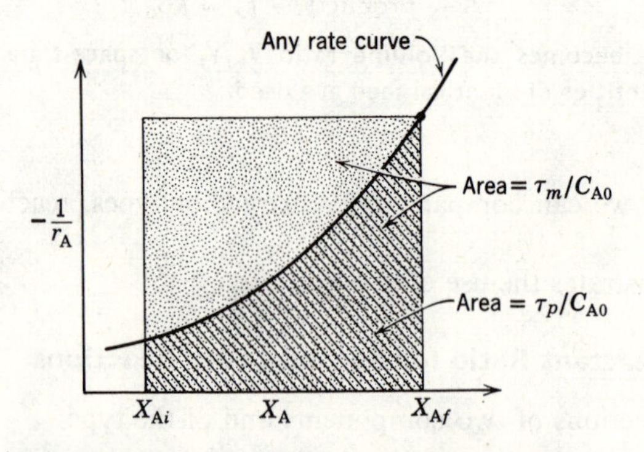

FIGURE 2. Comparison of performance of mixed flow and plug flow reactors for any reaction kinetics.

EXAMPLE 1. Reactor performance from design charts

The aqueous reaction A + B → products with known kinetics

$$-r_A = (500 \text{ liter/mol}\cdot\text{min})C_A C_B$$

is to take place in an experimental tubular reactor (assume plug flow) under the following conditions:

volume of reactor $V = 0.1$ liter
volumetric feed rate $v = 0.05$ liter/min
concentration of reactants in feed, $C_{A0} = C_{B0} = 0.01$ mol/liter

(*a*) What fractional conversion of reactants can be expected?
(*b*) For the same conversion as in part (*a*), what size of stirred tank reactor (assume mixed flow) is needed?
(*c*) What conversion can be expected in a mixed reactor equal in size to the plug flow reactor?

SOLUTION

For an equimolar feed we may treat this as a second-order reaction of the type

$$A \rightarrow \text{products}, \qquad -r_A = kC_A^2$$

in which the performance charts, Figs. 1 and 6, apply. The sketch of Fig. E1 shows how these charts are used.

(***a***) *Conversion in plug flow.* The space-time is

$$\tau = \frac{VC_{A0}}{F_{A0}} = \frac{V}{v} = \frac{0.1 \text{ liter}}{0.05 \text{ liter/min}} = 2 \text{ min}$$

therefore, the value of the second-order reaction rate group is

$$kC_{A0}\tau = (500 \text{ liter/mol}\cdot\text{min})(2 \text{ min})(0.01 \text{ mol/liter}) = 10$$

From the plug flow line of Fig. 6 and the calculated value for $kC_{A0}\tau$ we find

$$\text{Conversion:} \quad X_A = X_B = 0.91$$

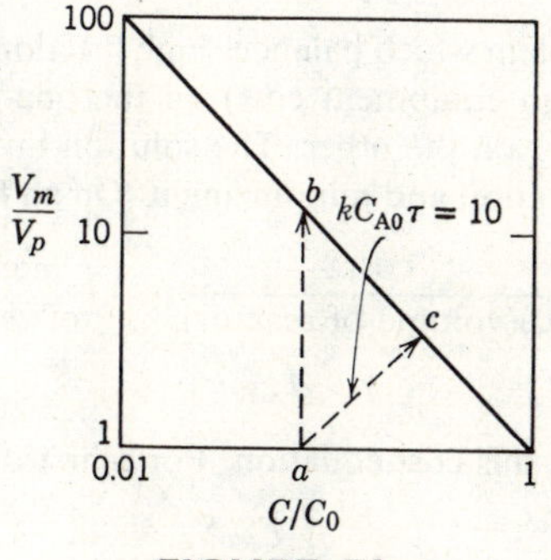

FIGURE E1.

(*b*) *Size of mixed reactor for identical treatment rate of feed.* For identical C_{A0} and F_{A0} the ordinate of Fig. 1 or 6 reduces to the volume ratio of the two reactors, V_m/V_p. Thus for the same X_A we find

$$\frac{V_m}{V_p} = 11$$

in which case

$$V_m = (11)(0.1 \text{ liter}) = 1.1 \text{ liter}$$

(*c*) *Conversion in a mixed reactor of the same size.* For the same size reactor we must remain on the same $kC_{A0}\tau$ line. Thus moving along this line from plug to mixed flow in Fig. 6 we find

$$\text{Conversion:} \quad X_A = 0.73$$

The graphical method was used in this example to illustrate the use of these charts. Analytic solution is also possible. The following examples illustrate the latter approach in searching for optimum operating conditions.

EXAMPLE 2. Finding the optimum reactor size

One hundred gram moles of R are to be produced hourly from a feed consisting of a saturated solution of A (C_{A0} = 0.1 mol/liter) in a mixed flow reactor. The reaction is

$$A \rightarrow R, \qquad r_R = (0.2 \text{ hr}^{-1})C_A$$

Cost of reactant at C_{A0} = 0.1 mol/liter is

$$\$_A = \$0.50/\text{mol A}$$

Cost of reactor including installation, auxiliary equipment, instrumentation, overhead, labor, depreciation, etc., is

$$\$_m = \$0.01/\text{hr}\cdot\text{liter}$$

What reactor size, feed rate, and conversion should be used for optimum operations? What is the unit cost of R for these conditions if unreacted A is discarded?

SOLUTION

This is an optimization problem which balances high fractional conversion (low reactant cost) in a large reactor (high equipment cost) on the one hand versus low fractional conversion in a small reactor on the other. The solution involves finding an expression for the total cost of the operation and minimizing it. On an hourly basis the total cost is

$$\$_t = \begin{pmatrix}\text{volume of}\\ \text{reactor}\end{pmatrix}\left(\frac{\text{cost}}{(\text{hr})(\text{volume of reactor})}\right) + \begin{pmatrix}\text{feed rate}\\ \text{of reactant}\end{pmatrix}\begin{pmatrix}\text{unit cost}\\ \text{of reactant}\end{pmatrix}$$

$$= V\$_m + F_{A0}\$_A$$

Let us evaluate the terms in this cost equation. For a first-order reaction Eq. 5.11 gives

$$V = \frac{F_{A0}X_A}{kC_{A0}(1 - X_A)}$$

Noting that the rate of production of R

$$F_R = F_{A0}X_A = 100 \text{ mol/hr}$$

we can eliminate F_{A0} and can write the total cost expression in terms of one variable alone, X_A. Thus

$$\$_t = \frac{F_R}{kC_{A0}(1 - X_A)}\$_m + \frac{F_R}{X_A}\$_A$$

$$= \frac{100}{(0.2)(0.1)(1 - X_A)}(0.01) + \frac{100}{X_A}(0.5)$$

$$= \frac{50}{1 - X_A} + \frac{50}{X_A}$$

To find the condition for minimum cost, differentiate this expression and set to zero. Thus

$$\frac{d(\$_t)}{dX_A} = 0 = \frac{50}{(1 - X_A)^2} - \frac{50}{X_A^2}$$

or

$$X_A = 0.5$$

Hence the optimum conditions are

Conversion:

$$X_A = 0.5$$

Feed rate:

$$F_{A0} = \frac{F_R}{X_A} = \frac{100}{0.5} = 200 \text{ mol A/hr} \qquad \text{or} \qquad v = \frac{F_{A0}}{C_{A0}} = 2000 \text{ liter/hr}$$

Reactor size:

$$V = \frac{F_{A0}X_A}{kC_{A0}(1 - X_A)} = \frac{100}{(0.2)(0.1)(0.5)} = 10{,}000 \text{ liter}$$

Cost of product:

$$\frac{\$_t}{F_R} = \frac{V\$_m + F_{A0}\$_A}{F_R} = \frac{10{,}000(0.01) + 200(0.5)}{100} = \$2.00/\text{mol R}$$

EXAMPLE 3. Finding the best size for a reactor and reclaiming unit

Suppose all unreacted A of the product stream of Example 2 can be reclaimed and brought up to the initial concentration $C_{A0} = 0.1$ mol/liter at a total cost $\$_r =$ \$0.125/mol A processed (Fig. E3). With this reclaimed A as a recycle stream, find the new optimum operating conditions and unit cost of producing R.

SOLUTION

The solution involves finding a balance between low reactor and high recycle cost on the one hand and high reactor and low recycle cost on the other. Referring to the

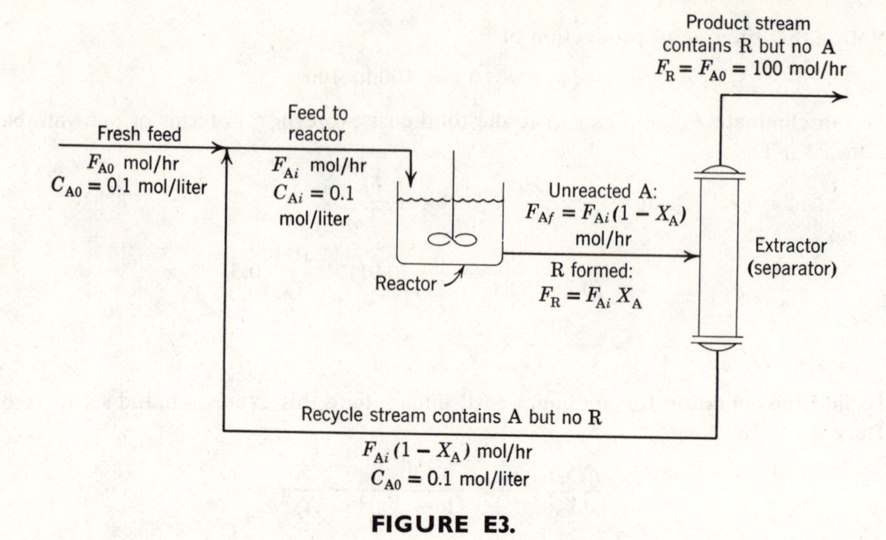

FIGURE E3.

accompanying sketch which indicates all flowing streams in terms of F, the hourly molar flow rate, we find by material balance that

$$F_R = F_{Ai}X_A = F_{A0} = 100 \text{ mol/hr}$$

The total hourly cost is then given by

$$\$_t = \begin{pmatrix}\text{volume of}\\ \text{reactor}\end{pmatrix}\begin{pmatrix}\text{hourly cost per unit}\\ \text{volume of reactor}\end{pmatrix} + \begin{pmatrix}\text{feed rate of}\\ \text{fresh reactant}\end{pmatrix}\begin{pmatrix}\text{unit cost of}\\ \text{fresh reactant}\end{pmatrix}$$

$$+ \begin{pmatrix}\text{feed rate of}\\ \text{reclaimed reactant}\end{pmatrix}\begin{pmatrix}\text{unit cost of}\\ \text{reclaimed reactant}\end{pmatrix}$$

$$= V\$_m + F_{A0}\$_A + F_{Ai}(1 - X_A)\$_r$$

With feed rate to the reactor F_{Ai} we have

$$V = \frac{F_{Ai}X_A}{kC_{A0}(1 - X_A)}$$

Eliminating F_{Ai} by the material balance, the total cost expression can be then written as a function of a single variable X_A. Thus

$$\$_t = \frac{F_{A0}}{kC_{A0}(1 - X_A)}\$_m + F_{A0}\$_A + \frac{F_{A0}}{X_A}(1 - X_A)\$_r$$

$$= \frac{100}{(0.2)(0.1)(1 - X_A)}(0.01) + 100(0.5) + 100\left(\frac{1 - X_A}{X_A}\right)(0.125)$$

$$= \frac{50}{1 - X_A} + 50 + 12.5\left(\frac{1 - X_A}{X_A}\right)$$

By differentiating and setting to zero, we obtain the condition for minimum cost. Thus

$$\frac{d\$_t}{dX_A} = 0 = \frac{50}{(1 - X_A)^2} - \frac{12.5}{X_A{}^2}$$

or

$$X_A = 0.33$$

Hence for optimum operations:

Conversion within reactor:

$$X_A = 0.33$$

Reactor size:

$$V = \frac{100}{(0.2)(0.1)(0.67)} = 7500 \text{ liters}$$

Flow rate into reactor:

$$F_{Ai} = \frac{F_{A0}}{X_A} = 300 \text{ mol A/hr} \qquad \text{or} \qquad v = 3000 \text{ liter/hr}$$

Recycle rate:

$$F_{Ai} - F_{A0} = 200 \text{ mol A/hr} \qquad \text{or} \qquad v' = 2000 \text{ liter/hr}$$

Cost of product:

$$\frac{\$_t}{F_R} = \frac{50/0.67 + 50 + 12.5(0.67/0.33)}{100} = \$1.50/\text{mol R}$$

MULTIPLE-REACTOR SYSTEMS

Plug Flow Reactors in Series and/or in Parallel

Consider N plug flow reactors connected in series, and let $X_1, X_2, \ldots, X_N$ be the fractional conversion of component A leaving reactor $1, 2, \ldots, N$. Basing the material balance on the feed rate of A to the first reactor, we find for the ith reactor from Eq. 5.18

$$\frac{V_i}{F_0} = \int_{X_{i-1}}^{X_i} \frac{dX}{-r}$$

or for the N reactors in series

$$\frac{V}{F_0} = \sum_{i=1}^{N} \frac{V_i}{F_0} = \frac{V_1 + V_2 + \cdots + V_N}{F_0}$$

$$= \int_{X_0=0}^{X_1} \frac{dX}{-r} + \int_{X_1}^{X_2} \frac{dX}{-r} + \cdots + \int_{X_{N-1}}^{X_N} \frac{dX}{-r} = \int_0^{X_N} \frac{dX}{-r_A}$$

Hence N plug flow reactors in series with a total volume V gives the same conversion as a single plug flow reactor of volume V.

For plug flow reactors connected in parallel or in any parallel-series combination, we can treat the whole system as a single plug flow reactor of volume equal to the total volume of the individual units if the feed is distributed in such a manner that fluid streams which meet have the same composition. Thus for reactors in parallel V/F or τ must be the same for each parallel line. Any other way of feeding is usually less efficient.

EXAMPLE 4. Operating a number of plug flow reactors

The reactor setup shown in Fig. E4 consists of three plug flow reactors in two parallel branches. Branch D has a reactor of volume 50 liters followed by a reactor of volume 30 liters. Branch E has a reactor of volume 40 liters. What fraction of the feed should go to branch D?

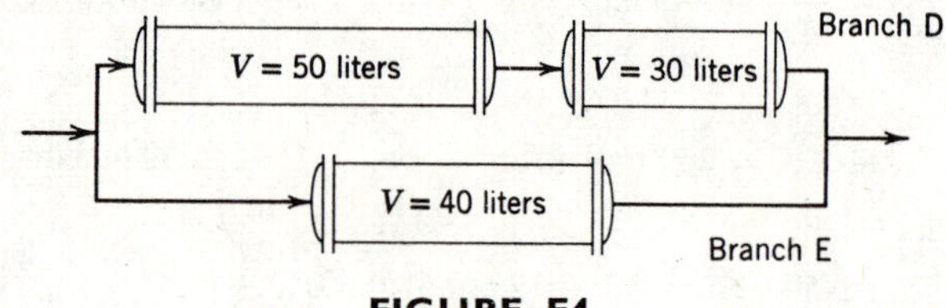

FIGURE E4.

SOLUTION

Branch D consists of two reactors in series; hence it may be considered to be a single reactor of volume

$$V_D = 50 + 30 = 80 \text{ liters}$$

Now for reactors in parallel V/F must be identical if the conversion is to be the same in each branch. Therefore

$$\left(\frac{V}{F}\right)_D = \left(\frac{V}{F}\right)_E$$

or

$$\frac{F_D}{F_E} = \frac{V_D}{V_E} = \frac{80}{40} = 2$$

Therefore two-thirds of the feed must be fed to branch D.

Equal-size Mixed Reactors in Series

In plug flow the concentration of reactant decreases progressively through the system; in mixed flow the concentration drops immediately to a low value. Because of this fact a plug flow reactor is more efficient than a mixed reactor for reactions whose rates increase with reactant concentration, such as nth-order irreversible reactions, $n > 0$.

Consider a system of N mixed reactors connected in series. Though the concentration is uniform in each reactor, there is nevertheless a change in concentration as fluid moves from reactor to reactor. This stepwise drop in concentration, illustrated in Fig. 3, suggests that the larger the number of units in series the closer should the behavior of the system approach plug flow. This will be shown to be so.

Let us now quantitatively evaluate the behavior of a series of N equal-size mixed reactors. Density changes will be assumed to be negligible; hence $\varepsilon = 0$ and $t = \tau$. As a rule, with mixed reactors it is more convenient to develop the necessary equations in terms of concentrations rather than fractional conversions; therefore we use this approach. The nomenclature used is shown in Fig. 4 with subscript i referring to the ith vessel.

First-order Reactions. From Eq. 5.12 a material balance for component A about vessel i gives

$$\tau_i = \frac{C_0 V_i}{F_0} = \frac{V_i}{v} = \frac{C_0(X_i - X_{i-1})}{-r_A}$$

Because $\varepsilon = 0$ this may be written in terms of concentrations. Hence

$$\tau_i = \frac{C_0[(1 - C_i/C_0) - (1 - C_{i-1}/C_0)]}{kC_i} = \frac{C_{i-1} - C_i}{kC_i}$$

or

$$\frac{C_{i-1}}{C_i} = 1 + k\tau_i \tag{5}$$

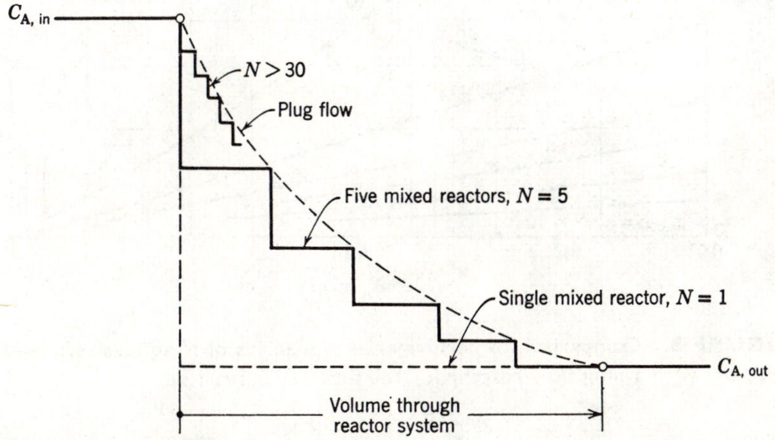

FIGURE 3. Concentration profile through an N-stage mixed reactor system compared with single flow reactors.

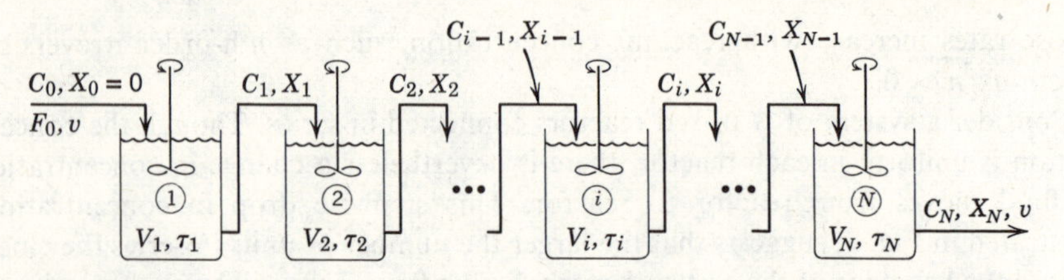

FIGURE 4. Notation for a system of N equal-size mixed reactors in series.

Now the space-time τ (or mean residence time t) is the same in all the equal-size reactors of volume V_i. Therefore

$$\frac{C_0}{C_N} = \frac{1}{1 - X_N} = \frac{C_0}{C_1}\frac{C_1}{C_2}\cdots\frac{C_{N-1}}{C_N} = (1 + k\tau_i)^N \tag{6a}$$

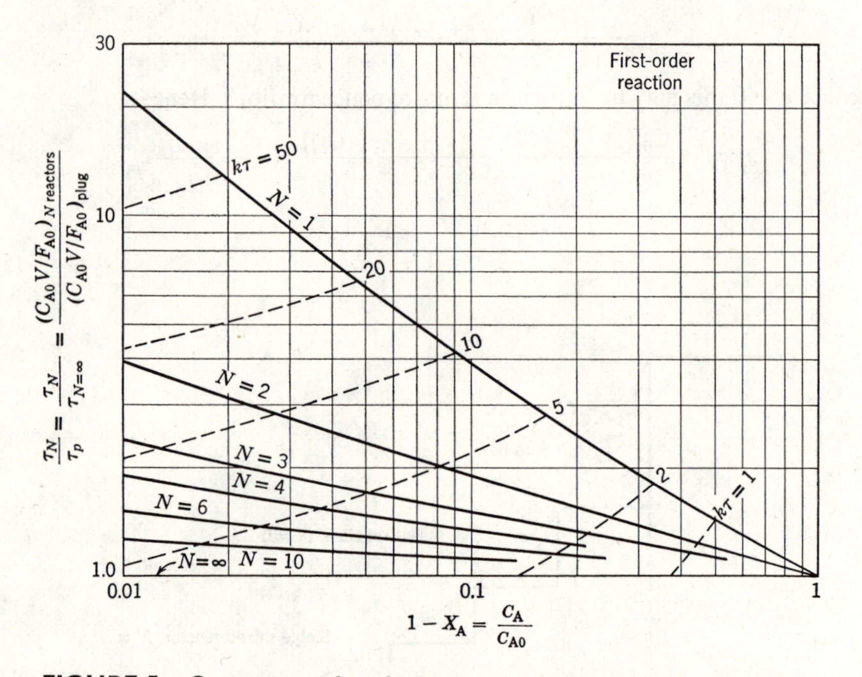

FIGURE 5. Comparison of performance of a series of N equal-size mixed reactors with a plug flow reactor for the first-order reaction

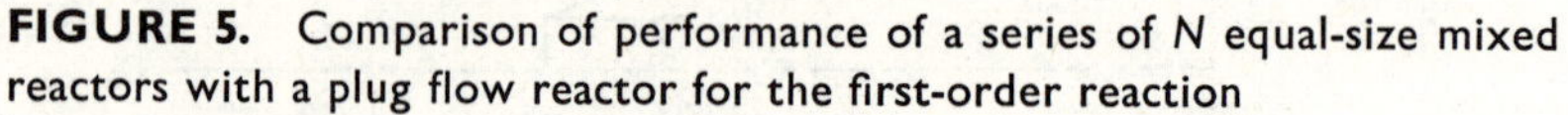

$$A \rightarrow R, \quad \varepsilon = 0$$

For the same processing rate of identical feed the ordinate measures the volume ratio V_N/V_p directly.

Rearranging, we find for the system as a whole

$$\tau_{N\text{ reactors}} = N\tau_i = \frac{N}{k}\left[\left(\frac{C_0}{C_N}\right)^{1/N} - 1\right] \tag{6b}$$

In the limit, for $N \to \infty$, this equation reduces to the plug flow equation

$$\tau_p = \frac{1}{k}\ln\frac{C_0}{C} \tag{7}$$

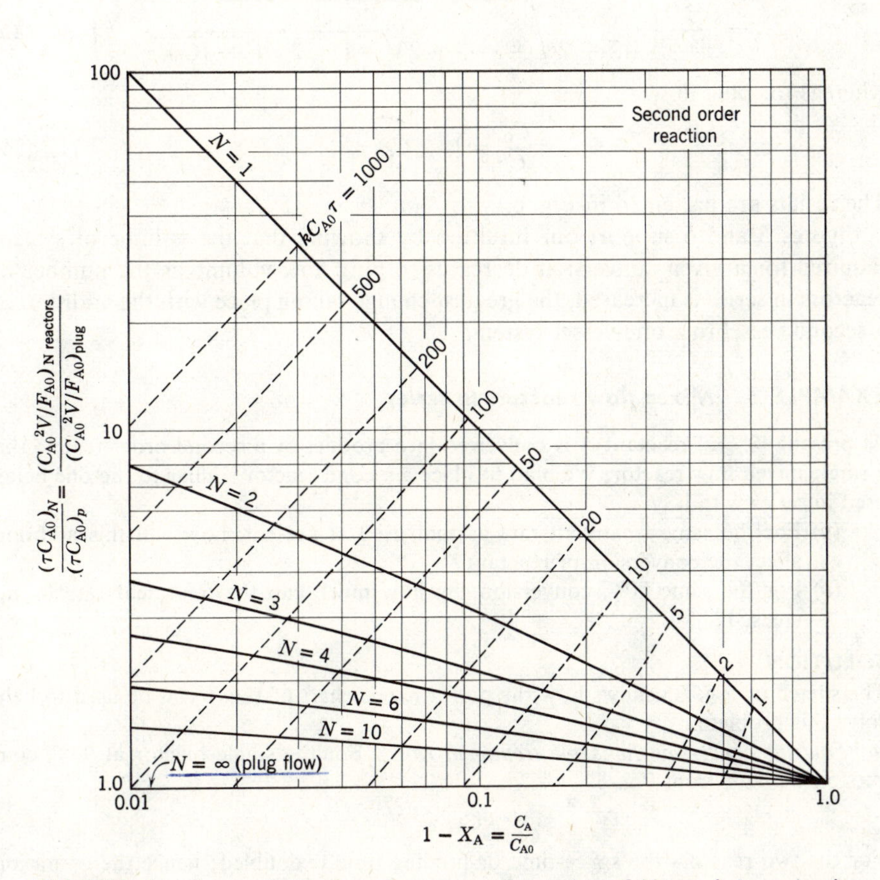

FIGURE 6. Comparison of performance of a series of N equal-size mixed reactors with a plug flow reactor for elementary second-order reactions

$$2A \rightarrow \text{products}$$
$$A + B \rightarrow \text{products}, \qquad C_{A0} = C_{B0}$$

with negligible expansion. For the same processing rate of identical feed the ordinate measures the volume ratio V_N/V_p or space-time ratio τ_N/τ_p directly.

With Eqs. 6*b* and 7 we can compare performance of N reactors in series with a plug flow reactor or with a single mixed reactor. This comparison is shown in Fig. 5 for first-order reactions in which density variations are negligible.

Second-order Reactions. We may evaluate the performance of a series of mixed reactors for a second-order, bimolecular-type reaction, no excess of either reactant, by a procedure similar to that for a first-order reaction. Thus for N reactors in series we find

$$C_N = \frac{1}{4k\tau_i}\left(-2 + 2\sqrt{-1\cdots + 2\sqrt{-1 + 2\sqrt{1 + 4C_0k\tau_i}}}\Bigg\}N\right) \tag{8a}$$

whereas for plug flow

$$\frac{C_0}{C} = 1 + C_0k\tau_p \tag{8b}$$

The results are presented in Fig. 6.

Figures 5 and 6 support our intuition by showing that the volume of system required for a given conversion decreases to plug flow volume as the number of reactors in series is increased, the greatest change taking place with the addition of a second vessel to a one-vessel system.

EXAMPLE 5. Mixed flow reactors in series

At present 90% of reactant A is converted into product by a second-order reaction in a single mixed flow reactor. We plan to place a second reactor similar to the one being used in series with it.

(*a*) For the same treatment rate as that used at present, how will this addition affect the conversion of reactant?

(*b*) For the same 90% conversion, by how much can the treatment rate be increased?

SOLUTION

The sketch of Fig. E5 shows how the performance chart of Fig. 6 can be used to help solve this problem.

(*a*) *Find the conversion for same treatment rate.* For the single reactor at 90% conversion we have from Fig. 6

$$kC_0\tau = 90$$

For the two reactors the space-time or holding time is doubled; hence the operation will be represented by the dashed line where

$$kC_0\tau = 180$$

This line cuts the $N = 2$ line at a conversion $X = 97.4\%$.

(*b*) *Find the treatment rate for the same conversion.* Staying on the 90% conversion line we find for $N = 2$ that

$$kC_0\tau = 27.5$$

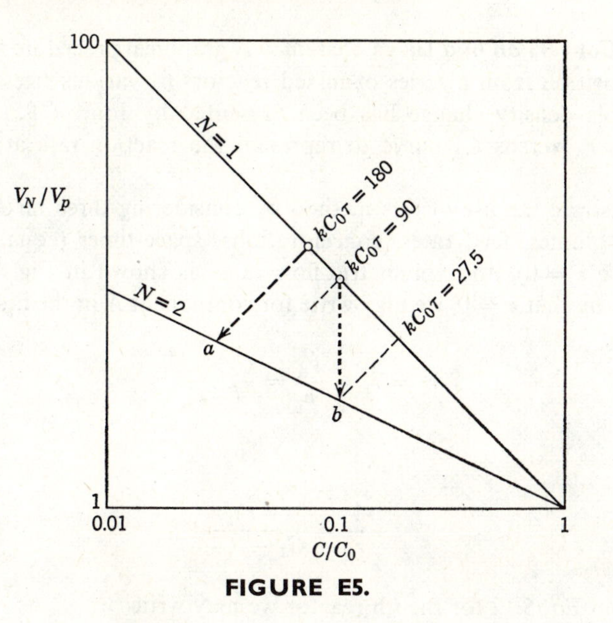

FIGURE E5.

Comparing the value of the reaction rate group for $N = 1$ and $N = 2$ we find

$$\frac{(kC_0\tau)_{N=2}}{(kC_0\tau)_{N=1}} = \frac{\tau_{N=2}}{\tau_{N=1}} = \frac{(V/v)_{N=2}}{(V/v)_{N=1}} = \frac{27.5}{90}$$

Since $V_{N=2} = 2V_{N=1}$ the ratio of flow rates becomes

$$\frac{v_{N=2}}{v_{N=1}} = \frac{90}{27.5}(2) = 6.6$$

Thus the treatment rate can be raised to 6.6 times the original.
Note. If the second reactor had been operated in parallel with the original unit then the treatment rate could only be doubled. Thus there is a definite advantage in operating these two units in series. This advantage becomes much more pronounced at higher conversions.

Performance charts different in kind from Figs. 1, 5, and 6 are given by MacMullin and Weber (1935), Jenney (1955), Lessells (1957), and Schoenemann (1958).

Mixed Flow Reactors of Different Sizes in Series

For arbitrary kinetics in mixed reactors of different size two types of questions may be asked: how to find the outlet conversion from a given reactor system, and the inverse question, how to find the best setup to achieve a given conversion. Different procedures are convenient for these two problems. We treat them in turn.

Finding the Conversion in a Given System. A graphical procedure for finding the outlet composition from a series of mixed reactors of various sizes for reactions with negligible density change has been presented by Jones (1951). All that is needed is an r_A versus C_A curve to represent the reaction rate at various concentrations.

Let us illustrate the use of this method by considering three mixed reactors in series with volumes, feed rates, concentrations, space-times (equal to residence times because $\varepsilon = 0$), and volumetric flow rates as shown in Fig. 7. Now from Eq. 5.11, noting that $\varepsilon = 0$, we may write for component A in the first reactor

$$\tau_1 = \bar{t}_1 = \frac{V_1}{v} = \frac{C_0 - C_1}{(-r)_1}$$

or

$$-\frac{1}{\tau_1} = \frac{(-r)_1}{C_1 - C_0} \tag{9}$$

Similarly from Eq. 5.12 for the ith reactor we may write

$$-\frac{1}{\tau_i} = \frac{(-r)_i}{C_i - C_{i-1}} \tag{10}$$

Plot the C versus r curve for component A and suppose that it is as shown in Fig. 8. To find the conditions in the first reactor note that the inlet concentration C_0 is known (point L), that C_1 and $(-r)_1$ correspond to a point on the curve to be found (point M), and that the slope of the line $LM = MN/NL = (-r)_1/(C_1 - C_0) = -(1/\tau_1)$ from Eq. 9. Hence from C_0 draw a line of slope $-(1/\tau_1)$ until it cuts the rate curve. This gives C_1. Similarly we find from Eq. 10 that a line of slope $-(1/\tau_2)$ from point N cuts the curve at P, giving the concentration C_2 of material leaving the second reactor. This procedure is then repeated as many times as needed.

With slight modification this graphical method can be extended to reactions in which density changes are appreciable.

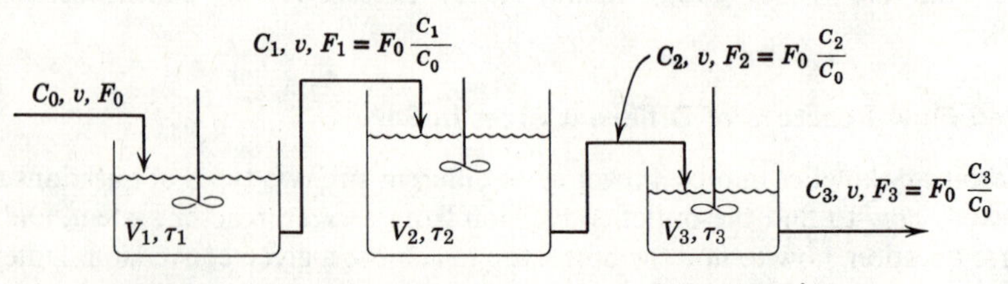

FIGURE 7. Notation for a series of unequal-size mixed reactors.

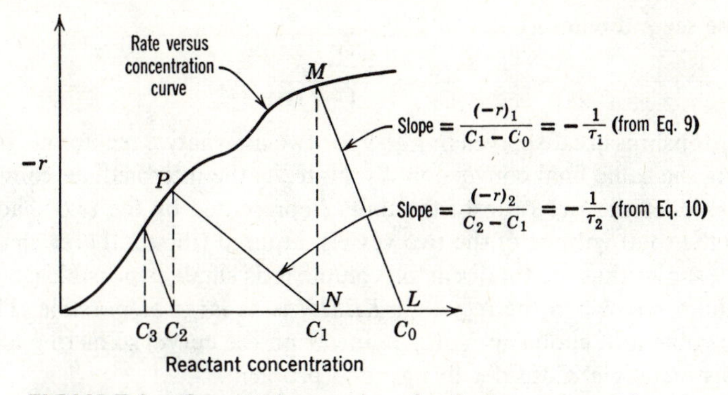

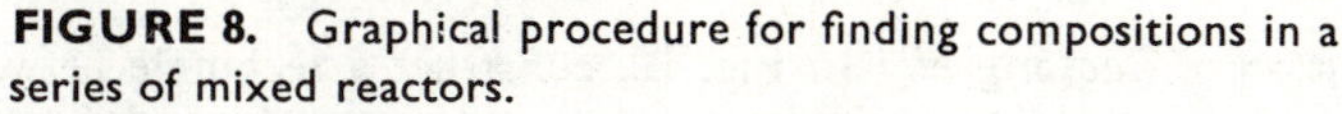

FIGURE 8. Graphical procedure for finding compositions in a series of mixed reactors.

Determining the Best System for a Given Conversion. Suppose we want to find the minimum size of two mixed reactors in series to achieve a specified conversion of feed which reacts with arbitrary but known kinetics. The basic performance expressions, Eqs. 5.11 and 5.12, then give, in turn, for the first reactor

$$\frac{\tau_1}{C_0} = \frac{X_1}{(-r)_1} \tag{11}$$

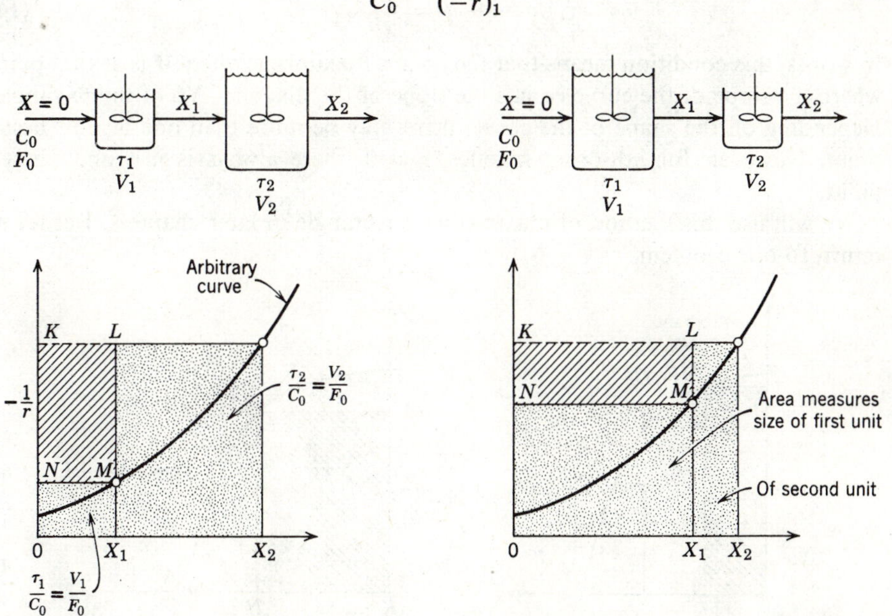

FIGURE 9. Graphical representation of the variables for two mixed reactors in series.

and for the second reactor

$$\frac{\tau_2}{C_0} = \frac{X_2 - X_1}{(-r)_2} \tag{12}$$

These relationships are displayed in Fig. 9 for two alternative reactor arrangements, both giving the same final conversion X_2. Note, as the intermediate conversion X_1 changes, so does the size ratio of the units (represented by the two shaded areas) as well as the total volume of the two vessels required (the total area shaded).

Figure 9 shows that the total reactor volume is as small as possible (total shaded area is minimized) when the rectangle $KLMN$ is as large as possible. This brings us to the problem of choosing X_1 (or point M on the curve) so as to maximize the area of this rectangle. Consider this general problem.

Maximization of Rectangles. In Fig. 10, construct a rectangle between the x-y axes and touching the arbitrary curve at point $M(x, y)$. The area of the rectangle is then

$$A = xy \tag{13}$$

This area is maximized when

$$dA = 0 = y\,dx + x\,dy$$

or where

$$-\frac{dy}{dx} = \frac{y}{x} \tag{14}$$

In words, this condition means that the area is maximized when M is at that point where the slope of the curve equals the slope of the diagonal NL of the rectangle. Depending on the shape of the curve, there may be more than one or no "best" point. However, for nth-order kinetics, $n > 0$, there always is just one "best" point.

We will use this method of maximizing a rectangle in later chapters. But let us return to our problem.

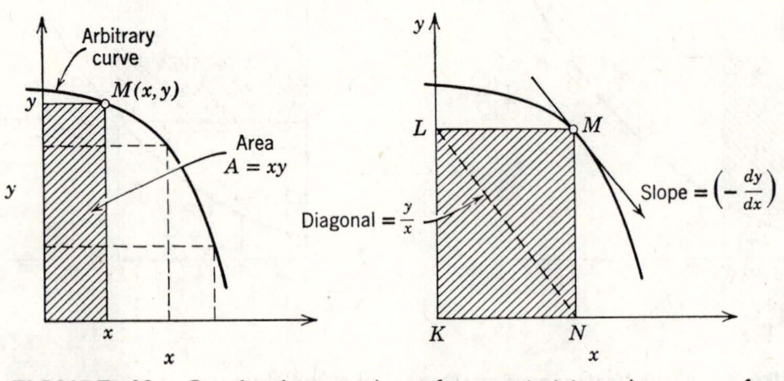

FIGURE 10. Graphical procedure for maximizing the area of a rectangle.

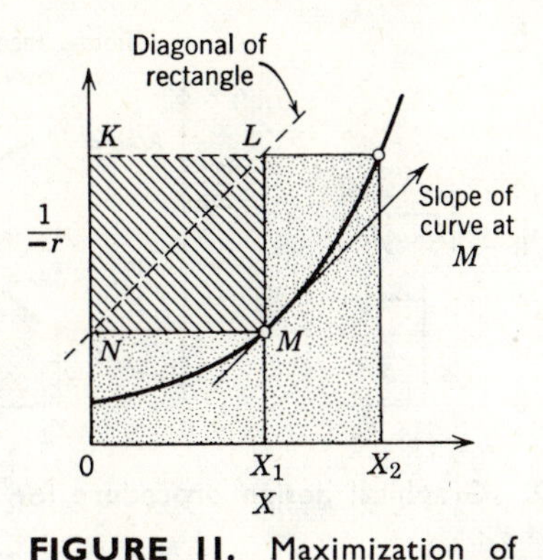

FIGURE 11. Maximization of rectangles applied to find the optimum intermediate conversion and optimum sizes of two mixed reactors in series.

The optimum size ratio of the two reactors is achieved where the slope of the rate curve at M equals the diagonal NL. The best value of M is shown in Fig. 11, and this determines the intermediate conversion X_1 as well as the size of units needed.

The optimum size ratio for two mixed reactors in series is found in general to be dependent on the kinetics of the reaction and on the conversion level. For the special case of first-order reactions equal-size reactors are best; for reaction orders $n > 1$ the smaller reactor should come first; for $n < 1$ the larger should come first (see Problem 3).

Charts showing the optimum size ratio for various n values and conversion levels are given by Szepe and Levenspiel (1964). Since the advantage of the minimum size system over the equal-size system is found to be quite small, only a few percent at most, over-all economic consideration would nearly always recommend using equal-size units.

The above procedure can be extended directly to multistage operations (see Problem 14); however, here the argument for equal-size units is stronger still than for the two-stage system.

Reactors of Different Types in Series

If reactors of different types are put in series, such as a mixed reactor followed by a plug flow reactor which in turn is followed by another mixed reactor, we may write for the three reactors

$$\frac{V_1}{F_0} = \frac{X_1 - X_0}{(-r)_1}, \qquad \frac{V_2}{F_0} = \int_{X_1}^{X_2} \frac{dX}{-r}, \qquad \frac{V_3}{F_0} = \frac{X_3 - X_2}{(-r)_3}$$

These relationships are represented in graphical form in Fig. 12. This allows us to predict the over-all conversions for such systems, or conversions at intermediate

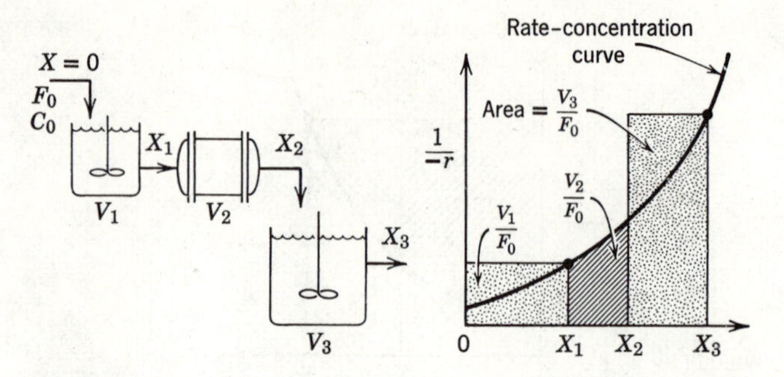

FIGURE 12. Graphical design procedure for reactors in series.

points between the individual reactors. These intermediate conversions may be needed to determine the duty of interstage heat exchangers.

Best Arrangement of a Set of Ideal Reactors. For the most effective use of a given set of ideal reactors we have the following general rules:

1. For a reaction whose rate-concentration curve rises monotonically (any nth-order reaction, $n > 0$) the reactors should be connected in series. They should be ordered so as to keep the concentration of reactant as high as possible if the rate-concentration curve is concave ($n > 1$), as low as possible if the curve is convex ($n < 1$). As an example, for the case of Fig. 12 the ordering of units should be plug, small mixed, large mixed, for $n > 1$; the reverse ordering should be used when $n < 1$.

2. For reactions where the rate-concentration curve passes through a maximum or minimum the arrangement of units depends on the actual shape of curve, the conversion level desired, and the units available. No simple rules can be suggested.

3. Whatever may be the kinetics and the reactor system, an examination of the $1/(-r_A)$ versus C_A curve is a good way to find the best arrangement of units.

The problems at the end of this chapter illustrate these findings.

RECYCLE REACTOR

In certain situations it is found to be advantageous to divide the product stream from a plug flow reactor and return a portion of it to the entrance of the reactor. Let the *recycle ratio* **R** be defined as

$$\mathbf{R} = \frac{\text{volume of fluid returned to the reactor entrance}}{\text{volume leaving the system}} \tag{15}$$

This recycle ratio can be made to vary from zero to infinity. Reflection suggests that as the recycle ratio is raised the behavior shifts from plug flow ($\mathbf{R} = 0$) to mixed flow ($\mathbf{R} = \infty$). Thus recycling provides a means for obtaining various degrees of backmixing with a plug flow reactor. Let us develop the performance equation for the recycle reactor.

Consider a recycle reactor with nomenclature as shown in Fig. 13. Across the reactor itself Eq. 5.18 for plug flow gives

$$\frac{V}{F'_{A0}} = \int_{X_{A1}}^{X_{A2}=X_{Af}} \frac{dX_A}{-r_A} \tag{16}$$

where F'_{A0} would be the feed rate of A if the stream entering the reactor (fresh feed plus recycle) were unconverted. Since F'_{A0} and X_{A1} are not known directly they must be written in terms of known quantities before Eq. 16 can be used. Let us now do this.

The flow entering the reactor includes both fresh feed and the recycle stream. Measuring the flow split at point L (point K will not do if $\varepsilon \neq 0$) we then have

$$F'_{A0} = \begin{pmatrix}\text{A which would enter in an}\\ \text{unconverted recycle stream}\end{pmatrix} + \begin{pmatrix}\text{A entering in}\\ \text{fresh feed}\end{pmatrix}$$

$$= \mathbf{R}F_{A0} + F_{A0} = (\mathbf{R} + 1)F_{A0} \tag{17}$$

Now to the evaluation of X_{A1}: from Eq. 3.72 we may write

$$X_{A1} = \frac{1 - C_{A1}/C_{A0}}{1 + \varepsilon_A C_{A1}/C_{A0}} \tag{18}$$

Because the pressure is taken to be constant, the streams meeting at point K may be added directly. This gives

$$C_{A1} = \frac{F_{A1}}{v_1} = \frac{F_{A0} + F_{A3}}{v_0 + \mathbf{R}v_f} = \frac{F_{A0} + \mathbf{R}F_{A0}(1 - X_{Af})}{v_0 + \mathbf{R}v_0(1 + \varepsilon_A X_{Af})} = C_{A0}\left(\frac{1 + \mathbf{R} - \mathbf{R}X_{Af}}{1 + \mathbf{R} + \mathbf{R}\varepsilon_A X_{Af}}\right) \tag{19}$$

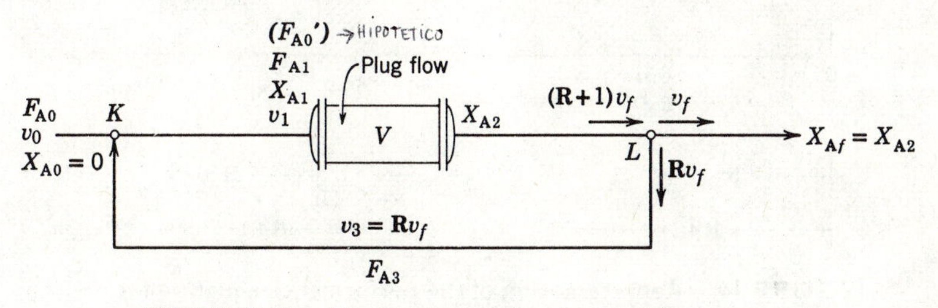

FIGURE 13. Nomenclature for the recycle reactor.

Combining Eqs. 18 and 19 gives X_{A1} in terms of measured quantities, or

$$X_{A1} = \left(\frac{\mathbf{R}}{\mathbf{R}+1}\right) X_{Af} \tag{20}$$

Finally, on replacing Eqs. 17 and 20 in Eq. 16 we obtain the useful form for the performance equation for recycle reactors, good for any kinetics, and any ε value

$$\frac{V}{F_{A0}} = (\mathbf{R}+1) \int_{\left(\frac{\mathbf{R}}{\mathbf{R}+1}\right) X_{Af}}^{X_{Af}} \frac{dX_A}{-r_A} \tag{21}$$

For the special case where density changes are negligible we may also write this equation in terms of concentrations, or

$$\tau = \frac{C_{A0}V}{F_{A0}} = -(\mathbf{R}+1) \int_{\frac{C_{A0}+\mathbf{R}C_{Af}}{\mathbf{R}+1}}^{C_{Af}} \frac{dC_A}{-r_A} \tag{22}$$

These expressions are represented graphically in Fig. 14.

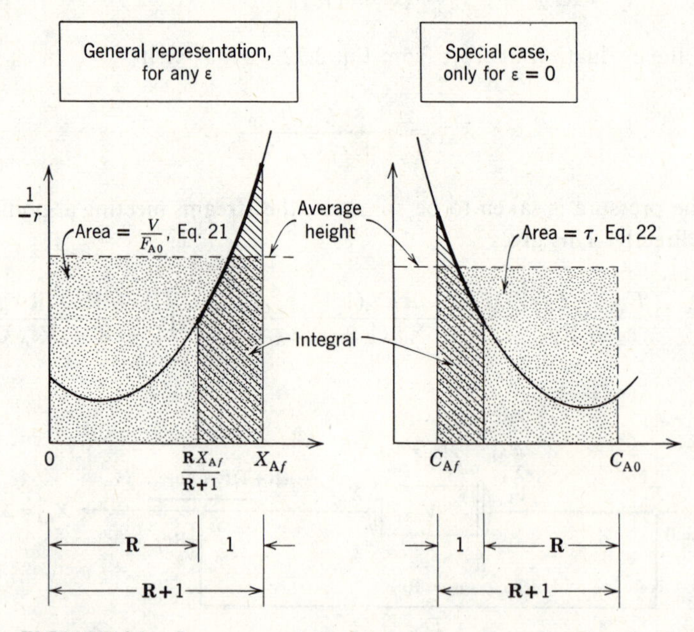

FIGURE 14. Representation of the performance equation for recycle reactors.

For the extremes of negligible and infinite recycle the system approaches plug flow and mixed flow, or

$$\frac{V}{F_{A0}} = (\mathbf{R}+1)\int_{\frac{\mathbf{R}}{\mathbf{R}+1}X_{Af}}^{X_{Af}} \frac{dX_A}{-r_A}$$

$\mathbf{R}=0$ ↓ ↓ $\mathbf{R}=\infty$

$$\frac{V}{F_{A0}} = \int_0^{X_{Af}} \frac{dX_A}{-r_A} \qquad \frac{V}{F_{A0}} = \frac{X_{Af}}{-r_{Af}}$$

plug flow mixed flow

The approach to these extremes is shown in Fig. 15.

Integration of the recycle equation gives, for *first-order reaction*, $\varepsilon_A = 0$,

$$\frac{k\tau}{\mathbf{R}+1} = \ln\left[\frac{C_{A0}+\mathbf{R}C_{Af}}{(\mathbf{R}+1)C_{Af}}\right] \tag{23}$$

and for *second-order reaction*, $2A \rightarrow$ products, $-r_A = kC_A^2$, $\varepsilon_A = 0$,

$$\frac{kC_{A0}\tau}{\mathbf{R}+1} = \frac{C_{A0}(C_{A0}-C_{Af})}{C_{Af}(C_{A0}+\mathbf{R}C_{Af})} \tag{24}$$

The expressions for $\varepsilon_A \neq 0$ and for other reaction orders can be evaluated, but are more cumbersome.

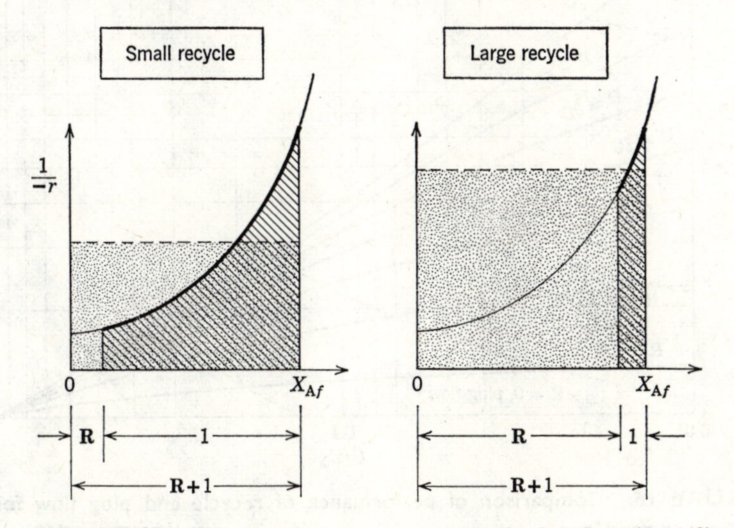

FIGURE 15. The recycle extremes approach plug flow ($\mathbf{R} \rightarrow 0$) and mixed flow ($\mathbf{R} \rightarrow \infty$).

Figures 16 and 17 show the transition from plug to mixed flow as **R** increases, and a match of these curves with those for N tanks in series (Figs. 5 and 6) gives the following rough comparison for equal performance:

N	**R** for first-order reaction, at $X_A = 0.5$,	$= 0.90$,	$= 0.99$	**R** for second-order reaction, at $X_A = 0.5$,	$= 0.90$,	$= 0.99$
1	∞	∞	∞	∞	∞	∞
2	1.0	2.2	5.4	1.0	2.8	7.5
3	0.5	1.1	2.1	0.5	1.4	2.9
4	0.33	0.68	1.3	0.33	0.90	1.7
10	0.11	0.22	0.36	0.11	0.29	0.5
∞	0	0	0	0	0	0

The recycle reactor is a convenient way for approaching mixed flow with what is essentially a plug flow device. Its particular usefulness is with solid catalyzed reactions with their fixed bed contactors. We meet this and other applications of recycle reactors in later chapters.

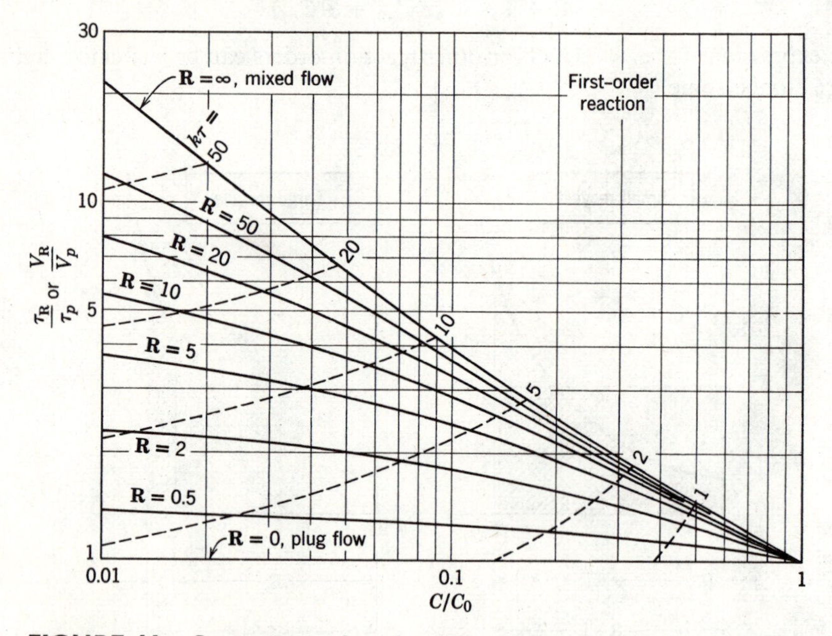

FIGURE 16. Comparison of performance of recycle and plug flow for first-order reactions

$$A \rightarrow R, \quad \varepsilon = 0.$$

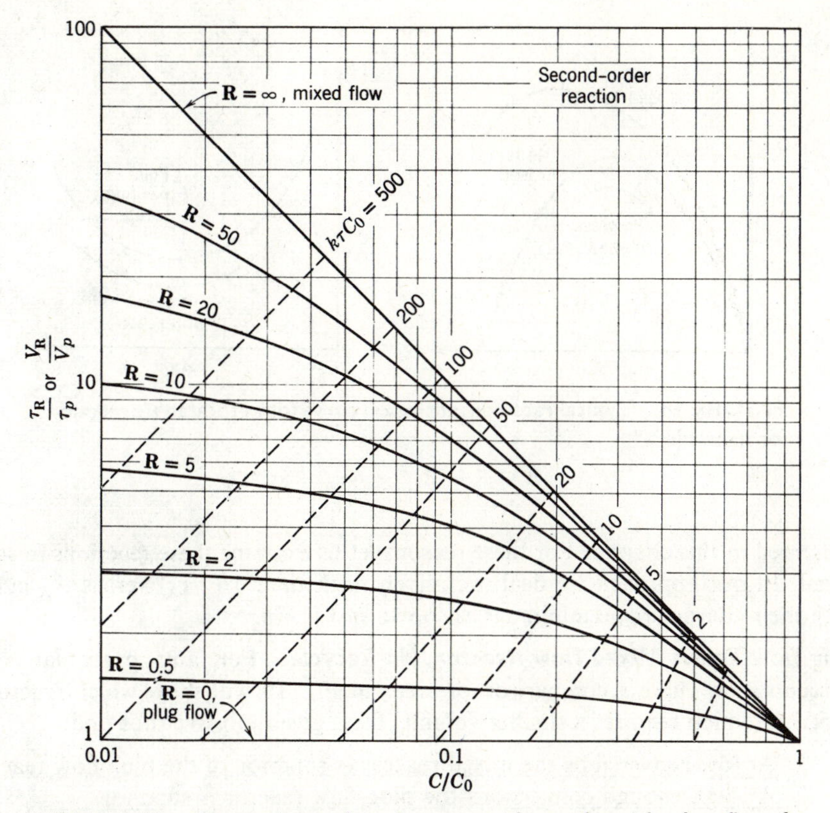

FIGURE 17. Comparison of performance of recycle with plug flow for elementary second-order reactions

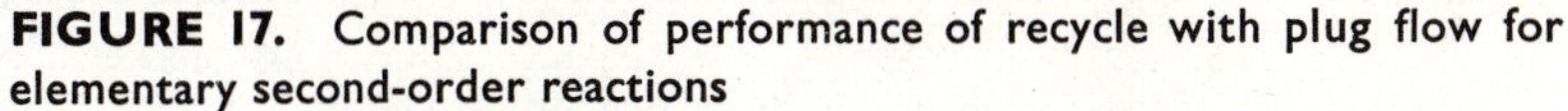

$$2A \rightarrow \text{products}, \qquad \varepsilon = 0$$
$$A + B \rightarrow \text{products}, \qquad C_{A0} = C_{B0} \text{ with } \varepsilon = 0$$

(Personal communication, from T. J. Fitzgerald and P. Fillesi.)

AUTOCATALYTIC REACTIONS

When a material reacts away by a first- or second-order rate in a batch reactor, its rate of disappearance is rapid at the start when the concentration of reactant is high. This rate then slows progressively as reactant is consumed. In an autocatalytic reaction, however, the rate at the start is low because little product is present; it increases to a maximum as product is formed and then drops again to a low value as reactant is consumed. Figure 18 shows a typical situation.

Reactions with such rate-concentration curves lead to interesting optimization problems. In addition, they provide a good illustration of the general design method

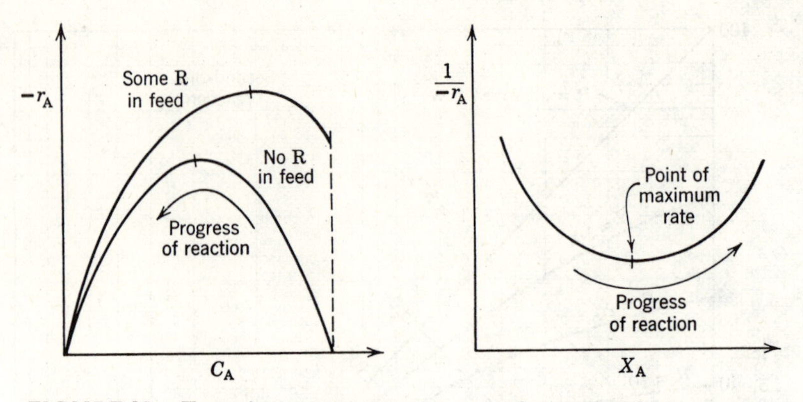

FIGURE 18. Typical rate-concentration curve for autocatalytic reactions, for example

$$A + R \rightarrow R + R, \qquad -r_A = kC_A^a C_R^r$$

presented in this chapter. For these reasons let us examine these reactions in some detail. In our approach we deal exclusively with their $1/(-r_A)$ versus X_A curves with their characteristic minimum as shown in Fig. 18.

Plug Flow Versus Mixed Flow Reactor, No Recycle. For any particular rate-concentration curve a comparison of areas in Fig. 19 will show which reactor is superior (which requires a smaller volume) for a given job. We thus find:

1. At low conversions the mixed reactor is superior to the plug flow reactor.
2. At high enough conversions the plug flow reactor is superior.

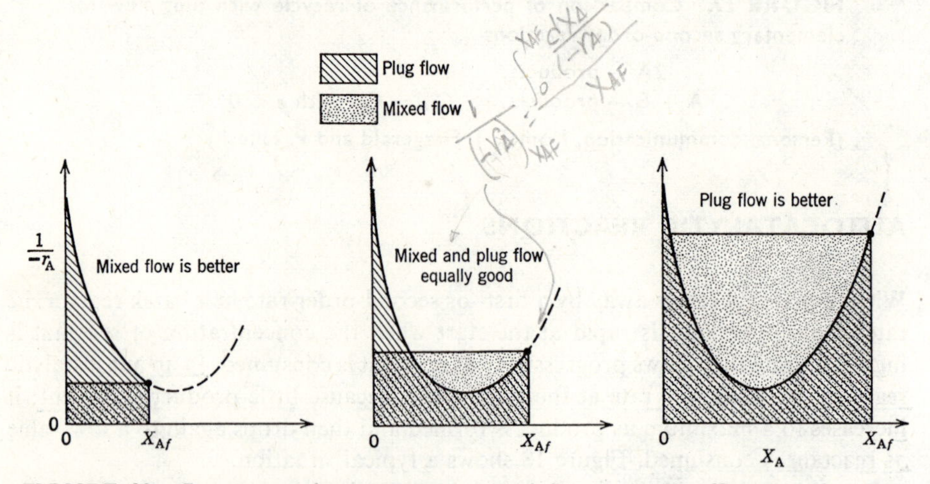

FIGURE 19. For autocatalytic reactions mixed flow is more efficient at low conversions, plug flow is more efficient at high conversions.

These findings differ from ordinary nth-order reactions ($n > 0$) where the plug flow reactor is always more efficient than the mixed reactor. In addition, we should note that since some product must normally be present in the feed for the autocatalytic reaction to proceed, a plug flow reactor will not operate at all with a feed of pure reactant. In such a situation the feed must be continually primed with product, an ideal opportunity for using a recycle reactor.

Recycle Reactor. When material is to be processed to some fixed final conversion X_{Af} in a recycle reactor, reflection suggests that there must be a particular recycle ratio which is optimum in that it minimizes the reactor volume or space-time. Let us determine this value of **R**.

The *optimum recycle ratio* is found by differentiating Eq. 21 with respect to **R** and setting to zero, thus

$$\text{take} \quad \frac{d(\tau/C_{A0})}{d\mathbf{R}} = 0 \quad \text{for} \quad \frac{\tau}{C_{A0}} = \int_{X_{Ai}=\frac{\mathbf{R}X_{Af}}{\mathbf{R}+1}}^{X_{Af}} \frac{\mathbf{R}+1}{(-r_A)}\, dX_A \tag{25}$$

This operation requires differentiating under an integral sign. From the theorems of calculus if

$$F(\mathrm{R}) = \int_{a(\mathrm{R})}^{b(\mathrm{R})} f(x, \mathrm{R})\, dx \tag{26}$$

then

$$\frac{dF}{d\mathrm{R}} = \int_{a(\mathrm{R})}^{b(\mathrm{R})} \frac{\partial f(x,\mathrm{R})}{\partial \mathrm{R}}\, dx + f(b, \mathrm{R})\frac{db}{d\mathrm{R}} - f(a, \mathrm{R})\frac{da}{d\mathrm{R}} \tag{27}$$

For our case, Eq. 25, we then find

$$\frac{d(\tau/C_{A0})}{d\mathbf{R}} = 0 = \int_{X_{Ai}}^{X_{Af}} \frac{dX_A}{(-r_A)} + 0 - \left.\frac{\mathbf{R}+1}{(-r_A)}\right|_{X_{Ai}} \frac{dX_{Ai}}{d\mathbf{R}}$$

where

$$\frac{dX_{Ai}}{d\mathbf{R}} = \frac{X_{Af}}{(\mathbf{R}+1)^2}$$

Combining and rearranging then gives

$$\left.\frac{1}{-r_A}\right|_{X_{Ai}} = \frac{\displaystyle\int_{X_{Ai}}^{X_{Af}} \frac{dX_A}{-r_A}}{(X_{Af} - X_{Ai})} \tag{28}$$

In words, the optimum recycle ratio introduces to the reactor a feed whose $1/(-r_A)$ value (KL in Fig. 20) equals the average $1/(-r_A)$ value in the reactor as a whole (PQ in Fig. 20). Figure 20 compares this optimum with conditions where the recycle is either too high or too low.

Nota: $F(a) = \frac{\int_a^b F(x)dx}{b-a}$;

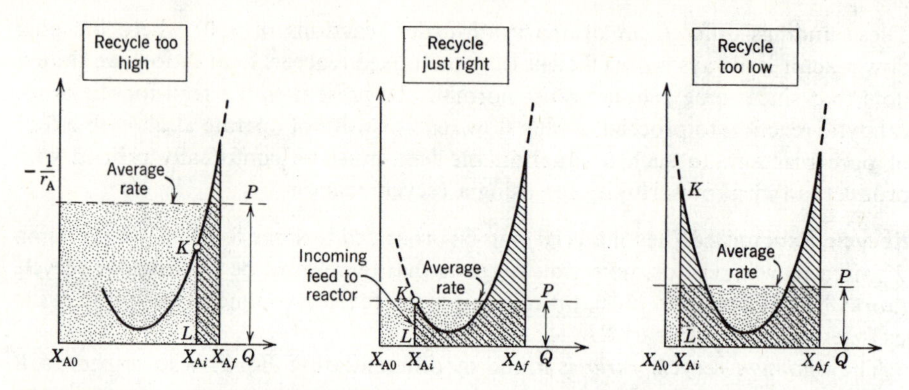

FIGURE 20. Correct recycle ratio for an autocatalytic reaction compared with recycle ratios which are too high and too low.

On comparing with the single plug flow or single mixed reactor we find:

1. For final conversions smaller than the point of maximum rate the mixed reactor is better than any recycle reactor.
2. For conversions higher than the point of maximum rate the recycle reactor with the proper recycle ratio is superior to either the plug flow or mixed reactor.

Multiple Reactor System. To minimize the total volume we should choose a system where most of the processing occurs at or near the composition of highest rate. Now this point can be reached in one jump and without going through the intermediate compositions of lower rate by using a mixed reactor. If we desire still higher conversions, we should move progressively from this composition to the final composition. This requires a plug flow reactor. Thus in processing material past the point of maximum rate the best setup is a mixed reactor operating at the maximum rate followed by a plug flow reactor. This is shown in Fig. 21.

This scheme is superior to the single plug flow reactor, the single mixed reactor, or the recycle reactor.

Reactors with Separation and Recycle of Unconverted Reactant. If unconverted reactant can be separated from the product stream and returned to the reactor then all the processing can be made to occur at one composition, for any conversion level. In such a case the best scheme is to use a mixed reactor operating at the point of maximum reaction rate as shown in Fig. 21.

The volume required is now the very minimum, less than any of the previous ways of operating. However, the over-all economics, including the cost of separation and of recycle, will determine which scheme is the optimum overall.

Occurrence of Autocatalytic Reactions. The most important examples of autocatalytic reactions are the broad class of fermentation reactions which result from

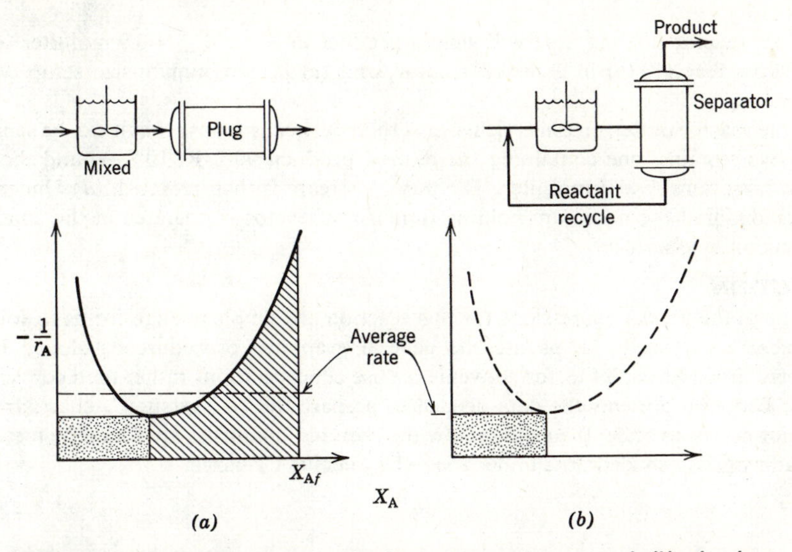

FIGURE 21. (*a*) The best multiple reactor scheme, and (*b*) the best scheme when unconverted reactant can be separated and recycled.

the action of a microorganism on an organic feed. When they can be treated as single reactions, then the methods of this chapter can be applied directly. Often, however, various microorganisms compete to produce different products. The methods of the next chapter treat this problem. Fermentations in general are discussed there. Another type of reaction which has autocatalytic behavior is the exothermic reaction (say the combustion of fuel gas) proceeding in an adiabatic manner with cool reactants entering the system. In such a reaction, called autothermal, heat may be considered to be the product which sustains the reaction. Thus with plug flow the reaction will die. With backmixing the reaction will be self-sustaining because the heat generated by the reaction can raise fresh reactants to a temperature at which they will react. Autothermal reactions are of great importance in solid catalyzed gas phase systems and are treated later in the book.

EXAMPLE 6. Plug and mixed flow for an autocatalytic reaction

We wish to explore various reactor setups for the transformation of A into R. The feed contains 99% A, 1% R; the desired product is to consist of 10% A, 90% R. The transformation takes place by means of the elementary reaction

$$A + R \rightarrow R + R$$

with rate constant $k = 1$ liter/mol·min. The concentration of active materials is

$$C_{A0} + C_{R0} = C_A + C_R = C_0 = 1 \text{ mol/liter}$$

throughout.

What reactor holding time will yield a product in which $C_R = 0.9$ mol/liter (*a*) in a plug flow reactor, (*b*) in a mixed reactor, and (*c*) in a minimum-size setup without recycle.

If the reactor outlet stream is richer in A than desired, suppose that it can be separated into two streams, one containing the desired product, 90% R, 10% A, and the other containing pure A at 1 mol/liter. The pure A stream is then recycled. (*d*) Under these conditions find the minimum holding time if the reactor is operated at the conditions of maximum efficiency.

SOLUTION

Although the kinetic expressions for this reaction are simple enough for us to solve the problem analytically, let us use the general graphical procedure developed in this chapter. In addition, let us for convenience use concentrations rather than conversions. Thus Table E6 presents the data needed to prepare the $1/r_A$ versus C_A chart, and Fig. E6, not drawn to scale, then shows how the required quantities are found by measuring the appropriate shaded area under the $-1/r_A$ versus C_A curve.

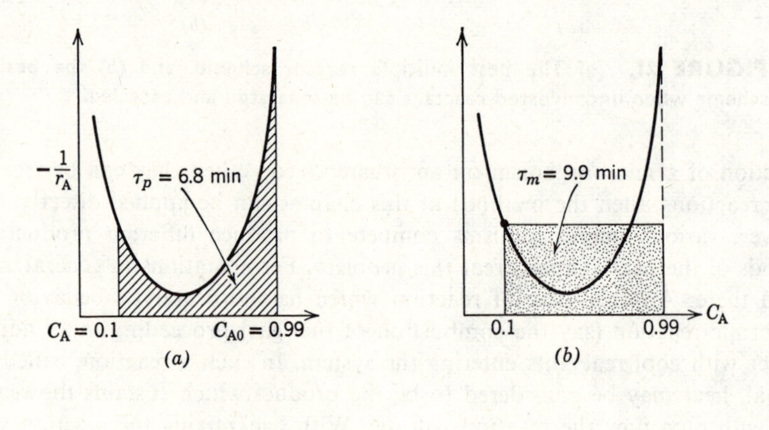

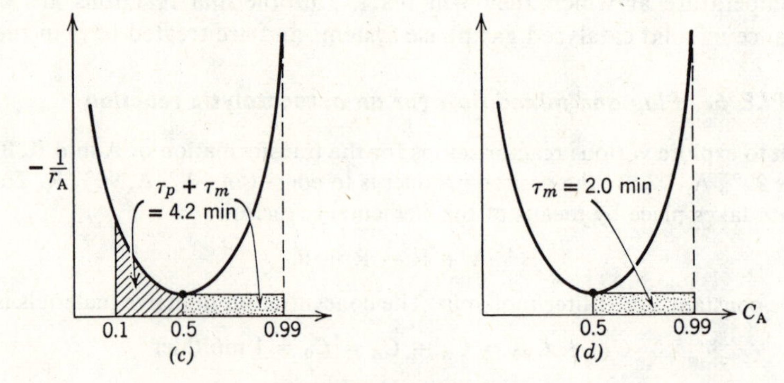

FIGURE E6.

TABLE E6

C_A	C_R	$-r_A = C_A C_R$	$-1/r_A$
0.99	0.01	0.0099	101.01
0.95	0.05	0.0475	21.05
0.90	0.10	0.09	11.11
0.70	0.30	0.21	4.76
0.50	0.50	0.25	4.00
0.30	0.70	0.21	4.76
0.10	0.90	0.09	11.11

For a product where $C_A = 0.1$ mol/liter and $C_R = 0.9$ mol/liter we find from Fig. E6:

(*a*) For plug flow

$$\tau_p = 6.8 \text{ min}$$

(*b*) For mixed flow

$$\tau_m = 9.9 \text{ min}$$

(*c*) For the most efficient setup without recycle we need a mixed reactor ($\tau_m = 2.0$ min) followed by a plug flow reactor ($\tau_p = 2.2$ min) with total space-time

$$\tau_{m+p} = 4.2 \text{ min}$$

(*d*) With product separation and recycle the most efficient setup has

$$\tau_{\min} = 2.0 \text{ min}$$

EXAMPLE 7. Optimum recycle ratio for an autocatalytic reaction

Substance A reacts according to the elementary autocatalytic reaction

$$A + R \rightarrow R + R, \qquad k = 1 \text{ liter/mol}\cdot\text{min}$$

We plan to process $F_{A0} = 1$ mol/min of a feed consisting of A alone ($C_{A0} = 1$ mol/liter, $C_{R0} = 0$) to 99% conversion in a recycle reactor. (*a*) Find the recycle rate which will minimize the size of reactor needed and determine this size. Compare this optimum size (*b*) with a reactor with recycle ratio, $\mathbf{R} = 4$, (*c*) with a mixed reactor, or $\mathbf{R} = \infty$ (*d*) with a plug flow reactor, or $\mathbf{R} = 0$.

SOLUTION

(*a*) *At optimum recycle.* For this reaction in a recycle reactor Eq. 21 becomes

$$\frac{V}{F_{A0}} = (\mathbf{R} + 1)\int_{X_{Ai}=\frac{\mathbf{R}X_{Af}}{\mathbf{R}+1}}^{X_{Af}} \frac{dX_A}{kC_{A0}^2 X_A(1 - X_A)}$$

Integrating and simplifying gives the performance equation

$$\frac{V}{F_{A0}} = \frac{\mathbf{R} + 1}{kC_{A0}^2} \ln \frac{1 + \mathbf{R}(1 - X_{Af})}{\mathbf{R}(1 - X_{Af})} \tag{i}$$

Now to find the optimum recycle ratio minimize V with respect to **R** in Eq. (i); in other words, choose **R** to satisfy the condition of Eq. 28. For our reaction this gives

$$\ln \frac{1 + \mathbf{R}(1 - X_{Af})}{\mathbf{R}(1 - X_{Af})} = \frac{\mathbf{R} + 1}{\mathbf{R}[1 + \mathbf{R}(1 - X_{Af})]}$$

Putting $X_{Af} = 0.99$ and simplifying, this condition reduces to

$$\ln \frac{1 + 0.01\mathbf{R}}{0.01\mathbf{R}} = \frac{\mathbf{R} + 1}{\mathbf{R}(1 + 0.01\mathbf{R})}$$

Solving by trial and error gives $\mathbf{R} = 0.19$. Thus the volume of reactor needed, found from Eq. (i), is

$$V = 1.19 \ln \frac{1.0019}{0.0019} = 7.46 \text{ liters}$$

(*b*) *For a recycle ratio*, $\mathbf{R} = 4$. For this condition the performance equation, Eq. (i), gives

$$V = 5 \ln \frac{1.04}{0.04} = 16.3 \text{ liters}$$

(*c*) *For mixed flow.* From Chapter 5 the performance equation for these particular kinetics becomes

$$\frac{V}{F_{A0}} = \frac{X_{Af}}{kC_{Af}C_{Rf}} = \frac{X_{Af}}{kC_{A0}^2 X_{Af}(1 - X_{Af})}$$

and on replacing known values we find

$$V = \frac{0.99}{0.99(0.01)} = 100 \text{ liters}$$

(*d*) *For plug flow.* From Chapter 5 the performance equation becomes

$$\frac{V}{F_{A0}} = \int_0^{X_{Af}} \frac{dX_A}{kC_{A0}^2 X_A(1 - X_A)}$$

On replacing values we find that

$$V = \infty \text{ liters}$$

This result shows that this reaction will not proceed at all in a reactor where the fluid strictly follows the ideal of plug flow.

RELATED READINGS

Denbigh, K. G., *Trans. Faraday Soc.*, **40**, 352 (1944).
Jenney, T. M., *Chem. Eng.*, **62**, 198 (Dec. 1955).

REFERENCES

Jones, R. W., *Chem. Eng. Progr.*, **47**, 46 (1951).
Leclerc, V. R., *Chem. Eng. Sci.*, **2**, 213 (1953).
Lessells, G. A., *Chem. Eng.*, **64**, 251 (Aug. 1957).
MacMullin, R. B., and M. Weber, Jr., *Trans. Am. Inst. Chem. Engrs.*, **31**, 409 (1935).
Schoenemann, K., *Chem. Eng. Sci.*, **8**, 161 (1958).
Szepe, S., and O. Levenspiel, *Ind. Eng. Chem. Process Design Develop.*, **3**, 214 (1964).

PROBLEMS

1. Substance A reacts according to second-order kinetics and conversion is 95% from a single flow reactor. We buy a second unit identical to the first. For the same conversion, by how much is the capacity increased if we operate these two units in parallel or in series?

(*a*) The reactors are both plug flow.

(*b*) The reactors are both mixed flow.

2. (*a*) Derive an expression for the concentration of reactant in the effluent from a series of mixed reactors of different sizes. Let the reaction follow first-order kinetics and let the holding time in the *i*th reactor be τ_i.
(*b*) Show that this expression reduces to the appropriate equation in this chapter when the reactors are all the same size.

3. Your company has two mixed reactors of unequal size for producing a specified product formed by homogeneous first-order reaction. How should these reactors be connected to achieve a maximum production rate?

4. A first-order liquid-phase reaction, 92% conversion, is taking place in a mixed reactor. It has been suggested that a fraction of the product stream, with no additional treatment, be recycled. If the feed rate remains unchanged, in what way would this affect conversion?

5. 100 liters/hr of radioactive fluid having a half-life of 20 hr is to be treated by passing it through two ideal stirred tanks in series, $V = 40{,}000$ liters each. In passing through this system, how much has the activity decayed?

6. At present the elementary liquid-phase reaction $A + B \rightarrow R + S$ takes place in a plug flow reactor using equimolar quantities of A and B. Conversion is 96%, $C_{A0} = C_{B0} = 1$ mol/liter.

(*a*) If a mixed reactor ten times as large as the plug flow reactor were hooked up in series with the existing unit, which unit should come first and by what fraction could production be increased for that setup?

(*b*) Does the concentration level of the feed affect the answer, and if so, in what way?

Note: Conversion is to remain unchanged.

7. The elementary reaction $A + B \rightarrow R + S$ is effected in a setup consisting of a mixed reactor into which the two reactant solutions are introduced followed by a plug flow reactor. A large enough excess of B is used so that the reaction is first order with respect to A. Various ways of increasing production have been suggested, one of which is to reverse the order of the two units. How would this change affect conversion?

8. A certain material polymerizes at high temperature. With temperature above 105°C a product with undesirable properties is obtained; therefore an operating temperature of 102°C is selected. At this temperature the polymerization proceeds by means of a reaction which can be represented adequately by a 1.5-order rate law with respect to the monomer. At present the monomer is being treated in two equal-size mixed reactors in series yielding a product in which the monomer content is about 20%. An increase in production is contemplated by adding a reactor similar to those being used. By what percentage can we increase the feed rate and still obtain a product containing no more than 20% monomer if the third reactor is connected to receive the effluent of the second reactor, thus giving three reactors in series?

9. The kinetics of the aqueous-phase decomposition of A is investigated in two mixed reactors in series, the second having twice the volume of the first reactor. At steady state with a feed concentration of 1 mol A/liter and mean residence time of 96 sec in the first reactor, the concentration in the first reactor is 0.5 mol A/liter and in the second is 0.25 mol A/liter. Find the kinetic equation for the decomposition.

10. Using a color indicator which shows when the concentration of A falls below 0.1 mol/liter, the following scheme is devised to explore the kinetics of the decomposition of A. A feed of 0.6 mol A/liter is introduced into the first of the two mixed reactors in series, each having a volume of 400 cm^3. The color change occurs in the first reactor for a steady-state feed rate of 10 cm^3/min, and in the second reactor for a steady-state feed rate of 50 cm^3/min. Find the rate equation for the decomposition of A from this information.

11. The elementary irreversible aqueous-phase reaction $A + B \rightarrow R + S$ is carried out isothermally as follows. Equal volumetric flow rates of two liquid streams are introduced into a 4-liter mixing tank. One stream contains 0.020 mol A/liter, the other 1.400 mol B/liter. The mixed stream is then passed through a 16-liter plug flow reactor. We find that some R is formed in the mixing tank, its concentration being 0.002 mol/liter. Assuming that the mixing tank acts as a mixed reactor, find the concentration of R at the exit of the plug flow reactor as well as the fraction of initial A that has been converted in the system.

12. Two unequal-size mixed reactors are connected in series. It is suspected that the order in which these vessels are connected as well as the reaction type may affect the degree of conversion for a given feed rate. Determine in general for reactions with concave or convex rate-concentration curves which reactor should come first. How does this apply to *n*th-order reactions?

13. At present conversion is 2/3 for our elementary second-order liquid reaction $2A \rightarrow 2R$ when operating in an isothermal plug flow reactor with a recycle ratio of unity. What will be the conversion if the recycle stream is shut off?

14. What operating condition minimizes the volume of recycle reactor needed to achieve $C_R = 0.9$ mol/liter for the reaction and feed of Example 6? Compare this volume with that for plug flow and mixed flow.

15. Reactant A decomposes according to a first-order reaction $A \rightarrow R$ and we are given two feed streams

$$\text{Feed 1:} \quad C_{A0} = 1, \qquad F_{A0} = 1$$

$$\text{Feed 2:} \quad C_{A0} = 2, \qquad F_{A0} = 2$$

which we want to process down to $C_{A,out} = 0.25$. For stream 1 alone in a plug flow reactor the volume needed is V_1. Now that we have two streams how should we contact them so as to minimize the total volume and still maintain $C_{A,out} = 0.25$. What is this minimum volume?

16. (*a*) Referring to Fig. P16, show that the optimum size ratio of three mixed reactors in series is achieved when X_1 and X_2 are located such that

$$\left.\begin{aligned}(\text{tangent at } P) &= (\text{slope } LM)\\(\text{tangent at } Q) &= (\text{slope } MN)\end{aligned}\right\}$$

(*b*) Present a procedure to evaluate X_1 and X_2 for a fixed final conversion X_3.
(*c*) Briefly indicate how to extend this procedure to N reactors in series.
Note: This solution only represents the optimum where the rate increases monotonically with conversion.

17. A first-order reaction is to be treated in a series of two mixed reactors. Show that the total volume of the two reactors is minimum when the reactors are equal in size.

18. Repeat Example 6.2 with two modifications: first, we plan to use a plug flow reactor and its cost per unit volume is identical to that for the mixed reactor; second, the rate of reaction is given by

$$-r_A = (2 \text{ liter/mol}\cdot\text{hr})C_A{}^2$$

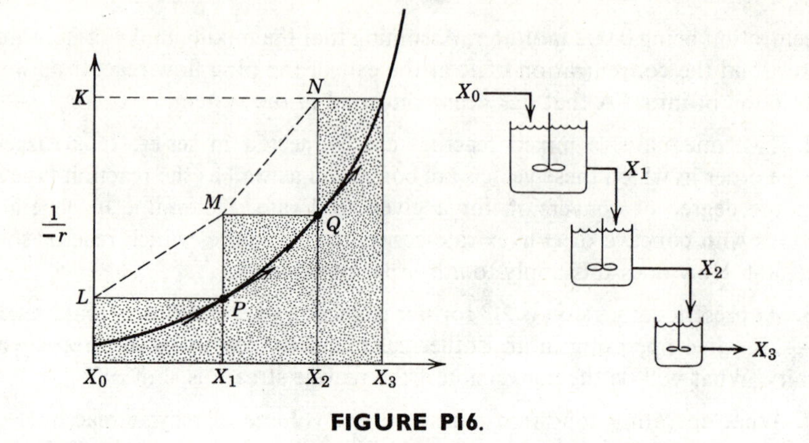

FIGURE P16.

19. At room temperature the aqueous second-order reaction proceeds as follows:

$$A \rightarrow R, \qquad -r_A = (1\ \text{liter/mol}\cdot\text{hr})C_A^2, \qquad C_{A0} = 1\ \text{mol/liter}$$

We plan to make this product batch after batch day and night in a vat. How should we operate the unit:

(*a*) For maximum production rate of R.
(*b*) For maximum rate of profit, if unreacted A is discarded.
(*c*) For maximum rate of profit, if unreacted A is reclaimed and reused at no cost.
(*d*) For maximum rate of profit, if unreacted A is reclaimed at a cost of 50% of fresh A.

What is the daily profit for these four cases?

Data: Shutdown time between batches is 1 hr. Cost of reactants is \$100/batch, and the value of the product fluid is given by \$200 X_A/batch.

20. One hundred moles of A per hour are available in a concentration of 0.1 mol/liter by a previous process. This stream is to be reacted with B to produce R and S. The reaction proceeds by the aqueous-phase elementary reaction,

$$A + B \xrightarrow{k} R + S, \qquad k = 5\ \text{liters/mol}\cdot\text{hr}$$

The amount of R required is 95 mol/hr. In extracting R from the reacted mixture A and B are destroyed, hence recycle of unused reactants is out of the question. Calculate the optimum reactor size and type as well as feed composition for this process.

Data: B costs \$1.25/mol in crystalline form. It is highly soluble in the aqueous solution and even when present in large amounts does not change the concentration of A in solution. Capital and operating costs are \$0.015/hr·liter for plug flow reactors, \$0.004/hr·liter for mixed flow reactors.

21. A commercial installation produces 40 kmol R/hr by hydrolysis in a mixed reactor of a feed stream containing 1 kmol/m^3 of reactant A. Because of the large excess of water used the reaction may be considered to be first order, or $A \rightarrow R$ even though it is bimolecular. The effluent stream from the reactor goes to a countercurrent extraction column in which R is quantitatively extracted. Two percent of the incoming A passes through the system unreacted. Fixed and operating costs for this process are $20/hr, reactant cost is $1.00/kmol, and R can be sold at $1.32/kmol. It is suspected that the plant is not being operated at optimum conditions. Therefore you have been asked to study the operations with the aim of optimizing them.

(*a*) What are the present profits on an hourly basis?

(*b*) How should the installation be operated (feed rate of A, conversion of A, and production rate of R) to maximize the profits? What are these profits on an hourly basis?

Note: All R produced may be sold. Separation equipment is flexible since it has been designed to adapt to large changes in capacity.

22. We are considering getting something of value from a waste stream of a process. This stream (20,000 liters/day) contains chemical A (0.01 kg/liter) which can be hydrolyzed in aqueous solution to give chemical R (value = $1.00/kg transformed). Product R can be recovered from the solution at negligible cost while unreacted A goes to waste. From the information below calculate the size of mixed reactor and conversion level which will maximize

(*a*) the profits,

(*b*) the rate of return on investment.

Data: On an annual basis the cost of reactor and supporting equipment, including depreciation and interest charges is

$$\$_m = \$225V^{1/2}, \quad \text{volume in liters}$$

Labor and operating costs

$$= \$20/\text{operating day.}$$

300 operating days/yr. The hydrolysis reaction is first order with respect to A with a rate constant $k = 0.25$/hr.

23. From a feed $C_{A0} = 1$ mol/liter we plan to produce product R in a vat. This is a liquid-phase reaction which at room temperature proceeds as follows:

$$A \rightarrow R, \qquad -r_A = (1\ \text{hr}^{-1})C_A$$

(*a*) If we run batch after batch day and night, what conversion and reaction time will give the maximum production rate of R. What is the profit (net income/day) under these conditions?

(*b*) What should be the operating conditions for maximum profit per batch, and what are these profits on a daily basis.

(*c*) How should we operate the batch reactor for maximum rate of profit, and what is this daily profit?

(*d*) We could use this vat as a mixed flow reactor. Would this be more profitable than running the unit as a batch reactor (give the optimum operating conditions and daily profit)?

Data: In batch operations the shutdown time to empty, clean, and refill is 1 hr. There is no shutdown time in operating the unit as a mixed flow reactor. Cost of reactant fluid is \$100/batch. Value of product fluid is dependent on the conversion level and is given by \$200 X_A/batch. Operating cost: for batch \$5/hr, for mixed flow \$5/day.

24. The reaction by which R is formed is $A \rightarrow R$ with rate equation

$$-r_A = (0.01 \text{ liter/mol}\cdot\text{hr})C_A C_R$$

From a feed of pure A (100 mol/liter, \$0.10/mol) 1000 mol R/hr are to be produced using either a mixed reactor alone or a mixed reactor followed by an A-R separator, in which case unreacted A may be recycled and reused.

The separator operates by an extraction process which, because of a favorable phase equilibrium, yields streams of essentially pure A and pure R. Its cost is \$8/hr + \$0.01/liter of fluid treated. The hourly cost of the reactor is \$8 + \$0.01/liter of reactor. Consider the density of all A-R mixtures to be constant. What system, reactor alone, or with separator, is most economical, and what is the unit cost of R produced by this setup?

25. Present facilities for the production of R cannot keep up with the demand for the material; hence you are asked to make an exploratory study to see whether production can be increased.

Product R is formed by the elementary irreversible reaction of A with B in a mixed flow reactor. Because a large excess of B is used the reaction may be considered to be first order with respect to A with rate constant k. R is quantitatively separated from the reactor effluent stream which is then discarded. The separation equipment is rather flexible and can easily handle greatly different flow rates.

(*a*) In general if feed A at $\$\alpha$/mol enters a mixed reactor of volume V liters at a concentration C_{A0} mol/liter and rate F_{A0} mols/hr, find the conversion of A at which unit cost of product R is a minimum. Let fixed and operating costs be $\$y$/hr.

(*b*) Under present operating conditions, what is the unit cost of producing R?

(*c*) Have you any suggestions for how the plant should be operated (conversion of reactant and production rate of R) so as to maximize production but still maintain the present unit cost of product?

(*d*) What is the minimum unit cost of producing R, and at what conversion of reactant and at what production rate will this occur?

Data: A is supplied at \$4/mol at a concentration of 0.1 mol/liter. The cost of B is negligible. Fixed and operating charges for the reactor and separation system are \$20/hr. At present the production rate $F_R = 25$ mol R/hr at conversion $X_A = 0.95$.

7

DESIGN FOR MULTIPLE REACTIONS

The preceding chapter on single reactions showed that the performance (size) of a reactor was influenced by the pattern of flow within the vessel. We now extend the discussion to multiple reactions and show that for these reactions both the size requirement and the distribution of reaction products are affected by the pattern of flow within the vessel. We may recall at this point that the distinction between a *single* reaction and *multiple* reactions is that the single reaction requires only one rate expression to describe its kinetic behavior whereas multiple reactions require more than one rate expression.

Since multiple reactions are so varied in type and seem to have so little in common, we may despair of finding general guiding principles for design. Fortunately this is not so because all multiple reactions can be considered to be combinations of two primary types: *parallel* reactions and *series* reactions. Because these two primary reaction types are the building blocks or components for more involved reaction schemes which we may call *series-parallel* reactions, we study them first and discover their characteristics. Then we select a particular series-parallel scheme and show how its behavior is governed by its component parallel and series reactions.

Let us consider the general approach and nomenclature. First of all, we find it more convenient to deal with concentrations than conversions. Secondly, in examining product distribution the procedure is to eliminate the time variable by dividing one rate equation by another. We end up then with equations relating the rates of change of certain components with respect to other components of the systems. Such relationships are relatively easy to treat. Thus we use two distinct analyses, one for determination of reactor size and the other for the study of product distribution.

The two requirements, small reactor size and maximization of desired product, may run counter to each other in that a good design with respect to one requirement may be poor with respect to the other. In such a situation an economic analysis will yield the best compromise. In general, however, product distribution controls; consequently this chapter concerns primarily optimization with respect to product distribution, a factor which plays no role in single reactions.

Finally, we ignore expansion effects in this chapter; thus we take $\varepsilon = 0$ throughout. This means that we may use the terms mean residence time, reactor holding time, space time, and reciprocal space velocity interchangeably.

REACTIONS IN PARALLEL

Qualitative Discussion About Product Distribution. Consider the decomposition of A by either one of two paths:

$$A \xrightarrow{k_1} R \quad \text{(desired product)} \tag{1a}$$

$$A \xrightarrow{k_2} S \quad \text{(unwanted product)} \tag{1b}$$

with corresponding rate equations

$$r_R = \frac{dC_R}{dt} = k_1 C_A^{a_1} \tag{2a}$$

$$r_S = \frac{dC_S}{dt} = k_2 C_A^{a_2} \tag{2b}$$

Dividing Eq. 2*b* by Eq. 2*a* gives a measure of the relative rates of formation of R and S. Thus

$$\frac{r_S}{r_R} = \frac{dC_S}{dC_R} = \frac{k_2}{k_1} C_A^{a_2 - a_1} \tag{3}$$

and we wish this ratio to be as small as possible.

Now C_A is the only factor in this equation which we can adjust and control (k_1, k_2, a_1, and a_2 are all constant for a specific system at a given temperature) and we can keep C_A low throughout the reactor by any of the following means: by using a mixed reactor, maintaining high conversions, increasing inerts in the feed, or decreasing the pressure in gas-phase systems. On the other hand, we can keep C_A high by using a batch or plug flow reactor, maintaining low conversions, removing inerts from the feed, or increasing the pressure in gas-phase systems.

For the reactions of Eq. 1 let us see whether the concentration of A should be kept high or low.

If $a_1 > a_2$, or the desired reaction is of higher order than the unwanted reaction, $a_2 - a_1$ is negative and from Eq. 3 we see that a high reactant concentration is desirable since it decreases the S/R ratio. As a result a batch or plug flow reactor would favor formation of product R and would require a minimum reactor size.

If $a_1 < a_2$, or the desired reaction is of lower order than the unwanted reaction, we need a low reactant concentration to favor formation of R. But this would also require large reactor size. Unfortunately, the demand for a desirable product distribution works against that of small reactor size. In such a case optimum design is an economic compromise between these conflicting demands.

If $a_1 = a_2$, or the two reactions are of the same order, Eq. 3 becomes

$$\frac{r_S}{r_R} = \frac{dC_S}{dC_R} = \frac{k_2}{k_1} = \text{constant}$$

Hence product distribution is fixed by k_2/k_1 alone and is unaffected by type of reactor used. Thus the reactor volume requirement will govern the design.

Besides this method, we may control product distribution by varying k_2/k_1. This can be done in two ways:

1. By changing the temperature level of operation. If the activation energies of the two reactions are different, k_2/k_1 can be made to vary. Chapter 8 considers this problem.
2. By using a catalyst. One of the most important features of a catalyst is its selectivity in depressing or accelerating specific reactions. This may be a much more effective way of controlling product distribution than any of the methods discussed so far.

For other reactions in parallel the reasoning is analogous to that presented. For example, consider the reactions

$$A + B \xrightarrow{k_1} R, \text{ desired} \qquad r_R = \frac{dC_R}{dt} = k_1 C_A^{a_1} C_B^{b_1} \tag{4}$$

$$A + B \xrightarrow{k_2} S, \text{ unwanted} \qquad r_S = \frac{dC_S}{dt} = k_2 C_A^{a_2} C_B^{b_2} \tag{5}$$

Dividing Eq. 5 by Eq. 4 gives

$$\frac{r_S}{r_R} = \frac{dC_S}{dC_R} = \frac{k_2}{k_1} C_A^{a_2 - a_1} C_B^{b_2 - b_1} \tag{6}$$

and since R is the desired product, we want to minimize

$$\frac{dC_S}{dC_R} \qquad \text{or} \qquad \frac{k_2}{k_1} C_A^{a_2 - a_1} C_B^{b_2 - b_1}$$

Thus we must examine separately whether $a_2 - a_1$ and $b_2 - b_1$ are positive or negative. This will determine whether A or B is to be kept at low or high concentration.

We may summarize our qualitative findings as follows:

For reactions in parallel the concentration level of reactants is the key to proper control of product distribution. A high reactant concentration favors the reaction of higher order, a low concentration favors the reaction of lower order, while the concentration level has no effect on the product distribution for reactions of the same order.

From this discussion we see that certain combinations of high and low reactant concentrations are desirable depending on the kinetics of the competing reactions. These combinations can be obtained by controlling the concentration of feed materials, by having certain components in excess, and by using the correct contacting pattern of reacting fluids. Figures 1 and 2 illustrate methods of contacting two reacting fluids in continuous and noncontinuous operations that keep the concentrations of these components both high, both low, one high the other low. In general, the number of reacting fluids involved, the possibility of recycle, and the cost of possible alternative setups must all be considered before the most desirable contacting pattern can be arrived at.

In any case the use of the proper contacting pattern is the critical factor in obtaining a favorable distribution of products for multiple reactions.

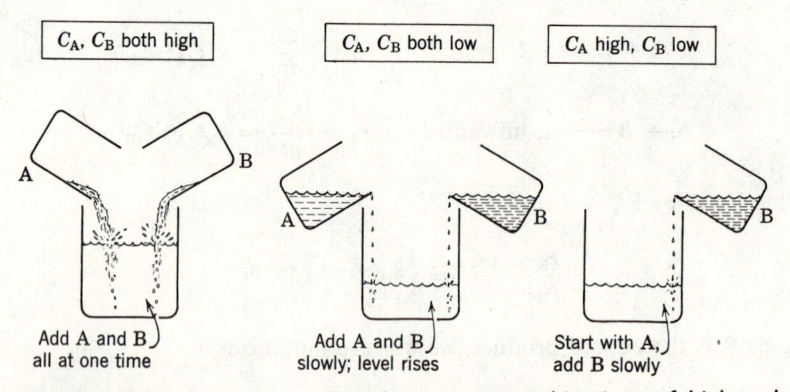

FIGURE 1. Contacting patterns for various combinations of high and low concentration of reactants in noncontinuous operations.

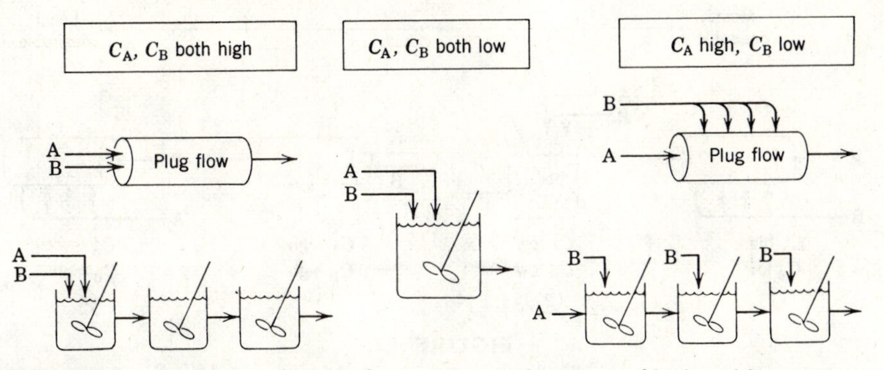

FIGURE 2. Contacting patterns for various combinations of high and low concentration of reactants in continuous flow operations.

EXAMPLE 1. Contacting patterns for reactions in parallel

The desired liquid-phase reaction

$$A + B \xrightarrow{k_1} R + T \qquad \frac{dC_R}{dt} = \frac{dC_T}{dt} = k_1 C_A C_B^{0.3} \tag{i}$$

is accompanied by the unwanted side reaction

$$A + B \xrightarrow{k_2} S + U \qquad \frac{dC_S}{dt} = \frac{dC_U}{dt} = k_2 C_A^{0.5} C_B^{1.8} \tag{ii}$$

From the standpoint of favorable product distribution, order the contacting schemes of Fig. 2, from the most desirable to the least desirable.

SOLUTION

Dividing Eq. (ii) by Eq. (i) gives the ratio

$$\frac{r_S}{r_R} = \frac{k_2}{k_1} C_A^{-0.5} C_B^{1.5}$$

which is to be kept as small as possible. According to the rule for reactions in parallel we want to keep C_A high, C_B low, and since the concentration dependency of B is more pronounced than of A, it is more important to have low C_B than high C_A. The contacting schemes are therefore ordered as shown in Fig. E1.

Comment. Example 2 verifies these qualitative findings. We should also note that there are still other contacting schemes which are superior to the best found in this example. For example, if we can use an excess of a reactant, or if it is practical to separate and recycle unconverted reactant, then vastly improved product distribution is possible.

Quantitative Treatment of Product Distribution and of Reactor Size. If rate equations are known for the individual reactions, we can quantitatively determine product distribution and reactor-size requirements. For convenience in evaluating

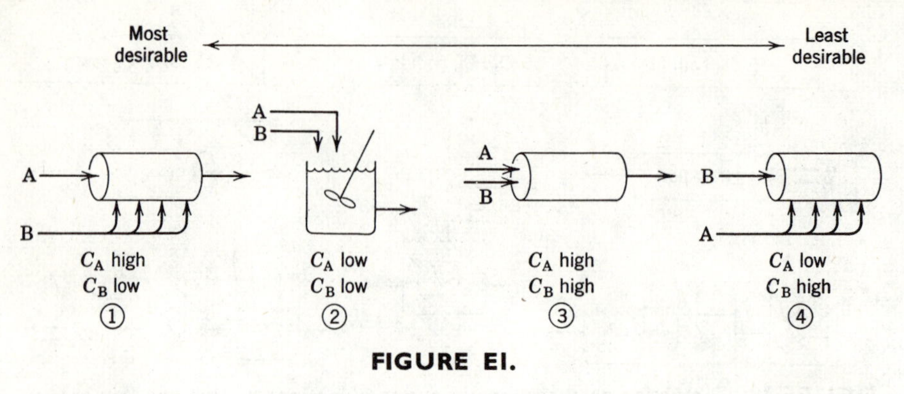

FIGURE E1.

product distribution we introduce two terms, φ and Φ. First, consider the decomposition of one reactant A, and let φ be the fraction of A disappearing at any instant which is transformed into desired product R. We call this the *instantaneous fractional yield of* R. Thus at any C_A

$$\varphi = \left(\frac{\text{moles R formed}}{\text{moles A reacted}}\right) = \frac{dC_R}{-dC_A} \tag{7}$$

For any particular set of reactions and rate equations φ is a specific function of C_A, and since C_A in general varies through the reactor φ will also change with position in the reactor. So let us define Φ as the fraction of all the reacted A which has been converted into R, and let us call this the *over-all fractional yield of* R. The over-all fractional yield is then the mean of the instantaneous fractional yields at all points within the reactor, thus we may write

$$\Phi = \left(\frac{\text{all R formed}}{\text{all A reacted}}\right) = \frac{C_{Rf}}{C_{A0} - C_{Af}} = \frac{C_{Rf}}{(-\Delta C_A)} = \bar{\varphi}_{\text{in reactor}} \tag{8}$$

It is the over-all fractional yield which really concerns us for it represents the product distribution at the reactor outlet. Now the proper averaging for φ depends on the type of flow within the reactor. Thus for *plug flow,* where C_A changes progressively through the reactor, we have

$$\Phi_p = \frac{-1}{C_{A0} - C_{Af}} \int_{C_{A0}}^{C_{Af}} \varphi \, dC_A = \frac{1}{\Delta C_A} \int_{C_{A0}}^{C_{Af}} \varphi \, dC_A \tag{9}$$

For *mixed flow* the composition is C_{Af} everywhere, so φ is likewise constant throughout the reactor, and we have

$$\Phi_m = \varphi_{\text{evaluated at } C_{Af}} \tag{10}$$

The fractional yields from mixed and plug flow reactors processing A from C_{A0} to C_{Af} are related by

$$\Phi_m = \left(\frac{d\Phi_p}{dC_A}\right)_{\text{at } C_{Af}} \quad \text{and} \quad \Phi_p = \frac{1}{\Delta C_A} \int_{C_{A0}}^{C_{Af}} \Phi_m \, dC_A \tag{11}$$

These expressions allow us to predict the yields from one type of reactor given the yields from the other.

From a series of $1, 2, \ldots, N$ mixed reactors in which the concentration of A is $C_{A1}, C_{A2}, \ldots, C_{AN}$ the over-all fractional yield is obtained by summing the fractional yields in each of the N vessels and weighting by the amount of reaction occurring in each vessel. Thus

$$\varphi_1(C_{A0} - C_{A1}) + \cdots + \varphi_N(C_{A,N-1} - C_{AN}) = \Phi_{N\,\text{mixed}}(C_{A0} - C_{AN})$$

from which

$$\Phi_{N\,\text{mixed}} = \frac{\varphi_1(C_{A0} - C_{A1}) + \varphi_2(C_{A1} - C_{A2}) + \cdots + \varphi_N(C_{A,N-1} - C_{AN})}{C_{A0} - C_{AN}} \tag{12}$$

For any reactor type the exit concentration of R is obtained directly from Eq. 8. Thus

$$C_{Rf} = \Phi(C_{A0} - C_{Af}) \tag{13}$$

and Fig. 3 shows how C_R is found for different types of reactors. For mixed reactors, or mixed reactors in series, the best outlet concentration to use, that which maximizes C_R, may have to be found by maximization of rectangles (see Chapter 6).

Now the shape of the φ versus C_A curve determines which type of flow gives the best product distribution, and Fig. 4 shows typical shapes of these curves for which plug flow, mixed flow, and mixed followed by plug flow are best.

These fractional yield expressions allow us to relate the product distribution from different types of reactors, and to search for the best contacting scheme. However, one condition must be satisfied before we can safely use these relationships: we must truly have parallel reactions in which no product influences the rate

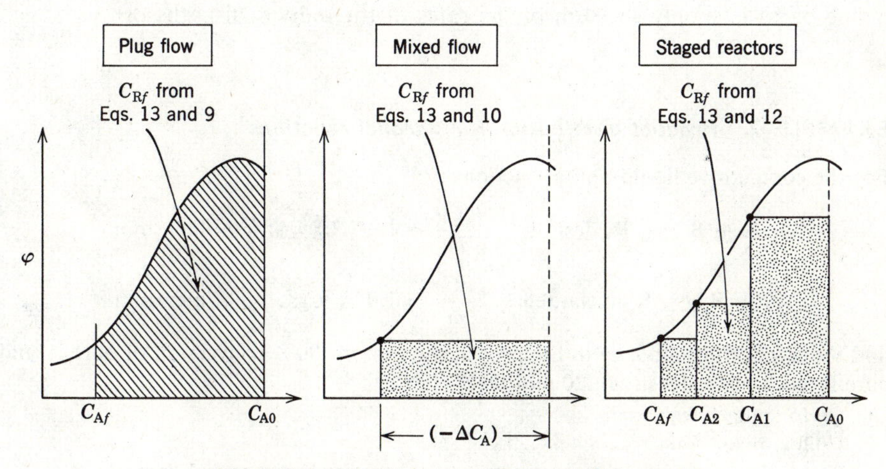

FIGURE 3. Shaded area gives total R formed.

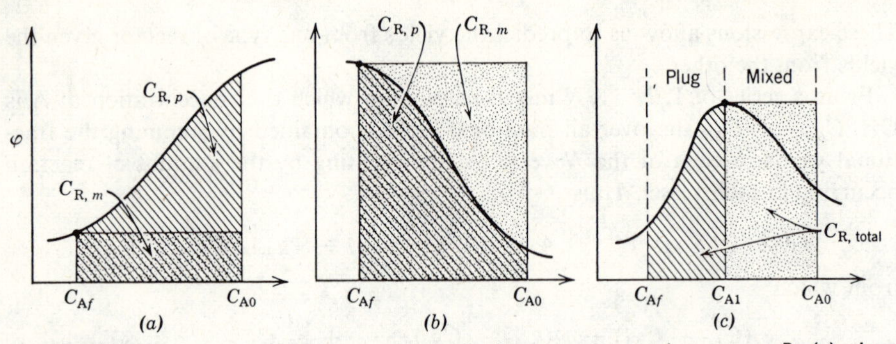

FIGURE 4. The contacting pattern with the largest area produces most R: (*a*) plug flow is best, (*b*) mixed flow is best, (*c*) mixed flow up to C_{A1} followed by plug flow is best.

to change the product distribution. The easiest way to test this is to add products to the feed and verify that the product distribution is in no way altered.

So far, the fractional yield of R has been taken as a function of C_A alone, and has been defined on the basis of the amount of this component consumed. More generally, when there are two or more reactants involved, the fractional yield can be based on one of the reactants consumed, on all reactants consumed, or on products formed. It is simply a matter of convenience which definition is used. Thus, in general, we define $\varphi(M/N)$ as the instantaneous fractional yield of M, based on the disappearance or formation of N.

The use of fractional yields to determine the product distribution for parallel reactions was developed by Denbigh (1944, 1961).

The determination of reactor volume is no different from that for a single reaction if we recall that the over-all rate of disappearance of reactant by a number of competing paths is simply the sum of the rates of the individual paths or

$$r = r_1 + r_2 + \cdots$$

EXAMPLE 2. Product distribution for parallel reactions

For the competitive liquid-phase reactions

$$A + B \xrightarrow{k_1} R, \text{ desired} \qquad \frac{dC_R}{dt} = 1.0C_A C_B^{0.3}, \quad \text{mol/liter}\cdot\text{min}$$

$$A + B \xrightarrow{k_2} S, \text{ unwanted} \qquad \frac{dC_S}{dt} = 1.0C_A^{0.5} C_B^{1.8}, \quad \text{mol/liter}\cdot\text{min}$$

find the fraction of impurity in the product stream for 90% conversion of pure A and pure B (each has a density of 20 mol/liter)

(*a*) for plug flow,
(*b*) for mixed flow,
(*c*) for the better of the two plug-mixed contacting schemes of Example 7.1.

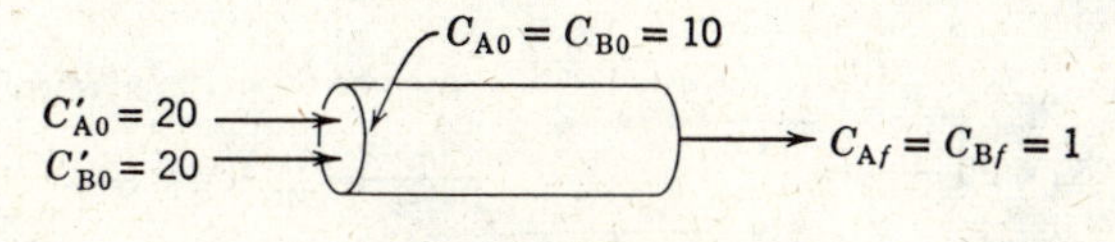

FIGURE E2a

SOLUTION

For this reaction the instantaneous fractional yield of R is given by

$$\varphi = \frac{dC_R}{dC_R + dC_S} = \frac{k_1 C_A C_B^{0.3}}{k_1 C_A C_B^{0.3} + k_2 C_A^{0.5} C_B^{1.8}} = \frac{1}{1 + C_A^{-0.5} C_B^{1.5}}$$

(*a*) *Plug flow.* Referring to the accompanying sketch, Fig. E2*a*, noting that the starting concentration of each reactant in the combined entering feed is $C_{A0} = C_{B0} = 10$ mol/liter and that $C_A = C_B$ everywhere, we find from Eq. 9 that

$$\Phi_p = \frac{-1}{C_{A0} - C_{Af}} \int_{C_{A0}}^{C_{Af}} \varphi \, dC_A = \frac{-1}{10 - 1} \int_{10}^{1} \frac{dC_A}{1 + C_A} = \frac{1}{9} \ln (1 + C_A) \Big|_1^{10} = 0.19$$

Therefore impurities in the R − S product = 81%.

(*b*) *Mixed flow.* Referring to the accompanying sketch, Fig. E2*b*, we have from Eq. 10

$$\Phi_m = \varphi_{\text{at exit conditions}} = \frac{1}{1 + C_{Af}} = \frac{1}{1 + 1} = 0.5$$

Therefore impurities in the R − S product = 50%.

(*c*) *Plug (A)—mixed (B) flow.* Assuming that B is introduced into the reactor in such a way that $C_B = 1$ mol/liter throughout the reactor, we find concentrations as shown in the accompanying sketch, Fig. E2*c*. Then accounting for the variation of C_A through the reactor we find

$$\Phi = \frac{-1}{C_{A0} - C_{Af}} \int_{C_{A0}}^{C_{Af}} \varphi \, dC_A = \frac{-1}{19 - 1} \int_{19}^{1} \frac{dC_A}{1 + C_A^{-0.5}(1)^{1.5}} = \frac{1}{18} \int_{1}^{19} \frac{dC_A}{1 + C_A^{-0.5}}$$

$$= \frac{1}{18} \left[(19 - 1) - 2(\sqrt{19} - 1) + 2 \ln \frac{1 + \sqrt{19}}{2} \right] = 0.741$$

Therefore impurities in the R − S product = 25.9%.

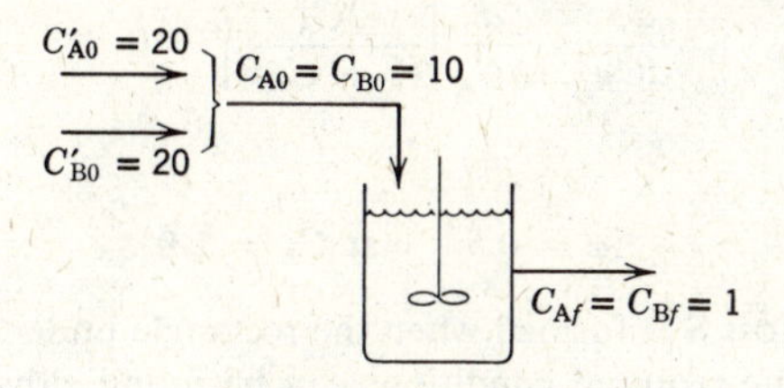

FIGURE E2b.

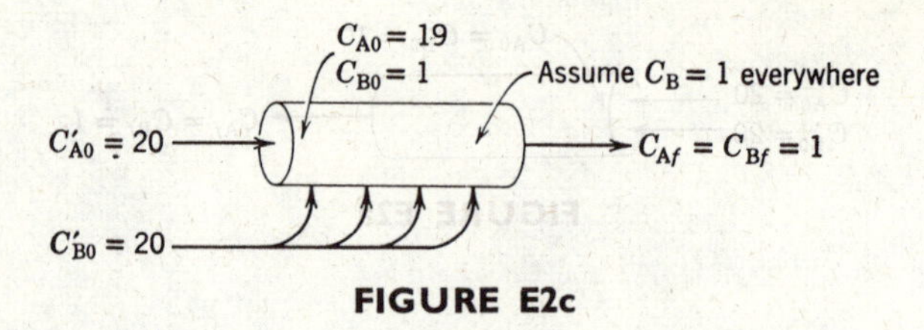

FIGURE E2c

Note. These results verify the qualitative findings of Example 7.1. Also in case (*c*) when C_B varies along the reactor length the whole analysis becomes considerably more complicated. For this situation see Kramers and Westerterp (1963).

EXAMPLE 3. Best operating conditions for parallel reactions

Often a desired reaction is accompanied by a variety of undesired side reactions, some of higher order, some of lower order. To see which type of operation gives the best product distribution consider the simplest typical case, the parallel decompositions of A, $C_{A0} = 2$,

$$A \rightarrow R \qquad r_R = 1$$
$$A \rightarrow S \qquad r_S = 2C_A$$
$$A \rightarrow T \qquad r_T = C_A^2$$

Find the maximum expected C_S for isothermal operations

(*a*) in a mixed reactor.
(*b*) in a plug flow reactor.
(*c*) in a reactor of your choice if unreacted A can be separated from the product stream and returned to the feed at $C_{A0} = 2$.

SOLUTION

Since S is the desired product let us write fractional yields in terms of S. Thus

$$\varphi(S/A) = \frac{dC_S}{dC_R + dC_S + dC_T} = \frac{2C_A}{1 + 2C_A + C_A^2} = \frac{2C_A}{(1 + C_A)^2}$$

Plotting this function we find the curve of Fig. E3 whose maximum occurs where

$$\frac{d\varphi}{dC_A} = \frac{d}{dC_A}\left[\frac{2C_A}{(1 + C_A)^2}\right] = 0$$

Solving we find

$$\varphi = 0.5 \qquad \text{at } C_A = 1.0$$

(*a*) *Mixed reactor.* Most S is formed when the rectangle under the φ versus C_A curve has the largest area. The required conditions can be found either by maximization of rectangles or analytically. Since simple explicit expressions are available for the various

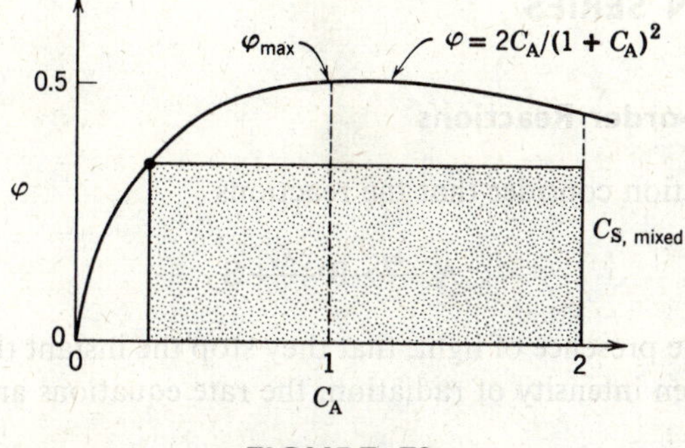

FIGURE E3.

quantities let us use the latter approach. Then from Eqs. 10 and 13 we find for the area of the rectangle

$$C_{Sf} = \Phi(S/A)\cdot(-\Delta C_A) = \varphi(S/A)\cdot(-\Delta C_A) = \frac{2C_A}{(1+C_A)^2}(C_{A0} - C_A)$$

Differentiating and setting to zero to find the conditions at which most S is formed gives

$$\frac{dC_{Sf}}{dC_A} = \frac{d}{dC_A}\left[\frac{2C_A}{(1+C_A)^2}(2 - C_A)\right] = 0$$

Evaluating this quantity gives the optimum operating conditions of a mixed reactor as

$$C_{Sf} = \tfrac{2}{3} \qquad \text{at } C_{Af} = \tfrac{1}{2}$$

(*b*) *Plug flow reactor.* The production of S is maximum when the area under the φ versus C_A curve is maximum. This occurs at 100% conversion of A. Thus from Eqs. 9 and 13

$$C_{Sf} = -\int_{C_{A0}}^{C_{Af}} \varphi(S/A)\, dC_A = \int_0^2 \frac{2C_A}{(1+C_A)^2}\, dC_A$$

Evaluating this integral gives, for the plug flow optimum,

$$C_{Sf} = 0.867 \qquad \text{at } C_{Af} = 0$$

(*c*) *Any reactor with separation and recycle of unused reactant.* Since no reactant leaves the system unconverted what is important is to operate at conditions where the fractional yield is at its highest value. This is at $C_A = 1$, where $\varphi(S/A) = 0.5$. Thus we should use a mixed reactor operating at $C_A = 1$. We would then have 50% of reactant A fed to the system forming product S.

Comment. Summarizing, we find

$$\left(\frac{\text{moles S formed}}{\text{moles A fed}}\right) \begin{aligned} &= 0.33 \text{ for part } (a) \\ &= 0.43 \text{ for part } (b) \\ &= 0.50 \text{ for part } (c) \end{aligned}$$

Thus a mixed reactor operating at conditions of highest φ with separation and recycle of unused reactant gives the highest product distribution. This result is quite general for parallel reactions of different order.

REACTIONS IN SERIES

Successive First-order Reactions

For easy visualization consider that the reactions

$$A \xrightarrow{k_1} R \xrightarrow{k_2} S \tag{14}$$

proceed only in the presence of light, that they stop the instant the light is shut off, and that for a given intensity of radiation, the rate equations are

$$r_A = -k_1 C_A \tag{15}$$

$$r_R = k_1 C_A - k_2 C_R \tag{16}$$

$$r_S = k_2 C_R \tag{17}$$

Our discussion will center about these reactions.

Qualitative Discussion About Product Distribution. Consider the following two ways of treating a beaker containing A: first, the contents are irradiated all at one time; second, a small stream is continuously withdrawn from the beaker, irradiated, and returned to the beaker; the rate of absorption of radiant energy is the same in the two cases. The two schemes are shown in Figs. 5 and 6. During this process A disappears and products are formed. Is the product distribution of R and S different in the two beakers? Let us see whether we can answer this question qualitatively for all values of the rate constants.

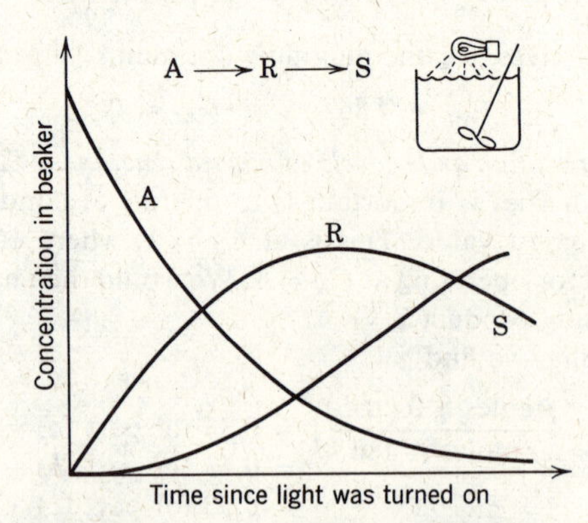

FIGURE 5. Concentration-time curves if the contents of the beaker are irradiated uniformly.

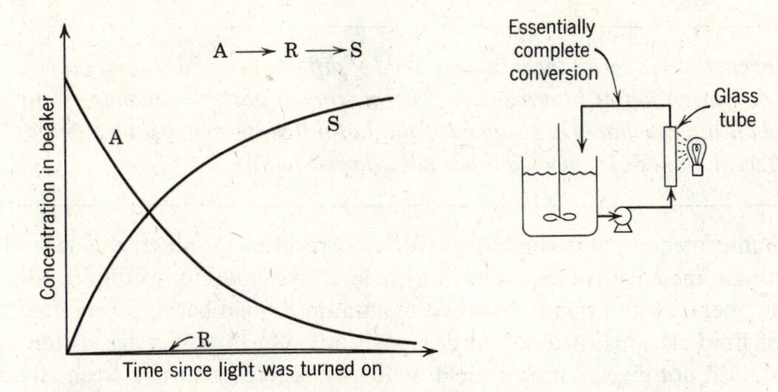

FIGURE 6. Concentration-time curves for the contents of the beaker if only a part of the fluid is irradiated at any instant.

In the first beaker, when the contents are being irradiated all at the same time, the first bit of light will attack A alone because only A is present at the start. The result is that R is formed. With the next bit of light both A and R will compete; however, A is in very large excess so it will preferentially absorb the radiant energy to decompose and form more R. Thus the concentration of R will rise while the concentration of A will fall. This process will continue until R is present in high enough concentration so that it can compete favorably with A for the radiant energy. When this happens, a maximum R concentration is reached. After this the decomposition of R becomes more rapid than its rate of formation, and its concentration drops. A typical concentration time curve is shown in Fig. 5.

In the alternate way of treating A a small fraction of the beaker's contents is continuously removed, irradiated, and returned to the beaker. Since the total absorption rate is the same in the two cases, however, the intensity of radiation received by the fluid removed is greater, and it could well be, if the flow rate is not too high, that the fluid being irradiated reacts essentially to completion. In this case, then, A is removed and S is returned to the beaker. So as time proceeds the concentration of A slowly decreases in the beaker, S rises, while R is absent. This progressive change is shown in Fig. 6.

These two methods of reacting the contents of the beaker yield different product distributions and represent the two extremes in possible operations, one with a maximum possible formation of R and the other with a minimum, or no formation, of R. How can we best characterize this behavior? We note in the first method that the contents of the beaker remain homogeneous throughout, all changing slowly with time, whereas in the second a stream of reacted fluid is continually being mixed with fresh fluid. In other words, we are mixing two streams of different compositions. This discussion suggests the following rule governing product distribution for reactions in series.

For reactions in series the mixing of fluid of different composition is the key to the formation of intermediate. The maximum possible amount of any and all intermediates is obtained if fluid of different compositions and at different stages of conversion are not allowed to mix.

As the intermediate is frequently the desired reaction product, this rule allows us to evaluate the effectiveness of various reactor systems. For example, plug flow and batch operations should both give a maximum R yield because here there is no mixing of fluid streams of different compositions. On the other hand, the mixed reactor should not give as high a yield of R as possible because a fresh stream of pure A is being mixed continually with an already reacted fluid in the reactor.

The following examples illustrate the point just made. We then give a quantitative treatment which will verify these qualitative findings.

EXAMPLE 4. *Favorable contacting for reactions in series*

From each of the pair of reactor setups in Fig. E4 select the method of operation which will produce a larger amount of R and briefly explain why this should be so. The reaction is

$$A \xrightarrow{\text{light}} R \xrightarrow{\text{light}} S$$

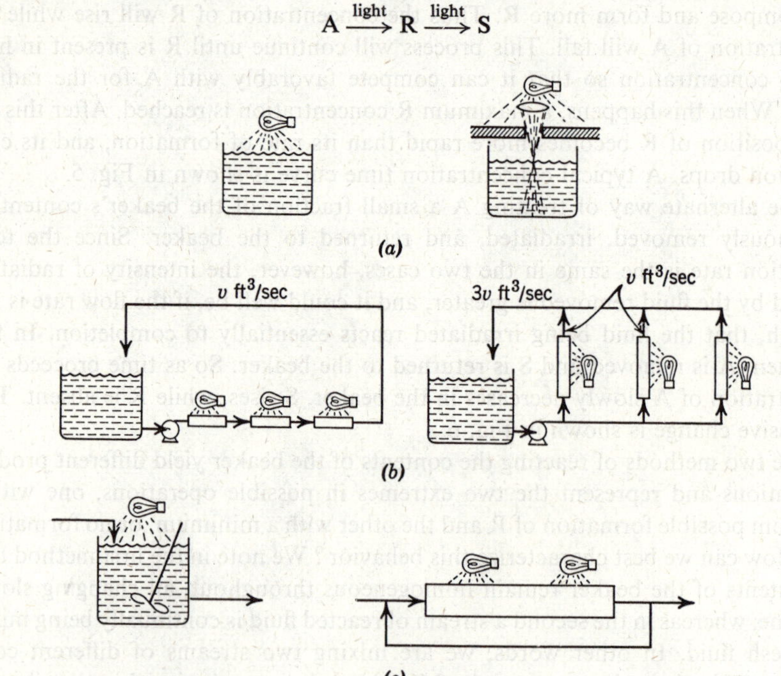

FIGURE E4.

SOLUTION

(*a*) For a relatively slow reaction either setup is satisfactory, yielding a maximum R concentration, but for a fast reaction and with insufficient fluid mixing the concentration of products in the beam of light of the second setup may rise above that in the surrounding fluid. The resulting heterogeneity results in a lowering in R yield. Thus for fast reactions the first setup is preferable.

(*b*) For reactions which are slow enough so that the conversion per pass is very small, either setup should be satisfactory; however, for a fast reaction the second setup is preferred because the conversion of reactant per pass is approximately one-third that for the first setup. Thus streams with smaller concentration differences are being mixed.

(*c*) To examine this situation first consider various extents of recycle in the second setup. With no recycle we have a plug flow system with no mixing of streams of different compositions, hence a maximum R yield. As the recycle rate is increased the second setup approaches the mixed reactor with a resultant decrease in R yield, hence the second setup with no recycle is preferred.

Quantitative Treatment, Plug Flow or Batch Reactor. In Chapter 3 we developed the equations relating concentration with time for all components of the unimolecular-type reactions

$$A \xrightarrow{k_1} R \xrightarrow{k_2} S$$

in batch reactors. The derivations assumed that the feed contained no reaction products R or S. If we replace reaction time by the space time, these equations apply equally well for plug flow reactors, thus

$$\frac{C_A}{C_{A0}} = e^{-k_1\tau} \tag{3.47}$$

$$\frac{C_R}{C_{A0}} = \frac{k_1}{k_2 - k_1}(e^{-k_1\tau} - e^{-k_2\tau}) \tag{3.49}$$

$$C_S = C_{A0} - C_A - C_R$$

The maximum concentration of intermediate and time at which it occurs is given by

---

$$\frac{C_{R,\max}}{C_{A0}} = \left(\frac{k_1}{k_2}\right)^{k_2/(k_2-k_1)} \tag{3.52}$$

$$\tau_{p,\text{opt}} = \frac{1}{k_{\text{log mean}}} = \frac{\ln(k_2/k_1)}{k_2 - k_1} \tag{3.51}$$

This is also the point at which the rate of formation of S is most rapid.

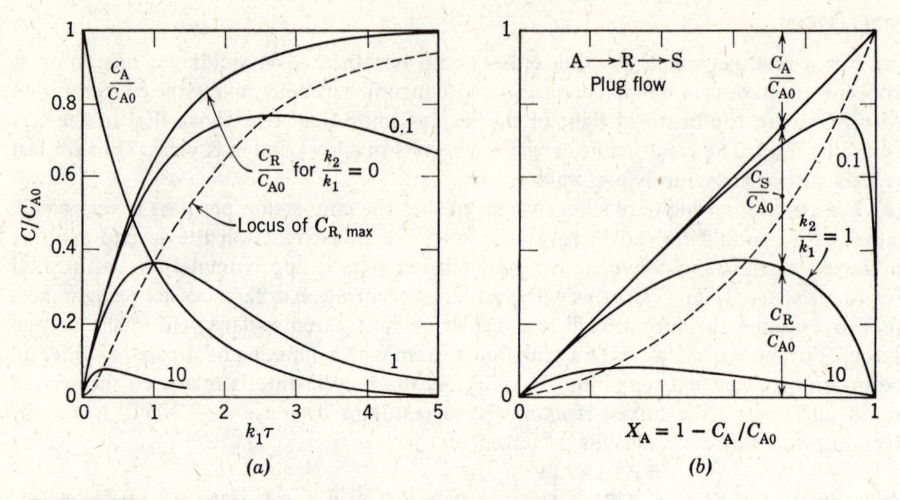

FIGURE 7. Behavior of unimolecular-type reactions

$$A \xrightarrow{k_1} R \xrightarrow{k_2} S$$

in a plug flow reactor: (*a*) concentration-time curves, and (*b*) relative concentration of the reaction components.

Figure 7*a*, prepared for various k_2/k_1 values, illustrates how this ratio governs the concentration-time curves of the intermediate R. Figure 7*b*, a time-independent plot, relates the concentration of all reaction components.

Quantitative Treatment, Mixed Flow Reactor. Let us develop the concentration-time curves for this reaction when it takes place in a mixed reactor. This may be done by referring to Fig. 8. Again the derivation will be limited to a feed which contains no reaction product R or S.

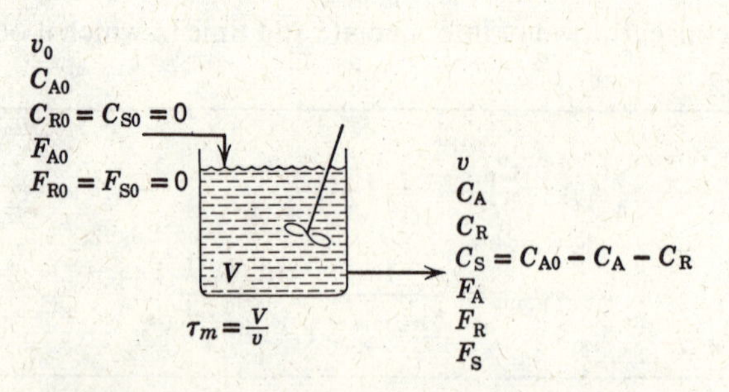

FIGURE 8. Variables for reactions in series (no R or S in the feed) occurring in a mixed reactor.

By the steady-state material balance we obtain for any component

$$\text{input} = \text{output} + \text{disappearance by reaction} \tag{4.1}$$

which for reactant A becomes

$$F_{A0} = F_A + (-r_A)V$$

or

$$vC_{A0} = vC_A + k_1 C_A V$$

Noting that

$$\frac{V}{v} = \tau_m = \bar{t} \tag{18}$$

we obtain, on rearranging,

$$\frac{C_A}{C_{A0}} = \frac{1}{1 + k_1 \tau_m} \tag{19}$$

For component R the material balance, Eq. 4.1, becomes

$$vC_{R0} = vC_R + (-r_R)V$$

or

$$0 = vC_R + (-k_1 C_A + k_2 C_R)V$$

With Eqs. 18 and 19 we obtain, on rearranging,

$$\frac{C_R}{C_{A0}} = \frac{k_1 \tau_m}{(1 + k_1 \tau_m)(1 + k_2 \tau_m)} \tag{20}$$

C_S is found by simply noting that at any time

$$C_A + C_R + C_S = C_{A0} = \text{constant}$$

hence

$$\frac{C_S}{C_{A0}} = \frac{k_1 k_2 \tau_m^2}{(1 + k_1 \tau_m)(1 + k_2 \tau_m)} \tag{21}$$

The location and maximum concentration of R are found by taking $dC_R/d\tau_m = 0$. Thus

$$\frac{dC_R}{d\tau_m} = 0 = \frac{C_{A0}k_1(1 + k_1\tau_m)(1 + k_2\tau_m) - C_{A0}k_1\tau_m[k_1(1 + k_2\tau_m) + (1 + k_1\tau_m)k_2]}{(1 + k_1\tau_m)^2(1 + k_2\tau_m)^2}$$

which simplifies neatly to give

$$\tau_{m,\text{opt}} = \frac{1}{\sqrt{k_1 k_2}} \tag{22}$$

The corresponding concentration of R is given by replacing Eq. 22 in Eq. 20. On rearranging this becomes

$$\frac{C_{R,\max}}{C_{A0}} = \frac{1}{[(k_2/k_1)^{1/2} + 1]^2} \tag{23}$$

Typical concentration-time curves for various k_2/k_1 values are shown in Fig. 9*a*. A time-independent plot, Fig. 9*b*, relates the concentrations of reactant and products.

Remarks on Performance Characteristics, Kinetic Studies, and Design. Figures 7*a* and 9*a* show the general time-concentration behavior for plug and mixed flow reactors and are an aid in visualizing the actual progress of the reaction. Comparison of these figures shows that except when $k_1 = k_2$ the plug flow reactor always requires a smaller time than does the mixed reactor to achieve the maximum concentration of R, the difference in times becoming progressively larger as k_2/k_1 departs from unity (see Eqs. 22 and 3.51). In addition, for any reaction the maximum obtainable concentration of R in a plug flow reactor is always higher than the maximum obtainable in a mixed reactor (see Eqs. 23 and 3.52). This verifies the conclusions arrived at by qualitative reasoning.

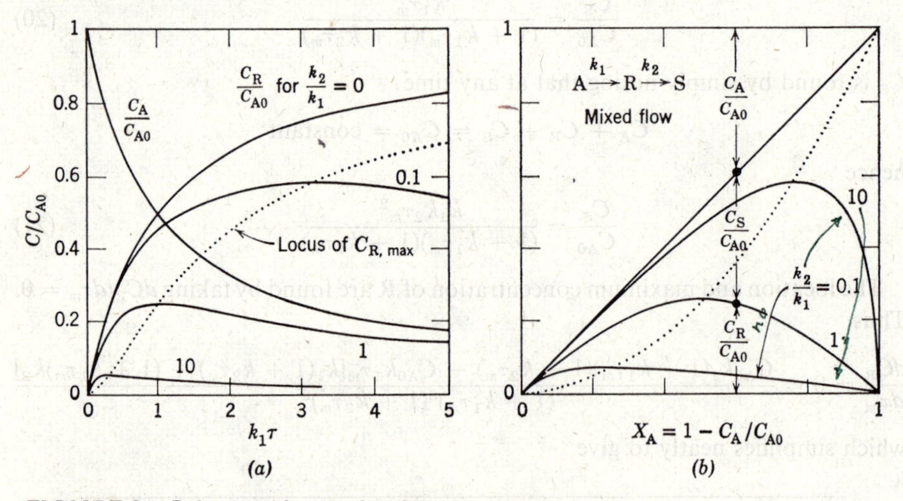

FIGURE 9. Behavior of unimolecular-type reactions

$$A \xrightarrow{k_1} R \xrightarrow{k_2} S$$

in a mixed reactor: (*a*) concentration-time curves, and (*b*) relative concentration of the reaction components.

Figures *7b* and *9b*, time-independent plots, show the distribution of materials during reaction. Such plots find most use in kinetic studies since they allow the determination of k_2/k_1 by matching the experimental points with one of the family of curves on the appropriate graph. Figures 15 and 16 are more detailed representations of these two figures.

Though not shown in the figures, C_S can be found by difference between C_{A0} and $C_A + C_R$. For small times, $\tau < \tau_{opt}$, the amount of S formed is small. For $\tau \gg \tau_{opt}$ the C_S curve approaches asymptotically that for a simple one-step reaction; thus the formation of S can be viewed as a one-step reaction with rate constant k_2 and having an induction period which becomes relatively unimportant as time increases.

Figure 10 presents the fractional yield curves for intermediate R as a function of the conversion level and the rate constant ratio. These curves clearly show that the fractional yield of R is always higher for plug flow than for mixed flow for any conversion level. A second important observation in this figure concerns the extent of conversion of A we should plan for. If for the reaction considered k_2/k_1 is much

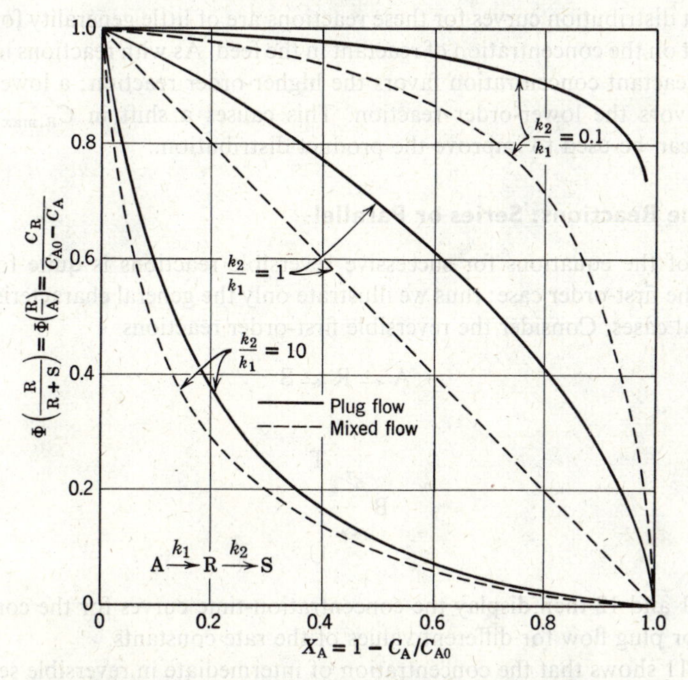

FIGURE 10. Comparison of the fractional yields of R in mixed and plug flow reactors for the unimolecular-type reactions

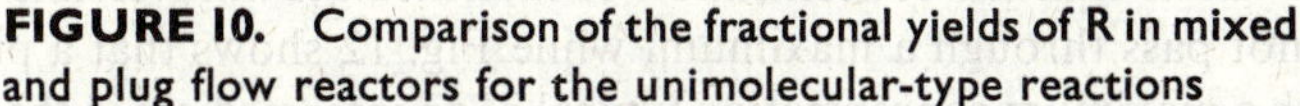

$$A \xrightarrow{k_1} R \xrightarrow{k_2} S$$

smaller than unity, we should design for a high conversion of A and probably dispense with recycle of unused reactant. However, if k_2/k_1 is greater than unity, the fractional yield drops very sharply even at low conversion. Hence to avoid obtaining unwanted S instead of R we must design for a very small conversion of A per pass, separation of R, and recycle of unused reactant. In such a case large quantities of material will have to be treated in the A-R separator and recycled, and this part of the process will figure prominently in cost considerations.

Successive Irreversible Reactions of Different Orders

In principle concentration-time curves can be constructed for successive reactions of different orders. For the plug flow or batch reactor this involves simultaneous solution of the governing differential equations, while for the mixed reactor we have only simultaneous algebraic equations to deal with. In both these cases explicit solutions are difficult to obtain; thus numerical methods provide the best tool for treating such reactions. In all cases these curves exhibit characteristics similar to successive first-order reactions; therefore we may generalize the conclusions for that reaction set to all irreversible reactions in series.

Product distribution curves for these reactions are of little generality for they are dependent on the concentration of reactant in the feed. As with reactions in parallel, a rise in reactant concentration favors the higher-order reaction; a lower concentration favors the lower-order reaction. This causes a shift in $C_{R,\max}$ and this property can be used to improve the product distribution.

Reversible Reactions: Series or Parallel

Solution of the equations for successive reversible reactions is quite formidable even for the first-order case; thus we illustrate only the general characteristics for a few typical cases. Consider the reversible first-order reactions

$$A \rightleftarrows R \rightleftarrows S \tag{24}$$

and

$$B \begin{array}{l} \nearrow\!\!\!\swarrow \; T \\ \searrow\!\!\!\nwarrow \; U \end{array} \tag{25}$$

Figures 11 and 12 then display the concentration-time curves for the components in batch or plug flow for different values of the rate constants.

Figure 11 shows that the concentration of intermediate in reversible series reactions need not pass through a maximum, while Fig. 12 shows that a product may pass through a maximum concentration typical of an intermediate in the irreversible series reactions; however, the reactions may be of a different kind. A comparison

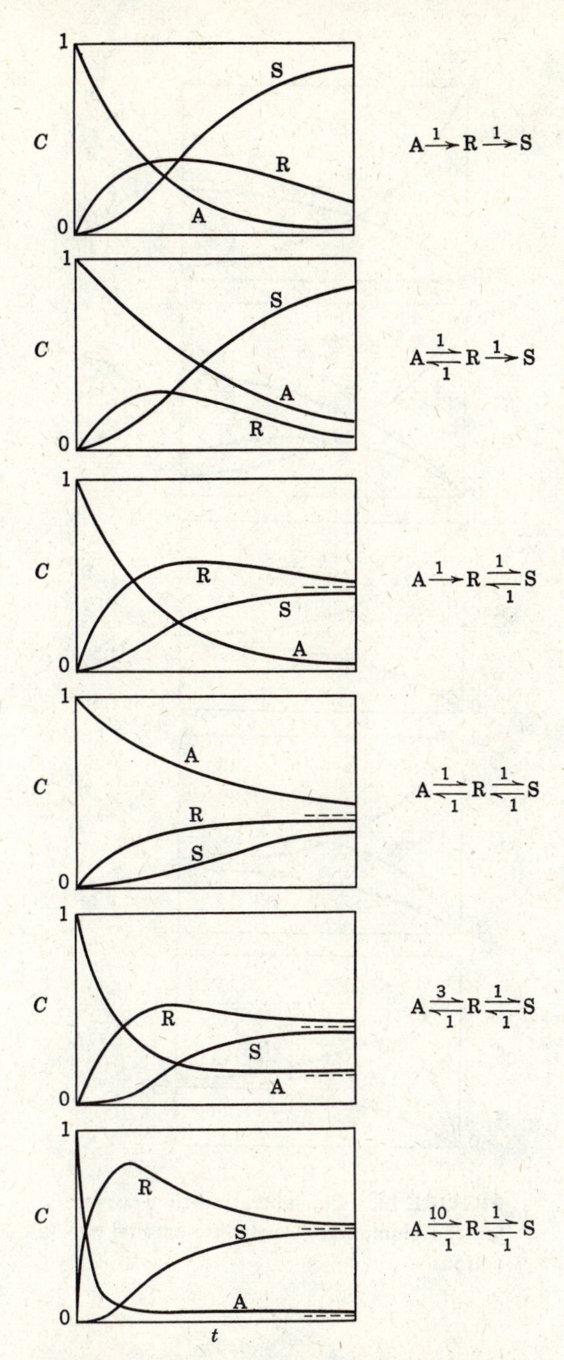

FIGURE 11. Concentration-time curves for the elementary reversible reactions

$$A \underset{k_2}{\overset{k_1}{\rightleftharpoons}} R \underset{k_4}{\overset{k_3}{\rightleftharpoons}} S$$

from Jungers *et al.* (1958), p. 207.

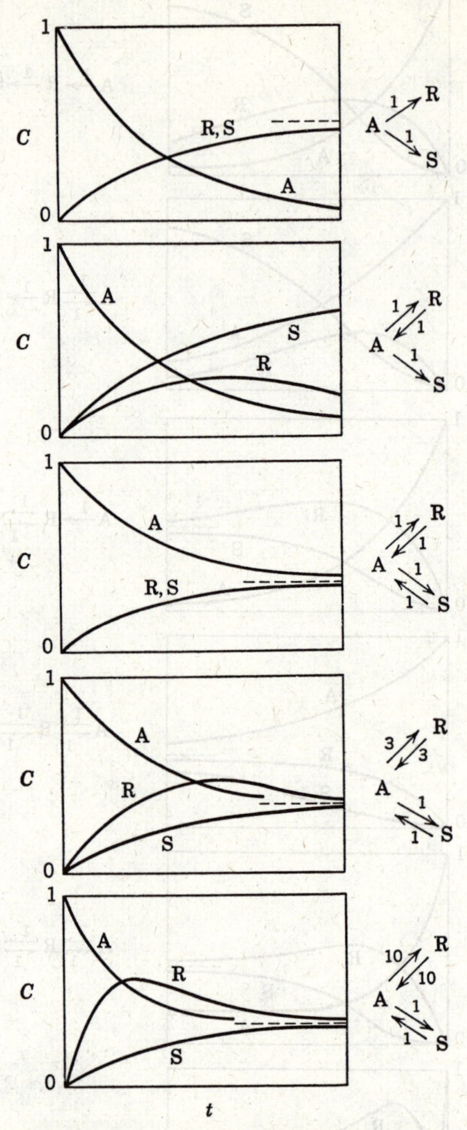

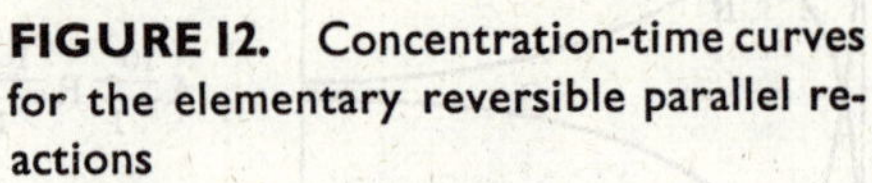

FIGURE 12. Concentration-time curves for the elementary reversible parallel reactions

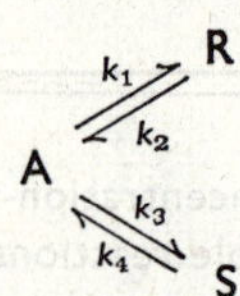

from Jungers *et al.* (1958), p. 207.

of these figures shows that many of the curves are similar in shape, making it difficult to select the correct mechanism of reaction by experiment, especially if the kinetic data are somewhat scattered. Probably the best clue in distinguishing between parallel and series reactions is to examine initial rate data—data obtained for very small conversion of reactant. For series reactions the time-concentration curve for S has a zero slope, whereas for parallel reactions this is not so.

SERIES-PARALLEL REACTIONS

Multiple reactions which consist of steps in series and steps in parallel are called series-parallel reactions. From the point of view of proper contacting, these reactions are more interesting than the simpler types already considered because a larger choice of contacting is usually possible, leading to much wider differences in product distribution. Thus the design engineer is dealing with a more flexible system and this affords him the opportunity to display his talents in devising the best of the wide variety of possible contacting patterns. Let us develop our ideas with a reaction type which represents a broad class of industrially important reactions. We will then generalize our findings to other series-parallel reactions.

For the reaction set consider the successive attack of a compound by a reactive material. The general representation of these reactions is

$$\left.\begin{aligned} A + B &\xrightarrow{k_1} R \\ R + B &\xrightarrow{k_2} S \\ S + B &\xrightarrow{k_3} T \\ &\text{etc.} \end{aligned}\right\} \qquad (26)$$

or

$$A \xrightarrow{+B,\, k_1} R \xrightarrow{+B,\, k_2} S \xrightarrow{+B,\, k_3} T$$

where A is the compound to be attacked, B is the reactive material, and R, S, T, etc., are the polysubstituted materials formed during reaction. Examples of such reactions may be found in the successive substitutive halogenation (or nitration) of hydrocarbons, say benzene or methane, to form monohalo, dihalo, trihalo, etc., derivatives as shown below:

$$C_6H_6 \xrightarrow{+Cl_2} C_6H_5Cl \xrightarrow{+Cl_2} \cdots \xrightarrow{+Cl_2} C_6Cl_6$$

$$C_6H_6 \xrightarrow{+HNO_3} C_6H_5NO_2 \xrightarrow{+HNO_3} \cdots \xrightarrow{+HNO_3} C_6H_3(NO_2)_3$$

$$CH_4 \xrightarrow{+Cl_2} CH_3Cl \xrightarrow{+Cl_2} \cdots \xrightarrow{+Cl_2} CCl_4$$

Another important example is the addition of alkene oxides, say ethylene oxide, to compounds of the proton donor class such as amines, alcohols, water, and hydrazine to form monoalkoxy, dialkoxy, trialkoxy, etc., derivatives, some examples of which are shown below:

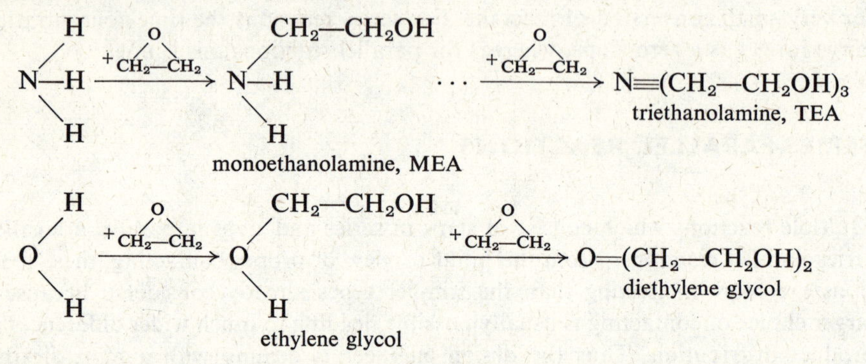

Such processes are frequently bimolecular, irreversible, hence second-order kinetically. When occurring in the liquid phase they are also essentially constant-density reactions.

We first consider the two-step reaction where the first substitution product is desired. Actually for an n-step reaction the third and succeeding reactions do not occur to any appreciable extent and may be ignored if the mole ratio of A to B is high (see qualitative treatment given next). The reaction set considered is thus

$$\left.\begin{aligned} A + B &\xrightarrow{k_1} R \\ R + B &\xrightarrow{k_2} S \end{aligned}\right\} \tag{27}$$

With the assumption that the reaction is irreversible, bimolecular, and of constant density, the rate expressions are given by

$$r_A = \frac{dC_A}{dt} = -k_1 C_A C_B \tag{28}$$

$$r_B = \frac{dC_B}{dt} = -k_1 C_A C_B - k_2 C_R C_B \tag{29}$$

$$r_R = \frac{dC_R}{dt} = k_1 C_A C_B - k_2 C_R C_B \tag{30}$$

$$r_S = \frac{dC_S}{dt} = k_2 C_R C_B \tag{31}$$

Qualitative Discussion About Product Distribution. To get the "feel" for what takes place when A and B react according to Eq. 27, imagine that we have two beakers, one containing A and the other containing B. Should it make any difference in the product distribution how we mix A and B? To find out, consider the following ways of mixing the reactants: (*a*) add A slowly to B, (*b*) add B slowly to A, and finally (*c*) mix A and B together rapidly.

For the first alternative pour A a little at a time into the beaker containing B, stirring thoroughly and making sure that all the A is used up and that the reaction stops before the next bit is added. With each addition a bit of R is produced in the beaker. But this R finds itself in an excess of B so it will react further to form S. The result is that at no time during the slow addition will A and R be present in any appreciable amount. The mixture becomes progressively richer in S and poorer in B. This continues until the beaker contains only S. Figure 13 shows this progressive change.

Now pour B a little at a time into the beaker containing A, again stirring thoroughly. The first bit of B will be used up, reacting with A to form R. This R cannot react further for there is now no B present in the mixture. With the next addition of B, both A and R will compete with each other for the B added, and since A is in very large excess it will react with most of the B, producing even more R. This process will be repeated with progressive buildup of R and depletion of A until the concentration of R is high enough so that it can compete favorably with A for the B added. When this happens, the concentration of R reaches a maximum, then drops. Finally, after addition of 2 moles of B for each mole of A, we end up with a solution containing only S. This progressive change is shown in Fig. 14.

Now consider the third alternative where the contents of the two beakers are rapidly mixed together, the reaction being slow enough so that it does not proceed

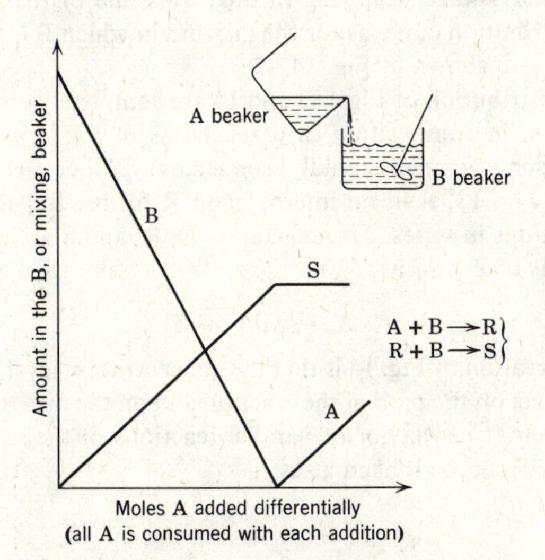

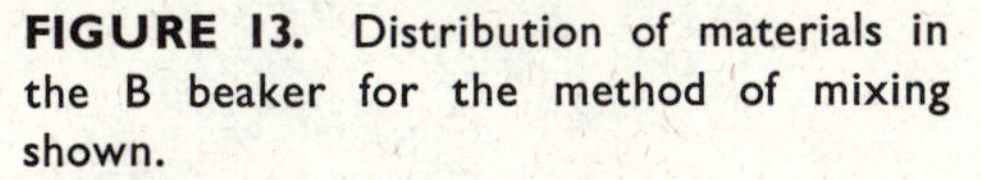

FIGURE 13. Distribution of materials in the B beaker for the method of mixing shown.

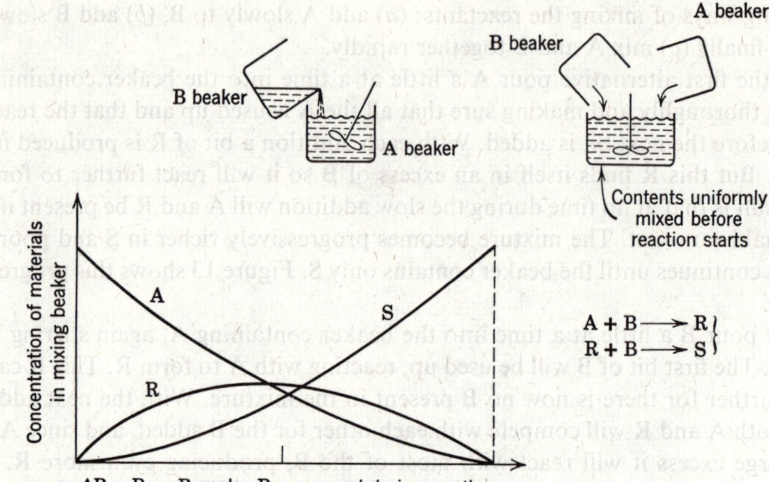

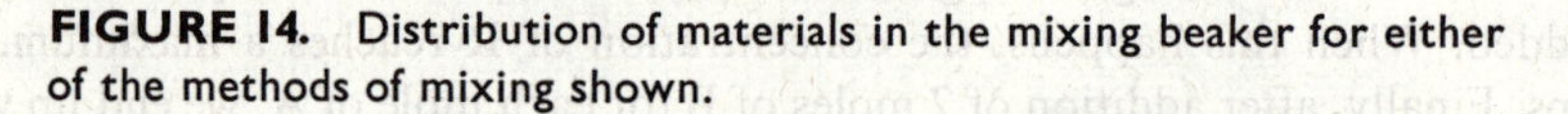

FIGURE 14. Distribution of materials in the mixing beaker for either of the methods of mixing shown.

to any appreciable extent before the mixture becomes uniform. During the first few reaction increments R finds itself competing with a large excess of A for B and hence is at a disadvantage. Carrying through this line of reasoning, we find the same type of distribution curve as for the mixture in which B is added slowly to A. This situation is also shown in Fig. 14.

The product distribution of Figs. 13 and 14 are completely different. Thus, when A is kept uniform in composition as it reacts, as in Fig. 14, then R is formed. However, when some A reacts rapidly, some slowly, when fresh A is mixed with reacted A, as in Fig. 13, then no intermediate R forms. But this is precisely the behavior of reactions in series. Thus, as far as A, R, and S are concerned, we may view the reactions of Eq. 27 as

$$A \xrightarrow{+B} R \xrightarrow{+B} S$$

A second observation of Fig. 14 is that the concentration level of B, whether high or low, has no effect on the path of the reaction and on the distribution of products. But this is precisely the behavior of parallel reactions of the same order. So with respect to B Eq. 27 can be looked at as

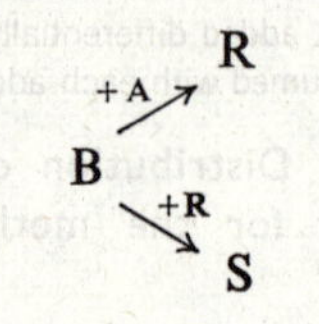

From this discussion we propose the general rule:

Series-parallel reactions can be analyzed in terms of their constituent series reactions and parallel reactions in that optimum contacting for favorable product distribution is the same as for the constituent reactions.

For the reactions of Eq. 27 where R is desired this rule shows that the best way of contacting A and B is to react A uniformly, while adding B any way convenient.

This is a powerful generalization which, without needing specific values for the rate constants, can already show in many cases which are the favorable contacting patterns. It is essential, however, to have the proper representation of the stoichiometry and form of rate equation. Example 6 applies to these generalizations.

Quantitative Treatment, Plug Flow or Batch Reactor. Here we quantitatively treat the reactions of Eq. 27 with the understanding that R, the intermediate, is the desired product, and that the reaction is slow enough so that we may ignore the problems of partial reaction during mixing of reactants.

In general, taking the ratio of two rate equations eliminates the time variable and gives information on the product distribution. So dividing Eq. 30 by Eq. 28 we obtain the first-order linear differential equation

$$\frac{r_R}{r_A} = \frac{dC_R}{dC_A} = -1 + \frac{k_2 C_R}{k_1 C_A} \tag{32}$$

whose method of solution is shown in Chapter 3. With some R present in the feed the limits of integration are C_{A0} to C_A for A and C_{R0} to C_R for R, and the solution of this differential equation is

---

$$\frac{C_R}{C_{A0}} = \frac{1}{1 - k_2/k_1}\left[\left(\frac{C_A}{C_{A0}}\right)^{k_2/k_1} - \frac{C_A}{C_{A0}}\right] + \frac{C_{R0}}{C_{A0}}\left(\frac{C_A}{C_{A0}}\right)^{k_2/k_1}, \qquad \frac{k_2}{k_1} \neq 1$$

$$\frac{C_R}{C_{A0}} = \frac{C_A}{C_{A0}}\left(\frac{C_{R0}}{C_{A0}} - \ln\frac{C_A}{C_{A0}}\right), \qquad \frac{k_2}{k_1} = 1 \tag{33}$$

This gives the relationship between C_R and C_A at any time in a batch or plug flow reactor. To find the concentrations of the other components simply make a material balance. An A balance gives

$$C_{A0} + C_{R0} + C_{S0} = C_A + C_R + C_S$$

or

$$\Delta C_A + \Delta C_R + \Delta C_S = 0 \tag{34}$$

from which C_S can be found as a function of C_A and C_R. Finally a balance about B gives

$$\Delta C_B + \Delta C_R + 2\Delta C_S = 0 \tag{35}$$

from which C_B can be found.

Quantitative Treatment, Mixed Flow. Writing the design equation for mixed flow in terms of A and R gives

$$\tau_m = \frac{C_{A0} - C_A}{-r_A} = \frac{C_{R0} - C_R}{-r_R}$$

or

$$\tau_m = \frac{C_{A0} - C_A}{k_1 C_A C_B} = \frac{C_{R0} - C_R}{k_2 C_R C_B - k_1 C_A C_B}$$

Rearranging, we obtain

$$\frac{C_{R0} - C_R}{C_{A0} - C_A} = -1 + \frac{k_2 C_R}{k_1 C_A}$$

which is the difference equation corresponding to the differential equation, Eq. 32. Writing C_R in terms of C_A then gives

$$C_R = \frac{C_A(C_{A0} - C_A + C_{R0})}{C_A + (k_2/k_1)(C_{A0} - C_A)} \tag{36}$$

Equations 34 and 35, material balances about A and B in plug flow, hold equally well for mixed flow and serve to complete the set of equations giving complete product distribution in this reactor.

Graphical Representation. Figures 15 and 16, time-independent plots, show the distribution of materials in plug and mixed flow and are prepared from Eqs. 33 to 36. As mentioned earlier, A, R, and S behave like the components in first-order reactions in series. Comparing Figs. 15 and 16 with Figs. 7*b* and 9*b*, we see that the distribution of these materials is the same in both cases, plug flow again giving a higher concentration of intermediate than mixed flow. As the reaction proceeds B is progressively consumed while the composition of A and R follows the corresponding k_2/k_1 curve from left to right. The lines of slope 2 on these charts show the amount of B consumed to reach any particular point on the curve. It makes no

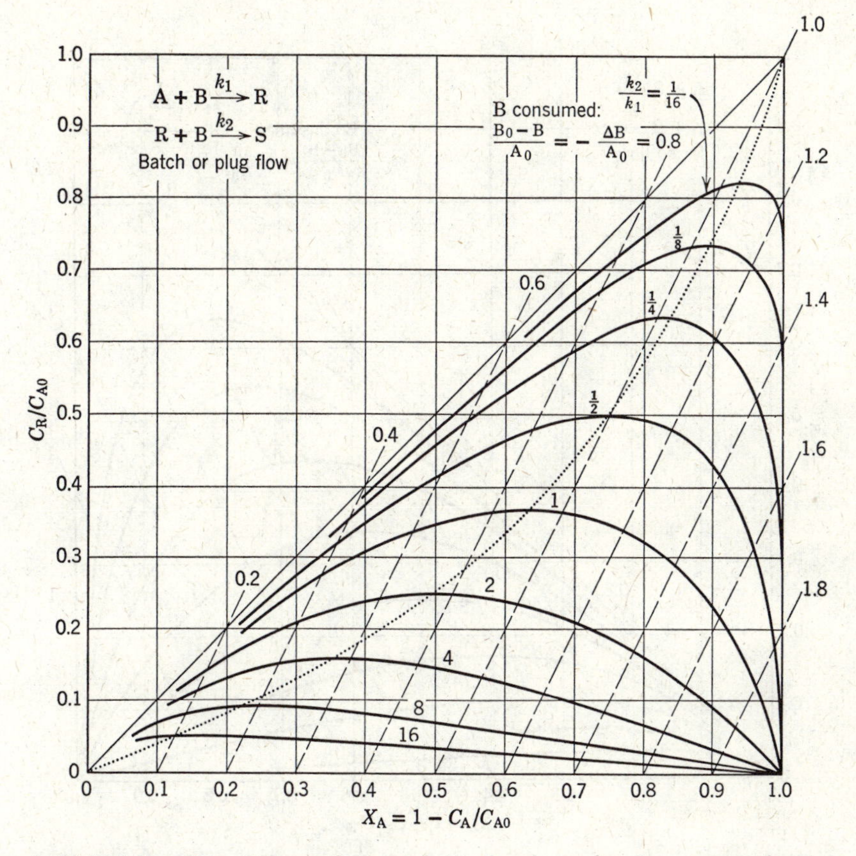

FIGURE 15. Distribution of materials in a batch or plug flow reactor for the elementary series-parallel reactions

$$A + B \xrightarrow{k_1} R$$

$$R + B \xrightarrow{k_2} S$$

difference whether B is added all at one time as in a batch reactor or a little at a time as in a semibatch reactor; in either case the same point on the chart will be reached when the same total amount of B is consumed.

These figures indicate that no matter what reactor system is selected, when the fractional conversion of A is low the fractional yield of R is high. Thus if it is possible to separate cheaply small amounts of R from a large reactant stream, the optimum setup for producing R is to have small conversions per pass coupled with a separation of R and recycle of unused A. The actual mode of operation will, as usual, depend on the economics of the system under study.

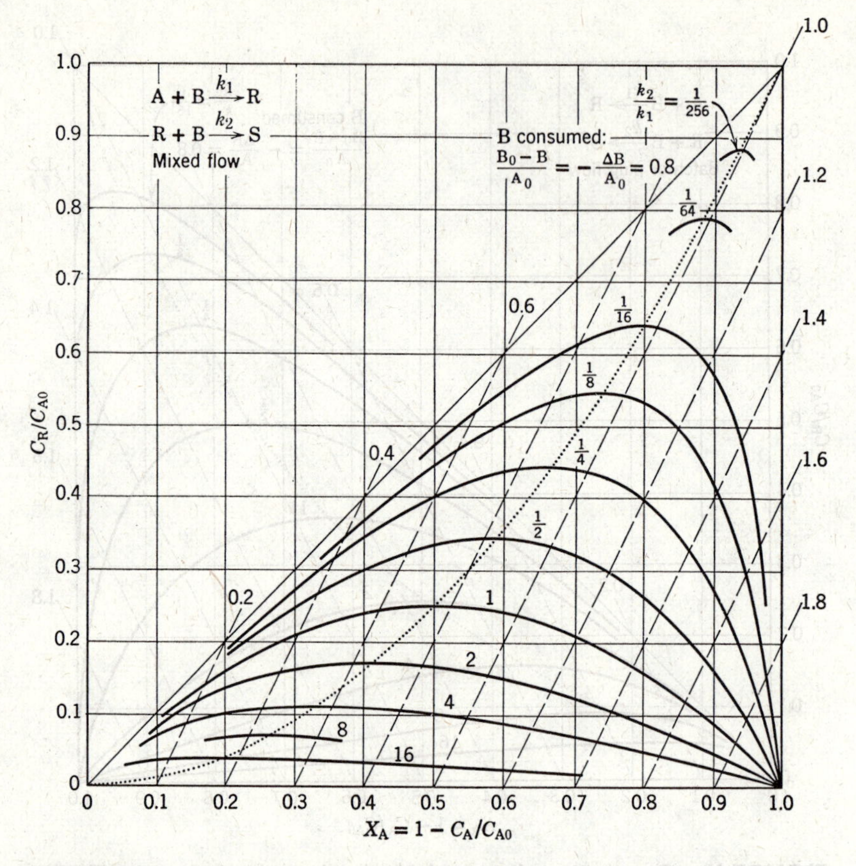

FIGURE 16. Distribution of materials in a mixed flow reactor for the elementary series-parallel reactions

$$A + B \xrightarrow{k_1} R$$
$$R + B \xrightarrow{k_2} S$$

Experimental Determination of the Kinetics of Reaction. The ratio k_2/k_1 may be found by analyzing the products of reaction from an experiment, and locating the corresponding point on the appropriate design chart. The simplest way to do this is to use different ratios of B to A in a batch reactor, allowing the reaction to go to completion each time. For each run a value of k_2/k_1 can be determined. The best mole ratios to use are those where the lines of constant k_2/k_1 are furthest apart, or where $-(\Delta B/A_0) \approx 1.0$, or close to equimolar ratios.

With k_2/k_1 known, all that is needed is k_1 which must be found by kinetic experiments. Since the rate of reaction between A and B is initially second order when R is absent in the feed, k_1 can be found by extrapolating rate data to zero holding time. In other words, k_1 is found from initial rate data. Another way is to deal directly with Eqs. 28–33, fitting the best values of the constants by a least-squares procedure. This may become somewhat difficult, however.

Frost and Pearson (1961) consider in detail the experimental determination of the rate constants of this reaction.

Intermediate in Feed or Recycle Stream. If R is present in the feed to the plug flow reactor, either from the initial feed or recycle stream, its effect can easily be found because the progress of the reaction will still be along the same k_2/k_1 line of Fig. 15 but starting from the point where this line cuts a second line of slope $-(C_R/C_A)_{\text{initial}}$ and emanating from $C_A = C_R = 0$. The effect of this R in the feed reduces the net fractional yield of R.

EXAMPLE 5. Kinetics of series-parallel reaction

From each of the following experiments, what can we say about the rate constants of the multiple reactions

$$A + B \xrightarrow{k_1} R$$

$$R + B \xrightarrow{k_2} S$$

(*a*) Half a mole of B is poured bit by bit, with stirring, into a flask containing a mole of A. The reaction proceeds slowly, and when completed B is entirely consumed; however, 0.67 mole of A remains unreacted.

(*b*) One mole of A and 1.25 moles of B are rapidly mixed, and the reaction is slow enough so that it does not proceed to any appreciable extent before homogeneity in composition is achieved. On completion of the reaction 0.5 mole of R is found to be present in the mixture.

(*c*) One mole of A and 1.25 moles of B are rapidly brought together. The reaction is slow enough so that it does not proceed to any appreciable extent before homogeneity in A and B is achieved. At the time when 0.9 mole of B is consumed 0.3 mole of S is present in the mixture.

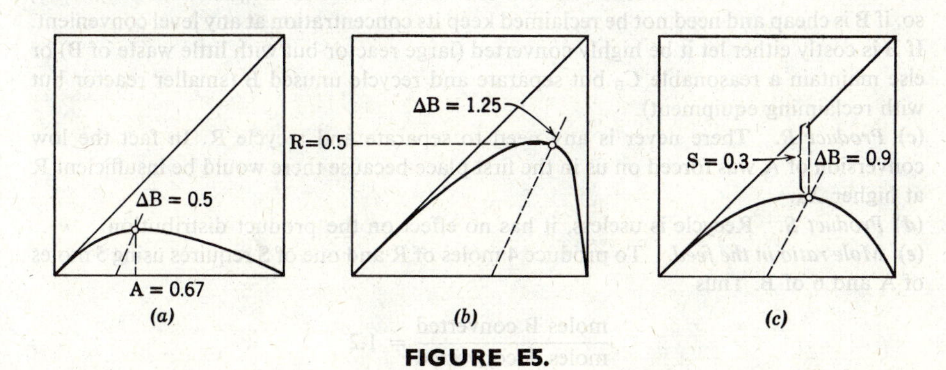

FIGURE E5.

SOLUTION

The sketches in Fig. E5 show how Fig. 15 is used to find the desired information. Thus we find:

$$(a)\ k_2/k_1 = 4, \qquad (b)\ k_2/k_1 = 0.4, \qquad (c)\ k_2/k_1 = 1.45$$

EXAMPLE 6. Design decisions for a given product distribution

We wish to produce continuously at least a 4:1 mixture of R and S from separate feed streams of A and of B which react according to the elementary reactions

$$\left.\begin{array}{l} A + B \xrightarrow{k_1} R \\ R + B \xrightarrow{k_2} S \end{array}\right\} \quad k_1 = k_2$$

We face numerous choices; the type of reactor, the mole ratio of feed, the conversion level of reactants, whether to separate and recycle any of the materials, and if so which? Let us briefly indicate the factors which lead to the various design alternatives.

SOLUTION

(*a*) *Reactant A.* From Figs. 15 and 16 we find that the desired product distribution occurs

$$\text{at } X_A = 0.36 \text{ in plug flow}$$

$$\text{at } X_A = 0.20 \text{ in mixed flow}$$

At conversions higher than these the ratio of R to S is lower than the required 4:1. Thus in one case we waste 80% of the incoming A, in the other case 64%. These are the conditions at which the reactor must operate.

If A is costly then it must be separated from the product stream and reused. Since conversion is higher for plug flow its recycle rate is smaller (45% that for mixed flow) and the reclaiming system needed is likewise smaller. On the other hand, if A is cheap enough that it need not be reclaimed (the production of ethylene glycol is an example, see p. 186: here chemical A is water) then there is no particular advantage to either ideal reactor.

(*b*) *Reactant B.* The concentration level of B has no effect on the product distribution; so, if B is cheap and need not be reclaimed keep its concentration at any level convenient. If B is costly either let it be highly converted (large reactor but with little waste of B) or else maintain a reasonable C_B but separate and recycle unused B (smaller reactor but with reclaiming equipment).

(*c*) *Product R.* There never is any need to separate and recycle R. In fact the low conversion of A was forced on us in the first place because there would be insufficient R at higher X_A.

(*d*) *Product S.* Recycle is useless, it has no effect on the product distribution.

(*e*) *Mole ratio in the feed.* To produce 4 moles of R and one of S requires using 5 moles of A and 6 of B. Thus

$$\frac{\text{moles B converted}}{\text{moles A converted}} = 1.2$$

The actual feed ratio may differ from this to account for any discarded A and B. For example, if A is discarded and B is very highly converted, then the feed ratio for mixed flow would be 0.24.

Summary. Without recycling A, X_A is fixed by the reaction kinetics and the R/S requirements. In this example X_A is unavoidably low, so if A is costly it must be reclaimed and recycled.

Any X_B can be used, so if B is costly use high X_B or reclaim and recycle B.

If the cost of A is a consideration then always use plug flow.

Do not recycle either of the products.

Extensions and Applications

Three or More Reactions. Analysis of three or more reactions can be made by procedures analogous to those presented. Of course, the mathematics becomes more involved; however, much of the extra labor can be avoided by selecting experimental conditions in which only two reactions need be considered at any time. Figure 17 shows product distribution curves for one such reaction set, the progressive chlorination of benzene. Note the similarity in shape with the time-concentration curves for successive first-order reactions in series, Fig. 3.13. The curves of Fig. 17 may be replotted in the manner of Figs. 15 and 16 for a three-step reaction [see Jungers *et al.* (1958)]. Unfortunately, because paper is only two-dimensional, a separate graph is needed for each value of k_2/k_1. Again, as with the two-reaction set, we find that a plug flow reactor yields a higher maximum concentration of intermediates than does a mixed reactor.

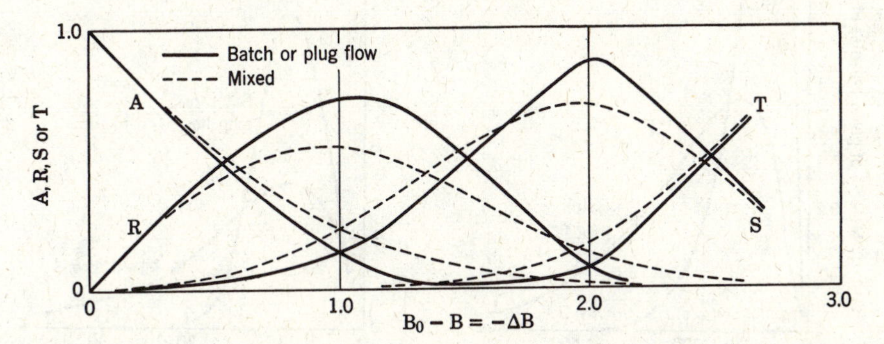

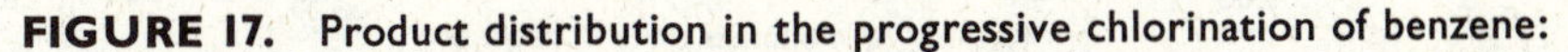

FIGURE 17. Product distribution in the progressive chlorination of benzene:

$$A + B \xrightarrow{k_1} R + U \qquad C_6H_6 + Cl_2 \xrightarrow{k_1} C_6H_5Cl + HCl$$

$$R + B \xrightarrow{k_2} S + U \quad \text{or} \quad C_6H_5Cl + Cl_2 \xrightarrow{k_2} C_6H_4Cl_2 + HCl$$

$$S + B \xrightarrow{k_3} T + U \qquad C_6H_4Cl_2 \xrightarrow{k_3} C_6H_3Cl_3 + HCl$$

with $k_2/k_1 = \frac{1}{8}$ and $k_3/k_1 = \frac{1}{240}$; from MacMullin (1948).

Polymerization. The field of polymerization affords an opportunity for a fruitful application of these ideas. Often hundreds or even thousands of reactions in series occur in the formation of polymers, and the type of crosslinking and molecular weight distribution of these products are what gives these materials their particular physical properties of solubility, density, flexibility, etc.

Since the mode of mixing of monomers with their catalysts profoundly affects product distribution, great importance must be paid to this factor if the product is to have the desired physical and chemical properties. Denbigh (1947, 1951) considered some of the many aspects of this problem, and Fig. 18 shows for various kinetics how reactor type influences molecular weight distribution of products.

Fermentation. From the simplest to the most complex, biological processes may be classed as fermentations, elementary physiological processes, and the action of living entities. Further, fermentations can be divided into two broad groups: those promoted and catalyzed by microorganisms or microbes (yeasts, bacteria, algae, molds, protozoa) and those catalyzed by enzymes (chemicals produced by micro-

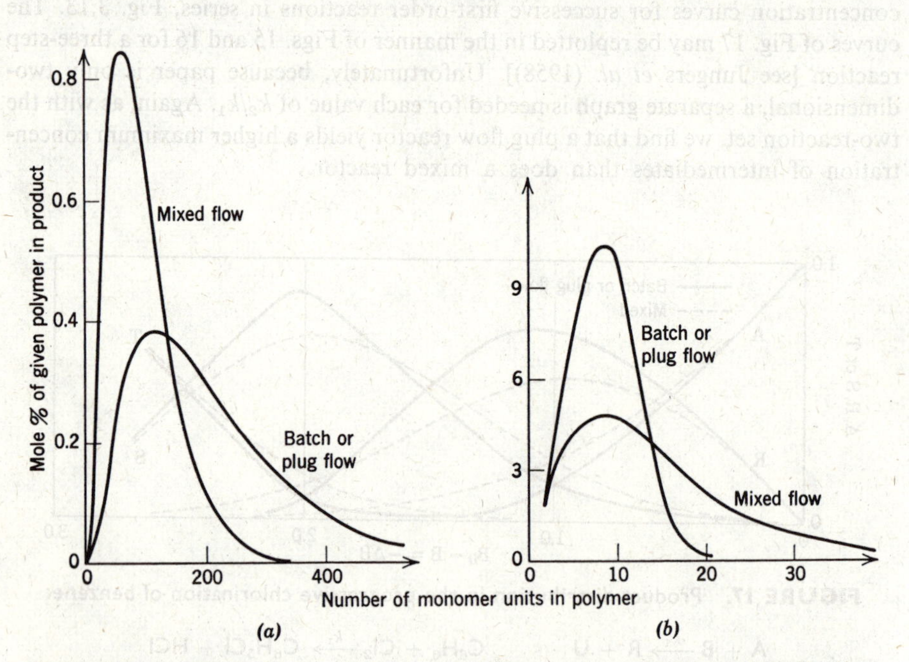

FIGURE 18. Type of flow and the kinetics influence the molecular weight distribution of polymer: (*a*) duration of polymerization reaction (life of active polymer) is short compared to the reactor holding time; (*b*) duration of polymerization reaction is long compared to the reactor holding time, or where polymerization has no termination reaction. Adapted from Denbigh (1947).

organisms). In general, then, fermentations are reactions wherein a raw organic feed is converted into product by the action of microorganisms or by the action of enzymes.

Consider *microbial fermentations*. Their action may be represented by

$$\text{(raw organic feed)} + \text{(microorganisms)} \rightarrow \begin{pmatrix}\text{product}\\ \text{chemicals}\end{pmatrix} + \begin{pmatrix}\text{more}\\ \text{microorganisms}\end{pmatrix}$$

or

$$\text{(organic feed: A)} \xrightarrow[\text{these cells are catalysts}]{\text{microorganisms: C,}} \text{(product: R)} + \text{(more cells: C)} \qquad (37)$$

Sometimes the product chemicals are the desired product (alcohol from sugar and starch), other times it is the microorganism or its enzyme (yeast, penicillin).

Typically, if a batch of organic feed is seeded with small amounts of microorganism we have the following progression of events. At first, since there is only a small amount of catalytic agent present, the conversion of feed is slow. However, as the microorganism propagates the reaction rate rises; it reaches a maximum; and with a gradual depletion of convertible feed and with an accumulation of products the rate then slows eventually to zero. Clearly this progression is autocatalytic and it is found that a reasonable first representation is given by the modified Michaelis–Menten expression (see Problem 2.19)

$$\text{A} \xrightarrow{\text{C}} r\text{R} + c\text{C}$$

where

$$-r_\text{A} = k_1 C_\text{C}\left(\frac{C_\text{A}}{k_2 + C_\text{A}}\right); \qquad -r_\text{A} = \frac{r_\text{C}}{c} = \frac{r_\text{R}}{r} \qquad (38)$$

If the stoichiometric coefficients are known and constant throughout the reaction (this means that each unit of feed consumed produces the same amount of new cells and product chemicals), then the system can be treated as a single autocatalytic reaction and the solution of problems of optimum design and reactor choice is direct, see Bischoff (1966). These assumptions are often made for they greatly simplify the analysis; however, whether they are justified requires a test with experiment. When they do not hold then we have a case of multiple reactions to treat, with all its attendant product distribution problems.

Consider a fermentation occurring in a mixed reactor. Often the feed is a mixture of organic materials (grape juice, carbohydrate mixture, raw sewage) and the injected cells consist of a variety of microorganisms (yeast starter, activated sludge) all competing for the feed. Since these processes are all autocatalytic, each particular temperature, mean residence time, and contacting pattern favors one of the reactions which then dominates and swamps all the other reactions. Changing the conditions such as in scale-up, or in shifting from batch to continuous operations may cause a different microorganism to dominate, giving a completely different product distribution. These questions of changing product are an interesting and extremely important application of the principles presented in this chapter.

Consider *enzyme fermentations*. These can be represented by

$$\text{(organic feed: A)} \xrightarrow{\text{enzyme: E, acting as catalyst}} \text{(product chemicals: R)} \tag{39}$$

where the simplest reasonable kinetics is given by the Michaelis–Menten equation (see Problems 2.19 and 3.17) as

$$-r_A = k_1 C_{E0}\left(\frac{C_A}{k_2 + C_A}\right), \qquad C_{E0} = \text{constant} \tag{40}$$

The key distinction between these two types of fermentations is as follows. In enzyme fermentation the catalytic agent (the enzyme) does not reproduce itself, while in microbial fermentation the catalytic agent (the cells with their enzyme content) does reproduce. Thus in a batch operation the cell concentration C_C in Eq. 38 is a changing quantity while the total enzyme concentration C_{E0} in Eq. 40 remains constant throughout. Also, in flow systems fresh enzyme must be fed continually to the reactor, while in microbial fermenters with sufficient recycle of product fluid no fresh microorganism need be added to the reactor.

The autocatalytic behavior of microbial fermentations suggests that mixed flow followed by plug flow likely gives the smallest reactor size. Because enzyme fermentations are not autocatalytic, plug flow is favored. The distinction between these systems is illustrated in Figs. 19 and 20.

Fermentations form a most important class of reactions. They are widely used by man to produce alcoholic beverages, antibiotics, vitamins, proteins, toxins, yeast, and many foodstuffs, such as tea, yoghurt, sauerkraut, and mouth-watering dill pickles; also to purify organic wastes such as sewage. They most likely rep-

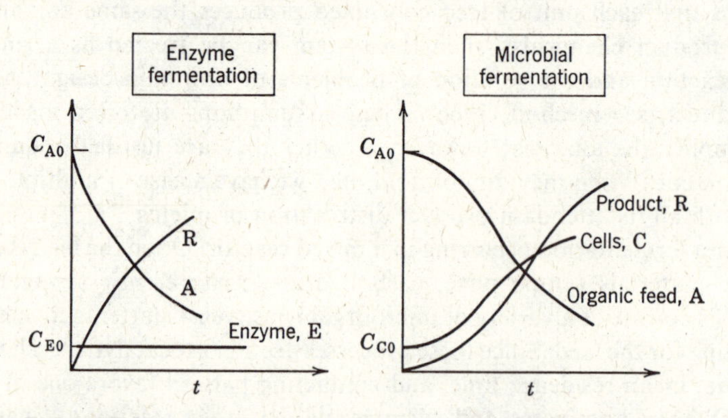

FIGURE 19. Distinction between the two types of fermentation, batch operations.

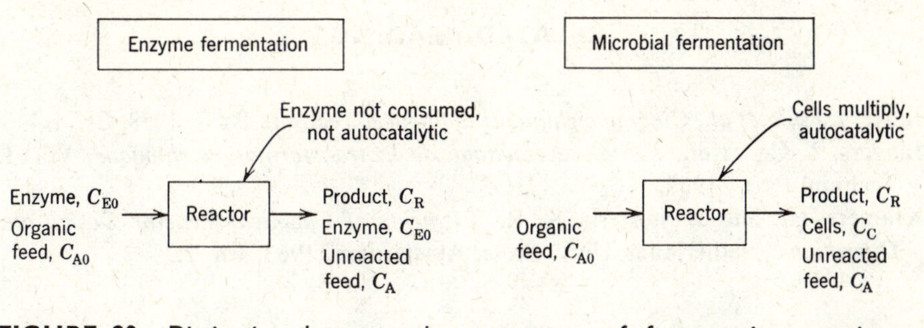

FIGURE 20. Distinction between the two types of fermentation, continuous operation.

resent man's earliest and happiest contacts with chemical technology. And in man's search for better foods and a cleaner environment these reactions most likely will become one of the most important applications of chemical reaction engineering.

CONCLUSION

The key to optimum design for multiple reactions is proper contacting and proper flow pattern of fluids within the reactor. These requirements are determined by the stoichiometry and observed kinetics.

To obtain high yields of a desired product we may have to maintain high concentrations or low concentrations (parallel reactions), or homogeneity of composition (series reactions) for the various reactants. With these requirements known desirable contacting patterns can then be devised by using the appropriate batch, semibatch, mixed, or plug flow unit with slow or rapid introduction of the various feeds, with or without separation and recycle of reactants.

Usually qualitative reasoning alone can already determine the correct contacting scheme. This is done by breaking the reaction stoichiometry into its component parallel and series reactions. Naturally, to determine the actual equipment size requires quantitative considerations.

When the kinetics and stoichiometry are not known, well-planned experimentation guided by the principles enunciated here should result in a reasonably close approach to optimum operations.

Additional discussions of how product distribution is affected by the contacting pattern and temperature of operations are given in Chapters 10 and 8 and in the chapters on heterogeneous systems. In particular, the discussion at the end of Chapter 13 shows how to vastly improve product distribution by deliberately adding a second immiscible phase to make the system heterogeneous.

RELATED READINGS

Jungers, J. C., *et al.*, *Cinétique chimique appliquée*, Technip, Paris, 1958, Ch. 4.
Jungers, J. C., *et al.*, *l'Analyse cinétique de la transformation chimique*, Vol. 1, Technip, Paris, 1967.
Kramers, H., and Westerterp, K. R., *Elements of Chemical Reactor Design and Operations*, Netherlands Univ. Press, Amsterdam, 1963, Ch. 2.

REFERENCES

Bischoff, K. B., *Can. J. Chem. Eng.*, **44**, 281 (1966).
Denbigh, K. G., *Trans. Faraday Soc.*, **40**, 352 (1944).
———, *Trans. Faraday Soc.*, **43**, 648 (1947).
———, *J. Appl. Chem.*, **1**, 227 (1951).
———, *Chem. Eng. Sci.*, **14**, 25 (1961).
Frost, A. A., and Pearson, R. G., *Kinetics and Mechanism*, 2nd ed., John Wiley & Sons, New York, 1961, Ch. 8.
Jungers, J. C., *et al.*, *Cinétique chimique appliquée*, Technip, Paris, 1958.
Kramers, H., and Westerterp, K. R., *Elements of Chemical Reactor Design and Operations*, Netherlands University Press, Amsterdam, 1963.
MacMullin, R. B., *Chem. Eng. Progr.*, **44**, 183 (1948).

PROBLEMS

These are grouped as follows:

Problems 1–6: Qualitative.
Problems 7–20: Reaction in parallel.
Problems 21–26: Series, and series-parallel reactions.
Problems 27–31: Miscellaneous.

1. In one sentence state the distinguishing characteristic of each of the following reactions: single, multiple, elementary, nonelementary.

2. Starting with separate feeds of reactants A and B of given concentration (no dilution with inerts permitted), for the competitive-consecutive reactions with stoichiometry and rate as shown

$$A + B \rightarrow R_{\text{desired}} \qquad \cdots r_1$$

$$R + B \rightarrow S_{\text{unwanted}} \qquad \cdots r_2$$

sketch the best contacting patterns for both continuous and noncontinuous operations:

(*a*) $r_1 = k_1 C_A C_B^2$, $r_2 = k_2 C_R C_B$ (*b*) $r_1 = k_1 C_A C_B$, $r_2 = k_2 C_R C_B^2$ (*c*) $r_1 = k_1 C_A C_B$, $r_2 = k_2 C_R^2 C_B$ (*d*) $r_1 = k_1 C_A^2 C_B$, $r_2 = k_2 C_R C_B$

3. The reaction of A and B produces the desired product R as well as the unwanted side products S, T, ... What reactant concentrations should we use for the following sets of elementary reactions so as to promote the conversion to R.

(*a*) $\left.\begin{array}{l} A + B \rightarrow R \\ A \rightarrow S \end{array}\right\}$ (*b*) $\left.\begin{array}{l} A + B \rightarrow R \\ 2A \rightarrow T \end{array}\right\}$ (*c*) $\left.\begin{array}{l} A + B \rightarrow R \\ A \rightarrow U \\ B \rightarrow V \end{array}\right\}$

Note: Costs of materials, desired conversion levels, possibility of recycle will all enter in selecting the best contacting pattern; however, these factors are not considered here.

4. Reactant A decomposes as follows:

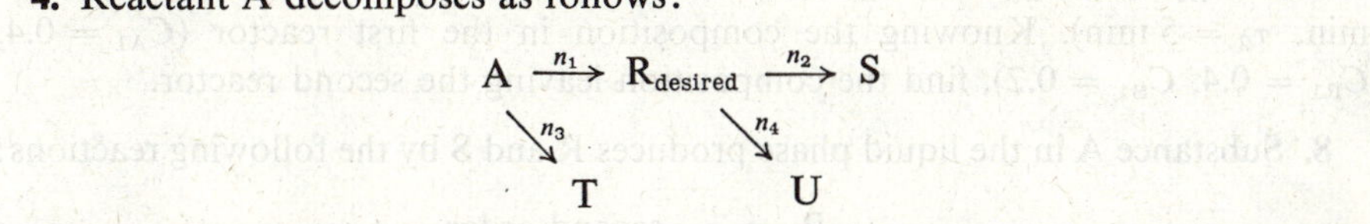

where n_i are reaction orders. Qualitatively find what flow (plug, mixed, or intermediate) and what C_{A0} (high, low, or intermediate) should be used for a high conversion to R.

	n_1	n_2	n_3	n_4
Case (*a*)	1	1	1	1
(*b*)	1	2	1	1
(*c*)	2	1	1	1
(*d*)	1	2	1	0
(*e*)	1	1	2	1
(*f*)	1	0	2	1

5. Reactant A decomposes as follows and R is the desired product:

$$A \xrightarrow{k_1} R \xrightarrow{k_2} S \quad \cdots \text{both first order}$$

$$A + A \xrightarrow{k_3} T \quad \cdots \text{second order}$$

Qualitatively, how should we run this reaction so as to obtain a high conversion to R.

(*a*) Consider that C_{A0} is fixed and that either plug, mixed, or recycle flow can be used.

(*b*) Consider that C_{A0} can be set at will, and that either plug, mixed, or recycle flow can be used.

6. What feed concentration (high or low C_{A0}), contacting pattern (plug or mixed flow), and conversion level (high or low) will promote the formation of desired product R for the following sets of elementary reactions:

$$(a)\ \left.\begin{matrix} A + R \rightarrow 2R \\ A \rightarrow S \end{matrix}\right\} \qquad (b)\ \left.\begin{matrix} A \rightarrow R \\ A + R \rightarrow S \end{matrix}\right\} \qquad (c)\ \left.\begin{matrix} 2A \rightarrow R \\ R \rightarrow S \end{matrix}\right\}$$

7. Substance A in a liquid reacts to produce R and S as follows:

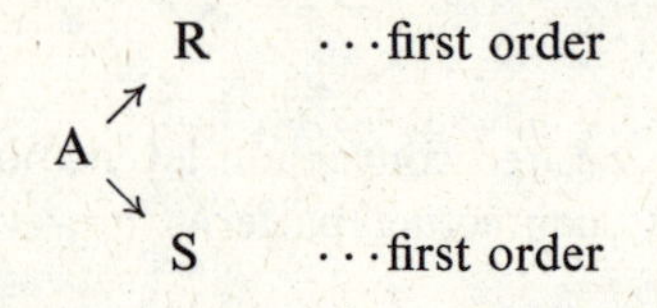

A feed ($C_{A0} = 1$, $C_{R0} = 0$, $C_{S0} = 0$) enters two mixed reactors in series ($\tau_1 = 2.5$ min, $\tau_2 = 5$ min). Knowing the composition in the first reactor ($C_{A1} = 0.4$, $C_{R1} = 0.4$, $C_{S1} = 0.2$), find the composition leaving the second reactor.

8. Substance A in the liquid phase produces R and S by the following reactions:

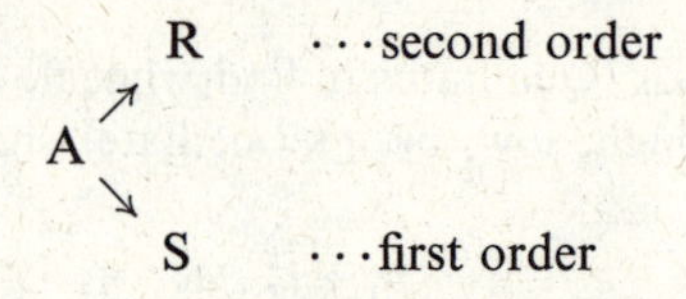

The feed ($C_{A0} = 1.0$, $C_{R0} = 0$, $C_{S0} = 0$) enters two mixed reactors in series ($\tau_1 = 2.5$ min, $\tau_2 = 10$ min). Knowing the composition in the first reactor ($C_{A1} = 0.4$, $C_{R1} = 0.4$, $C_{S1} = 0.2$), find the composition leaving the second reactor.

9. The stoichiometry of a liquid phase decomposition is known to be

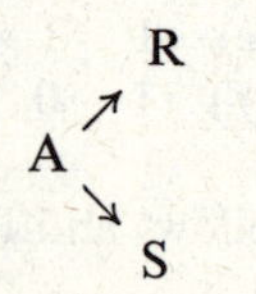

In a series of steady-state flow experiments ($C_{A0} = 100$, $C_{R0} = C_{S0} = 0$) in a laboratory mixed flow reactor the following results are obtained:

C_A	90	80	70	60	50	40	30	20	10	0
C_R	7	13	18	22	25	27	28	28	27	25

Further experiments indicate that the level of C_R and C_S have no effect on the progress of the reaction.

With a feed $C_{A0} = 100$ and exit concentration $C_{Af} = 20$, find the value of C_R at the exit from

(*a*) a plug flow reactor,
(*b*) a mixed flow reactor,
(*c*) and (*d*): repeat parts (*a*) and (*b*) for $C_{A0} = 200$.

10. For the reaction of Problem 9, how should we operate a mixed reactor so as to maximize the production of R? Separation and recycle of unused reactant is not possible, and $C_{A0} = 150$.

11. Reactant A decomposes in an isothermal batch reactor ($C_{A0} = 100$) to produce wanted R and unwanted S and the following progressive concentration readings are recorded:

C_A	(100)	90	80	70	60	50	40	30	20	10	(0)
C_R	(0)	1	4	9	16	25	35	45	55	64	(71)

Additional runs show that adding R or S does not affect the distribution of products formed and that only A does. Also it is noted that the number of moles of A, R, and S is constant.

(*a*) Find the φ versus C_A curve for this reaction.

With a feed $C_{A0} = 100$ and $C_{Af} = 10$, find C_R

(*b*) from a mixed flow reactor,
(*c*) from a plug flow reactor,
(*d*) and (*e*): repeat parts (*b*) and (*c*) with the modifications that $C_{A0} = 70$.

12. Reactant A in a liquid either isomerizes or dimerizes as follows

$$A \rightarrow R_{\text{desired}} \qquad r_R = k_1 C_A$$
$$A + A \rightarrow S_{\text{unwanted}} \qquad r_S = k_2 C_A^2$$

(*a*) Write $\varphi(R/A)$ and $\varphi(R/R + S)$.

With a feed stream of concentration C_{A0} find $C_{R,\max}$ which can be formed

(*b*) in a plug flow reactor,
(*c*) in a mixed flow reactor.

A quantity of A of initial concentration $C_{A0} = 1$ mol/liter is dumped into a batch reactor and is reacted to completion.

(*d*) If $C_S = 0.18$ mol/liter in the resultant mixture what does this tell of the kinetics of the reaction?

13. For the parallel decompositions of A, where R is desired,

$$A \rightarrow \begin{cases} R, & r_R = 1 \\ S, & r_S = 2C_A \\ T, & r_T = C_A^2 \end{cases} \quad \ldots \text{with } C_{A0} = 1$$

What is the maximum C_R we may expect in isothermal operations

(*a*) in a mixed reactor,

(*b*) in a plug flow reactor.

14. Often a desired reaction $A \rightarrow S$ is accompanied by alternative decompositions of higher or of lower order. In a rather generalized formulation this may be represented by

$$\begin{matrix} & & R, \\ & \nearrow & \\ A & \rightarrow & S, \\ & \searrow & \\ & & T, \end{matrix} \qquad \left.\begin{aligned} r_R &= k_0 C_A^{a_0} \\ r_S &= k_1 C_A^{a_1} \\ r_T &= k_2 C_A^{a_2} \end{aligned}\right\} \qquad \text{where } a_0 < a_1 < a_2$$

(*a*) Sketch the $\varphi(S/A)$ versus C_A curve for this reacting system and indicate any noteworthy features of this curve.

(*b*) Without recycle of unused reactant, what single reactor or combination of reactors will maximize production of S from a given feed of A?

(*c*) Repeat part (*b*) with the modification that separation and recycle of unused reactant is possible.

All operations are at a fixed temperature.

15. A and B react with each other as follows:

$$2A \rightarrow R, \qquad r_R = k_1 C_A^2$$
$$A + B \rightarrow S, \qquad r_S = k_2 C_A C_B$$
$$2B \rightarrow T, \qquad r_T = k_3 C_B^2$$

Find what ratio of A to B should be maintained in a mixed flow reactor so as to maximize the fractional yield of desired product S.

16. Given the reactions

$$\left.\begin{aligned} A + 2B \rightarrow R, \qquad r_R &= k_1 C_A C_B^2 \\ A + B \rightarrow S, \qquad r_S &= k_2 C_A C_B \end{aligned}\right\} \qquad \text{with } k_2 = 2k_1$$

(*a*) What are the fractional yield expressions $\varphi(R/A)$ and $\varphi(R/B)$ for this system.

(*b*) How should we operate a mixed reactor so as to maximize the production of R from a single feed consisting of $C_{A0} = C_{B0} = 1$.

17. We have a mixture consisting of 90 mole % A (45 mol/liter) and 10 mole % impurity B (5 mol/liter). To be of satisfactory quality the mole ratio of A to B in the mixture must be 100 to 1 or higher. D reacts with both A and B as follows:

$$A + D \rightarrow R, \qquad -r_A = 21 C_A C_D$$
$$B + D \rightarrow S, \qquad -r_B = 147 C_B C_D$$

Assuming that the reactions go to completion, how much D need be added to a batch of mixture to bring about the desired quality?

18. A and B react as follows:

$$A + 2B \rightarrow 2R + S, \qquad (r_A)_1 = -k_1 C_A C_B^2$$

$$A + B \rightarrow T + U, \qquad (r_A)_2 = -k_2 C_A C_B$$

Equimolar quantities of A and B are introduced into a batch reactor and are left to react to completion. With all B consumed, $C_{Af} = 0.1 C_{A0} = 0.1$. What information does this experiment give us about the rate constants for the reactions?

19. Equimolar quantities of A, B, and D are fed continuously to a mixed flow reactor where they combine by the elementary reactions

$$\left.\begin{array}{l} A + D \xrightarrow{k_1} R \\ B + D \xrightarrow{k_2} S \end{array}\right\} \quad \text{with } \frac{k_2}{k_1} = 0.2$$

(*a*) If 50% of the incoming A is consumed find what fraction of the products formed is R.

(*b*) If 50% of the incoming D is consumed find what fraction of the products formed is R.

20. The desired product R is formed as follows:

$$A + B \rightarrow R, \qquad r_R = k_1 C_A C_B$$

Under conditions which favor this reaction, however, B also dimerizes to form unwanted S:

$$2B \rightarrow S, \qquad r_S = k_2 C_B^2$$

At present R is produced in a mixed reactor using a large excess of A so as to depress formation of unwanted dimer. Feeds of A and B are so adjusted that the mole ratio of A to B in the reactor is kept at 40 to 1. Unreacted A is cleanly separated from the rest of the materials in the reactor effluent stream and is returned to the reactor. Unreacted B cannot easily be separated and is discarded. R and S are produced in equimolar quantities, and 50% of entering B is unreacted and is discarded unused.

(*a*) Find the fractional yield of R based on A consumed, on B consumed, on total entering B, and on the sum of all products formed.

This scheme seems to be inefficient because much B forms the wrong product or is wasted by remaining unreacted. To improve utilization of B, let us connect a second reactor in series with the present reactor, leaving the conditions within the first reactor unchanged. Both reactors are to be of the same size, and separation and recycle of A occurs after passage of fluid through both reactors. Because it is in great excess, assume that the concentration of A is constant throughout the system.

(*b*) Find the fraction of R present in the R-S product and the fraction of B in the feed which has been transformed into R.

(*c*) With unchanged feed rate of B and unchanged concentration of A within the system, repeat part (*b*) if the two mixed reactors are connected in parallel rather than in series.

21. Under appropriate conditions A decomposes as follows:

$$A \xrightarrow{k_1 = 0.1/\text{min}} R \xrightarrow{k_2 = 0.1/\text{min}} S$$

R is to be produced from 1000 liter/hr of feed in which $C_{A0} = 1$ mol/liter, $C_{R0} = C_{S0} = 0$.

(*a*) What size plug flow reactor will maximize the yield of R, and what is the concentration of R in the effluent stream from this reactor?

(*b*) What size mixed reactor will maximize the yield of R, and what is $C_{R,max}$ in the effluent stream from this reactor?

22. At Sandy's Rock and Gravel Company they want to shift a mountain of gravel, estimated at about 20,000 tons, from one side of their yard to the other. For this they intend to use a power shovel to fill a hopper, which in turn feeds a belt conveyor. The latter then transports the gravel to the new location.

The shovel scoops up large amounts of gravel at first; however, as the gravel supply decreases the handling capacity of the shovel also decreases because of the increased time required to move away from the hopper for a load and then return and dump. Roughly, then, we may estimate that the shovel's gravel-handling rate is proportional to the size of the pile still to be moved, its initial rate being 10 ton/min. The conveyer, on the other hand, transports the gravel at a uniform 5 ton/min. At first the shovel will work faster than the conveyor, then slower. Hence the storage bin will accumulate material, then empty.

(*a*) What will be the largest amount of gravel in the bin?

(*b*) When will this occur?

(*c*) When will the rates of bin input and output be equal?

(*d*) When will the bin become empty?

23. Consider the following elementary reactions:

$$A + B \xrightarrow{k_1} R$$

$$R + B \xrightarrow{k_2} S$$

(*a*) One mole A and 3 moles B are rapidly mixed together. The reaction is very slow, allowing analysis of compositions at various times. When 2.2 moles B remain unreacted, 0.2 mole S is present in the mixture. What should be the composition of the mixture (A, B, R, and S) when the amount of S present is 0.6 mole?

(*b*) One mole A is added bit by bit with constant stirring to 1 mole B. Left overnight, then analyzed, 0.5 mole S is found. What can we say about k_2/k_1?

(*c*) One mole A and 1 mole B are thrown together and mixed in a flask. The reaction is very rapid and goes to completion before any rate measurements can be made. On analysis of the products of reaction 0.25 mole S is found to be present. What can we say about k_2/k_1?

24. For single reactions we have seen that if the reactor system consists of a number of mixed reactors in series, its performance (volume or capacity) lies between plug flow and the single mixed reactor; the larger the number of reactors in series, the closer the approach to plug flow. We may expect the same behavior to occur with multiple reactions with respect not only to capacity but to product distribution as well. Let us see if it is true for one particular case. Consider the reaction

$$A \xrightarrow{k_1 = 1,\, n_1 = 1} R \xrightarrow{k_2 = 1,\, n_2 = 1} S$$

For a single plug flow reactor (see Eq. 3.52)

$$\frac{C_{R,max}}{C_{A0}} = \frac{1}{e} = 0.368 \quad \text{and} \quad \tau_{opt} = \frac{1}{k_{log\ mean}} = 1$$

For a single mixed reactor (see Eq. 23)

$$\frac{C_{R,max}}{C_{A0}} = 0.25 \quad \text{and} \quad \tau_{opt} = \frac{1}{\sqrt{k_1 k_2}} = 1$$

For two mixed reactors in series find $C_{R,max}/C_{A0}$ and τ_{opt}, and see whether these lie between the values found for plug flow and mixed flow.

25. For the elementary reactions

$$A \xrightarrow{k_1} R \xrightarrow{k_2} S, \qquad k_2 = k_1 + k_3$$
$$A \xrightarrow{k_3} T$$

find $C_{R,max}/C_{A0}$ and τ_{opt} in a plug flow reactor.

26. For the elementary reactions

$$A \xrightarrow{k_1} R \xrightarrow{k_2} S$$
$$A \xrightarrow{k_3} T$$

(*a*) Show for plug flow that

$$\frac{C_{R,max}}{C_{A0}} = \frac{k_1}{k_1 + k_3}\left(\frac{k_2}{k_1 + k_3}\right)^{k_2/(k_1 - k_2 + k_3)} \quad \text{at } \tau_{opt} = \frac{\ln\,[(k_1 + k_3)/k_2]}{k_1 - k_2 + k_3}$$

(*b*) Show for mixed flow that

$$\frac{C_{R,max}}{C_{A0}} = \frac{k_1}{(\sqrt{k_1 + k_3} + \sqrt{k_2})^2} \quad \text{at } \tau_{opt} = \frac{1}{\sqrt{k_2(k_1 + k_3)}}$$

27. The great naval battle, to be known to history as the battle of Trafalgar (1805), was soon to be joined. Admiral Villeneuve proudly surveyed his powerful fleet of 33 ships stately sailing in single file in the light breeze. The British fleet under Lord Nelson was now in sight, 27 ships strong. Estimating that it would still be two hours before the battle, Villeneuve popped open another bottle of burgundy and point by point reviewed his carefully thought out battle strategy. As was the custom of naval battles at that time, the two fleets would sail in single file parallel to each other and in the same direction, firing their cannons madly. Now, by long experience in battles of this kind, it was a well-known fact that the rate of destruction of a fleet was proportional to the fire power of the opposing fleet. Considering his ships to be on a par, one for one, with the British, Villeneuve was confident of victory. Looking at his sundial, Villeneuve sighed and cursed the light wind—he'd never get it over with in time for his afternoon snooze. "Oh well," he sighed, "c'est la vie." He could see the headlines next morning. "British fleet annihilated, Villeneuve's losses are . . ." Villeneuve stopped short. How many ships would he lose? Villeneuve called over his chief bottle cork popper, Monsieur Dubois, and asked this question. What answer did he get?

At this very moment, Nelson, who was enjoying the air on the poop deck of the *Victory*, was struck with the realization that all was ready except for one detail—he had forgotten to formulate his battle plan. Commodore Archibald Forsythe-Smythe, his trusty trusty, was hurriedly called over for a conference. Being familiar with the firepower law, Nelson was loathe to fight the whole French fleet (he could see the headlines too). Now certainly it was no disgrace for Nelson to be defeated in battle by superior forces, so long as he did his best and played the game; however, he had a sneaking suspicion that maybe he could pull a fast one. With a nagging conscience whether it was cricket or not, he proceeded to investigate this possibility.

It was possible to "break the line," in other words, to start parallel to the French fleet, and then cut in and divide the enemy fleet into two sections. The rear section could be engaged and disposed of before the front section could turn around and rejoin the fray. Now to the question. Should he split the French fleet and if so then where? Commodore Forsythe-Smythe, who was so rudely taken from his grog, grumpily agreed to consider this question and to advise Nelson at what point to split the French fleet so as to maximize their chance of success. He also agreed to predict the outcome of the battle using this strategy. What did he come up with?

28. Pure A ($C_{A0} = 100$) is fed to a mixed reactor, R and S are formed, and the following outlet concentrations are recorded:

	C_A	C_R	C_S
Run 1	75	15	10
Run 2	25	25	50

(*a*) Find a kinetic scheme to fit this data.

(*b*) What conversion level should be maintained in a mixed reactor to maximize C_R? What is $C_{R,max}$?

(*c*) In what type of single reactor, plug or mixed flow, would you expect to find the highest possible C_R? What is this C_R?

29. A 20-liter mixed reactor is to treat a reactant which decomposes as follows:

$$A \rightarrow R, \qquad r_R = k_1 C_A = (4/\text{hr}) C_A$$

$$A \rightarrow S, \qquad r_S = k_2 C_A = (1/\text{hr}) C_A$$

Find the feed rate and conversion of reactant so as to maximize profits. What are these on an hourly basis?

Data: Feed material A costs \$1.00/mol at $C_{A0} = 1$ mol/liter, product R sells for \$5.00/mol, and S has no value. The total operating cost of reactor and product separation equipment is \$25/hr + \$1.25/mol A fed to the reactor. Unconverted A is not recycled.

30. Chemical R is to be produced by the decomposition of A in a given mixed reactor. The reaction proceeds as follows:

$$A \rightarrow R, \qquad r_R = k_1 C_A$$

$$2A \rightarrow S, \qquad r_S = k_2 C_A^2$$

Let the molar cost ratio $\$_R/\$_A = M$ (S is waste material of no value), and for convenience let $k_1 = N k_2 C_{A0}$. In the feed C_{A0} is fixed.

(*a*) Ignoring operating costs, find what conversion of A should be maintained in the reactor to maximize the gross earnings and therefore the profits.

(*b*) Repeat part (*a*) with the hourly operating cost dependent on the feed rate and given by $\alpha + \beta F_{A0}$.

31. Chemicals A and B react as follows:

$$A + B \rightarrow R, \qquad r_R = k_1 C_A C_B = (68.8 \text{ liter/hr}\cdot\text{mol}) C_A C_B$$

$$2B \rightarrow S, \qquad r_S = k_2 C_B^2 = (34.4 \text{ liter/hr}\cdot\text{mol}) C_B^2$$

In this reaction 100 mol R/hr are to be produced at minimum cost in a mixed reactor. Find the feed rates of A and B to be used and the size of reactor required.

Data: Reactants are available in separate streams at $C_{A0} = C_{B0} = 0.1$ mol/liter and both cost \$0.50/mol. The cost of reactor is \$0.01/hr·liter.

8 TEMPERATURE AND PRESSURE EFFECTS

In our search for favorable conditions for reaction we have already considered how reactor type and size influence the extent of conversion and distribution of products. The reaction temperature and pressure also influence the progress of reactions, and it is the role of these variables that we now consider.

We follow a three-step procedure: First of all, we must find how equilibrium yield, rate of reaction, and product distribution are affected by changes in operating temperatures and pressures. This will allow us to determine the optimum temperature progression: in time for batch reactors, along the length of plug flow reactors, and from reactor to reactor in a series of reactors. This progression represents the ideal which we strive to approximate with a real design.

Secondly, chemical reactions are usually accompanied by heat effects, and we must know how these will change the temperature of the reacting mixture. With this information we are able to propose a number of favorable reactor and heat exchange systems—those which closely approach the optimum.

Finally, economic considerations will select one of these favorable systems as the best.

So with the emphasis on finding the optimum conditions and then seeing how best to approach them in actual design rather than determining what specific reactors will do, let us start with discussions of single reactions and follow this with the special considerations of multiple reactions.

SINGLE REACTIONS

With single reactions we are concerned with conversion level and reactor stability. Questions of product distribution do not occur.

Thermodynamics gives two important pieces of information related to conversion and reactor stability, the first being the heat liberated or absorbed for a given extent of reaction, the second being the maximum expected conversion. Let us now briefly summarize these findings. A justification of the expressions to follow as well as their many special forms and extensions can be found in the standard thermodynamics texts for chemical engineers.

Heats of Reaction from Thermodynamics

The heat liberated or absorbed during reaction depends on the nature of the reacting system, the amount of material reacting, and the temperature and pressure of the reacting system, and is calculated from the heat of reaction ΔH_r for the reaction in question. When this is not known, it can in most cases be calculated from known and tabulated thermochemical data on heats of formation ΔH_f or heats of combustion ΔH_c of the reacting materials. As a brief reminder consider the reaction

$$aA \rightarrow rR + sS$$

By convention we define the heat of reaction at temperature T as the heat transferred to the reacting system from the surroundings when a moles of A disappear to produce r moles of R and s moles of S with the system measured at the same temperature and pressure before and after the change. Thus

$$aA \rightarrow rR + sS, \qquad \Delta H_{rT} \begin{cases} \text{positive, endothermic} \\ \text{negative, exothermic} \end{cases} \tag{1}$$

Heats of Reaction and Temperature. The heat of reaction at temperature T_2 in terms of the heat of reaction at temperature T_1 is found by the law of conservation of energy as follows:

$$\begin{pmatrix} \text{heat absorbed} \\ \text{during reaction} \\ \text{at temperature} \\ T_2 \end{pmatrix} = \begin{pmatrix} \text{heat added to} \\ \text{reactants to} \\ \text{change their} \\ \text{temperature} \\ \text{from } T_2 \text{ to } T_1 \end{pmatrix} + \begin{pmatrix} \text{heat absorbed} \\ \text{during reaction} \\ \text{at temperature} \\ T_1 \end{pmatrix} + \begin{pmatrix} \text{heat added to} \\ \text{products to} \\ \text{bring them} \\ \text{back to } T_2 \\ \text{from } T_1 \end{pmatrix} \tag{2}$$

In terms of enthalpies of reactants and products this becomes

$$\Delta H_{r2} = -(H_2 - H_1)_{\text{reactants}} + \Delta H_{r1} + (H_2 - H_1)_{\text{products}} \tag{3}$$

where subscripts 1 and 2 refer to quantities measured at temperatures T_1 and T_2 respectively. In terms of specific heats

$$\Delta H_{r2} = \Delta H_{r1} + \int_{T_1}^{T_2} \nabla C_p \, dT \tag{4}$$

where

$$\nabla C_p = rC_{pR} + sC_{pS} - aC_{pA} \tag{5}$$

When the specific heats are functions of temperature as follows,

$$\begin{aligned} C_{pA} &= \alpha_A + \beta_A T + \gamma_A T^2 \\ C_{pR} &= \alpha_R + \beta_R T + \gamma_R T^2 \\ C_{pS} &= \alpha_S + \beta_S T + \gamma_S T^2 \end{aligned} \tag{6}$$

we obtain

$$\begin{aligned} \Delta H_{r2} &= \Delta H_{r1} + \int_{T_1}^{T_2} (\nabla\alpha + \nabla\beta T + \nabla\gamma T^2) \, dT \\ &= \Delta H_{r1} + \nabla\alpha(T_2 - T_1) + \frac{\nabla\beta}{2}(T_2^2 - T_1^2) + \frac{\nabla\gamma}{3}(T_2^3 - T_1^3) \end{aligned} \tag{7}$$

where

$$\begin{aligned} \nabla\alpha &= r\alpha_R + s\alpha_S - a\alpha_A \\ \nabla\beta &= r\beta_R + s\beta_S - a\beta_A \\ \nabla\gamma &= r\gamma_R + s\gamma_S - a\gamma_A \end{aligned} \tag{8}$$

Knowing the heat of reaction at any one temperature as well as the specific heats of reactants and products in the temperature range concerned allows us to calculate the heat of reaction at any other temperature. From this the heat effects of the reaction can be found.

Equilibrium Constants from Thermodynamics

From the second law of thermodynamics, equilibrium constants, hence equilibrium compositions of reacting systems, may be calculated. We must remember, however, that real systems do not necessarily achieve this conversion; therefore the conversions calculated from thermodynamics are only suggested attainable values.

As a brief reminder, the standard free energy ΔG° for the reaction of Eq. 1 at temperature T is defined as

$$\Delta G^\circ = rG^\circ_{\mathrm{R}} + sG^\circ_{\mathrm{S}} - aG^\circ_{\mathrm{A}} = -RT\ln K = -RT\ln \frac{\left(\dfrac{f}{f^\circ}\right)_{\mathrm{R}}^{r}\left(\dfrac{f}{f^\circ}\right)_{\mathrm{S}}^{s}}{\left(\dfrac{f}{f^\circ}\right)_{\mathrm{A}}^{a}} \tag{9}$$

where f is the fugacity of the component at equilibrium conditions, f° is the fugacity of the component at the arbitrarily selected standard state at temperature T, the same one used in calculating ΔG°, G° is the standard free energy of a reacting component, tabulated for many compounds, and K is the thermodynamic equilibrium constant for the reaction. Standard states at given temperature are commonly chosen as follows:

Gases—pure component at one atmosphere, at which pressure ideal gas behavior is closely approximated.
Solid—pure solid component at unit pressure.
Liquid—pure liquid at its vapor pressure.
Solute in liquid—1 molar solution; or at such dilute concentrations that the activity is unity.

For convenience define

$$K_f = \frac{f_{\mathrm{R}}^{\,r} f_{\mathrm{S}}^{\,s}}{f_{\mathrm{A}}^{\,a}}, \qquad K_p = \frac{p_{\mathrm{R}}^{\,r} p_{\mathrm{S}}^{\,s}}{p_{\mathrm{A}}^{\,a}}, \qquad K_y = \frac{y_{\mathrm{R}}^{\,r} y_{\mathrm{S}}^{\,s}}{y_{\mathrm{A}}^{\,a}}, \qquad K_C = \frac{C_{\mathrm{R}}^{\,r} C_{\mathrm{S}}^{\,s}}{C_{\mathrm{A}}^{\,a}} \tag{10}$$

and

$$\Delta n = r + s - a$$

Simplified forms of Eq. 9 can be obtained for various systems. For *gas reactions* standard states are usually chosen at a pressure of 1 atm. At this low pressure the deviation from ideality invariably is small; hence fugacity and pressure are identical and $f^\circ = p^\circ = 1$ atm. Thus

$$K = e^{-\Delta G^\circ/RT} = K_f\{p^\circ = 1\ \mathrm{atm}\}^{-\Delta n} \tag{11}$$

The term in braces in this equation and in Eq. 13 is always unity but is retained to keep the equations dimensionally correct.

For *ideal gases*

$$f_i = p_i = y_i\pi = C_iRT \tag{12}$$

for any component i. Hence

$$K_f = K_p$$

and

$$K = \frac{K_p}{\{p^\circ = 1\ \mathrm{atm}\}^{\Delta n}} = \frac{K_y\pi^{\Delta n}}{\{p^\circ = 1\ \mathrm{atm}\}^{\Delta n}} = \frac{K_C(RT)^{\Delta n}}{\{p^\circ = 1\ \mathrm{atm}\}^{\Delta n}} \tag{13}$$

For a *solid component* taking part in a reaction, fugacity variations with pressure are small and can usually be ignored. Hence

$$\left(\frac{f}{f^\circ}\right)_{\text{solid component}} = 1 \tag{14}$$

Equilibrium Conversion. The equilibrium composition, as governed by the equilibrium constant, changes with temperature, and from thermodynamics the rate of this change is given by

$$\frac{d(\ln K)}{dT} = \frac{\Delta H_r}{RT^2} \tag{15}$$

On integrating Eq. 15, we see how the equilibrium constant changes with temperature. When the heat of reaction ΔH_r can be considered to be constant in the temperature interval, integration yields

$$\ln\frac{K_2}{K_1} = -\frac{\Delta H_r}{R}\left(\frac{1}{T_2} - \frac{1}{T_1}\right) \tag{16}$$

When the variation of ΔH_r must be accounted for in the integration we have

$$\ln\frac{K_2}{K_1} = \frac{1}{R}\int_{T_1}^{T_2}\frac{\Delta H_r}{T^2}\,dT \tag{17}$$

Where ΔH_r is given by a special form of Eq. 4 in which subscript 0 refers to the base temperature,

$$\Delta H_r = \Delta H_{r0} + \int_{T_0}^{T} \nabla C_p\,dT \tag{18}$$

Replacing Eq. 18 in Eq. 17 and integrating, while using the temperature dependency for C_p given by Eq. 8, gives

$$R\ln\frac{K_2}{K_1} = \nabla\alpha \ln\frac{T_2}{T_1} + \frac{\nabla\beta}{2}(T_2 - T_1) + \frac{\nabla\gamma}{6}(T_2{}^2 - T_1{}^2)$$

$$+\left(-\Delta H_{r0} + \nabla\alpha T_0 + \frac{\nabla\beta}{2}T_0{}^2 + \frac{\nabla\gamma}{3}T_0{}^3\right)\left(\frac{1}{T_2} - \frac{1}{T_1}\right) \tag{19}$$

These expressions allow us to find the variation of the equilibrium constant, hence equilibrium conversion, with temperature.

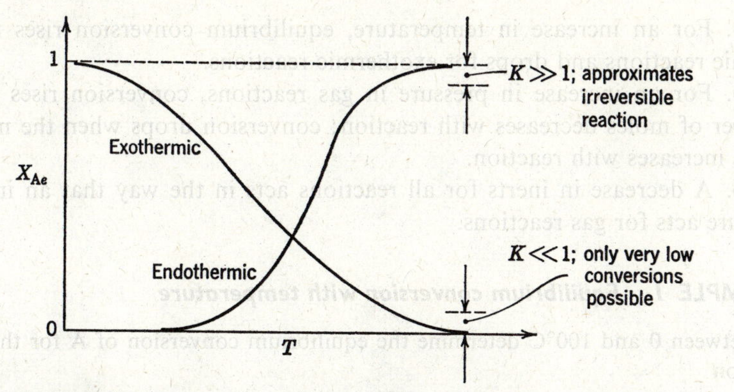

FIGURE 1. Effect of temperature on equilibrium conversion as predicted by thermodynamics (pressure fixed).

The following conclusions may be drawn from thermodynamics. These are illustrated in part by Figs. 1 and 2.

1. The thermodynamic equilibrium constant is unaffected by the pressure of the system, by the presence or absence of inerts, or by the kinetics of the reaction, but is affected by the temperature of the system.

2. Though the thermodynamic equilibrium constant is unaffected by pressure or inerts, the equilibrium concentration of materials and equilibrium conversion of reactants can be influenced by these variables.

3. $K \gg 1$ indicates that practically complete conversion is possible and that the reaction can be considered to be irreversible. $K \ll 1$ indicates that reaction will not proceed to any appreciable extent.

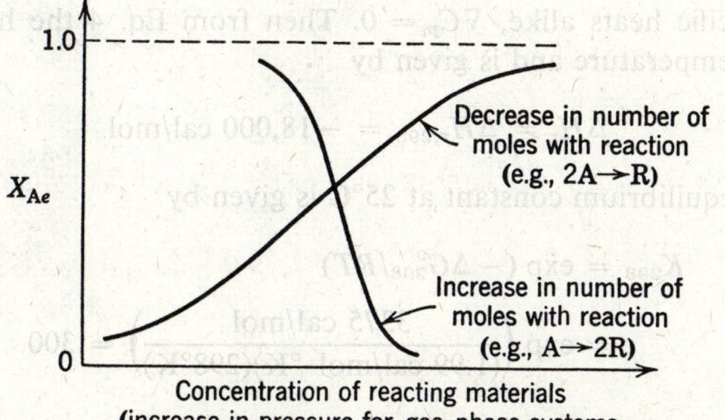

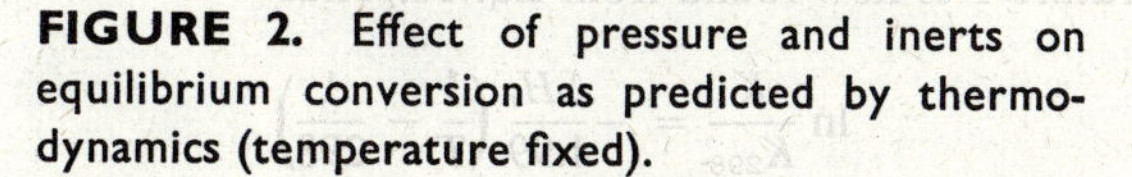

FIGURE 2. Effect of pressure and inerts on equilibrium conversion as predicted by thermodynamics (temperature fixed).

4. For an increase in temperature, equilibrium conversion rises for endothermic reactions and drops for exothermic reactions.

5. For an increase in pressure in gas reactions, conversion rises when the number of moles decreases with reaction; conversion drops when the number of moles increases with reaction.

6. A decrease in inerts for all reactions acts in the way that an increase in pressure acts for gas reactions.

EXAMPLE 1. Equilibrium conversion with temperature

(*a*) Between 0 and 100°C determine the equilibrium conversion of A for the aqueous reaction

$$A \rightleftharpoons R \qquad \begin{aligned} \Delta G^\circ_{298} &= -3375 \text{ cal/mol} \\ \Delta H_{r,298} &= -18{,}000 \text{ cal/mol} \end{aligned}$$

Present the result in the form of a plot of conversion versus temperature.

(*b*) What restrictions should be placed on a reactor operating isothermally if we are to obtain fractional conversions of 75% or higher?

Data. The reported ΔG° is based on the following standard states of reactants and products:

$$C^\circ_R = C^\circ_A = 1 \text{ mol/liter}$$

Assume ideal solution, in which case

$$K = \frac{C_R/C^\circ_R}{C_A/C^\circ_A} = \frac{C_R}{C_A} = K_C$$

In addition, assume specific heats of all solutions are equal to that of water.

SOLUTION

(*a*) With all specific heats alike, $\nabla C_p = 0$. Then from Eq. 4 the heat of reaction is independent of temperature and is given by

$$\Delta H_r = \Delta H_{r,298} = -18{,}000 \text{ cal/mol} \tag{i}$$

From Eq. 9 the equilibrium constant at 25°C is given by

$$\begin{aligned} K_{298} &= \exp\left(-\Delta G^\circ_{298}/RT\right) \\ &= \exp\left(\frac{3375 \text{ cal/mol}}{(1.99 \text{ cal/mol}\cdot{}^\circ\text{K})(298^\circ\text{K})}\right) = 300 \end{aligned} \tag{ii}$$

Since the heat of reaction does not change with temperature, the equilibrium constant K at any temperature T is now found from Eq. 16. Thus

$$\ln \frac{K}{K_{298}} = -\frac{\Delta H_r}{1.99}\left(\frac{1}{T} - \frac{1}{298}\right)$$

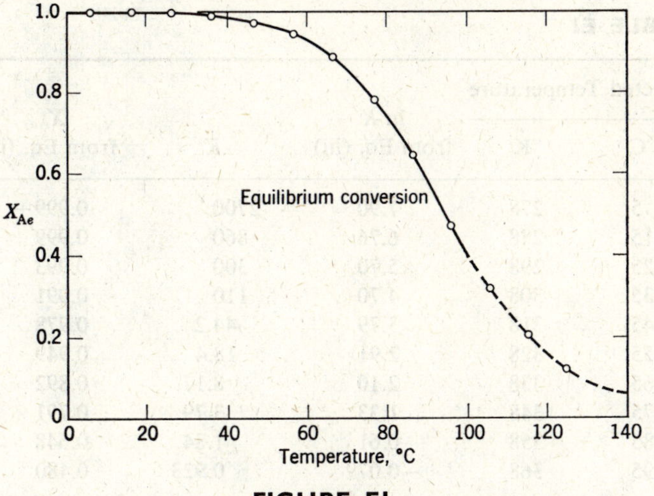

FIGURE E1.

Combining with Eqs. (i) and (ii) gives

$$\ln \frac{K}{300} = \frac{18{,}000}{1.99}\left(\frac{1}{T} - \frac{1}{298}\right)$$

or

$$\ln K = \frac{18{,}000}{RT} - 24.7 \tag{iii}$$

Values of ln K and K, calculated from Eq. (iii), are shown for 10°C intervals in Table E1. From thermodynamics

$$K = \frac{C_{Re}}{C_{Ae}} = \frac{C_{A0} - C_{Ae}}{C_{Ae}} = \frac{X_{Ae}}{1 - X_{Ae}}$$

Hence the fractional conversion at equilibrium is given by

$$X_{Ae} = \frac{K}{1 + K} \tag{iv}$$

Table E1 also gives the values of X_{Ae} calculated from Eq. (iv), and Fig. E1 shows the changing equilibrium conversion as a function of temperature in the range of 0 to 100°C.

(*b*) From the graph we see that the temperature must stay below 78°C if conversions of 75% or higher are to be obtained.

TABLE EI

Selected Temperature °C	°K	ln K from Eq. (iii)	K	X_{Ae} from Eq. (iv)
5	278	7.90	2700	0.999+
15	288	6.76	860	0.999
25	298	5.70	300	0.993
35	308	4.70	110	0.991
45	318	3.79	44.2	0.978
55	328	2.91	18.4	0.949
65	338	2.10	8.17	0.892
75	348	1.33	3.79	0.791
85	358	0.61	1.84	0.648
95	368	−0.079	0.923	0.480

General Graphical Design Procedure

Temperature, composition, and reaction rate are uniquely related for any single homogeneous reaction, and this may be represented graphically in one of three ways, as shown in Fig. 3. The first of these, the composition-temperature plot, is the most convenient so we will use it throughout to represent data, to calculate reactor sizes, and to compare design alternatives.

For a given feed (fixed C_{A0}, C_{B0}, . . .) and using conversion of key component as a measure of the composition and extent of reaction, the X_A versus T plot has the general shape as shown in Fig. 4. This plot can be prepared either from a thermodynamically consistent rate expression for the reaction (the rate must be zero at equilibrium) or by interpolating from a given set of kinetic data in con-

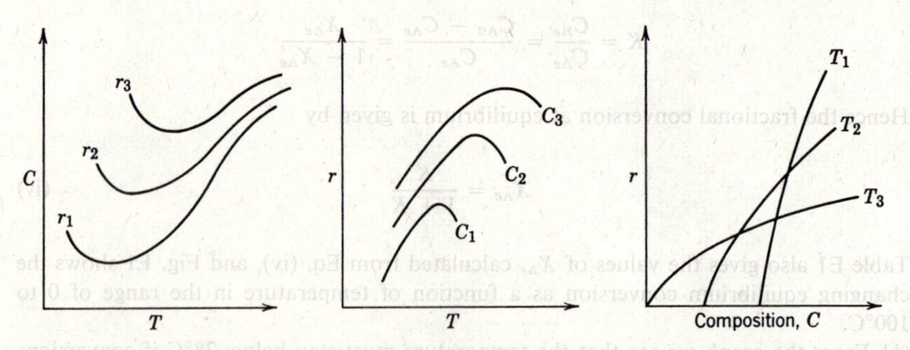

FIGURE 3. Different ways of representing the relationship of temperature, composition, and rate for a single homogeneous reaction.

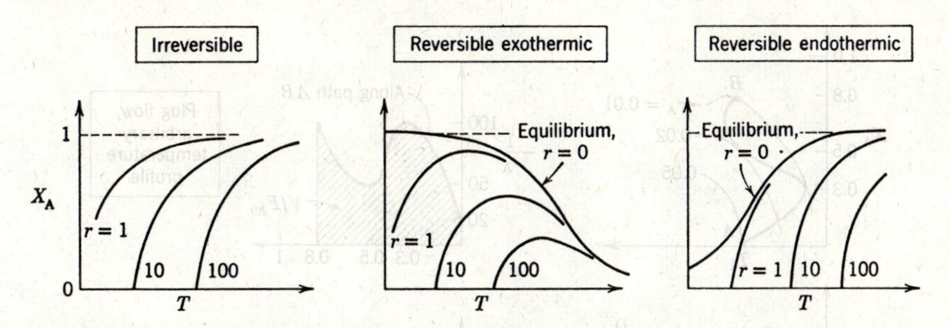

FIGURE 4. General shape of the temperature-conversion plot for different reaction types.

junction with thermodynamic information on the equilibrium. Naturally the reliability of all the calculations and predictions which follow are directly dependent on the accuracy of this chart. Hence it is imperative to obtain good kinetic data to construct this chart.

The size of reactor required for a given duty and for a given temperature progression is found as follows:

1. Draw the reaction path on the X_A versus T plot. This is the *operating line* for the operation.
2. Find the rate at various X_A along this path.
3. Plot the $1/(-r_A)$ versus X_A curve for this path.
4. Find the area under this curve. This gives V/F_{A0}.

This procedure is illustrated in Fig. 5 for three paths: path AB for plug flow with an arbitrary temperature profile, path CD for nonisothermal plug flow with 50% recycle, and point E for mixed flow. Note that for mixed flow the operating line reduces to a single point.

This procedure is quite general, applicable for any kinetics, any temperature progression, and any reactor type or any series of reactors. So, once the operating line is known, the reactor size can be found by the procedure just outlined.

Optimum Temperature Progression

We define the optimum temperature progression to be that progression which minimizes V/F_{A0} for a given conversion of reactant. This optimum may be an isothermal or it may be a changing temperature: in time for a batch reactor, along the length of a plug flow reactor, or from stage to stage for a series of mixed reactors. It is important to know what is this progression because it is the ideal which we try to approach with a real system. It also allows us to estimate how far any real system departs from this ideal.

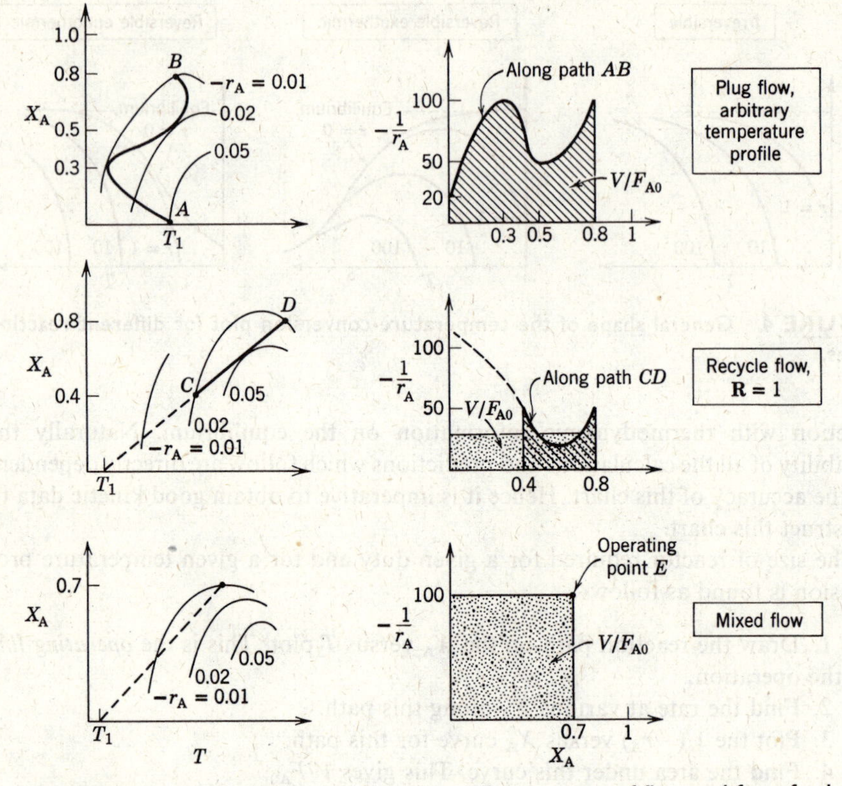

FIGURE 5. Finding the reactor size for different types of flow and for a feed temperature T_1.

The condition for the optimum temperature progression in a given type of reactor is as follows: Whatever the composition, always have the system at the temperature where the rate is a maximum. The locus of maximum rates is found by examining the $r(T, C)$ curves of Fig. 4; Fig. 6 shows this progression.

For irreversible reactions, the rate always increases with temperature at any composition, so the highest rate occurs at the highest allowable temperature. This temperature is set by the materials of construction or by the possible increasing importance of side reactions.

For endothermic reactions a rise in temperature increases both equilibrium conversion and the rate of reaction. Thus, as with irreversible reactions, the highest allowable temperature should be used.

For exothermic reversible reactions the situation is different for here two opposing factors are at work when the temperature is raised—the rate of forward reaction speeds up but the maximum attainable conversion decreases. Because of this,

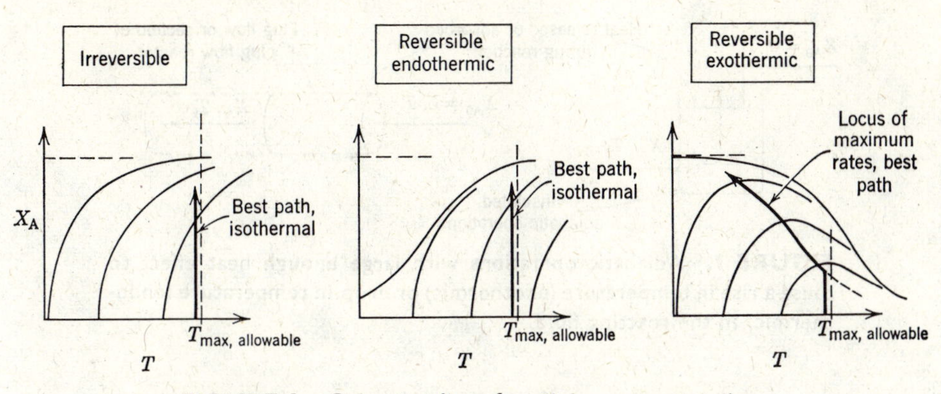

FIGURE 6. Operating lines for minimum reactor size.

where the system is far from equilibrium use a high temperature to take advantage of the high rate; where equilibrium is approached and dominates, lower the temperature to shift the equilibrium to a more favorable value. Thus in general, for reversible exothermic reactions the optimum progression is a changing temperature; a high starting temperature which decreases as conversion rises. Figure 6 shows this progression and its precise values are found by connecting the maxima of the different rate curves. We call this line the *locus of maximum rates*.

Heat Effects

When the heat absorbed or released by reaction can markedly change the temperature of the reacting fluid, this factor must be accounted for in design. Thus we need to use both the material and energy balance expressions, Eqs. 4.1 and 4.2, rather than the material balance alone, which was the starting point of all the analyses of isothermal operations of Chapters 5 and 6.

First of all if the reaction is exothermic and if heat transfer is unable to remove all of the liberated heat, then the temperature of the reacting fluid will rise as conversion rises. By similar arguments, for endothermic reactions the fluid cools as conversion rises. Let us relate this temperature change with extent of conversion.

We start with adiabatic operations, later extending the treatment to account for heat interchange with the surroundings.

Adiabatic Operations

Consider either a mixed flow reactor, a plug flow reactor, or a section of plug flow reactor, in which conversion is X_A, as shown in Fig. 7. In Chapters 5 and 6 one component, usually the limiting reactant, was selected as the basis for all material

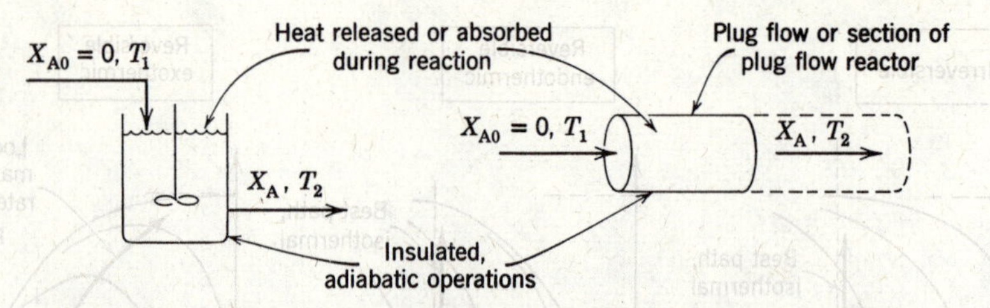

FIGURE 7. Adiabatic operations with large enough heat effect to cause a rise in temperature (exothermic) or drop in temperature (endothermic) in the reacting fluid.

balance calculations. The same procedure is followed here, with limiting reactant A taken as the basis. Let

Subscripts 1, 2 refer to temperatures of entering and leaving streams.
$\mathbf{C}_p'$, $\mathbf{C}_p''$ = mean specific heat of unreacted feed stream and of completely converted product stream per mole of entering reactant A.
$\mathbf{H}'$, $\mathbf{H}''$ = enthalpy of unreacted feed stream and of completely converted product stream per mole of entering reactant A.
$\Delta\mathbf{H}_r$ = heat of reaction per mole of entering reactant A.

With T_1 as the reference temperature on which enthalpies and heats of reaction are based we have

Enthalpy of entering feed:

$$\mathbf{H}_1' = \mathbf{C}_p'(T_1 - T_1) = 0 \text{ cal/mol A}$$

Enthalpy of leaving stream:

$$\mathbf{H}_2''X_A + \mathbf{H}_2'(1 - X_A) = \mathbf{C}_p''(T_2 - T_1)X_A + \mathbf{C}_p'(T_2 - T_1)(1 - X_A) \text{ cal/mol A}$$

Energy absorbed by reaction:

$$\Delta\mathbf{H}_{r1}X_A \text{ cal/mol A}$$

Replacing these quantities in the energy balance,

$$\text{input} = \text{output} + \text{accumulation} + \text{disappearance by reaction} \tag{4.2}$$

we obtain at steady state

$$0 = [\mathbf{C}_p''(T_2 - T_1)X_A + \mathbf{C}_p'(T_2 - T_1)(1 - X_A)] + \Delta\mathbf{H}_{r1}X_A \tag{20}$$

By rearranging,

$$X_A = \frac{C'_p(T_2 - T_1)}{-\Delta H_{r1} - (C''_p - C'_p)(T_2 - T_1)} = \frac{C'_p \, \Delta T}{-\Delta H_{r1} - (C''_p - C'_p)\, \Delta T} \tag{21}$$

or, with Eq. 18

operación adiabática:

$$X_A = \frac{C'_p \, \Delta T}{-\Delta H_{r2}} = \left(\frac{\text{heat needed to raise feed stream to } T_2}{\text{heat released by reaction at } T_2} \right) \tag{22}$$

which for complete conversion becomes

$$-\Delta H_{r2} = C'_p \, \Delta T, \qquad \text{for } X_A = 1 \tag{23}$$

The last form of the equation simply states that the heat released by reaction just balances the heat necessary to raise the reactants from T_1 to T_2.

The relation between temperature and conversion as given by the energy balances of Eq. 21 or 22 is shown in Fig. 8. The resulting lines are straight for all practical purposes since the variation of the denominator term of these equations is relatively

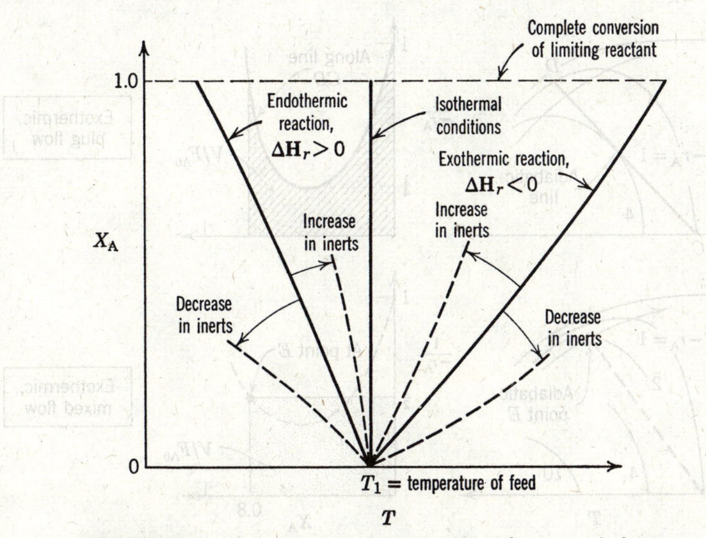

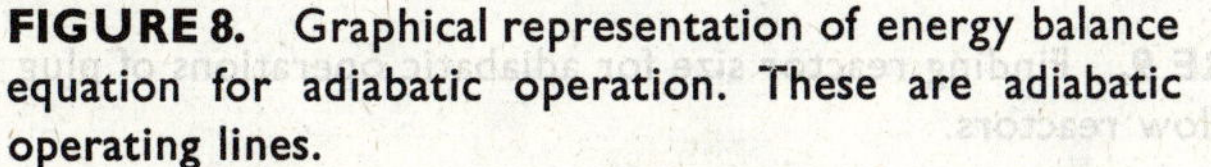

FIGURE 8. Graphical representation of energy balance equation for adiabatic operation. These are adiabatic operating lines.

small. When $\mathbf{C}_p'' - \mathbf{C}_p' = 0$, the heat of reaction is independent of temperature and Eqs. 21 and 22 reduce to

$$X_A = \frac{\mathbf{C}_p \, \Delta T}{-\Delta \mathbf{H}_r} \tag{24}$$

which is a straight line in Fig. 8.

This figure illustrates the shape of the energy balance curve for both endothermic and exothermic reactions for both mixed and plug flow reactors. This representation shows that whatever the conversion at any point in the reactor, the temperature is at its corresponding value on the curve. For plug flow the fluid in the reactor moves progressively along the curve, for mixed flow the fluid immediately jumps to

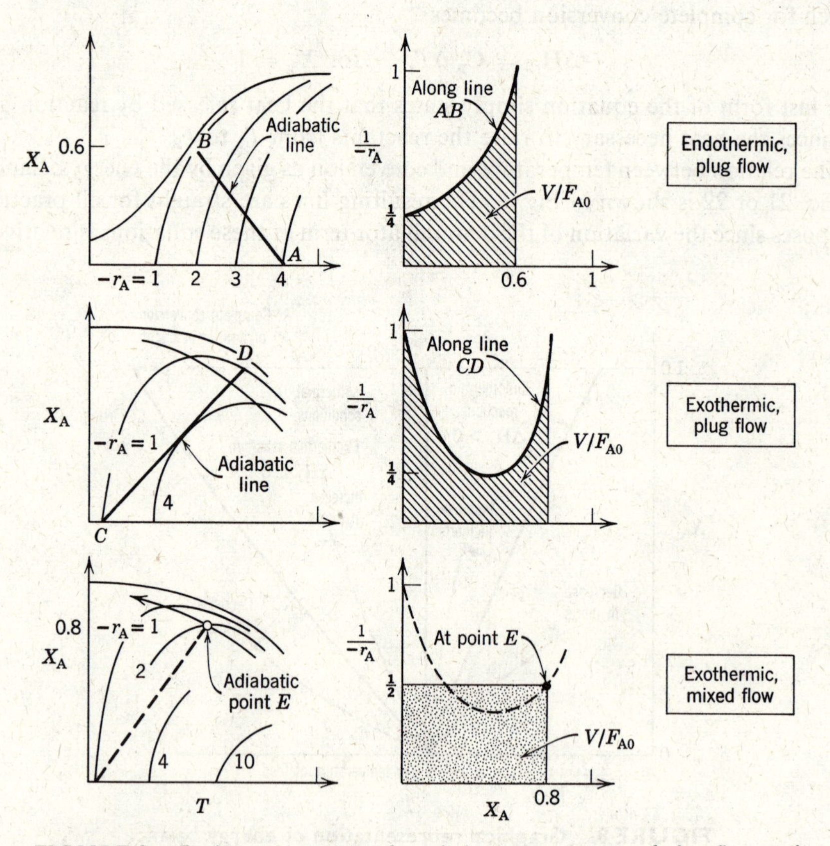

FIGURE 9. Finding reactor size for adiabatic operations of plug flow and mixed flow reactors.

its final value on the curve. These are the *adiabatic operating lines* for the reactor. With increased inerts $\mathbf{C_p}$ rises, and these curves become more closely vertical. A vertical line indicates that temperature is unchanged as reaction proceeds. This then is the special case of isothermal reactions treated in Chapters 5–7.

The *size of reactor* needed for a given duty is found as follows. For plug flow tabulate the rate for various X_A along this adiabatic operating line, prepare the $1/(-r_A)$ versus X_A plot and integrate. For mixed flow simply use the rate at the conditions within the reactor. Figure 9 illustrates this procedure.

Note that this procedure in fact simultaneously solves the governing material and energy balance equations. Moving along the adiabatic operating line satisfies the energy balance; finding the rates along this line and evaluating V/F_{A0} satisfies the material balance.

The *best adiabatic operations* of a single plug flow reactor is found by shifting the operating line (changing the inlet temperature) to where the rates have the highest mean value. For endothermic operations this means starting at the highest allowable temperature. For exothermic reactions this means straddling the locus of maximum rates as shown in Fig. 10. A few trials will locate the best inlet temperature, that which minimizes V/F_{A0}. For mixed flow the reactor should operate on the locus of maximum rates, again shown in Fig. 10.

The best reactor type, that which minimizes V/F_{A0}, is found directly from this X_A versus T graph. If the rate progressively decreases with conversion then use plug flow. This is the case for endothermic reactions (Fig. 9*a*) and close to isothermal exothermic reactions. For exothermic reactions which have a large

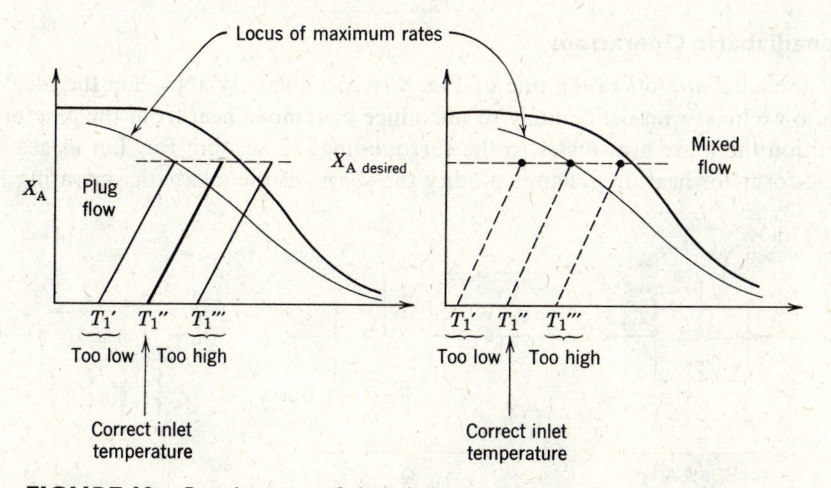

FIGURE 10. Best location of the adiabatic operating line. For plug flow a trial and error search is needed to find this line; for mixed flow no search is needed.

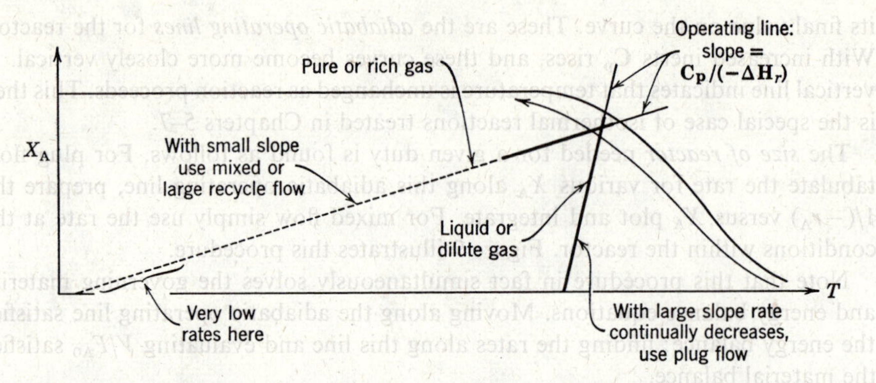

FIGURE 11. For exothermic reactions mixed flow is best where the temperature rise is large; plug flow is best for close to isothermal systems.

temperature rise during reaction the rate rises from a very low value to a maximum at some intermediate X_A, then falls. This behavior is characteristic of autocatalytic reactions, thus recycle operations are best. Figure 11 illustrates two situations, one where plug flow is best, the other where large recycle or mixed flow is best. The slope of the operating line, $\mathbf{C_p}/-\Delta\mathbf{H}_r$, will determine which case one has at hand. Thus:

1. for small $\mathbf{C_p}/-\Delta\mathbf{H}_r$ (pure gaseous reactants) mixed flow is best.
2. for large $\mathbf{C_p}/-\Delta\mathbf{H}_r$ (gas with much inerts, or liquid systems) plug flow is best.

Nonadiabatic Operations

For the adiabatic operating line of Fig. 8 to more closely approach the ideals of Fig. 6 we may want deliberately to introduce or remove heat from the reactor. In addition there are heat losses to the surroundings to account for. Let us see how these forms of heat interchange modify the shape of the adiabatic operating line.

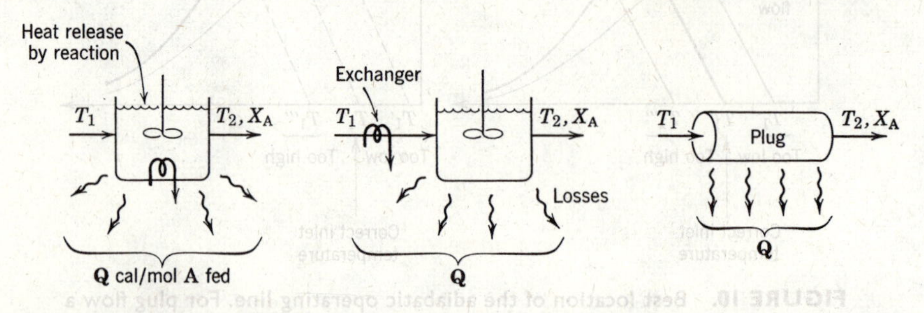

FIGURE 12. Nonadiabatic operations account for heat losses, deliberate heat exchange, and heat release by reaction.

Referring to Fig. 12 let **Q** be the total heat *added* to a mixed reactor, a plug flow reactor, or section of plug flow reactor per mole of entering reactant A, and let this heat also include the losses to the surroundings. Then Eq. 20, the energy balance about the system, is modified to account for this heat interchange and becomes

$$\mathbf{Q} = \mathbf{C}_p''(T_2 - T_1)X_A + \mathbf{C}_p'(T_2 - T_1)(1 - X_A) + \Delta\mathbf{H}_{r1}X_A$$

which on rearrangement and with Eq. 18 gives

$$X_A = \frac{\mathbf{C}_p' \, \Delta T - \mathbf{Q}}{-\Delta\mathbf{H}_{r2}} = \left(\frac{\text{net heat still needed after heat transfer to raise feed to } T_2}{\text{heat released by reaction at } T_2}\right) \tag{25}$$

and for $\mathbf{C}_p'' = \mathbf{C}_p'$, which often is a reasonable approximation,

$$X_A = \frac{\mathbf{C}_p \, \Delta T - \mathbf{Q}}{-\Delta\mathbf{H}_r} \tag{26}$$

With heat input proportional to $\Delta T = T_2 - T_1$ the energy balance line rotates about T_1. This change is shown in Fig. 13. Other modes of heat addition or removal yield corresponding shifts in the energy balance line.

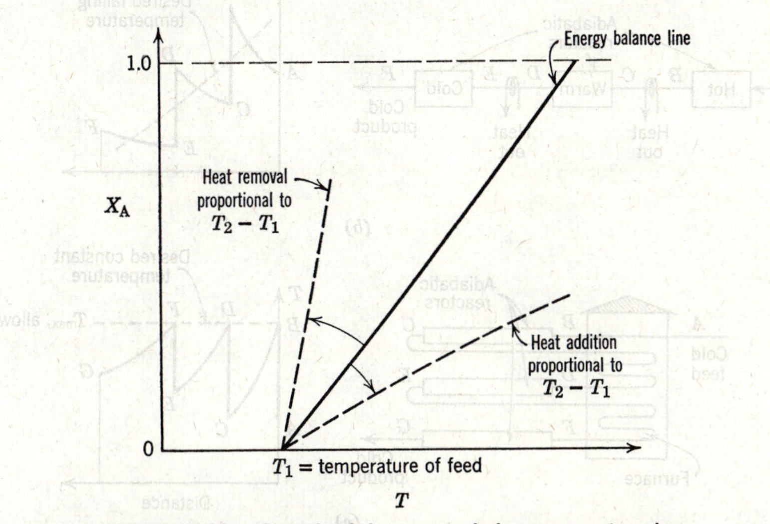

FIGURE 13. Sketch of the energy balance equation showing the shift in adiabatic line caused by heat exchange with surroundings.

Using this modified operating line the procedure for finding the reactor size and optimum operations follows directly from the discussion on adiabatic operations.

Comments and Extensions

Adiabatic operations of an exothermic reaction give a rising temperature with conversion. However, the desired progression is one of falling temperature. Thus very drastic heat removal may be needed to make the operating line approach the ideal, and many schemes may be proposed to do this. As an example we may have heat exchange with the incoming fluid (see Fig. 14*a*), a case treated by van Heerden (1953, 1958). Another alternative is to have multistage operations with interstage cooling between adiabatic sections (see Fig. 14*b*). In general, multistaging is used

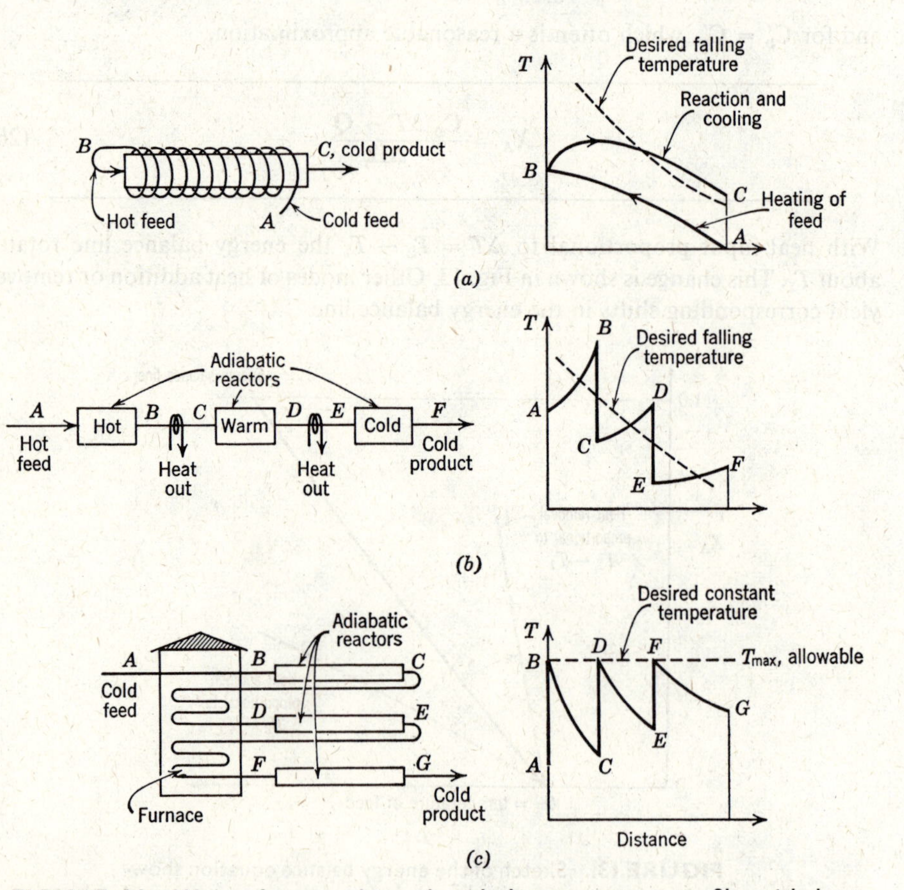

FIGURE 14. Ways of approaching the ideal temperature profile with heat exchange: (*a*) and (*b*) exothermic reaction; (*c*) endothermic reaction.

when it is impractical to effect the necessary heat exchange within the reactor itself. This is usually the case with gas-phase reactions with their relatively poor heat transfer characteristics. For endothermic reactions multistaging with reheat between stages is commonly used to keep the temperature from dropping too low (see Fig. 14*c*).

Since the main use of these and many other forms of multistage operations is with solid catalyzed gas-phase reactions we discuss these operations in Chap. 14. Design for homogeneous reactions parallels that for catalytic reactions, so the reader is referred to Chap. 14 for the development.

Exothermic Reactions in Mixed Reactors—a Special Problem

For exothermic reactions in mixed flow (or close to mixed flow) an interesting situation may develop in that more than one reactor composition may satisfy the governing material and energy balance equations. This means that we may not know which conversion level to expect. Let us examine this problem.

First, consider reactant fluid fed at a given rate (fixed τ or V/F_{A0}) to a mixed reactor. At each reactor temperature there will be some particular conversion which satisfies the material balance equation, Eq. 5.11. At low temperature the rate is low so this conversion is low. At higher temperature the conversion rises and approaches the equilibrium. At a still higher temperature we enter the region of falling equilibrium so the conversion for given τ will likewise fall. Figure 15 illustrates this behavior for different τ values. Note that these lines do not represent an operating line or a reaction path. Actually each point on these curves represents a particular solution of the material balance equations, thus represents an operating point or condition for the mixed reactor.

Now, for a given feed temperature T_1 the intersection of the energy balance line with the S-shaped material balance line for the operating τ gives the conditions

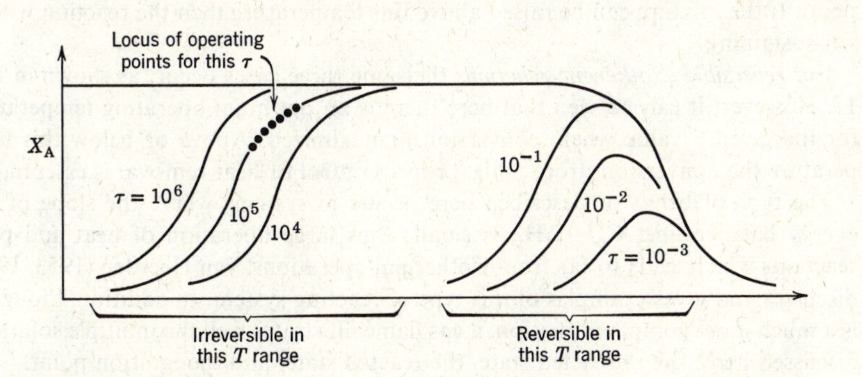

FIGURE 15. Conversion in a mixed reactor as a function of T and τ; from the material balance equation, Eq. 5.11.

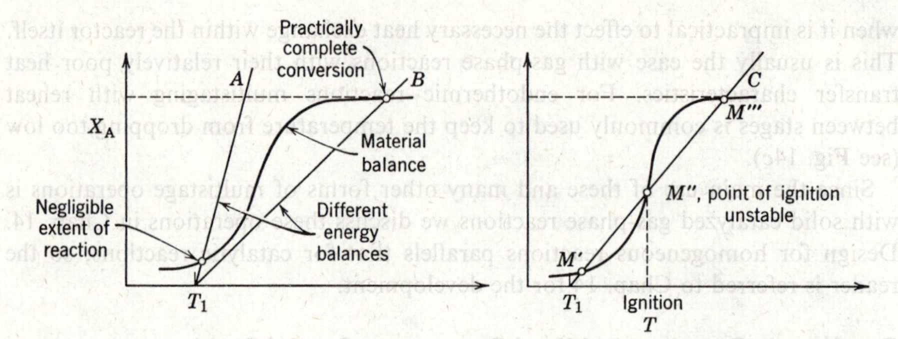

FIGURE 16. **Three types of solutions to the energy and material balances for exothermic irreversible reactions.**

within the reactor. Here three cases may be distinguished. These are shown in Fig. 16 for *irreversible reactions.*

First, the energy balance line T_1A represents the situation where insufficient heat is liberated by reaction to raise the temperature to a high enough level for the reaction to be self-sustaining. Hence conversion is negligible. At the other extreme if we have more than enough heat liberated the fluid will be hot and conversion essentially complete. This is shown as line T_1B. Finally line T_1C indicates an intermediate situation which has three solutions to the material and energy balance equations, points M', M'', and M'''. However, point M'' is an unstable state because with a small rise in temperature the heat produced (with the rapidly rising material balance curve) is greater than the heat consumed by the reacting mixture (energy balance curve). The excess heat produced will make the temperature rise until M''' is reached. By similar reasoning, if the temperature drops slightly below M'' it will continue to drop until M' is reached. Thus we look upon M'' as the ignition point. If the mixture can be raised above this temperature then the reaction will be self-sustaining.

For *reversible exothermic reactions* the same three cases occur, as shown in Fig. 17. However, it can be seen that here there is an optimum operating temperature for the given τ value where conversion is maximized. Above or below this temperature the conversion drops; thus proper control of heat removal is essential.

The type of behavior described here occurs in systems where the slope of the energy balance line, $\mathbf{C}_p/-\Delta\mathbf{H}_r$, is small; thus large liberation of heat and pure reactants which leads to far from isothermal operations. van Heerden (1953, 1958) discusses and gives examples of this type of reacting system. In addition, though it is a much more complex situation, a gas flame illustrates well the multiple solutions discussed here: the unreacted state, the reacted state, and the ignition point.

Reactor dynamics, stability, and start-up procedures are particularly important for auto-induced reactions such as these. For example a small change in feed rate

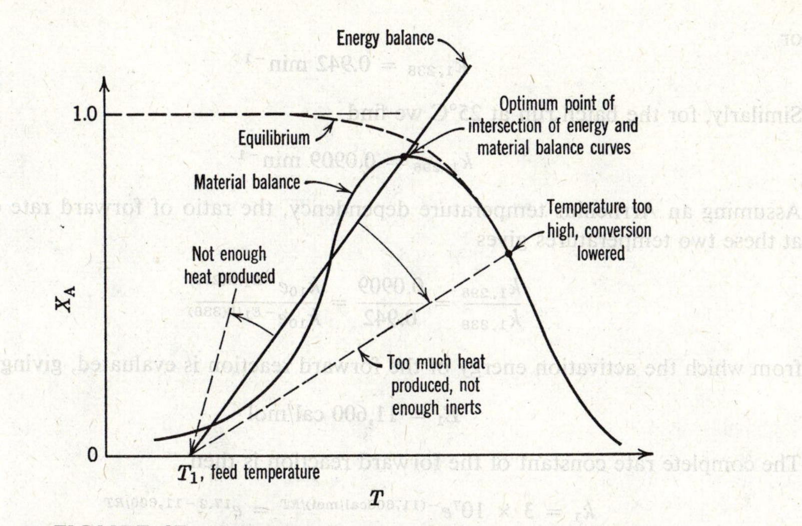

FIGURE 17. Solution of energy and material balances for reversible exothermic reaction.

(τ value), feed composition or temperature, or heat transfer rate may cause the reactor output to jump from one operating point to the other.

EXAMPLE 2. Construction of the rate-conversion-temperature chart from kinetic data

With the system of Example 1 and starting with an R-free solution, kinetic experiments in a batch reactor give 58.1% conversion in 1 min at 65°C, 60% conversion in 10 min at 25°C. Assuming reversible first-order kinetics find the rate expression for this reaction and prepare the conversion-temperature chart with reaction rate as parameter.

SOLUTION

Integrate the performance equation. For a reversible first-order reaction the performance equation for a batch reactor is

$$t = C_{A0}\int \frac{dX_A}{-r_A} = C_{A0}\int \frac{dX_A}{k_1 C_A - k_2 C_R} = \frac{1}{k_1}\int_0^{X_A} \frac{dX_A}{1 - X_A/X_{Ae}}$$

According to Eq. 3.54 this integrates to give

$$\frac{k_1 t}{X_{Ae}} = -\ln\left(1 - \frac{X_A}{X_{Ae}}\right) \qquad \text{(i)}$$

Calculate the Forward Rate Constant. From the batch run at 65°C, noting from Example 1 that $X_{Ae} = 0.89$, we find with Eq. (i)

$$\frac{k_1(1\ \text{min})}{0.89} = -\ln\left(1 - \frac{0.581}{0.89}\right)$$

or

$$k_{1,338} = 0.942 \text{ min}^{-1} \tag{ii}$$

Similarly, for the batch run at 25°C we find

$$k_{1,298} = 0.0909 \text{ min}^{-1} \tag{iii}$$

Assuming an Arrhenius temperature dependency, the ratio of forward rate constants at these two temperatures gives

$$\frac{k_{1,298}}{k_{1,338}} = \frac{0.0909}{0.942} = \frac{k_{10}e^{-E_1/R(298)}}{k_{10}e^{-E_1/R(338)}}$$

from which the activation energy of the forward reaction is evaluated, giving

$$E_1 = 11{,}600 \text{ cal/mol}$$

The complete rate constant of the forward reaction is then

$$k_1 = 3 \times 10^7 e^{-(11{,}600\text{cal/mol})/RT} = e^{17.2 - 11{,}600/RT}$$

Noting that $k_1/k_2 = K$ where K is given in Example 1, we then can find the value for k_2.

Summary. For the reversible first-order reaction of Examples 1 and 2 we have

---

$$A \underset{2}{\overset{1}{\rightleftarrows}} R; \qquad K = \frac{C_{Re}}{C_{Ae}}; \qquad -r_A = r_R = k_1C_A - k_2C_R$$

Equilibrium: $K = e^{18{,}000/RT - 24.7}$

Rate constants: $k_1 = e^{17.2 - 11{,}600/RT}$, min^{-1}

$k_2 = e^{41.9 - 29{,}600/RT}$, min^{-1}

From these values the X_A versus T chart for any specific C_{A0} can be prepared, and for this purpose the electronic computer is a great timesaver. Figure E2 is such a plot prepared for $C_{A0} = 1$ mol/liter and $C_{R0} = 0$. Also see the dust jacket of this book.

Since we are dealing with first-order reactions this plot can be used for any C_{A0} value by properly relabeling the rate curves. Thus for $C_{A0} = 10$ mol/liter simply multiply all the rate values on this graph by a factor of 10.

EXAMPLE 3. *Performance for the optimal temperature progression*

Using the optimal temperature progression in a plug flow reactor for the reaction of the previous examples

(*a*) Calculate the space time needed for 80% conversion of a feed where $C_{A0} = 1$ mol/liter.

(*b*) Plot the temperature and conversion profile along the length of the reactor.

Let the maximum allowable operating temperature be 95°C.

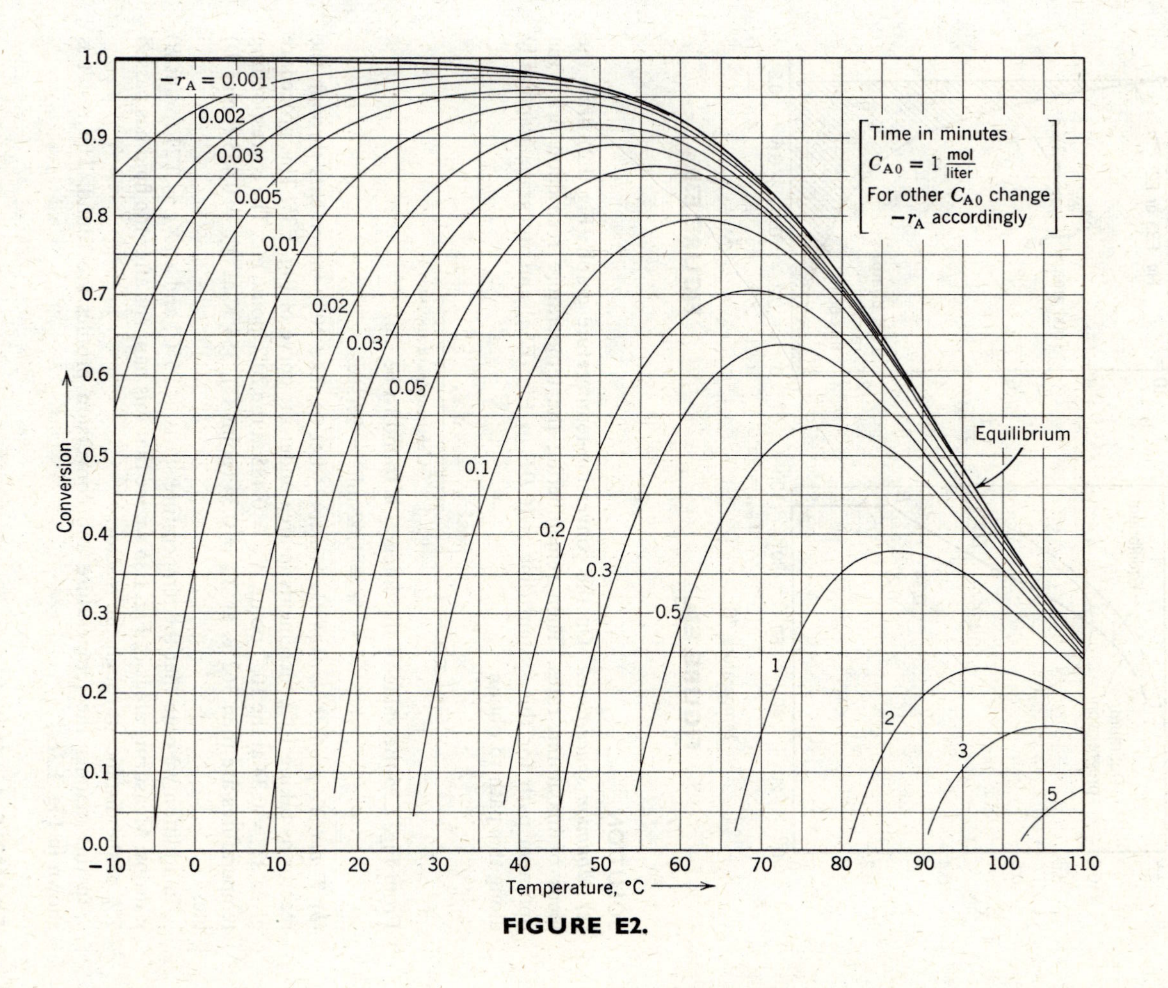

$-r_A = 0.001$
0.002
0.003
0.005
0.01
0.02
0.03
0.05
0.1
0.2
0.3
0.5
1
2
3
5
Time in minutes
$C_{A0} = 1 \frac{mol}{liter}$
For other C_{A0} change $-r_A$ accordingly
Equilibrium
Conversion
Temperature, °C
1.0
0.9
0.8
0.7
0.6
0.5
0.4
0.3
0.2
0.1
0.0
−10
0
10
20
30
40
50
60
70
80
90
100
110

FIGURE E2.

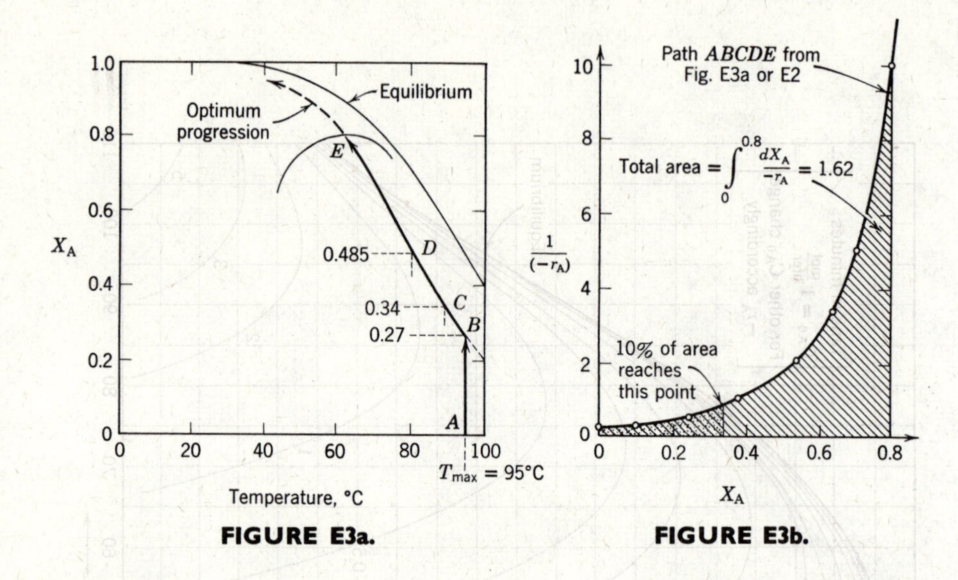

FIGURE E3a. **FIGURE E3b.**

SOLUTION

(*a*) *Minimum space-time.* On the conversion-temperature graph (Fig. E2) draw the locus of maximum rates. Then remembering the temperature restriction draw the optimum path for this system (line *ABCDE* in Fig. E3*a*) and integrate graphically along this path to obtain

$$\tau_{opt} = C_{A0} \int_0^{0.8} \frac{dX_A}{(-r_A)_{\text{optimum path } ABCDE}}$$

From Fig. E3*b* the value of this integral is found to be

$$\tau = 1.62 \text{ min} = 97 \text{ sec}$$

(*b*) *T and X_A profiles through the reactor.* Let us take 10% increments through the reactor by taking 10% increments in area under the curve of Fig. E3*b*. This procedure gives $X_A = 0.34$ at the 10% point, $X_A = 0.485$ at the 20% point, etc. The corresponding temperatures are then 362°K at $X_A = 0.34$ (point *C*), 354°K at $X_A = 0.485$ (point *D*) etc.

In addition we note that the temperature starts at 95°C, and at $X_A = 0.27$ (point *B*) it drops. Measuring areas in Fig. E3*b* we see that this happens after the fluid has passed 7% of the distance through the reactor.

In this manner the temperature and conversion profiles are found. The result is shown in Fig. E3*c*.

EXAMPLE 4. Mixed reactor performance

A concentrated aqueous A-solution of the previous examples ($C_{A0} = 4$ mol/liter, $F_{A0} = 1000$ mol/min, $C_p = C_{p,H_2O}$) is to be 80% converted in a mixed reactor.

(*a*) What size of reactor is needed?

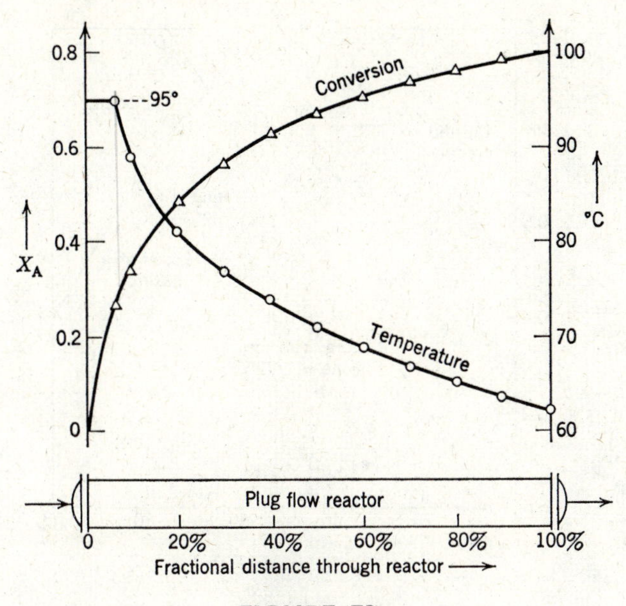

FIGURE E3c.

(*b*) What is the heat duty if feed enters at 25°C and product is to be withdrawn at this temperature?

(*c*) Compare this reactor size with the minimum size when the optimum temperature progression is used.

SOLUTION

(***a***) *Reactor volume.* For C_{A0} = 4 mol/liter we may use the X_A versus T chart of Fig. E2 as long as we multiply all rate values on this chart by 4.

Following Fig. 10 the mixed flow operating point should be located where the locus of optima intersects the 80% conversion line (point A on Fig. E4*a*). Here the reaction rate has the value

$$-r_A = 0.4 \text{ mol A converted/min}\cdot\text{liter}$$

From the performance equation for mixed reactors, Eq. 5.11, the volume required is given by

$$V = \frac{F_{A0}X_A}{(-r_A)} = \frac{(1000 \text{ mol/min})(0.80)}{0.4 \text{ mol/min}\cdot\text{liter}} = 2000 \text{ liters}$$

(***b***) *Heat duty.* The slope of the energy balance line is

$$\text{slope} = \frac{\mathbf{C_p}}{-\Delta \mathbf{H}_r} = \frac{(1000 \text{ cal/4 mol A}\cdot{}^\circ\text{K})}{(18{,}000 \text{ cal/mol A})} = \frac{1}{72}\ {}^\circ\text{K}^{-1}$$

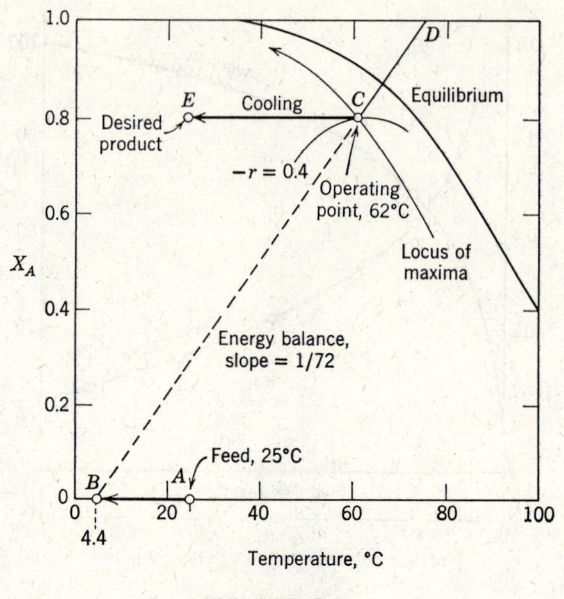

FIGURE E4a.

Drawing this line through point C (line BCD) we see that the feed must be cooled 20° (from point A to point B) before it enters and reacts adiabatically. Also, the product must be cooled 37°C (from point C to point E). Thus the heat duty is

$$\text{Precooler:} \quad Q_{AB} = (250\ \text{cal/mol A}\cdot{}^\circ\text{K})(20^\circ\text{K}) = 5000\ \text{cal/mol A fed}$$
$$= (5000\ \text{cal/mol A})(1000\ \text{mol A/min}) = 5{,}000{,}000\ \text{cal/min}$$

$$\text{Postcooler:} \quad Q_{CE} = (250)(37) = 9250\ \text{cal/mol A fed}$$
$$= (9250)(1000) = 9{,}250{,}000\ \text{cal/min}$$

Fig. E4b shows two alternate arrangements for the coolers.

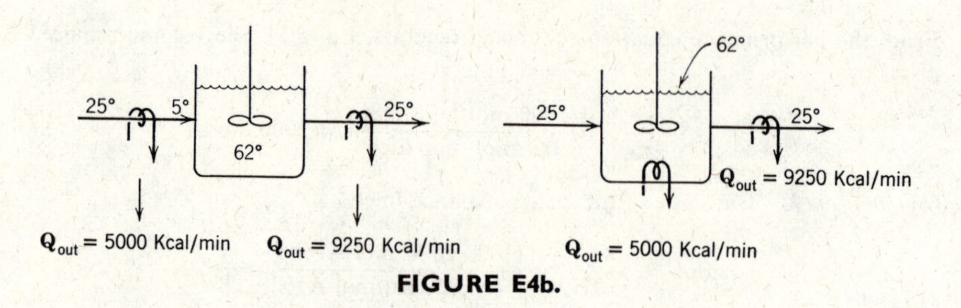

FIGURE E4b.

(*c*) *Comparison with minimum volume.* For the optimum temperature profile, $C_{A0} =$ 1 mol/liter, Example 3 gave

$$\tau = C_{A0} \int_0^{0.8} \frac{dX_A}{-r_A} = 1.62 \text{ min}$$

Here C_{A0} and $-r_A$ are both multiplied by a factor of 4 so τ remains unchanged, and the optimum path and temperature progression is also unchanged. Thus we have

$$V = \frac{\tau F_{A0}}{C_{A0}} = \frac{(1.62 \text{ min})(1000 \text{ mol/min})}{4 \text{ mol/liter}} = 405 \text{ liters}$$

Comparing with part (*a*) we see that the mixed reactor is 5 times the minimum size.

EXAMPLE 5. Plug flow reactor performance

Find the size of adiabatic plug flow reactor to react the feed of Example 4 to 80% conversion.

SOLUTION

Following the procedure of Fig. 10 draw trial operating lines (see Fig. E5*a*) with a slope of $\frac{1}{72}$ (from Example 4) and for each evaluate the integral

$$\int_0^{0.8} \frac{dX_A}{-r_A}$$

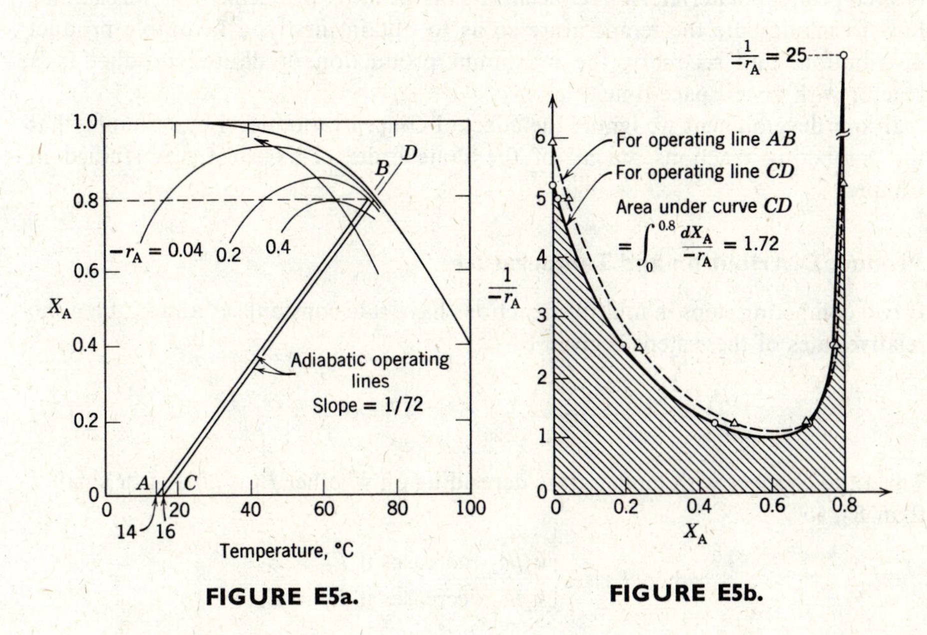

FIGURE E5a. **FIGURE E5b.**

to find which is smallest. Figures E5*a* and E5*b* show this procedure for lines *AB* and *CD*. Line *CD* has the smaller area, is in fact close to the minimum, and is therefore the desired adiabatic operating line. So

$$V = F_{A0}\int_0^{0.8} \frac{dX_A}{-r_A} = F_{A0}(\text{Area under curve } CD)$$

$$= (1000 \text{ mol/min})(1.72 \text{ liter}\cdot\text{min/mol})$$

$$= 1720 \text{ liters}$$

This volume is somewhat smaller than the volume of mixed reactor (from Example 4) but it is still four times as large as the minimum required.

Regarding temperatures: Figure E5*a* shows that the feed must first be cooled to 16.0°C, it then passes through the adiabatic reactor and leaves at 73.6°C and 80% conversion.

MULTIPLE REACTIONS

As pointed out in the introduction to Chapter 7, in multiple reactions both reactor size and product distribution are influenced by the processing conditions. Since the problems of reactor size are no different in principle than those for single reactions and are usually less important than the problems connected with obtaining the desired product material, let us concentrate on the latter problem. Thus we examine how to manipulate the temperature so as to obtain, firstly, a favorable product distribution, and secondly, the maximum production of desired product in a reactor with given space-time.

In our development we ignore the effect of concentration level by assuming that the competing reactions are all of the same order. This effect was studied in Chapter 7.

Product Distribution and Temperature

If two competing steps in multiple reactions have rate constants k_1 and k_2, then the relative rates of these steps is given by

$$\frac{k_1}{k_2} = \frac{k_{10}e^{-E_1/RT}}{k_{20}e^{-E_2/RT}} = \frac{k_{10}}{k_{20}}\, e^{(E_2-E_1)/RT} \propto e^{(E_2-E_1)/RT} \tag{27}$$

This ratio changes with temperature depending on whether E_1 is greater or smaller than E_2, so

$$\text{when } T \text{ rises}\begin{cases} k_1/k_2 \text{ increases if } E_1 > E_2 \\ k_1/k_2 \text{ decreases if } E_1 < E_2 \end{cases}$$

Thus the reaction with larger activation energy is the more temperature-sensitive of the two reactions. This finding leads to the following general rule on the influence of temperature on the relative rates of competing reactions.

A high temperature favors the reaction of higher activation energy, a low temperature favors the reaction of lower activation energy.

Let us apply this rule to find the proper temperature of operations for various types of multiple reactions.

For *parallel reactions*

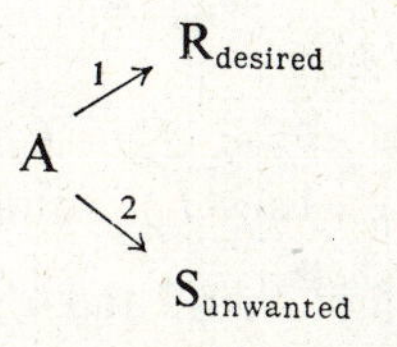

Step 1 is to be promoted, step 2 depressed, and k_1/k_2 made as large as possible. Thus from the above rule

$$\left.\begin{array}{l}\text{if } E_1 > E_2 \text{ use high } T\\ \text{if } E_1 < E_2 \text{ use low } T\end{array}\right\} \tag{28}$$

For *reactions in series*

$$A \xrightarrow{1} R_{\text{desired}} \xrightarrow{2} S \tag{29a}$$

the production of R is favored if k_1/k_2 is increased. Thus

$$\left.\begin{array}{l}\text{if } E_1 > E_2 \text{ use high } T\\ \text{if } E_1 < E_2 \text{ use low } T\end{array}\right\} \tag{29b}$$

For the general series-parallel reaction we introduce two additional considerations. First of all for *parallel steps* if one requirement is for a high temperature and another is for a low temperature then a particular intermediate temperature is best in that it gives the most favorable product distribution. As an example consider the reactions

$$A \begin{array}{l} \overset{1}{\nearrow} R_{\text{desired}} \\ \xrightarrow{2} S \\ \underset{3}{\searrow} T \end{array} \qquad \text{where } E_1 > E_2,\ E_1 < E_3 \tag{30a}$$

Now, $E_1 > E_2$ requires a high T, $E_1 < E_3$ requires a low T, and it can be shown that the most favorable product distribution is obtained when the temperature satisfies the following condition

$$\frac{1}{T} = \frac{R}{E_3 - E_2} \ln \left[\frac{E_3 - E_1}{E_1 - E_2} \frac{k_{30}}{k_{20}}\right] \tag{30b}$$

Secondly for *steps in series* if an early step needs a high temperature and a later step needs a low temperature, then a falling progression of temperatures should be used. Analogous arguments hold for other progressions. As an example consider the reaction

$$\begin{array}{ccccccc} A & \xrightarrow{1} & R & \xrightarrow{3} & S & \xrightarrow{5} & T_{\text{desired}} \\ & \searrow^{2} & & \searrow^{4} & & \searrow^{6} & \\ & U & & V & & W & \end{array} \tag{31}$$

where:

$$\begin{array}{ccc} E_1 > E_2 & E_3 < E_4 & E_5 > E_6 \\ (\text{high } T) & (\text{low } T) & (\text{high } T) \end{array}$$

Noting that steps 1 and 2 occur first, then 3 and 4, then 5 and 6—of course with some overlap—we see that the best sequence is to start with a high temperature, let this fall to a minimum where R is high, and then let it rise as S becomes appreciable.

Table 1 summarizes the favorable temperature levels and progressions for a number of rather general reaction schemes. These results follow directly from the arguments just presented, and extension to other schemes is straightforward. The problems at the end of this chapter verify some of the quantitative findings on T_{opt} and also show some possible extensions.

Temperature and Vessel Size (or τ) for Maximum Production

Maximum production rate requires both a favorable product distribution and high conversions. If the activation energies are such that a favorable product distribution is obtained at a high temperature then we should use the highest allowable temperature, because then the rates of reaction are high and the required reactor size is small.

If the activation energies of the various reaction steps are equal the temperature has no effect on the product distribution; however, we should still operate at the highest temperature because this minimizes τ or maximizes production.

The only problem is when the favorable product distribution occurs at low temperature where very little of anything is formed. In this situation it is preferable to use an intermediate temperature, for although product distribution suffers somewhat this may be more than offset by the increase in rate and conversion. The precise temperature level or progression to use for any particular space-time τ can

TABLE I. Operating temperature for favorable product distribution

---

$$\begin{array}{l} A \xrightarrow{1} R_{\text{desired}} \xrightarrow{3} S \\ A \xrightarrow{2} T \end{array} \tag{32}$$

(*a*) if $E_1 > E_2$ and E_3 use high T
(*b*) if $E_1 > E_2(\text{high } T)$, $E_1 < E_3(\text{low } T)$ use falling T
(*c*) if $E_1 < E_2$ and E_3 use low T
(*d*) if $E_1 < E_2(\text{low } T)$, $E_1 > E_3(\text{high } T)$ use rising T

$$\begin{array}{l} A \xrightarrow{1} R_{\text{desired}} \xrightarrow{2} S \\ R_{\text{desired}} \xrightarrow{3} T \end{array} \tag{33}$$

(*a*) if $E_1 > E_2$ and E_3 use high T
(*b*) if $E_1 > E_2(\text{high } T)$, $E_1 < E_3(\text{low } T)$ use intermediate T level
(*c*) if $E_1 < E_2$ and E_3 use low T
(*d*) if $E_1 < E_2(\text{low } T)$, $E_1 > E_3(\text{high } T)$ use intermediate T level

$$\begin{array}{l} A \xrightarrow{1} R \xrightarrow{3} S_{\text{desired}} \\ A \xrightarrow{2} T \end{array} \tag{34}$$

(*a*) if $E_1 > E_2$ use high T
(*b*) if $E_1 < E_2$ use low T

$$\begin{array}{l} A \xrightarrow{1} R \xrightarrow{2} S_{\text{desired}} \\ R \xrightarrow{3} T \end{array}$$

(*a*) if $E_2 > E_3$ use high T
(*b*) if $E_2 < E_3$ use low T

$$\begin{array}{l} A \xrightarrow{1} R \xrightarrow{3} S_{\text{desired}} \\ A \xrightarrow{2} T \\ R \xrightarrow{4} U \end{array} \tag{36}$$

(*a*) if $E_1 > E_2$, $E_3 > E_4$ use high T
(*b*) if $E_1 > E_2(\text{high } T)$, $E_3 < E_4(\text{low } T)$ use falling T
(*c*) if $E_1 < E_2$, $E_3 < E_4$ use low T
(*d*) if $E_1 < E_2(\text{low } T)$, $E_3 > E_4(\text{high } T)$ use rising T
Note: See Denbigh (1958) for discussion of this reaction.

---

be found either analytically or by search procedures. This then is the problem we deal with in this section.

Consider the following cases, first the *parallel reactions*,

$$A \overset{1}{\nearrow} R_{\text{desired}}, \quad A \overset{2}{\searrow} S \qquad \text{where } E_1 < E_2 \tag{37a}$$

(*a*) If τ is of no concern use the lowest allowable temperature with very large space-time.

(*b*) For given τ the temperature level for maximum production of R in a mixed reactor is found to be

$$T = \frac{E_2}{R} \Big/ \ln\left[k_{20}\tau\left(\frac{E_2}{E_1} - 1\right)\right] \tag{37b}$$

Note that the optimum temperature depends on both the kinetics and τ.

(*c*) If a temperature progression is permitted, say in a plug flow reactor, then for given τ a rising temperature progression is better still than the best isothermal optimum.

(*d*) In all cases, an economic balance of the two conflicting factors, increased production versus increased reactor size, will tell which pair of T and τ values should be used.

For the *reactions in series*

$$A \xrightarrow{1} R_{\text{desired}} \xrightarrow{2} S \qquad \text{where } E_1 < E_2 \tag{38}$$

(*a*) If τ is of no concern use the lowest allowable temperature with the proper corresponding τ to reach $C_{R,\max}$.

(*b*) For given τ and isothermal operations there is one temperature which is optimum and gives $C_{R,\max}$. This temperature is found from Eq. 3.51 for plug flow.

(*c*) If a temperature progression is permitted in a plug flow reactor then a falling temperature progression is better still than the isothermal optimum. Figure 18 illustrates this temperature progression for the situation where the maximum and minimum allowable temperatures are specified. Note that this progression is the reverse of that for reactions in parallel, treated above.

(*d*) In all cases the optimum temperature of operations is found by balancing the increase in reactor cost for the lower temperature of operations with the rising profits from an increase in product.

Aris (1960) deals with various aspects of this problem. We also should note that plug flow always is the most desirable of contacting patterns if an intermediate is desired. Naturally, if the final product S is desired then we should always use a high temperature, since this increases both k_1 and k_2.

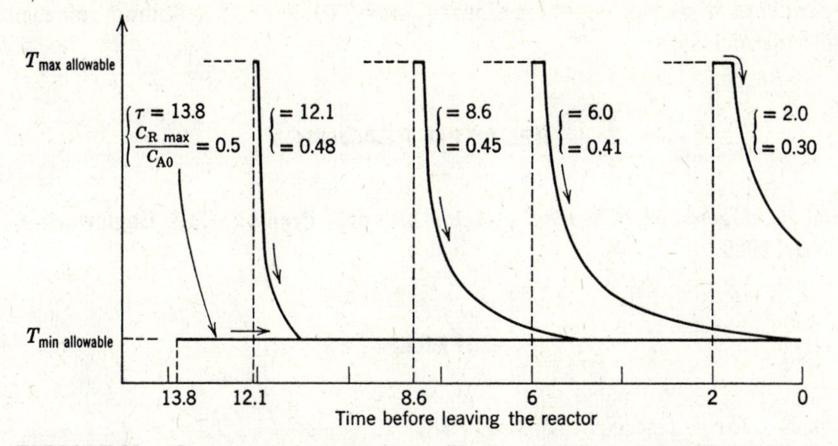

FIGURE 18. Optimum temperature profile in a plug flow reactor for various reactor sizes for the elementary reactions

$$A \longrightarrow R \longrightarrow S, \qquad \varepsilon_A = 0, \qquad E_1 < E_2$$

Adapted from Aris (1960).

The type of reasoning for these two types of reactions can be extended to other multiple reaction schemes.

Comments

This discussion of multiple reactions shows that the relative size of the activation energies will tell which temperature level or progression is favored, just as Chapter 7 showed what concentration level or progression and what state of mixing is best. Although the general pattern of low, high, falling, or rising temperature can usually be determined without much difficulty, calculation of the optimum profile is a different matter and requires rather sophisticated mathematics.

In experimentation we usually meet with the inverse of the situation outlined here in that we observe product distributions from experiment, and from this we wish to find the stoichiometry, kinetics, and the most favorable operating conditions. The generalizations of this chapter should be helpful in this inductive search.

Finally, when the reactions are of different order and of different activation energies then we must combine the methods of Chapters 7 and 8. Jackson *et al.* (1971) treat a particular system of this type and find that the optimum policy requires adjusting only one of the two factors, temperature or concentration, while keeping the other at its extreme. Which factor to adjust depends on whether the change in product distribution is more temperature dependent or concentration dependent. It would be interesting to know whether this finding represents a general conclusion.

RELATED READINGS

Aris, R., *Elementary Chemical Reactor Analysis*, Prentice Hall, Englewood Cliffs, N.J., 1969.

REFERENCES

Aris, R., *Chem. Eng. Sci.*, **13**, 18 (1960).
———, *Can. J. of Chem. Eng.*, **40**, 87 (1962).
———, *The Optimal Design of Chemical Reactors*, Academic Press, New York, 1961, Ch. 1.
Denbigh, K. G., *Chem. Eng. Sci.*, **8**, 125 (1958).
———, *Chem. Eng. Sci.*, **8**, 133 (1958).
Jackson, R., Obando, R., and Senior, M. G., *Chem. Eng. Sci.* **26**, 853 (1971).
Kramers, H., and Westerterp, K. R., *Chem. Eng. Sci.*, **17**, 423 (1962).
van Heerden, C., *Ind. Eng. Chem.*, **45**, 1242 (1953).

PROBLEMS

These are grouped as follows:

Problems 1–10: Equilibrium conversion of fast reactions.
Problems 11–20: Conversion for single reactions.
Problems 21–30: Conversion for multiple reactions.

1. The gases 20 cm^3 A, 40 cm^3 B, and 20 cm^3 C measured at standard conditions (0°C, 1 atm) are forced into a 20-cm^3 reaction vessel containing air at standard conditions. A, B, and C react according to the following stoichiometric equation

$$\tfrac{1}{2}A + \tfrac{3}{2}B = C$$

At 0°C equilibrium is attained when the pressure in the reaction vessel is 4 atm. Find the standard free-energy change $\Delta G°$ at 200 atm and 0°C for the reaction

$$2C = A + 3B$$

Assume that the perfect gas law holds throughout.

2. At 1000°K and 1 atm substance A is 2 mole % dissociated according to the gas-phase reaction 2A = 2B + C.

(*a*) Calculate the mole % dissociated at 200°K and 1 atm.
(*b*) Calculate the mole % dissociated at 200°K and 0.1 atm.

Data: Average C_p of A = 12 cal/mol·°K
Average C_p of B = 9 cal/mol·°K
Average C_p of C = 6 cal/mol·°K

At 25°C and 1 atm 2000 cal are released when 1 mole A is formed from the reactants B and C.

3. Given the gas-phase reaction A = B + C. Start with pure A. Suppose that 50% of the original A is dissociated at 1000°K and 10 atm as well as at 500°K and 0.1 atm.

(*a*) Calculate the percent dissociation at 250°K and 1 atm.
(*b*) Calculate the percent dissociation at 250°K and 0.01 atm.

Data: Average C_p of A = 12 cal/mol·°K
Average C_p of B = 7 cal/mol·°K
Average C_p of C = 5 cal/mol·°K

4. Given the gas-phase reaction 2A = B + C. Start with 1 mole A and 2 moles B in 5 moles inert gas. Suppose that at 1000°K and 10 atm 66.7% of original A is dissociated and at 500°K and 0.1 atm 50% of original A is dissociated.

(*a*) Calculate the percent dissociation at 250°K and 1 atm.
(*b*) Calculate the percent dissociation at 250°K and 0.01 atm.
(*c*) Find ΔG° for the reaction A = $\frac{1}{2}$B + $\frac{1}{2}$C at 250°K and 1 atm.
(*d*) Find ΔG° for the reaction 2A = B + C at 250°K and 0.01 atm.
(*e*) Find $\Delta H_{r,298}$ for the reaction 2A = B + C.

Data: Average C_p of A = 7 cal/mol·°K
Average C_p of B = 9 cal/mol·°K
Average C_p of C = 5 cal/mol·°K

5. At 1000°K and 1 atm substance A is 2 mole % dissociated according to the following reaction

$$2A = 2B + C$$

(*a*) Calculate the mole % dissociated at 200°K and 1 atm.
(*b*) Calculate the mole % dissociated at 200°K and 0.1 atm.

Data: Average C_p of A = 8 cal/mol·°K
Average C_p of B = 8 cal/mol·°K
Average C_p of C = 8 cal/mol·°K

At 25°C and 1 atm 2000 cal are released when 1 mole A is formed from the reactants B and C.

6. Given the following gas-phase reaction

$$A = B + C, \qquad \Delta H_{r,298} = 1000 \text{ cal/mol}$$

Start with 1 mole A in $4\frac{1}{2}$ moles inert gas and suppose that 50% of the original A is dissociated at 1000°K and 100 atm.

(*a*) Find ΔG° at 500°K and 1 atm.
(*b*) Find ΔG° at 500°K and 0.01 atm.
(*c*) Find the % A dissociated at 500°K and 1 atm.
(*d*) Find the % A dissociated at 500°K and 0.01 atm.

Data:

Substance	$\bar{C}_p$, cal/mol·°K	Critical Values T_c	p_c
A	13	1000	100
B	8	400	50
C	5	800	20

Assume that the system is an ideal solution:

$$f_A = y_A f_{A,\text{pure substance}} = y_A \pi \left(\frac{f_{A,\text{pure substance}}}{\pi} \right)$$

The quantity in brackets is given by the Newton charts which may be found in most thermodynamics texts for chemical engineers.

7. The rates for two different gas-phase reactions at 1028°C are given by

$r_1 = 670C_A{}^2$ measured in metric (MKS) units

$r_2 = 8388C_A^{0.5}C_B$ measured in English units with time in hours

Which reaction is more greatly influenced by
(*a*) a change in temperature,
(*b*) a change in pressure?

8. The following *very* fast reaction

$$A + B = 2R, \qquad K = 4$$

takes place isothermally in a plug flow reactor. Find the ratio of A to B which would minimize the total amount of reactants fed to the reactor to produce a given amount of R. Assume that unused reactants are discarded.

9. The following *very* fast reaction

$$A + B = 2R, \qquad K = 4$$

takes place in a plug flow reactor. Find the ratio of A to B to be fed to the reactor which would minimize the total reactant cost to produce a given amount of R. Assume that unconverted reactants are discarded.

Data: Reactant A costs $\$\alpha$/mol
Reactant B costs $\$\beta$/mol

10. The homogeneous gas-phase reaction $A + B = R + S$ yields 40% conversion of A in a given plug flow reactor when equimolar quantities of reactants are fed into the reactor at 2 atm and 25°C.

(*a*) At what temperature level should this reactor be operated so as to maximize conversion of reactant? Find this conversion.

(*b*) At this optimum temperature would it be possible to lower the pressure to slightly below atmospheric without a drop in conversion? Such a change is desirable since possible leaks in the system would then not result in contamination of the atmosphere or loss of reactants.

Data: For the reaction as written and at 25°C

$$\Delta H_r = 1800 \text{ cal}$$
$$\Delta G^\circ = 0$$

Possible operating temperatures are between 0°C and 500°C, in which range

$$\bar{C}_{pA} = 8 \text{ cal/mol}\cdot{}^\circ\text{K}$$
$$\bar{C}_{pB} = 10 \text{ cal/mol}\cdot{}^\circ\text{K}$$
$$\bar{C}_{pR} = 7 \text{ cal/mol}\cdot{}^\circ\text{K}$$
$$\bar{C}_{pS} = 5 \text{ cal/mol}\cdot{}^\circ\text{K}$$

Assume the usual Arrhenius temperature dependency for the rate.

11. For the system of the examples in this chapter find τ needed to achieve 60% conversion in a plug flow reactor having the optimal temperature profile if

(*a*) $C_{A0} = 1$ mol/liter

(*b*) $C_{A0} = 7.2$ mol/liter.

12. If the optimum temperature progression in a plug flow reactor were used for Examples 4 and 5 ($C_{A0} = 4$ mol/liter, $F_{A0} = 1000$ mol A/min) and feed and product are both to be at 25°C how much heating and cooling would be needed

(*a*) for the feed stream

(*b*) for the reactor

(*c*) for the stream leaving the reactor?

13. For the reacting system of the examples in this chapter, with $C_{A0} = 10$ mol/liter, what is the maximum conversion attainable

(*a*) in a mixed reactor where heat exchange within the reactor or with fluid before it enters the reactor is allowed (*b*) in an adiabatic plug flow reactor where heat exchange during reaction is not possible?

The temperature of reacting fluid is to remain between 5°C and 95°C.

14. For the system of the examples in this chapter, but with $C_{A0} = 5$ mol/liter

(*a*) find the space-time in a mixed reactor needed to achieve 90% conversion of A to R

(*b*) with a sketch show the location and determine the necessary heat exchange if feed enters at 25°C, product leaves at 25°C, and the allowable temperature range is from 5 °C to 95°C.

15. To increase the processing rate for the conditions of Example 5 it is proposed that we recycle a portion of the product stream from the plug flow reactor. For the operating line *CD* (feed enters at 16.0 °C) find the best recycle ratio and the increase, if any, in processing rate resulting from this modification.

16. Find the minimum volume of isothermal plug flow reactor for 80% conversion of the feed of Examples 4 and 5. At what temperature should we operate this reactor?

17. For the liquid reacting system of the examples in this chapter and with a more concentrated feed ($C_{A0} = 10$ mol A/liter)

(*a*) what range of adiabatic operations between plug and mixed flow can be used to obtain 80% conversion from a single reactor? Note that the allowable operating temperature range is from 5°C to 95°C.

(*b*) For a feed rate $F_{A0} = 1000$ mol A/min present a reasonable design giving the type and size of reactor, and sketch the location and duty of the necessary heat exchangers. Feed is available at 25 °C.

18. The fluid of Example 4 is to be processed to 80% conversion in two mixed reactors.

(*a*) What should be the size of the two reactors to minimize the total volume?

(*b*) What should be the temperature of the two reactors?

(*c*) Devise a flow scheme and find the duty of the heat exchangers for this scheme if the feed and product are both to be at 25 °C.

19. The reversible first-order gas reaction

$$A \underset{2}{\overset{1}{\rightleftarrows}} R$$

is to be carried out in a mixed reactor. For operations at 300°K the volume of reactor required is 100 liters for 60% conversion of A.

(*a*) What should be the volume of reactor for the same feed rate and conversion but with operations at 400°K?

(*b*) State or show with a sketch how to find the temperature of operations which would minimize the size of mixed reactor needed for this conversion and feed rate.

Data: $k_1 = 10^3 \exp[-4800/RT]$
$\Delta C_p = C_{pR} - C_{pA} = 0$
$\Delta H_r = -8000$ cal/mol at 300°K
$K = 10$ at 300°K
Feed consists of pure A
Total pressure stays constant

20. Repeat the previous problem with the following two modifications:

$$\Delta C_p = 5 \text{ cal/mol}\cdot°\text{K}$$
$$\Delta H_r = -4000 \text{ cal/mol at } 300°\text{C}$$

21. Qualitatively find the optimum temperature progression to maximize C_S for the reaction scheme

$$\begin{array}{ccccccc} A & \xrightarrow{1} & R & \xrightarrow{3} & S_{\text{desired}} & \xrightarrow{5} & T \\ \searrow^{2} & & \searrow^{4} & & \searrow^{6} & & \\ U & & V & & W & & \end{array}$$

Data: $E_1 = 10$, $E_2 = 25$, $E_3 = 15$, $E_4 = 10$, $E_5 = 20$, $E_6 = 25$

22. For all combinations of high and low E values for the different steps of the reaction scheme

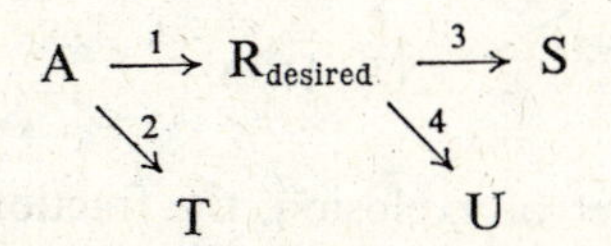

qualitatively find the optimum temperature of operations so as to maximize C_R assuming that τ can be adjusted to whatever value desired.

23. For the parallel decompositions of A, all of the same order,

$$A \xrightarrow{1} R_{\text{desired}}, \quad A \xrightarrow{2} S, \quad A \xrightarrow{3} T$$

find the temperature which maximizes the fractional yield of R.
Note: Equation 30*b* represents the solution to one of the cases.

24. For the parallel first order decompositions of A

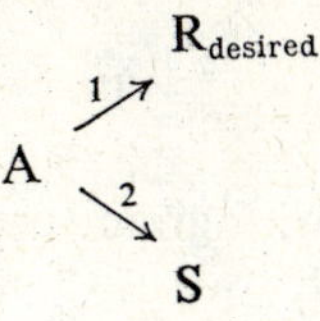

occurring in a mixed reactor with given space-time, find the temperature level which maximizes production of R.

Note: Equation 37*b* represents the solution to one case.

25. For a given space-time in a mixed reactor find the temperature which will maximize C_R for the elementary reactions

$$A \xrightarrow{1} R \xrightarrow{2} S$$

Note: We can exceed this $C_{R,max}$ by proper operations in a plug flow reactor.

26. The violent oxidation of xylene simply produces CO_2 and H_2O; however, when oxidation is gentle and carefully controlled, it can also produce useful quantities of valuable phthalic anhydride as follows

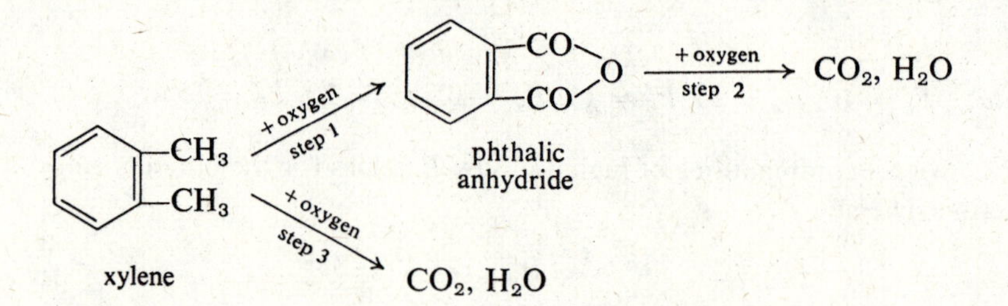

Also, because of the danger of explosion, the fraction of xylene in the reacting mixture must be kept below 1%.

Naturally the problem in this process is to obtain a favorable product distribution.

(*a*) In a plug flow reactor what circumstances would require that we operate at the maximum allowable temperature?

(*b*) Under what circumstances should the plug flow reactor have a falling temperature progression?

27. For the parallel first-order decompositions of A

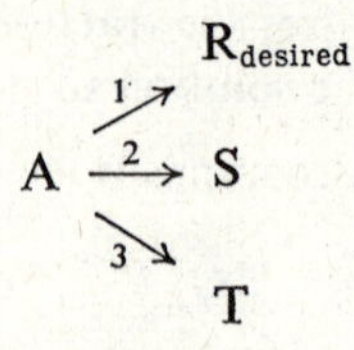

taking place in a mixed reactor with given space-time find the temperature level which maximizes production of R.

28. The first-order reactions

$$A \xrightarrow{1} R \xrightarrow{3} S_{\text{desired}} \qquad A \xrightarrow{2} T \qquad R \xrightarrow{4} U$$

$$k_1 = 10^9 e^{-6000/T}$$
$$k_2 = 10^7 e^{-4000/T}$$
$$k_3 = 10^8 e^{-9000/T}$$
$$k_4 = 10^{12} e^{-12{,}000/T}$$

are to be run in two mixed reactors in series anywhere between 10 and 90°C.

(*a*) If both reactors are kept at the same temperature, what should that temperature be for maximum fractional yield of desired product S? Find this fractional yield.

(*b*) If the reactors may be kept at different temperatures, what should these temperatures be for maximum fractional yield of S? Find this fractional yield.

(*c*) For part (*b*), what should the relative sizes of the two reactors be—approximately equal, first one larger, or first one smaller?

29. The aqueous reactions

$$A + B \xrightarrow{1} R + T$$

$$R + B \xrightarrow{2} S + T$$

proceed with elementary second-order kinetics. Find the optimum temperature of operation of a plug flow reactor, and find the corresponding maximum fractional yield of R obtainable, based on the amount of A consumed. R is the desired product, and any temperature level between 5 and 65°C may be used.

Data: Equimolar quantities of A and B are mixed and allowed to react in beakers at different temperatures. When all B is consumed, analysis shows that 75% A has reacted at 25°C, 60% A has reacted at 45°C.

30. For the series reactions of Fig. 18 let

$$T_{\text{min, allowable}} = 1000°\text{K}$$

$$E_1/R = 10{,}000°\text{K}$$

$$E_2/R = 20{,}000°\text{K}$$

If the minimum allowable temperature were lowered to 900°K

(*a*) what would be the highest obtainable $C_{R,\max}/C_{A0}$?

(*b*) what size plug flow reactor would be needed to obtain this product?

31. Consider the scheme of elementary reactions

$$A \xrightarrow{1} R \xrightarrow{2} S_{\text{desired}} \qquad R \xrightarrow{3} T$$

$$k_1 = 10 e^{-3500/T}, \text{ sec}^{-1}$$
$$k_2 = 10^{12} e^{-10{,}500/T}, \text{ sec}^{-1}$$
$$k_3 = 10^8 e^{-7000/T}, \text{ sec}^{-1}$$

Feed consists of A and inerts, $C_{A0} = 1$ mol/liter, and the operable temperature range is between 7 and 77°C.

(*a*) What is the maximum amount of S obtainable per mole of A, and at what temperature and in what type of reactor is this obtained?

(*b*) Find the minimum holding time to produce 99% of $C_{S,max}$.

(*c*) Repeat part (*b*) if $k_1 = 10^7 e^{-3500/T}$, all else remaining unchanged.

9

NONIDEAL FLOW

Thus far we have restricted ourselves to two idealized flow patterns, plug flow and mixed flow. Though real reactors never fully follow these flow patterns a large number of designs approximate these ideals with negligible error. In other cases deviation from ideality can be considerable. This deviation can be caused by channeling of fluid, by recycling of fluid, or by creation of stagnant regions in the vessel. Figure 1 shows this behavior. In all types of process equipment, such as heat exchangers, packed columns, and reactors this type of flow should be avoided since it always lowers the performance of the unit.

The problems of nonideal flow are intimately tied to those of scale-up because the question of whether to pilot-plant or not rests in large part on whether we are in control of all the major variables for the process. Often the uncontrolled factor in scale-up is the magnitude of the nonideality of flow, and unfortunately this very often differs widely between large and small units. Therefore ignoring this factor may lead to gross errors in design.

In this chapter we hope to consider enough about nonideal flow to acquire an intuitive feel for the magnitude of this phenomenon in reactors of various types, to know whether it needs to be treated, and if so to deal with it in a rational manner consistent with present-day knowledge.

RESIDENCE TIME DISTRIBUTION OF FLUID IN VESSELS

If we know precisely what is happening within the vessel, thus if we have a complete velocity distribution map for the fluid, then we are able to predict the behavior of a vessel as a reactor. Though fine in principle, the attendant complexities make it impractical to use this approach.

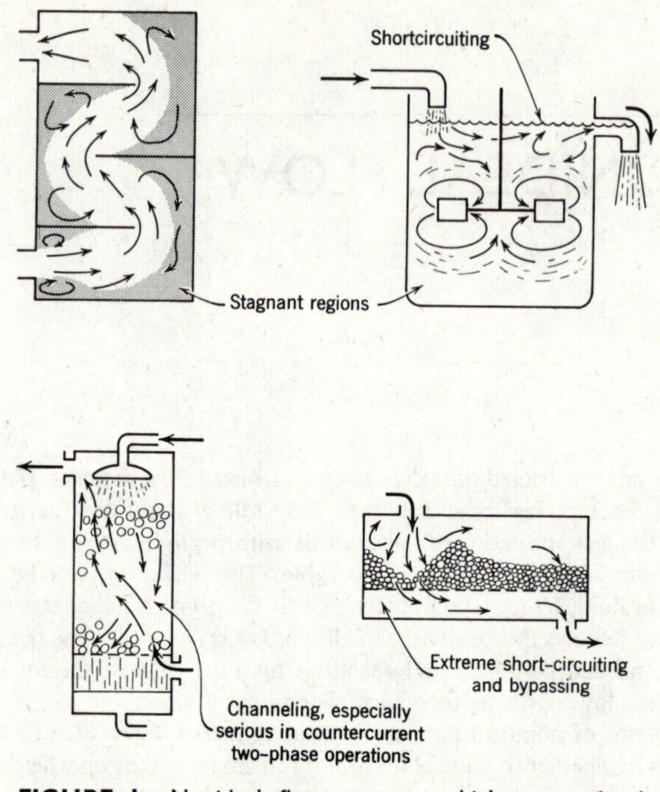

FIGURE 1. Nonideal flow patterns which may exist in process equipment.

Setting aside this goal of complete knowledge about the flow, let us be less ambitious and see what is the least we need to know about the flow which will be useful for design. In many cases we really do not need to know very much, simply how long the individual molecules stay in the vessel, or more precisely, the distribution of residence times of the flowing fluid. This information can be determined easily and directly by a widely used method of inquiry, the stimulus-response experiment.

This chapter deals in large part with the residence time distribution (or RTD) approach to nonideal flow. We show when it may legitimately be used, how to use it, and when it is not applicable what alternatives to turn to.

In developing the "language" for this treatment of nonideal flow (see Danckwerts (1953)), we will only consider the steady-state flow, without reaction and without density change, of a single fluid through a vessel.

E, the Age Distribution of Fluid Leaving a Vessel

It is evident that elements of fluid taking different routes through the reactor may require different lengths of time to pass through the vessel. The distribution of these times for the stream of fluid leaving the vessel is called the exit age distribution **E**, or the residence time distribution RTD of fluid.

We find it convenient to represent the RTD in such a way that the area under the curve is unity, or

$$\int_0^\infty \mathbf{E}\, dt = 1$$

This procedure is called normalizing the distribution, and Fig. 2 shows this distribution in normalized form.

With this representation the fraction of exit stream of age* between t and $t + dt$ is

$$\mathbf{E}\, dt$$

the fraction younger than age t_1 is

$$\int_0^{t_1} \mathbf{E}\, dt \tag{1}$$

whereas the fraction of material older than t_1, shown as the shaded area in Fig. 2, is

$$\int_{t_1}^\infty \mathbf{E}\, dt = 1 - \int_0^{t_1} \mathbf{E}\, dt \tag{2}$$

The **E** curve is the distribution needed to account for nonideal flow. Other related distributions are mentioned later in this chapter and have special uses.

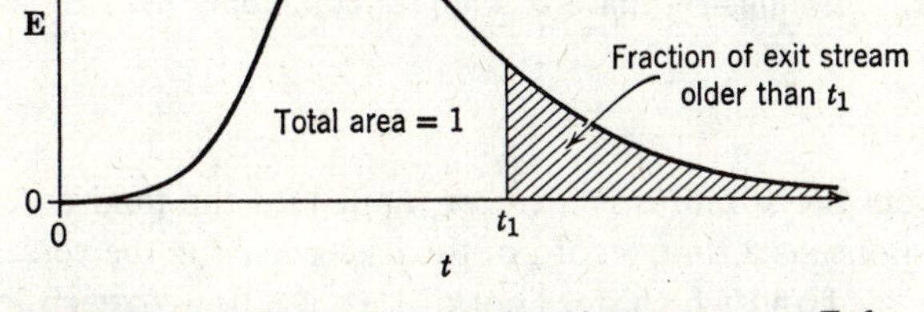

FIGURE 2. The exit age distribution curve **E** for fluid flowing through a vessel; also called the residence time distribution, or RTD.

* The term "age" for an element of the exit stream refers to the time spent by that element in the vessel.

Experimental Methods

Since we plan to characterize the extent of nonideal flow by means of the exit age distribution function, we should like to know how to evaluate **E** for any flow. For this we resort to one of a number of experimental techniques, all of which may be classed as stimulus-response techniques. In all such experimentation we disturb the system and then see how the system responds to this stimulus. An analysis of the response gives the desired information about the system. This method of experimentation is widely used in science.

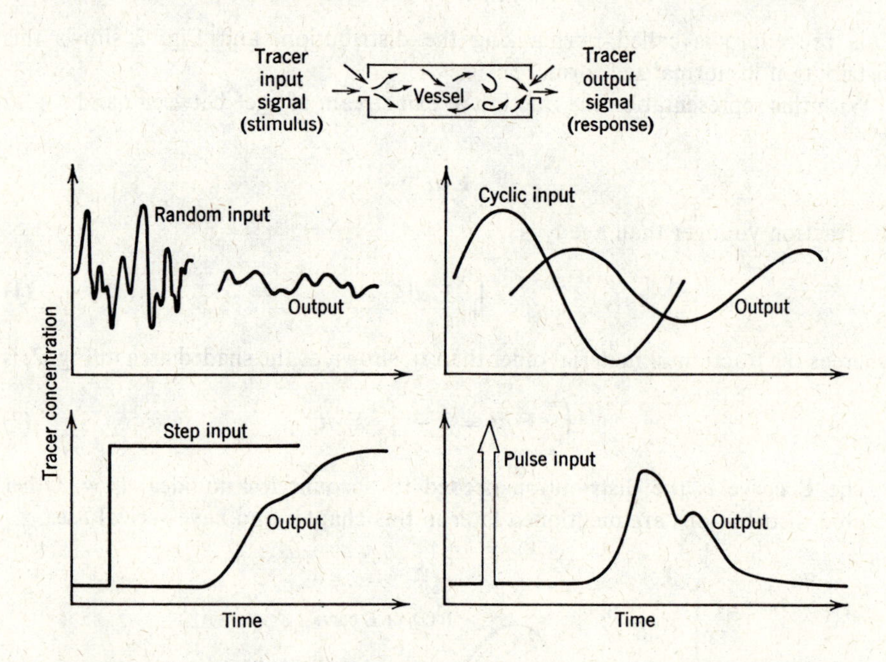

FIGURE 3. Stimulus-response techniques commonly used to study flow in vessels.

In our problem the stimulus is a tracer input into the fluid entering the vessel, whereas the response is a time record of the tracer leaving the vessel. Any material that can be detected and which does not disturb the flow pattern in the vessel can be used as tracer, and any type of input signal may be used—a random signal, a periodic signal, a step signal, or a pulse signal. These signals and their typical responses are shown in Fig. 3. Although exactly the same information can be extracted with these different inputs we will only consider the last two since they are simplest to treat.

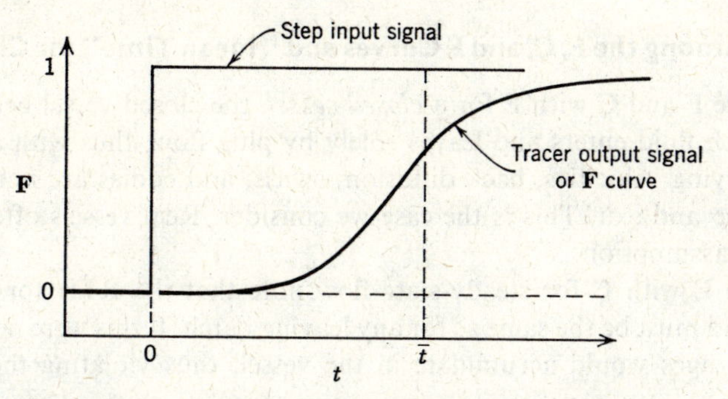

FIGURE 4. Typical downstream signal, called the **F** curve, in response to an upstream step input signal.

The F Curve

With no tracer initially present anywhere impose a step input of tracer of concentration C_0 on the fluid stream entering the vessel. Then a time record of tracer in the exit stream from the vessel, measured as C/C_0, is called the **F** curve. Figure 4 sketches this curve and shows that it always rises from 0 to 1.

The C Curve

With no tracer initially present anywhere impose an idealized instantaneous pulse of tracer on the stream entering the vessel. Such an input is often called a delta function or impulse. The normalized response is then called the **C** curve.

To perform this normalization we divide the measured concentration by Q, the area under the concentration-time curve. Thus we have on normalization

$$\int_0^\infty \mathbf{C}\,dt = \int_0^\infty \frac{C}{Q}\,dt = 1 \qquad \text{where} \qquad Q = \int_0^\infty C\,dt \tag{3}$$

Figure 5 shows the **C** curve and its properties.

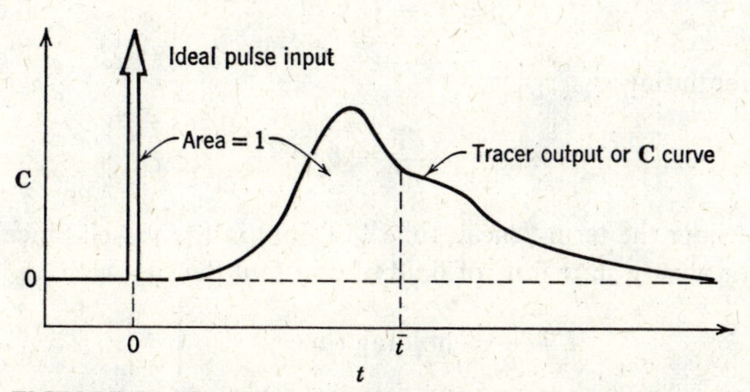

FIGURE 5. Typical downstream signal, called the **C** curve, in response to an upstream δ-function input signal.

Relation Among the F, C, and E Curves and "Mean Time" for Closed Vessels

Let us relate **F** and **C** with **E** for a *closed vessel*, the closed vessel being defined as one in which fluid enters and leaves solely by plug flow, thus with a flat velocity profile. Varying velocities, back diffusion, swirls, and eddies are not permitted at the entrance and exit. This is the case we consider. Real vessels often reasonably satisfy this assumption.

To relate **E** with **C** for steady-state flow note that the RTD for any batch of entering fluid must be the same as for any leaving batch. If this were not so, material of different ages would accumulate in the vessel, thus violating the steady-state assumption.

Now imagine the following experiment. At time $t = 0$ introduce a pulse of red tracer fluid into the stream of white flowing fluid. The **C** curve for the tracer then records when these molecules leave, in other words, their distribution of ages. Since the **C** curve represents the RTD for that particular batch of entering fluid, it must also be the RTD for any other batch, in particular, any batch in the exit stream. So we have

$$\mathbf{C} = \mathbf{E} \tag{4}$$

Thus the **C** curve gives directly the exit age distribution.

To relate **E** with **F** imagine a steady flow of white fluid. Then at time $t = 0$ switch to red and record the rising concentration of red fluid in the exit stream, the **F** curve. At any time $t > 0$ red fluid and only red fluid in the exit stream is younger than age t. Thus we have

$$\begin{pmatrix}\text{fraction of red fluid}\\ \text{in the exit stream}\end{pmatrix} = \begin{pmatrix}\text{fraction of exit stream}\\ \text{younger than age } t\end{pmatrix}$$

But the first term is simply the **F** value, while the second is given by Eq. 1. So we have, at time t,

$$\mathbf{F} = \int_0^t \mathbf{E}\, dt \tag{5a}$$

and on differentiating

$$\frac{d\mathbf{F}}{dt} = \mathbf{E} \tag{5b}$$

Finally consider the term "mean time" of fluid in the vessel. Since this chapter only treats the steady state flow of fluid of constant density we have

$$\left.\begin{aligned} \bar{t} = \frac{V}{v} &= \text{holding time} \\ &= \text{mean residence time} \\ &= \text{space-time} \end{aligned}\right\} \tag{6}$$

It may seem reasonable to expect that the mean of the **E** curve is given by $\bar{t}$ however this has yet to be shown. We do this by examining the contents of a vessel at time $t = 0$. We may thus write

$$\begin{pmatrix}\text{total volume of}\\ \text{fluid in the}\\ \text{vessel at } t = 0\end{pmatrix} = \sum_{\text{all time}} \begin{pmatrix}\text{volume of fluid}\\ \text{which had entered}\\ t \text{ to } t + dt\\ \text{seconds earlier}\end{pmatrix}\begin{pmatrix}\text{fraction of this fluid}\\ \text{which stays more than}\\ \text{about } t \text{ seconds in the}\\ \text{vessel}\end{pmatrix}$$

In symbols and using Eq. 2 this word equation becomes

$$V = \int_0^\infty (v\,dt)\left(\int_t^\infty \mathbf{E}\,dt\right)$$

or

$$\bar{t} = \frac{V}{v} = \int_0^\infty \left[\int_{t'}^\infty \mathbf{E}\,dt\right] dt'$$

By changing the order of integration we find the desired result, or

$$\bar{t} = \int_0^\infty \left[\int_0^t dt'\right] \mathbf{E}\,dt$$

$$= \int_0^\infty t\mathbf{E}\,dt = \bar{t}_{\mathbf{E}} \tag{7}$$

We may summarize our findings as follows: at any time t

$$\mathbf{E} = \mathbf{C} = \frac{d\mathbf{F}}{dt} \quad \text{or} \quad \mathbf{F} = \int_0^t \mathbf{E}\,dt = \int_0^t \mathbf{C}\,dt$$

and

$$\bar{t} = \bar{t}_{\mathbf{C}} = \bar{t}_{\mathbf{E}} \tag{8}$$

These relationships show how stimulus-response experiments, using either step or pulse inputs can conveniently give the RTD and mean flow rate of fluid in the vessel. We should remember that these relationships only hold for closed vessels. When these boundary conditions are not met then the **C** curve may differ appreciably from the **E** curve. In fact not even the mean times for these curves are equal: thus for open vessels

$$\bar{t} \neq \bar{t}_{\mathbf{C}} \stackrel{?}{=} \bar{t}_{\mathbf{E}}$$

In this case it would be interesting to decide how to define **E** usefully.

Figure 6 shows the shapes of these curves for various types of flow and also for a dimensionless measure of time θ, to be discussed later.

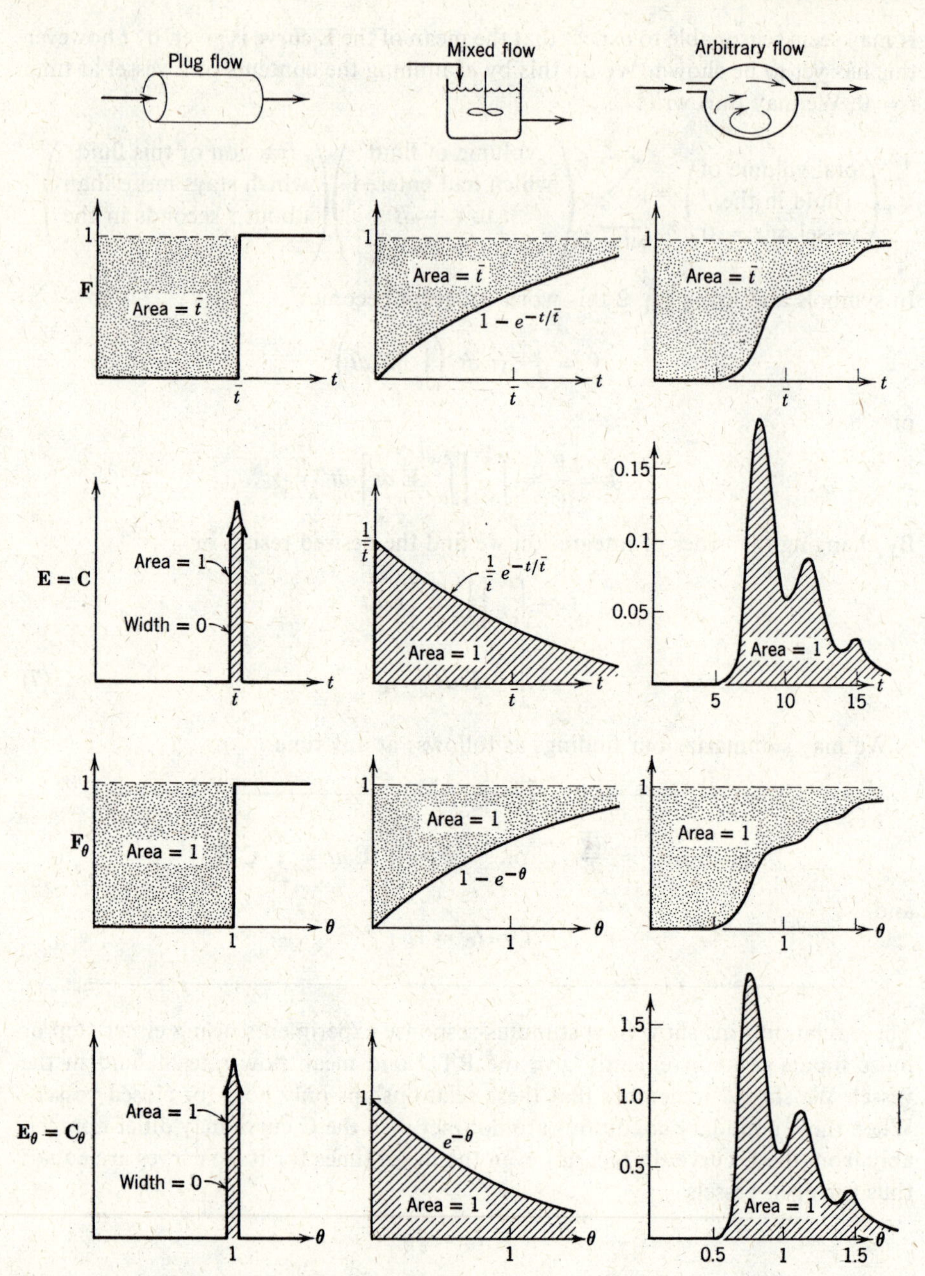

FIGURE 6. Properties of the **E**, **C**, and **F** curves for various flows. Curves are drawn in terms of ordinary and dimensionless time units. Relationship between curves is given by Eqs. 8 and 26.

Useful Mathematical Tools

Here we introduce three mathematical concepts which are widely used in tracer work. Many of the results in this chapter are obtained by their application.

The Mean and Variance. It is frequently desirable to characterize a distribution by a few numerical values. For this purpose the most important measure is the location of the distribution. This is called the mean value or the centroid of the distribution. Thus for a C versus t curve the mean is given by

$$\bar{t} = \frac{\int_0^\infty tC\,dt}{\int_0^\infty C\,dt} \tag{9a}$$

If the distribution curve is only known at a number of discrete time values t_i then

$$\bar{t} \cong \frac{\sum t_i C_i\,\Delta t_i}{\sum C_i\,\Delta t_i} \tag{9b}$$

The next most important descriptive quantity is the spread of the distribution. This is commonly measured by the variance σ^2, defined as

$$\sigma^2 = \frac{\int_0^\infty (t-\bar{t})^2 C\,dt}{\int_0^\infty C\,dt} = \frac{\int_0^\infty t^2 C\,dt}{\int_0^\infty C\,dt} - \bar{t}^2 \tag{10a}$$

Again, in discrete form

$$\sigma^2 \cong \frac{\sum (t_i-\bar{t})^2 C_i\,\Delta t_i}{\sum C_i\,\Delta t_i} = \frac{\sum t_i^2 C_i\,\Delta t_i}{\sum C_i\,\Delta t_i} - \bar{t}^2 \tag{10b}$$

The variance represents the square of the spread of the distribution and has units of (time)^2. It is particularly useful for matching experimental curves to one of a family of theoretical curves.

When used with normalized distributions for closed vessels these expressions simplify somewhat. Thus for a continuous curve or for discrete measurements at equal time intervals the mean becomes

$$\bar{t} = \int_0^\infty t\mathbf{E}\,dt \cong \frac{\sum t_i \mathbf{E}_i}{\sum \mathbf{E}_i} = \sum t_i \mathbf{E}_i\,\Delta t \tag{11}$$

and the variance becomes

$$\sigma^2 = \int_0^\infty (t-\bar{t})^2 \mathbf{E}\,dt = \int_0^\infty t^2 \mathbf{E}\,dt - \bar{t}^2$$

$$= \frac{\sum t_i^2 \mathbf{E}_i}{\sum \mathbf{E}_i} - \bar{t}^2 = \sum t_i^2 \mathbf{E}_i\,\Delta t - \bar{t}^2 \tag{12}$$

Figure 7 illustrates these terms.

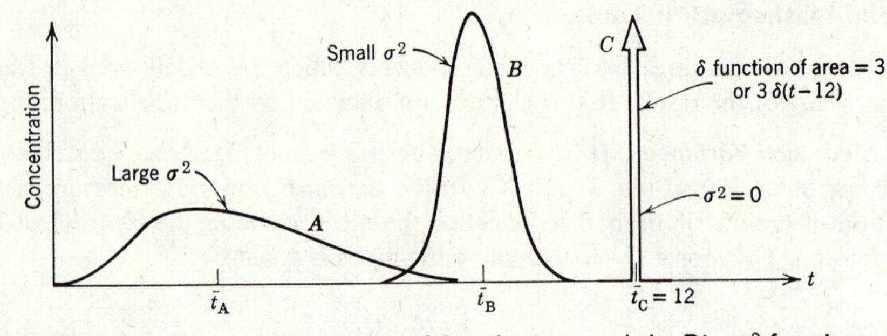

FIGURE 7. Means and variances of distributions, and the Dirac δ function.

The Dirac Delta Function. Ordinary calculus has difficulty dealing with discontinuous functions so a special function called the Dirac delta function, δ, is defined to treat the discontinuous pulse. Thus $\delta(t - t_0)$ is a distribution curve which is zero everywhere except at $t - t_0 = 0$ where it is infinite. The area under the curve is unity and the width of the pulse is zero. In symbols then

$$\left.\begin{aligned} \delta(t - t_0) &= \infty \quad \text{at } t = t_0 \\ \delta(t - t_0) &= 0 \quad \text{elsewhere} \end{aligned}\right\} \quad \text{such that} \int_{-\infty}^{\infty} \delta(t - t_0)\, dt = 1 \tag{13}$$

Figure 7 shows how we graphically represent this function.

The one useful property of this peculiar function is that integration with any other function $f(t)$ gives

$$\int_a^b \delta(t - t_0) f(t)\, dt = f(t_0) \qquad \text{if } a < t_0 < b \tag{14a}$$

$$= 0 \qquad \text{if the } ab \text{ interval does not contain } t_0 \tag{14b}$$

Thus integration is practically automatic; simply evaluate the function $f(t)$ at the location of the pulse. As examples of the use of the δ function

$$\int_2^4 \delta(t - 3) t^2\, dt = t^2 \Big|_3 = 9$$

$$\int_0^2 \delta(t - 3) t^2\, dt = 0$$

$$\int_{-\infty}^{\infty} \delta(t - 3) t^2\, dt = t^2 \Big|_3 = 9$$

$$\int_{-\infty}^{\infty} \delta(t) e^{-kt}\, dt = e^{-kt} \Big|_0 = 1$$

The Convolution Integral. Suppose we introduce into a vessel a one-shot tracer signal C_{in} versus t as shown in Fig. 8. In passing through the vessel the signal will

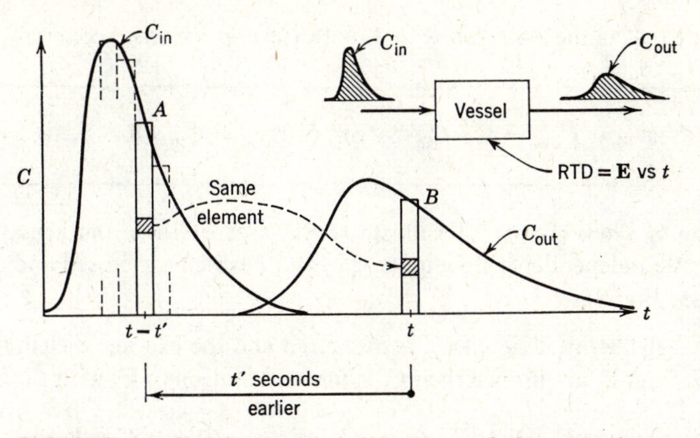

FIGURE 8. Sketch showing derivation of the convolution integral.

be modified to give an output signal C_{out} versus t. Since the flow with its particular RTD is responsible for this modification let us relate C_{in}, $\mathbf{E}$, and C_{out}.

Focus attention on tracer leaving at time about t. This is shown as the narrow rectangle B in Fig. 8. We may then write

$$\begin{pmatrix}\text{tracer leaving}\\ \text{in rectangle } B\end{pmatrix} = \begin{pmatrix}\text{all the tracer entering } t' \text{ seconds earlier than } t,\\ \text{and staying for time } t' \text{ in the vessel}\end{pmatrix}$$

We show the tracer which enters t' seconds earlier than t as the narrow rectangle A. In terms of this rectangle the above equation may be written

$$\begin{pmatrix}\text{tracer leaving}\\ \text{in rectangle } B\end{pmatrix} = \sum_{\substack{\text{all rectangles}\\ A \text{ which enter}\\ \text{earlier than}\\ \text{time } t}} \begin{pmatrix}\text{tracer in}\\ \text{rectangle}\\ A\end{pmatrix}\begin{pmatrix}\text{fraction of tracer in } A\\ \text{which stays for about}\\ t' \text{ seconds in the vessel}\end{pmatrix}$$

In symbols and taking limits (shrinking the rectangles) we obtain the desired relationship which is called the convolution integral

$$C_{out}(t) = \int_0^t C_{in}(t - t')\mathbf{E}(t')\,dt' \tag{15a}$$

In what can be shown to be equivalent form we also have

$$C_{out}(t) = \int_0^t C_{in}(t')\mathbf{E}(t - t')\,dt' \tag{15b}$$

We say that C_{out} is the *convolution* of $\mathbf{E}$ with C_{in} and we write concisely

$$C_{out} = \mathbf{E} * C_{in} \qquad \text{or} \qquad C_{out} = C_{in} * \mathbf{E} \tag{15c}$$

Application of These Tools. To illustrate the uses of these mathematical tools consider three independent† flow units a, b, and c which are closed and connected in series (see Fig. 9).

Problem 1. If the input signal C_{in} is measured and the exit age distribution functions $\mathbf{E}_a$, $\mathbf{E}_b$, and $\mathbf{E}_c$ are known then C_1 is the convolution of $\mathbf{E}_a$ with C_{in} and so on, thus

$$C_1 = C_{in} * \mathbf{E}_a, \qquad C_2 = C_1 * \mathbf{E}_b, \qquad C_{out} = C_2 * \mathbf{E}_c$$

and on combining

$$C_{out} = C_{in} * \mathbf{E}_a * \mathbf{E}_b * \mathbf{E}_c \tag{16}$$

Thus we can determine the output from a multiregion flow unit.

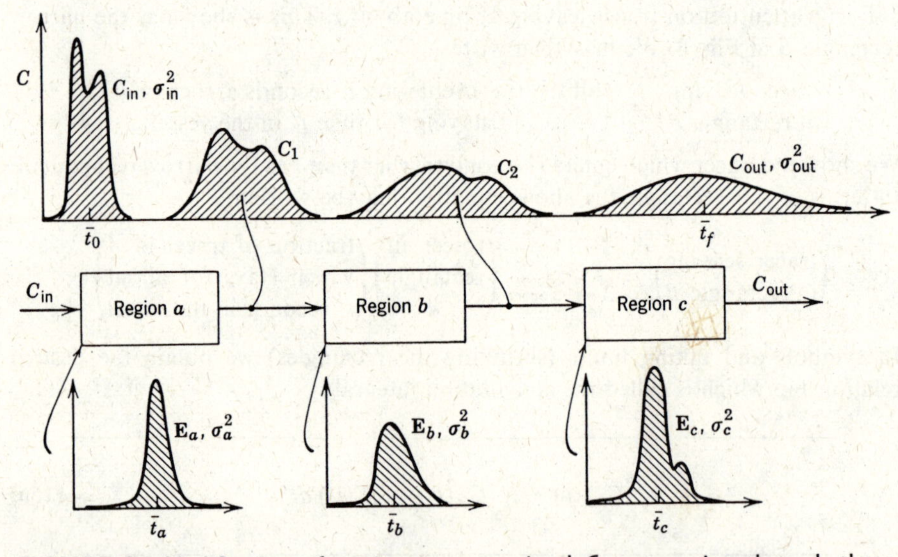

FIGURE 9. Modification of an input tracer signal C_{in} on passing through three successive regions.

† By independence we mean that the fluid loses its memory as it passes from vessel to vessel. Thus a faster moving fluid element in one vessel does not remember this fact in the next vessel and doesn't preferentially flow faster (or slower) there. Laminar flow often does not satisfy this requirement of independence; however, complete (or lateral) mixing of fluid between units satisfies this condition.

Problem 2. If we measure C_{in} and C_{out} and know $\mathbf{E}_a$ and $\mathbf{E}_c$ we can then extract the unknown $\mathbf{E}_b$. This type of problem is of particular importance in experimentation where the entrance region and collection region for tracer are both large compared with the experimental section.

It is a straightforward matter to convolute; however, to *deconvolute*, to find one of the distribution functions under the integral, is difficult and is best attacked by Fourier transforms. Thus Problem 2 is harder to treat than Problem 1. However a special case of Problem 2 is quite simple to treat, and that is when the input is a δ-function and the known regions are all in mixed flow. Fortunately this case often closely represents real systems with their end effects. We present a simple graphical method for treating this situation in Example 7 later in this chapter.

Problem 3. A useful property of the variance is its additivity for flow through independent vessels. Thus if σ_{in}^2 and σ_{out}^2 are the variances of the input and output tracer curves and σ_a^2, σ_b^2, σ_c^2 are the variance of the $\mathbf{E}$ curves for the three regions, as indicated in Fig. 9, then it can be shown that

$$\sigma_{\text{out}}^2 = \sigma_{\text{in}}^2 + \sigma_a^2 + \sigma_b^2 + \sigma_c^2 \tag{17}$$

This property allows us to extract the variance of the RTD for any region if the other variances are known.

This additivity property of variances parallels that of the means where

$$\bar{t}_{\text{out}} = \bar{t}_{\text{in}} + \bar{t}_a + \bar{t}_b + \bar{t}_c \tag{18}$$

This additivity of means does not require independence of regions, just closed vessels.

Ways of Using Age Distribution Information

Tracer information is used either directly or in conjunction with flow models to predict performance of real flow reactors, the method used depending in large part on whether the reactor can be considered to be a linear system or not.

Linear and Nonlinear Processes. A process is linear if any change in magnitude of the stimulus results in a corresponding proportional change in the magnitude of the response. In symbols

$$\frac{\Delta(\text{response})}{\Delta(\text{stimulus})} = \frac{d(\text{response})}{d(\text{stimulus})} = k_1 = \text{constant} \tag{19}$$

or integrating

$$(\text{response}) = k_1(\text{stimulus}) + k_2$$

Processes that do not satisfy these conditions are not linear.

Linear processes have the following highly desirable property. If a number of independent linear processes are occurring simultaneously in a system, their over-all effect is also a linear process. In addition, the over-all effect of these individual linear processes occurring simultaneously in a system can be analyzed by studying each of the processes separately. This property does not extend to nonlinear processes. Hence nonlinear systems must be studied "in the total situation," and their over-all behavior cannot be predicted by knowledge of each of the contributing processes. Because of this property solutions to problems involving linear processes are relatively simple and are of wide generality. On the other hand solutions for nonlinear processes are much more difficult and are specific to each problem.

Linear Systems without Flow Models. If the tracer has no unusual activity (adsorption at walls, disappearance by reaction), but simply passes through the vessel with the rest of the fluid, the stimulus-response experiment at steady state is linear in concentration. In other words, if we double the concentration of the stimulus we double the concentration of the response.

Now by the additivity property of linear processes stimulus-response information should be sufficient to account for the behavior of the nonideal flow vessel as a chemical reactor as long as the reaction rate is also linear in concentration. This includes ordinary first-order reactions of all types: irreversible, reversible, series, parallel, and in combination. Thus we can say

$$\begin{pmatrix}\text{tracer}\\ \text{information}\\ \text{for vessel}\end{pmatrix} + \begin{pmatrix}\text{kinetic data}\\ \text{for first}\\ \text{order reactions}\\ \text{of all kinds}\end{pmatrix} \rightarrow \begin{pmatrix}\text{behavior of}\\ \text{vessel as a}\\ \text{reactor}\end{pmatrix}$$

We treat such systems directly after Example 1.

Nonlinear Systems with Flow Models. Unfortunately if the reaction is not of first order then conversion cannot be found using age distribution information directly. To illustrate this point consider the two models of a flow reactor shown in Fig. 10. They both have identical tracer response curves and cannot be distinguished by tracer experiments. For reactions with rate linear in concentration, however, it is not necessary to know which is the true flow pattern because both give identical conversions (see Problem 6.3). On the other hand, since conversions will differ in these two systems for nonlinear reactions (see Problem 6.12 and Example 10.1), the flow pattern which actually exists must be known before predictions of performance can be made.

In the absence of the needed point-to-point information we hypothesize what we consider to be a reasonable model for the flow of fluid in the vessel and then calculate the conversion. Naturally the closeness of predicted conversion to actual conversion will depend on how well the model mirrors reality.

Linear Systems with Flow Models. In addition to predicting conversions in nonlinear reaction systems, flow models are often used to predict conversions in

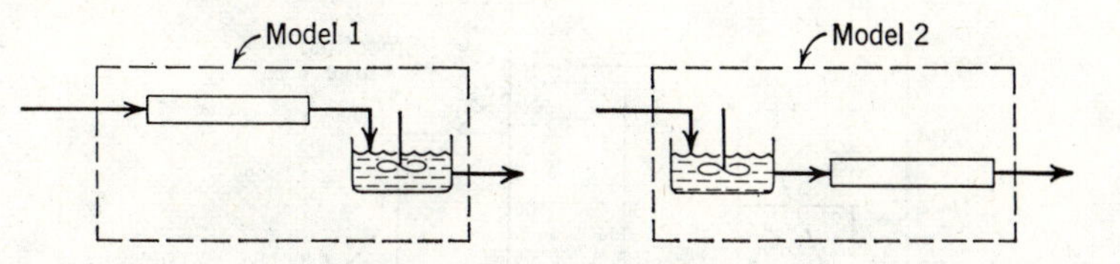

FIGURE 10. Both models give identical tracer response signals and therefore act alike for first order reactions, but act differently for reactions whose rates are nonlinear in concentration.

linear systems. This seemingly roundabout procedure is used because the parameters of these models often correlate with the variables of the system, such as Reynolds number, Schmidt number, etc. Such correlations can then be used to predict conversions without resorting to experiment. Such is the case with packed bed and tubular reactors. We take up this approach after showing how tracer data are used directly.

EXAMPLE 1. Finding the RTD by experiment

The concentration readings in Table E1 represent a continuous response to a delta-function input into a closed vessel which is to be used as a chemical reactor. Tabulate and plot the exit age distribution E.

TABLE E1

Time t, min	Tracer Output Concentration, gm/liter fluid
0	0
5	3
10	5
15	5
20	4
25	2
30	1
35	0

SOLUTION

The area under the concentration-time curve,

$$Q = \sum C\,\Delta t = (3 + 5 + 5 + 4 + 2 + 1)5$$
$$= 100 \text{ gm}\cdot\text{min/liter}$$

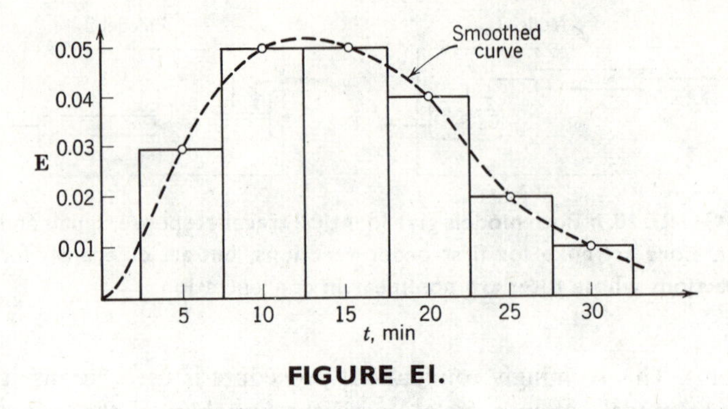

FIGURE E1.

gives the total amount of tracer introduced. To find **E**, the area under this curve must be unity; hence the concentration readings must each be divided by Q, giving

$$\mathbf{E} = \frac{C}{Q}$$

Thus we have

t, min	0	5	10	15	20	25	30
$\mathbf{E} = \dfrac{C}{Q}$, 1/min	0	0.03	0.05	0.05	0.04	0.02	0.01

Figure E1 is a plot of this distribution.

CONVERSION DIRECTLY FROM TRACER INFORMATION

Linear Process. A variety of flow patterns can give the same tracer output curve. For linear processes, however, these all result in the same conversion; consequently we may use any convenient flow pattern to determine conversions, as long as the pattern selected gives the same tracer response curve as the real reactor. The simplest pattern to use assumes that each element of fluid passes through the vessel with no intermixing with adjacent elements, the age distribution of material in the exit stream telling how long each of these individual elements remains within the reactor. Thus for reactant A in the exit stream

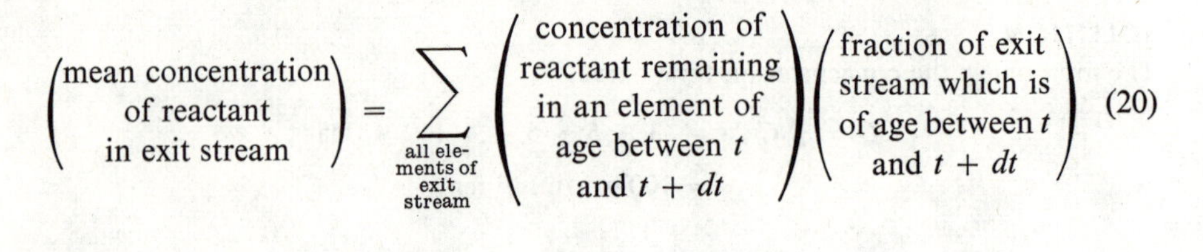

$$\begin{pmatrix}\text{mean concentration}\\ \text{of reactant}\\ \text{in exit stream}\end{pmatrix} = \sum_{\substack{\text{all ele-}\\ \text{ments of}\\ \text{exit}\\ \text{stream}}} \begin{pmatrix}\text{concentration of}\\ \text{reactant remaining}\\ \text{in an element of}\\ \text{age between } t\\ \text{and } t + dt\end{pmatrix}\begin{pmatrix}\text{fraction of exit}\\ \text{stream which is}\\ \text{of age between } t\\ \text{and } t + dt\end{pmatrix} \quad (20)$$

or in symbols

$$\bar{C}_A = \int_{t=0}^{\infty} C_{A,\text{element}} \mathbf{E}\, dt \tag{21}$$

For *irreversible first-order reactions* with no density change the concentration of reactant in any element changes with time as follows,

$$\ln \frac{C_{A,\text{element}}}{C_{A0}} = -kt \qquad \text{or} \qquad C_{A,\text{element}} = C_{A0} e^{-kt}$$

Hence Eq. 21 becomes

$$\bar{C}_A = C_{A0} \int_0^{\infty} e^{-kt} \mathbf{E}\, dt \tag{22}$$

For the *reversible first-order reactions* of Eq. 3.53, with no product in the feed, inserting Eq. 3.54 in Eq. 21 gives

$$\bar{C}_A = C_{A0} \int_0^{\infty} [1 - X_{Ae} + X_{Ae} e^{-kt/X_{Ae}}] \mathbf{E}\, dt \tag{23}$$

For intermediate R in the *successive first-order reactions* of Eq. 3.44 to 3.46 insertion of Eq. 3.49 in Eq. 21 gives

$$\bar{C}_R = \frac{C_{A0} k_1}{k_2 - k_1} \int_0^{\infty} [e^{-k_1 t} - e^{-k_2 t}] \mathbf{E}\, dt \tag{24}$$

Similar expressions can be obtained for any other system of first-order reactions.

These conversion equations can be solved either graphically or numerically for any pattern of flow. Thus the performance of nonideal flow reactors can be determined precisely given the residence time distribution and the rate constants for the first-order reactions.

Nonlinear Process. For reactions with nonlinear rates, conversions cannot be determined from tracer information alone. Equation 21, however, always gives one of the bounds to the conversion. Since it represents the latest possible mixing for any given RTD, we can show that it gives the lower bound to conversion for $n < 1$, upper bound for $n > 1$. The bound corresponding to the earliest possible mixing is difficult to calculate. This problem is discussed in Chapter 10.

EXAMPLE 2. Conversion in reactors having nonideal flow

The vessel of Example 1 is to be used as a reactor for a liquid decomposing with rate

$$-r_A = kC_A, \qquad k = 0.307 \text{ min}^{-1}$$

Find the fraction of reactant unconverted in the real reactor and compare this with the fraction unconverted in a plug flow reactor of the same size.

SOLUTION

With negligible density change we have from Example 1 for both reactors

$$\tau = \bar{t} = 15 \text{ min}$$

For the plug flow reactor then

$$\tau = C_{A0} \int_0^{X_A} \frac{dX_A}{-r_A} = -\frac{1}{k} \int_{C_{A0}}^{C_A} \frac{dC_A}{C_A} = \frac{1}{k} \ln \frac{C_{A0}}{C_A}$$

or

$$\frac{C_A}{C_{A0}} = e^{-k\tau} = e^{-(0.307)(15)} = e^{-4.6} = 0.01$$

Thus the fraction of reactant unconverted in a plug flow reactor equals 1.0%.

For the real reactor the fraction unconverted, given by Eq. 22, is found as shown in Table E2.

TABLE E2

t	E	kt	e^{-kt}	$e^{-kt}\mathbf{E}\,\Delta t$
5	0.03	1.53	0.2154	(0.2154)(0.03)(5) = 0.0323
10	0.05	3.07	0.0464	0.0116
15	0.05	4.60	0.0100	0.0025
20	0.04	6.14	0.0021	0.0004
25	0.02	7.68	0.0005	0.0001
30	0.01	9.21	0.0001	0
				$\sum e^{-kt}\mathbf{E}\,\Delta t = 0.0469$

Hence the fraction of reactant unconverted in the real reactor, C_A/C_{A0}, equals 4.7% From the table we see that the unconverted material comes mostly from the early portion of the E curve. This suggests that channeling and short-circuiting can seriously hinder attempts to achieve high conversion in reactors.

MODELS FOR NONIDEAL FLOW

Many types of models can be used to characterize nonideal flow within vessels. Some draw on the analogy between mixing in actual flow and a diffusional process. These are called *dispersion models*. Others build a chain or network of ideal mixers, while still others visualize various flow regions connected in series or parallel. Some models are useful in accounting for the deviation of real systems, such as tubular vessels or packed beds, from plug flow; others describe the deviation of real stirred tanks from the ideal of mixed flow; and still others attempt to account for fluidized beds and other contacting devices.

Models vary in complexity. For example, one-parameter models adequately represent packed beds or tubular vessels. On the other hand, two- to six-parameter models have been proposed to represent fluidized beds. Our plan is to first treat these simpler models, and then to consider the more complex models, and various related subjects.

Dimensionless Time Units. In treating models we often find it convenient to measure time in units of mean residence time. This then gives a dimensionless measure

$$\theta = \frac{t}{\bar{t}} \quad \text{and} \quad d\theta = \frac{dt}{\bar{t}} \tag{25}$$

The various distribution curves based on these dimensionless time units are denoted by the subscript θ, thus $\mathbf{E}_\theta$, $\mathbf{F}_\theta$, $\mathbf{C}_\theta$. Typical curves in both t and θ units are shown in Fig. 6.

To relate $\mathbf{E}$ and $\mathbf{E}_\theta$ pick the same point on these two curves. Then geometrical considerations require that

$$\theta \mathbf{E}_\theta = t\mathbf{E}$$

and on combining with Eq. 25 this gives

$$\mathbf{E}_\theta = \bar{t}\mathbf{E}$$

Similarly we can relate $\mathbf{F}_\theta$ and $\mathbf{F}$.

The relationship between distributions in both these time measures are summarized as follows:

$$\mathbf{E} = \mathbf{C} = \frac{d\mathbf{F}}{dt}, \qquad \mathbf{E}_\theta = \mathbf{C}_\theta = \frac{d\mathbf{F}_\theta}{d\theta}$$

$$\mathbf{E}_\theta = \bar{t}\mathbf{E}, \qquad \mathbf{C}_\theta = \bar{t}\mathbf{C}, \qquad \mathbf{F}_\theta = \mathbf{F}$$

$$\bar{\theta}_{\mathbf{C}} = \bar{\theta}_{\mathbf{E}} = 1, \qquad \bar{t}_{\mathbf{C}} = \bar{t}_{\mathbf{E}} = \bar{t} \tag{26}$$

$$\sigma_\theta^2 = \frac{\sigma^2}{\bar{t}^2} \text{ (dimensionless)}$$

DISPERSION MODEL (DISPERSED PLUG FLOW)

Consider the plug flow of a fluid, on top of which is superimposed some degree of backmixing or intermixing, the magnitude of which is independent of position within the vessel. This condition implies that there exist no stagnant pockets and no gross bypassing or short-circuiting of fluid in the vessel. This is called the dispersed plug flow model, or simply the dispersion model. Figure 11 shows the conditions visualized. Note that with varying intensities of turbulence or intermixing the predictions of this model should range from plug flow at one extreme to mixed flow at the other. As a result the reactor volume for this model will lie between those calculated for plug and mixed flow.

Since the mixing process involves a shuffling or redistribution of material either by slippage or eddies, and since this is repeated a considerable number of times during the flow of fluid through the vessel we can consider these disturbances to be statistical in nature, somewhat as in molecular diffusion. For molecular diffusion in the x direction the governing differential equation is given by Fick's law

$$\frac{\partial C}{\partial t} = \mathscr{D} \frac{\partial^2 C}{\partial x^2}$$

where $\mathscr{D}$, the coefficient of molecular diffusion, is a parameter which uniquely characterizes the process. In an analogous manner we may consider all the contributions to backmixing of fluid flowing in the x direction to be described by a similar form of expression, or

$$\frac{\partial C}{\partial t} = D \frac{\partial^2 C}{\partial x^2} \tag{27}$$

where the parameter D, which we call the *longitudinal* or *axial dispersion coefficient*, uniquely characterizes the degree of backmixing during flow. We use the terms "longitudinal" and "axial" because we wish to distinguish mixing in the direction of flow from mixing in the lateral or radial direction, which is not our primary

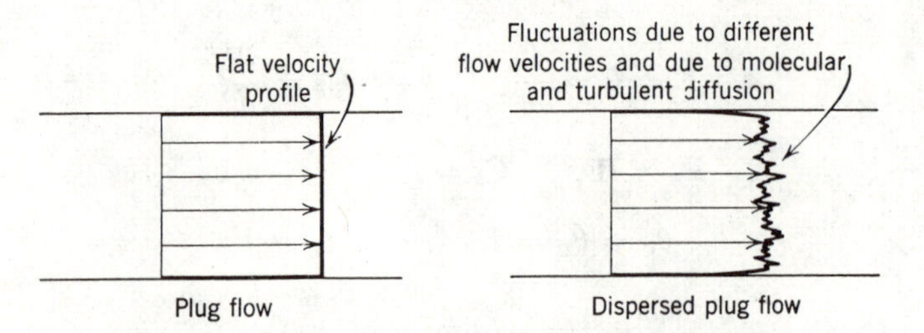

FIGURE 11. Representation of the dispersion (dispersed plug flow) model.

concern. These two quantities may be quite different in magnitude. For example, in streamline flow of fluids through pipes, axial mixing is mainly due to fluid velocity gradients whereas radial mixing is due to molecular diffusion alone.

In dimensionless form where $z = x/L$ and $\theta = t/\bar{t} = tu/L$ the basic differential equation representing this dispersion model becomes

$$\frac{\partial C}{\partial \theta} = \left(\frac{D}{uL}\right)\frac{\partial^2 C}{\partial z^2} - \frac{\partial C}{\partial z} \tag{28}$$

where the dimensionless group $\left(\frac{D}{uL}\right)$, called the vessel dispersion number, is the parameter which measures the extent of axial dispersion. Thus

$\frac{D}{uL} \rightarrow 0$ negligible dispersion, hence plug flow

$\frac{D}{uL} \rightarrow \infty$ large dispersion, hence mixed flow

This model usually represents quite satisfactorily flow that deviates not too greatly from plug flow, thus real packed beds and tubes (long ones if flow is streamline).

Fitting the Dispersion Model for Small Extents of Dispersion

If we impose an idealized pulse onto the flowing fluid then dispersion modifies this pulse as shown in Fig. 12. For small extents of dispersion (if D/uL is small) the spreading tracer curve does not significantly change in shape as it passes the measuring point (during the time it is being measured). Under these conditions the solution to Eq. 28 is not difficult, and gives the symmetrical **C** curve

$$\mathbf{C}_\theta = \frac{1}{2\sqrt{\pi(D/uL)}} \exp\left[-\frac{(1-\theta)^2}{4(D/uL)}\right] \tag{29}$$

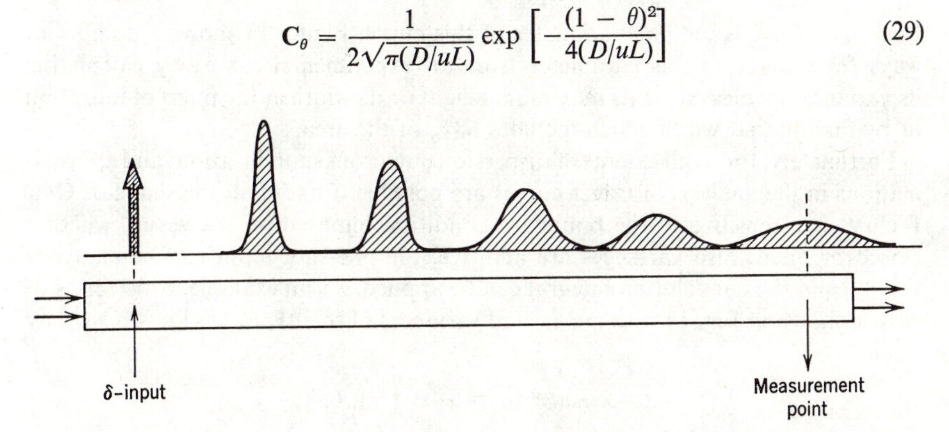

FIGURE 12. The dispersion model predicts a symmetrical distribution of tracer at any instant.

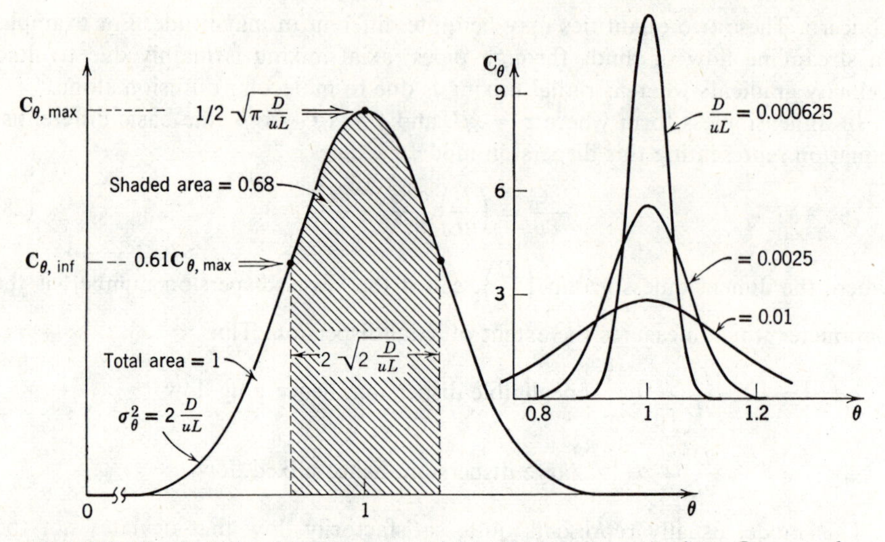

FIGURE 13. Relationship between D/uL and the dimensionless **C** curve for small extents of dispersion, Eq. 29.

which represents a family of gaussian, normal, or error curves with mean and variance

$$\bar{\theta}_{\mathbf{C}} = \frac{\bar{t}_{\mathbf{C}}}{\bar{t}} = 1 \tag{30}$$

$$\sigma_\theta^{\ 2} = \frac{\sigma^2}{\bar{t}^2} = 2\left(\frac{D}{uL}\right) \quad \text{or} \quad \sigma^2 = 2\left(\frac{DL}{u^3}\right) \tag{31}$$

Note that D/uL is the one parameter of this curve. Figure 13 shows a number of ways for evaluating this parameter from an experimental curve: by calculating its variance, by measuring its maximum height or its width at the point of inflection or by finding that width which includes 68% of the area.

Fortunately, for small extents of dispersion numerous simplifications and approximations in the analysis of tracer curves are possible. First of all the shape of **C** or **F** curve is insensitive to the boundary condition imposed on the vessel, whether closed or open; also variances are additive, the one-shot input can be analyzed simply, and the convolution integral can be applied. As an example, for a series of vessels shown in Fig. 14 with means and variances of their **E** curves known we may write

$$\bar{t}_{\text{overall}} = \bar{t}_a + \bar{t}_b + \cdots + \bar{t}_n$$

and

$$\sigma_{\text{overall}} = \sigma_a^{\ 2} + \sigma_b^{\ 2} + \cdots + \sigma_n^{\ 2} \tag{32}$$

FIGURE 14. Illustration of additivity of variances of the **E** curves of vessels $a, b, \ldots, n$.

This additivity property of variances also allows us to treat any one-shot tracer input, no matter what its shape, and to extract from it the variance of the **C** (or **E**) curve of the vessel. So, on referring to Fig. 15, if we write

$$\Delta\sigma^2 = \sigma^2_{\text{out}} - \sigma_{\text{in}}{}^2 \tag{33}$$

we then have with Eq. 31

$$\frac{\Delta\sigma^2}{\bar{t}^2} = \Delta\sigma_\theta{}^2 = 2\left(\frac{D}{uL}\right) \tag{34}$$

Thus no matter what the shape of the input curve, the corresponding output curve and D/uL value for the vessel can be found.

The goodness of fit for this simple treatment can only be evaluated by comparison with the more exact but much more complex solutions. From such a comparison we find that the maximum error in estimate of D/uL is given by

$$\text{error} < 5\% \text{ when } \frac{D}{uL} < 0.01$$

$$\text{error} < 0.5\% \text{ when } \frac{D}{uL} < 0.001$$

Fitting the Dispersion Model for Large Extents of Dispersion

When the tracer of Fig. 12 changes shape significantly during the time that it passes the recording point, then the measured curve is unsymmetrical with a somewhat extended tail. In this situation the flow conditions at the injection and

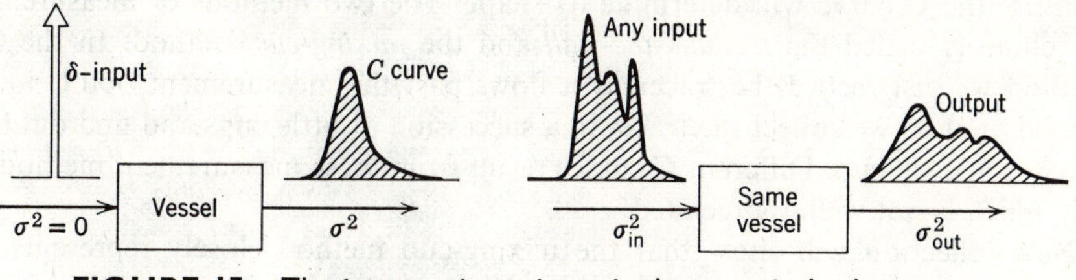

FIGURE 15. The increase in variance is the same in both cases, or

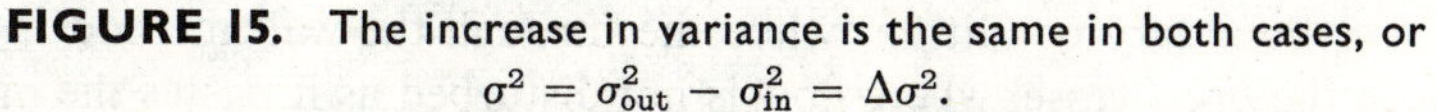

$$\sigma^2 = \sigma^2_{\text{out}} - \sigma^2_{\text{in}} = \Delta\sigma^2.$$

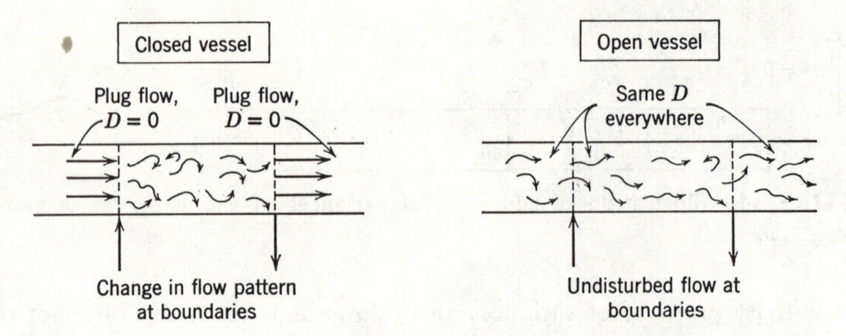

FIGURE 16. Two of the many possible boundary conditions for a flow vessel.

measurement point (called the boundary conditions) will influence the shape of the obtained **C** curve. Let us briefly consider just two of the many possible cases which have been treated by workers in this field, the closed vessel and the open vessel, as sketched in Fig. 16.

We should mention here that it has not been possible to obtain analytic expressions for **C** curves for any of the boundary conditions, except for one case, the open vessel. However, means and variances can be determined for all cases. Many of these values are reported by van der Laan (1958).

Closed vessel. Figure 17 shows the **C** curve for closed vessels as calculated by numerical methods. Note, as D/uL rises the curve becomes increasingly skewed. The mean and variance of this family of curves is found to be

$$\bar{\theta}_{\mathbf{C}} = \frac{\bar{t}_{\mathbf{C}}}{\bar{t}} = 1 \tag{35}$$

$$\sigma_\theta^2 = \frac{\sigma^2}{\bar{t}^2} = 2\frac{D}{uL} - 2\left(\frac{D}{uL}\right)^2(1 - e^{-uL/D}) \tag{36}$$

Open vessels. This is the only situation where the **C** curve can be derived analytically. However, here we encounter an additional complication in that the way we measure the **C** curve will determine its shape. The two methods of measurement are clumsily called the *through-the-wall* and the *mixing-cup* method. In the first method we just record the tracer as it flows past the measurement point; in the second method we collect the tracer in a succession of little cups and find out how much is in each cup. Different **C** curves result from these measurement methods, a fact which is not well appreciated.

Now reflection will show that the mixing-cup method closely represents the closed vessel boundary condition, while the through-the-wall method is more in accord with the open vessel where flow is not disturbed as it passes the measuring

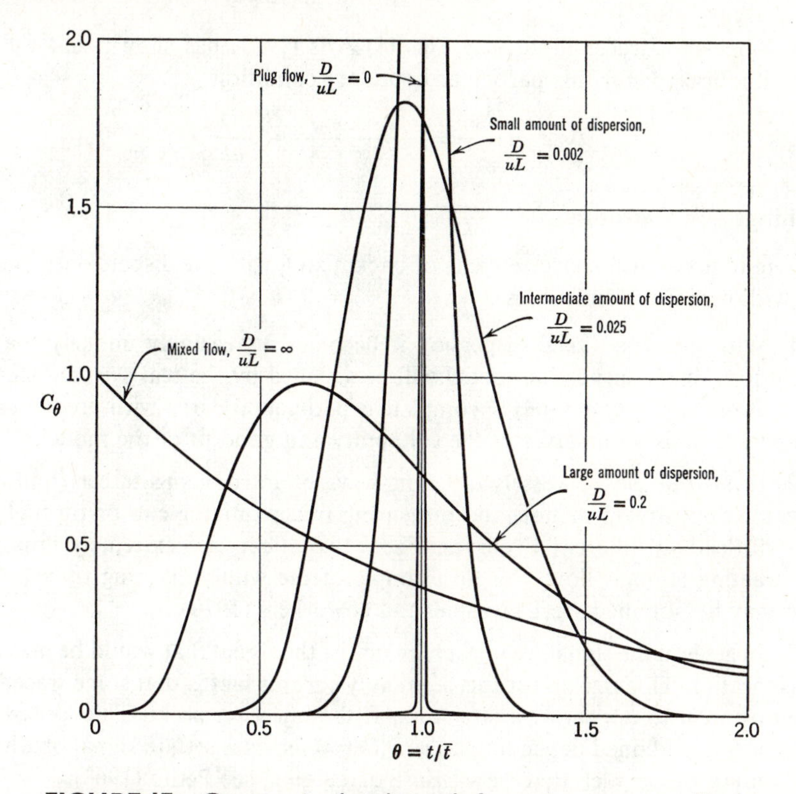

FIGURE 17. **C** curves in closed vessels for various extents of backmixing as predicted by the dispersion model.

point. Consequently we will only consider the equations for through-the-wall measurements for open vessels. These **C** curves are given by

$$\mathbf{C}_\theta = \frac{1}{2\sqrt{\pi\theta(D/uL)}} \exp\left[-\frac{(1-\theta)^2}{4\theta(D/uL)}\right] \tag{37}$$

with mean and variance

$$\bar{\theta}_{\mathbf{C}} = \frac{\bar{t}_{\mathbf{C}}}{\bar{t}} = 1 + 2\frac{D}{uL} \tag{38}$$

$$\sigma_\theta^2 = \frac{\sigma^2}{\bar{t}^2} = 2\frac{D}{uL} + 8\left(\frac{D}{uL}\right)^2 \tag{39}$$

The derivation of these expressions and shapes of these curves are given by Levenspiel and Smith (1957).

For a one-shot tracer input (see Fig. 15) Aris (1959) has shown that we may write without error for this particular boundary condition

$$\Delta\sigma_\theta{}^2 = \frac{\Delta\sigma^2}{\bar{t}^2} = \frac{\sigma_{out}^2 - \sigma_{in}{}^2}{\bar{t}^2} = 2\frac{D}{uL} \tag{34}$$

Warnings and Cautions

In trying to account for large extents of backmixing with the dispersion model we meet with numerous difficulties.

1. With increased axial dispersion it becomes increasingly unlikely that the assumptions of the dispersion model will be satisfied by the real system. Thus we should examine and compare the complete experimental curve with predictions of the model to satisfy ourselves of the suitability and good fit of the model.

2. In all but closed vessels different ways of introducing tracer (uniformly across the flow stream or not) and measuring tracer (mixing-cup or through-the-wall) will lead to different **C** curves. These two effects are extremely important when treating laminar flow. For an example of the wildly differing tracer curves which may be obtained see Levenspiel and coworkers (1970a,b).

3. In general the signal we inject may not be the signal that would be measured at this location. One reason for this seemingly curious fact is that some tracer may move upstream to appear at a later time at the injection point. Thus different **C** curves may be obtained depending on whether we *inject* a certain signal or whether we introduce tracer such that we *measure* that signal, see Petho (1968).

The literature in this field is profuse and often conflicting, primarily because of the unstated and unclear assumptions about what is happening at the vessel boundaries. The treatment of end conditions is full of mathematical subtleties as noted above, and the additivity of variances is questionable. Because of all this we should be very careful in using the dispersion model where backmixing is large, particularly if the system is not closed.

EXAMPLE 3. D/uL from a C curve

On the assumption that the closed vessel of Example 1, p. 267, is well represented by the dispersion model, calculate the vessel dispersion number D/uL.

SOLUTION

Since the **C** curve for this vessel is broad and unsymmetrical let us guess that dispersion is too large to allow use of the simplification leading to Fig. 13. We thus start with the variance matching procedure of Eq. 36. The variance of a continuous distribution measured at a finite number of equidistant locations is given by Eqs. 8 and 10 as

$$\sigma^2 = \frac{\sum t_i{}^2 C_i}{\sum C_i} - \bar{t}^2 = \frac{\sum t_i{}^2 C_i}{\sum C_i} - \left[\frac{\sum t_i C_i}{\sum C_i}\right]^2$$

Using the original tracer concentration-time data given in Example 1, we find

$$\sum C_i = 3 + 5 + 5 + 4 + 2 + 1 = 20$$

$$\sum t_i C_i = (5 \times 3) + (10 \times 5) + \cdots + (30 \times 1) = 300 \text{ min}$$

$$\sum t_i^2 C_i = (25 \times 3) + (100 \times 5) + \cdots + (900 \times 1) = 5450 \text{ min}^2$$

Therefore

$$\sigma^2 = \frac{5450}{20} - \left(\frac{300}{20}\right)^2 = 47.5 \text{ min}^2$$

and

$$\sigma_\theta^2 = \frac{\sigma^2}{\bar{t}^2} = \frac{47.5}{(15)^2} = 0.211$$

Now for a closed vessel Eq. 36 relates the variance to D/uL. Thus

$$\sigma_\theta^2 = 0.211 = 2\frac{D}{uL} - 2\left(\frac{D}{uL}\right)(1 - e^{-uL/D})$$

Ignoring the second term on the right, we have as a first approximation

$$\frac{D}{uL} \cong 0.106$$

Correcting for the term ignored we find by trial and error that

$$\frac{D}{uL} = 0.120$$

Our original guess was correct: This value of D/uL is much beyond the limit where the simple gaussian approximation should be used.

EXAMPLE 4. D/uL from an F curve

Von Rosenberg (1956) studied the displacement of benzene by *n*-butyrate in a $1\frac{1}{2}$-in.-diameter packed column 4 ft long, measuring the fraction of *n*-butyrate in the exit stream by refractive index methods. When graphed, the fraction of *n*-butyrate versus time was found to be *S*-shaped. This is the **F** curve, and it is shown in Fig. E4*a* for the run at the lowest flow rate where $u = 2.19 \times 10^{-5}$ ft/sec, which is about 2 ft/day.

Find the vessel dispersion number for this system.

SOLUTION

Instead of finding the **C** curve by taking the slopes of the **F** curve and then determining the spread of this curve, let us illustrate a short cut which can be used when D/uL is small.

When D/uL is small the **C** curve approaches the normal curve of Eq. 29, and the corresponding **F** curve, when plotted on probability paper, lies on a straight line. Plotting the original **F**-curve data on probability paper does actually give close to a straight line, as shown in Fig. E4*b*.

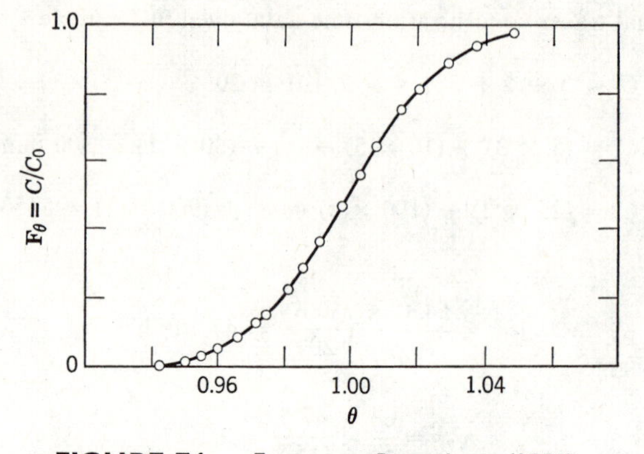

FIGURE E4a. From von Rosenberg (1956).

To find the variance and D/uL from a probability graph is a simple matter if we observe the following property of a normal curve: that one standard deviation σ on either side of the mean of the curve includes 68% of the total area under the curve. Hence the 16th and 84th percentile points of the **F** curve are two standard deviations apart. The 84th percentile intersects the straight line through the data at 187,750 sec and the 16th percentile intersects it at 178,550 sec, so the difference, 9200 sec, is taken as the value of two standard deviations. Thus the standard deviation is

$$\sigma = 4600 \text{ sec}$$

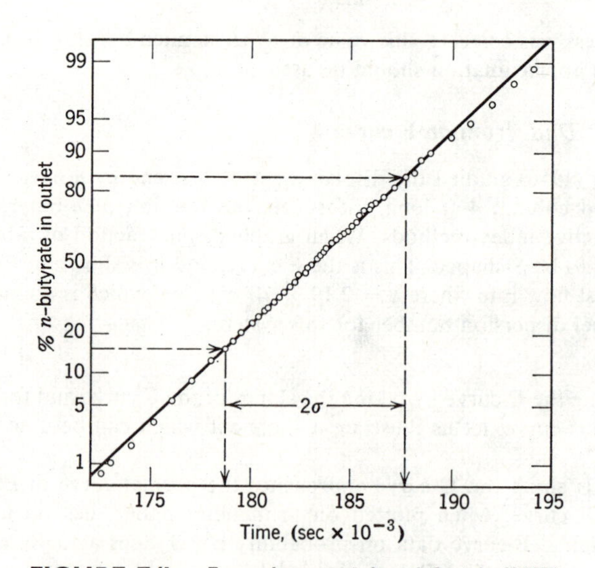

FIGURE E4b. From Levenspiel and Smith (1957).

We need this standard deviation in dimensionless time units if we are to find D. Therefore

$$\sigma_\theta = \frac{\sigma}{\bar{t}} = (4600\ \text{sec})\left(\frac{2.19 \times 10^{-5}\ \text{ft/sec}}{4\ \text{ft}}\right) = 0.0252$$

Hence the variance

$$\sigma_\theta^2 = (0.0252)^2 = 0.00064$$

and from Eq. 31

$$\frac{D}{uL} = \frac{\sigma_\theta^2}{2} = 0.00032$$

Note that the value of D/uL is well below 0.01, justifying the use of the Normal approximation to the C curve and this whole procedure.

EXAMPLE 5. D/uL from a one-shot input

Find the vessel dispersion number in a fixed-bed reactor packed with 0.625 cm catalyst pellets. For this purpose tracer experiments are run in equipment shown in Fig. E5.

Catalyst is laid down in a haphazard manner above a screen to a height of 120 cm, and fluid flows downward through this packing. A sloppy pulse of radioactive tracer is

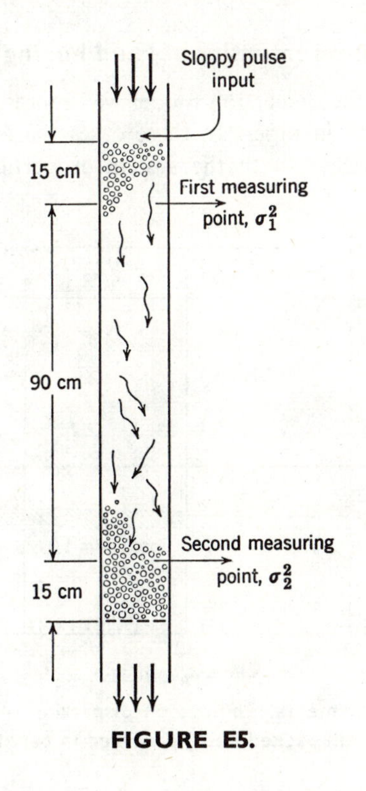

FIGURE E5.

injected directly above the bed, and output signals are recorded by Geiger counters at two levels in the bed 90 cm apart.

The following data apply to a specific experimental run. Bed voidage = 0.4, superficial velocity of fluid (based on an empty tube) = 1.2 cm/sec, and variances of output signals are found to be $\sigma_1^2 = 39\ \text{sec}^2$ and $\sigma_2^2 = 64\ \text{sec}^2$. Find D/uL.

SOLUTION

Bischoff and Levenspiel (1962) have shown that as long as the measurements are taken at least 2 or 3 particle diameters into the bed then the open vessel boundary conditions hold closely. This is the case here since the measurements are made 15 cm into the bed. As a result this experiment corresponds to a one-shot input to an open vessel for which Eq. 34 holds. Thus

$$\Delta\sigma^2 = \sigma_2^2 - \sigma_1^2 = 64 - 39 = 25\ \text{sec}^2$$

or in dimensionless form

$$\Delta\sigma_\theta^2 = \Delta\sigma^2\left(\frac{v}{V}\right)^2 = (25\ \text{sec}^2)\left[\frac{1.2\ \text{cm/sec}}{(90\ \text{cm})(0.4)}\right]^2 = \frac{1}{36}$$

from which the dispersion number is

$$\frac{D}{uL} = \frac{\Delta\sigma_\theta^2}{2} = \frac{1}{72}$$

Experimental Findings on Intensity of Fluid Mixing

Experiments show that the dispersion model well represents flow in packed beds as well as turbulent flow in pipes. In these cases the intensity of dispersion as measured by D/ud correlates with the system properties (see Figs. 18 and 20).

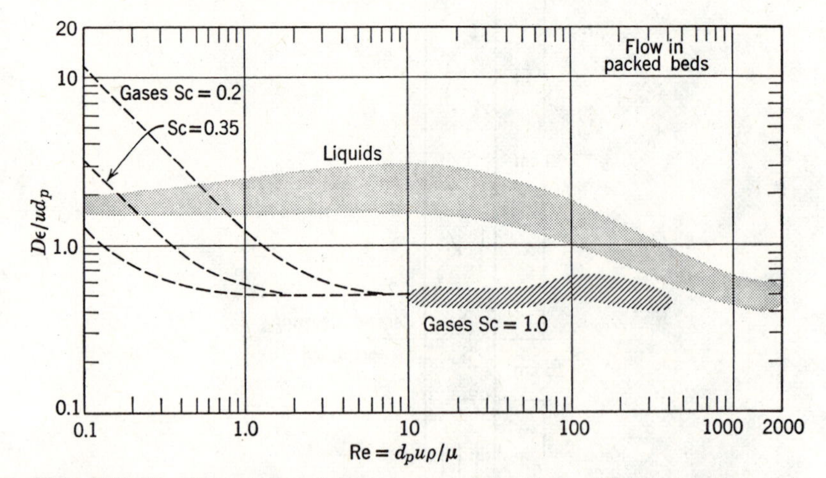

FIGURE 18. Experimental findings on dispersion of fluids flowing with mean axial velocity u in packed beds; prepared in part from Bischoff (1961).

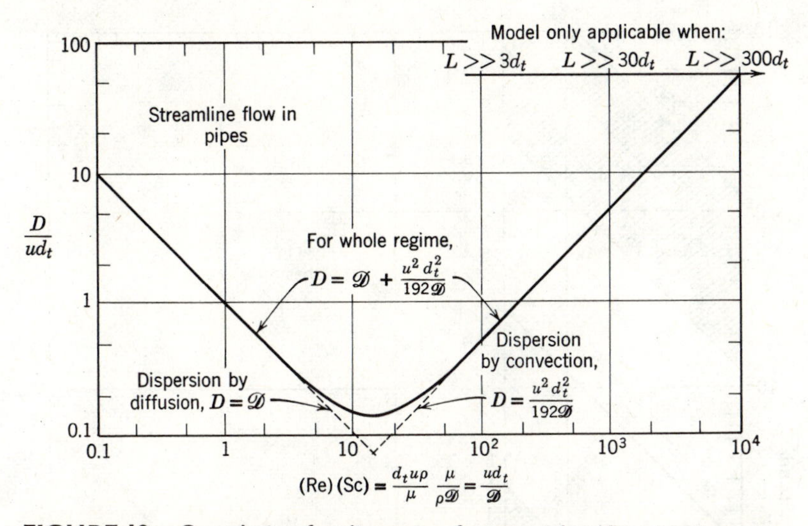

FIGURE 19. Correlation for dispersion for streamline flow in pipes; prepared from Taylor (1953, 1954a) and Aris (1956).

However, this model only represents streamline flow in pipes when the pipe is long enough to achieve radial uniformity of a pulse of tracer. For liquids this may require a rather long pipe, and Fig. 19 shows these results. Note that molecular diffusion strongly affects the rate of dispersion in laminar flow. At low flow rate it promotes dispersion; at higher flow rate it has the opposite effect.

Correlations similar to these are available or can be obtained for flow in beds of porous and/or adsorbing solids, in coiled tubes, in flexible channels, for pulsating flow, for non-newtonians, and so on.

The vessel or reactor dispersion number is simply the product of the intensity of dispersion, found from the charts, and the geometric factor for the reactor. Thus

$$\frac{D}{uL} = \begin{pmatrix}\text{intensity of}\\ \text{dispersion}\end{pmatrix}\begin{pmatrix}\text{geometric}\\ \text{factor}\end{pmatrix}$$

$$= \left(\frac{D}{ud}\right)\left(\frac{d}{L}\right)$$

where d is the characteristic length used in the charts.

Chemical Reaction and Dispersion

Our discussion has led to the measure of dispersion by a dimensionless group D/uL. Let us now see how this affects conversion in reactors.

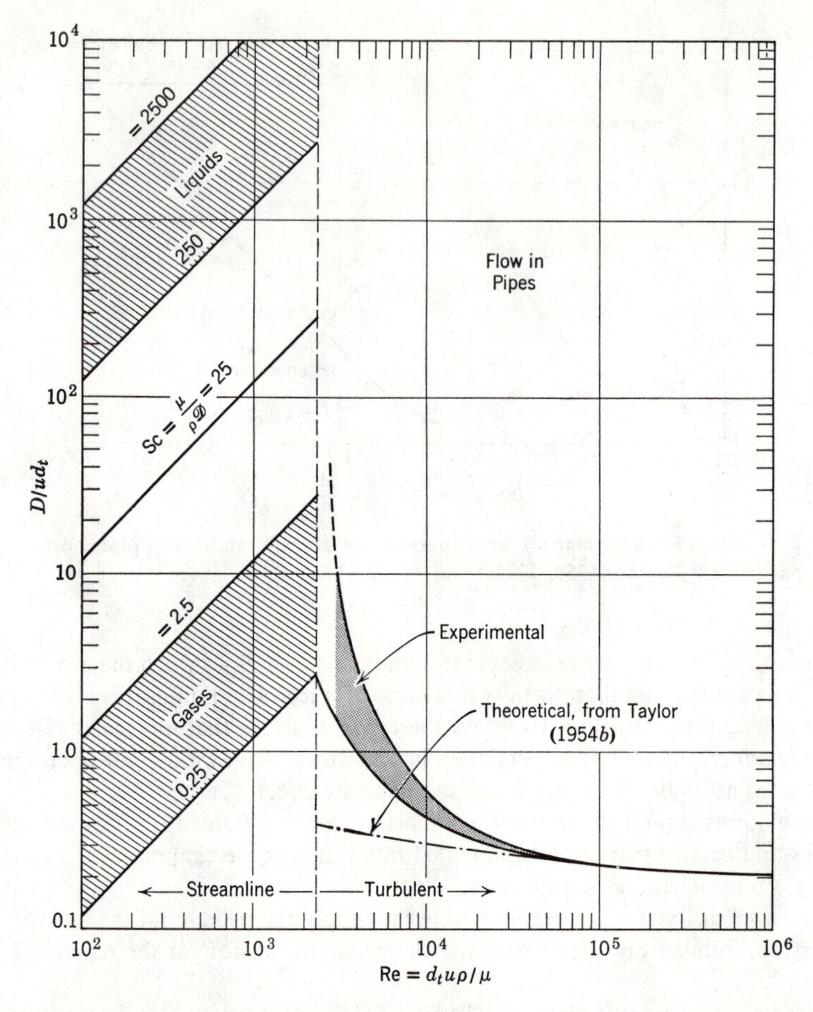

FIGURE 20. Correlation for the dispersion of fluids flowing in pipes, adapted from Levenspiel (1958).

Consider a steady-flow chemical reactor of length L through which fluid is flowing with a constant velocity u, and in which material is mixing axially with a dispersion coefficient D. Let the nth-order reaction be of the type

$$\mathrm{A} \rightarrow \text{products}, \qquad -r_{\mathrm{A}} = kC_{\mathrm{A}}^{n}$$

By referring to the elementary section of reactor as shown in Fig. 21, the basic material balance for any reaction component

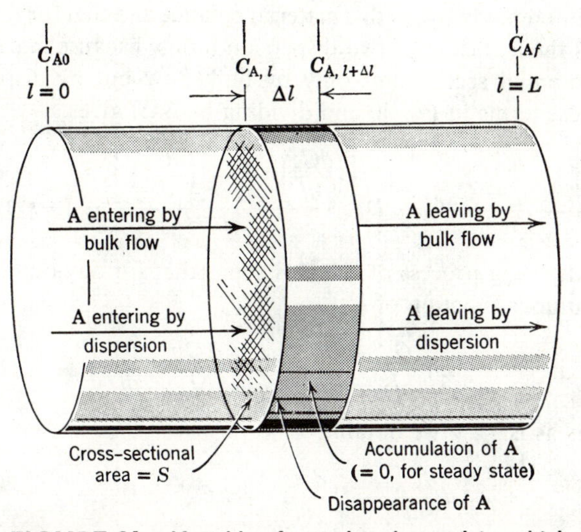

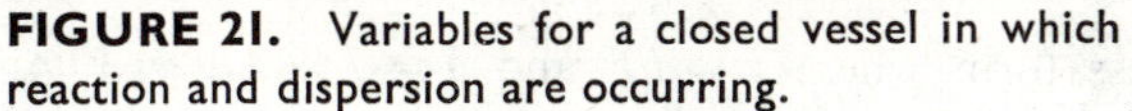

FIGURE 21. Variables for a closed vessel in which reaction and dispersion are occurring.

$$\text{input} = \text{output} + \text{disappearance by reaction} + \text{accumulation} \tag{4.1}$$

becomes for component A

$$(\text{out-in})_{\text{bulk flow}} + (\text{out-in})_{\text{axial dispersion}} + \begin{matrix}\text{disappearance}\\ \text{by reaction}\end{matrix} + \text{accumulation} = 0 \tag{40}$$

The individual terms (in moles A/time) are as follows:

$$\text{entering by bulk flow} = \left(\frac{\text{moles A}}{\text{volume}}\right)\left(\begin{matrix}\text{volumetric}\\ \text{flow rate}\end{matrix}\right)$$

$$= \left(\frac{\text{moles A}}{\text{volume}}\right)\left(\begin{matrix}\text{flow}\\ \text{velocity}\end{matrix}\right)\left(\begin{matrix}\text{cross-sectional}\\ \text{area}\end{matrix}\right)$$

$$= C_{A,l}uS$$

$$\text{leaving by bulk flow} = C_{A,l+\Delta l}uS$$

$$\text{entering by axial dispersion} = \frac{dN_A}{dt} = -\left(DS\frac{dC_A}{dl}\right)_l$$

$$\text{leaving by axial dispersion} = \frac{dN_A}{dt} = -\left(DS\frac{dC_A}{dl}\right)_{l+\Delta l}$$

$$\text{disappearance by reaction} = (-r_A)V = (-r_A)S\,\Delta l$$

Note that the difference between this material balance and that for the ideal reactors of Chapter 5 is the inclusion of two dispersion terms, because material enters and leaves the differential section not only by bulk flow but by dispersion as well. Entering all these terms in Eq. 40 and dividing by $S\,\Delta l$ gives

$$u\frac{(C_{A,l+\Delta l}-C_{A,l})}{\Delta l}-D\frac{\left[\left(\frac{dC_A}{dl}\right)_{l+\Delta l}-\left(\frac{dC_A}{dl}\right)_l\right]}{\Delta l}+(-r_A)=0$$

Now the basic limiting process of calculus states that for any quantity Q which is a smooth continuous function of l

$$\lim_{l_2\to l_1}\frac{Q_2-Q_1}{l_2-l_1}=\lim_{\Delta l\to 0}\frac{\Delta Q}{\Delta l}=\frac{dQ}{dl}$$

So taking limits as $\Delta l \to 0$ we obtain

$$u\frac{dC_A}{dl}-D\frac{d^2C_A}{dl^2}+kC_A{}^n=0 \qquad (41a)$$

In dimensionless form where $z = l/L$ and $\tau = \bar{t} = L/u = V/v$, this expression becomes

$$\frac{D}{uL}\frac{d^2C_A}{dz^2}-\frac{dC_A}{dz}-k\tau C_A{}^n=0 \qquad (41b)$$

or in terms of fractional conversion

$$\frac{D}{uL}\frac{d^2X_A}{dz^2}-\frac{dX_A}{dz}+k\tau C_{A0}^{n-1}(1-x_A)^n=0 \qquad (41c)$$

This expression shows that the fractional conversion of reactant A in its passage through the reactor is governed by three dimensionless groups: a reaction rate group $k\tau C_{A0}^{n-1}$, the dispersion group D/uL, and the reaction order n.

First-order Reaction. Equation 41 has been solved analytically by Wehner and Wilhelm (1956) for first-order reactions. For vessels with any kind of entrance and exit conditions the solution is

$$\frac{C_A}{C_{A0}}=1-X_A=\frac{4a\exp\left(\frac{1}{2}\frac{uL}{D}\right)}{(1+a)^2\exp\left(\frac{a}{2}\frac{uL}{D}\right)-(1-a)^2\exp\left(-\frac{a}{2}\frac{uL}{D}\right)} \qquad (42)$$

where

$$a=\sqrt{1+4k\tau(D/uL)}$$

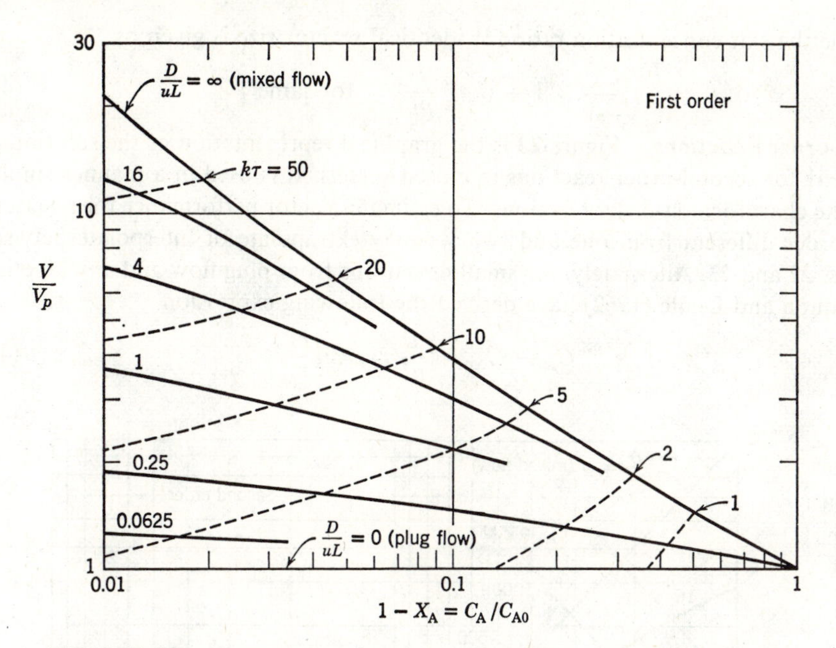

FIGURE 22. Comparison of real and plug flow reactors for the first-order reaction A → products, assuming negligible expansion; from Levenspiel and Bischoff (1959, 1961).

Figure 22 is a graphical representation of these results in useful form, prepared by combining Eqs. 42 and 5.17, and allows comparison of reactor sizes for plug and dispersed plug flow.

For *small deviations from plug flow* D/uL becomes small, the **E** curve approaches gaussian, hence on expanding the exponentials and dropping higher order terms Eq. 42 reduces to

$$\frac{C_A}{C_{A0}} = \exp\left[-k\tau + (k\tau)^2 \frac{D}{uL}\right] \tag{43}$$

$$= \exp\left[-k\tau + \frac{k^2\sigma^2}{2}\right] \tag{44*}$$

Equations 43 and 5.17 compare the performance of real reactors which are close to plug flow with plug flow reactors. Thus the size ratio needed for identical conversion is given by

$$\frac{L}{L_p} = \frac{V}{V_p} = 1 + (k\tau)\frac{D}{uL} \qquad \text{for same } C_{A,\text{out}} \tag{45}$$

* It should be noted that Eq. 44 applies to any gaussian RTD with variance σ^2.

while the exit concentration ratio for identical reactor size is given by

$$\frac{C_A}{C_{Ap}} = 1 + (k\tau)^2 \frac{D}{uL} \qquad \text{for same } V \tag{46}$$

nth-order Reactions. Figure 23 is the graphical representation of the solution of Eq. 41 for second-order reactions in closed vessels. It is used in a manner similar to the chart for first-order reactions. To estimate reactor performance for reactions of order different from one and two we may extrapolate or interpolate between Figs. 22 and 23. Alternately, for small deviations from plug flow and any kinetics, Pasquon and Dente (1962) have derived the following expression

$$C_A - C_{Ap} = \frac{D}{uL} r_{Ap} \tau \ln \frac{r_{Ap}}{r_{A0}} \tag{47}$$

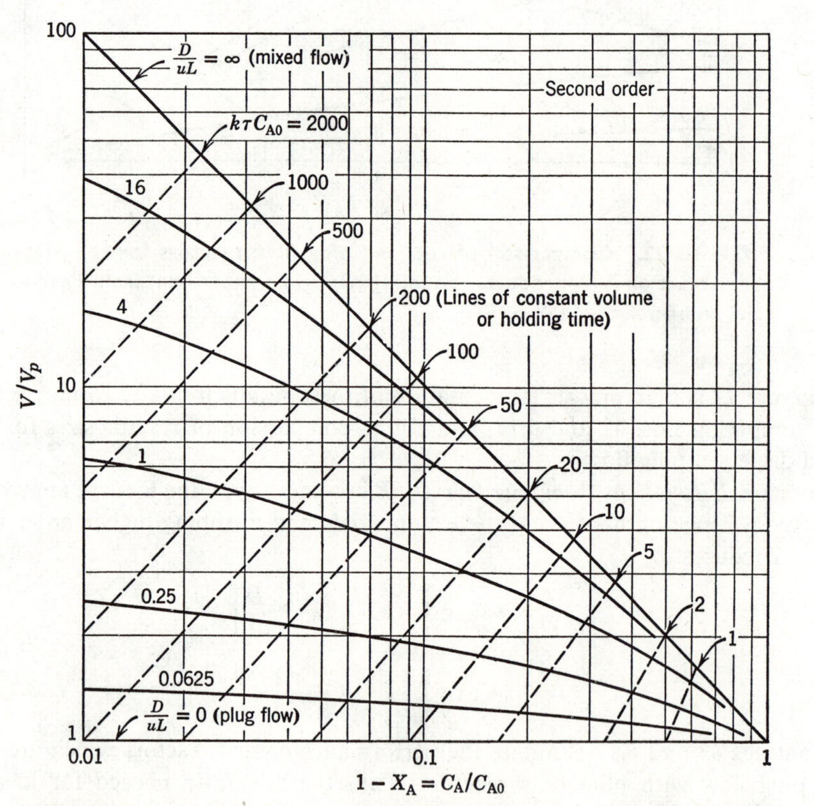

FIGURE 23. Comparison of real and plug flow reactors for the second-order reactions

$$A + B \rightarrow \text{products}, \qquad C_{A0} = C_{B0}$$
$$2A \rightarrow \text{products}$$

assuming negligible expansion; from Levenspiel and Bischoff (1959, 1961).

where r_{A0} and r_{Ap} are the rates at the inlet and outlet of a plug flow reactor with the same τ or L as the real reactor.

For nth order reactions this expression reduces to

$$\frac{C_A}{C_{Ap}} = 1 + n\left(\frac{D}{uL}\right)(kC_{A0}^{n-1}\tau) \ln \frac{C_{A0}}{C_{Ap}} \tag{48}$$

Note that backmixing does not affect performance for zero-order reactions.

Extensions. For multiple reactions of the consecutive type we do not at present have a simple graphical representation of the effects of dispersion on product distribution and reactor size. Tichacek (1963) presents equations accounting for this effect and shows that for systems which deviate slightly from plug flow, say for $D/uL < 0.05$, the fractional decrease in the maximum amount of intermediate formed is roughly given by the value of D/uL itself.

Large molar expansion or contraction influences the predicted reactor performance, and by numerical means Douglas and Bischoff (1964) have evaluated this effect for various reaction types.

Of the three reactor types generally used in industry, tubular, packed bed, and fluidized bed, only the first two exhibit marked nonisothermal behavior which need be considered in design. For these two reactor types the dispersion model, using axial and radial dispersion of both matter and heat, is the best way we now have for approximating the real situation. Himmelblau and Bischoff (1968) present and evaluate the various dispersion models which may be used here. They also present general correlations for axial and radial dispersion of matter and heat in various systems.

EXAMPLE 6. Conversion from the dispersion model

Redo Example 2 assuming that the dispersion model is a good representation of flow in the reactor. Compare the calculated conversion by the two methods and comment.

SOLUTION

Matching the experimentally found variance with that of the dispersion model, we find from Example 3

$$\frac{D}{uL} = 0.12$$

Conversion in the real reactor is found from Fig. 22. Thus moving along the $k\tau = (0.307)(15) = 4.6$ line from $C/C_0 = 0.01$ to $D/uL = 0.12$, we find that the fraction of reactant unconverted is approximately

$$\frac{C}{C_0} = 0.035, \qquad \text{or } 3.5\%$$

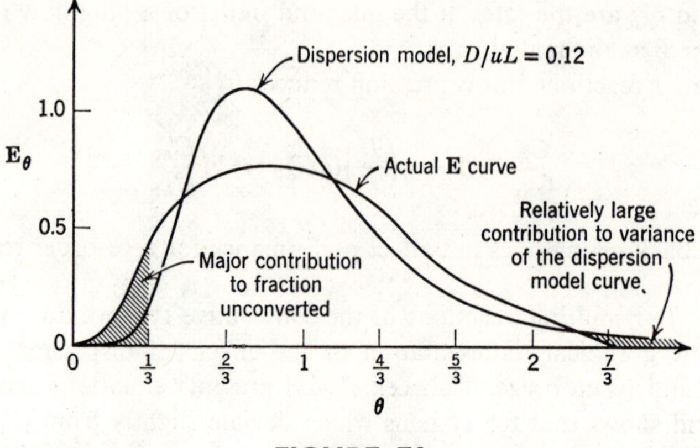

FIGURE E6.

Figure E6 shows that except for a long tail the dispersion model curve has for the most part a greater central tendency than the actual curve. On the other hand, the actual curve has more shortlived material leaving the vessel. Because this contributes most to the reactant remaining unconverted, the finding

$$\left(\frac{C}{C_0}\right)_{\text{actual}} = 4.7\% > \left(\frac{C}{C_0}\right)_{\substack{\text{dispersion}\\\text{model}}} = 3.5\%$$

is expected.

TANKS-IN-SERIES MODEL

Besides the dispersion model, the tanks-in-series model is the other one-parameter model widely used to represent nonideal flow. Here we view the fluid to flow through a series of equal-size ideal stirred tanks, and the one parameter of this model is the number of tanks in this chain.

The **C** or **E** curve and moments of this model are easy to obtain since problems of proper boundary conditions, method of tracer injection, and measurement do not intrude. Thus for one tank we have

$$\bar{t}_i\mathbf{E} = e^{-t/\bar{t}_i}, \qquad N = 1$$

for two tanks we find by any of a variety of methods, either by a material balance, by using the convolution integral, or by using Laplace transforms,

$$\bar{t}_i\mathbf{E} = \frac{t}{\bar{t}_i}\, e^{-t/\bar{t}_i}, \qquad N = 2 \tag{49}$$

See: Carberry pp 70–80

and similarly for N tanks in series we obtain in various forms

$$\bar{t}_i\mathbf{E} = \left(\frac{t}{\bar{t}_i}\right)^{N-1} \frac{1}{(N-1)!} e^{-t/\bar{t}_i} \tag{50a}$$

$$\mathbf{E}_{\theta_i} = \bar{t}_i\mathbf{E} = \frac{\theta_i^{N-1}}{(N-1)!} e^{-\theta_i} \tag{50b}$$

$$\mathbf{E}_\theta = (N\bar{t}_i)\mathbf{E} = \frac{N(N\theta)^{N-1}}{(N-1)!} e^{-N\theta} \tag{50c}$$

where

$\bar{t}_i$ = mean residence time in one tank

$\bar{t} = N\bar{t}_i$, mean residence time in the N tank system

$\theta_i = t/\bar{t}_i = Nt/\bar{t}$

$\theta = t/\bar{t} = t/N\bar{t}_i$

These curves are shown in Fig. 24, and their mean and variance are found to be

$$\bar{t} = N\bar{t}_i, \qquad \sigma^2 = N\bar{t}_i^{\,2} = \frac{\bar{t}^2}{N} \tag{51a}$$

$$\bar{t}_{\theta_i} = N, \qquad \sigma_{\theta_i}^{\,2} = N \tag{51b}$$

$$\bar{t}_\theta = 1, \qquad \sigma_\theta^{\,2} = \frac{1}{N} \tag{51c}$$

Figure 25 shows some of the properties of these curves which should be useful in estimating the value of the parameter. Thus N can be estimated from the maximum, the width at inflection, or the variance of the measured **C** curve.

For large N the RTD curve becomes increasingly symmetrical and approaches the normal curve of Fig. 13, and a comparison of these two curves allows us to relate the tanks in series and dispersion models.

With this model the additivity of variances and the convolution integral can safely be used. Thus if we have N_1 tanks and we add N_2 tanks to the chain then

$$\sigma_{N_1}^{\,2} + \sigma_{N_2}^{\,2} = \sigma_{N_1+N_2}^2 \tag{52}$$

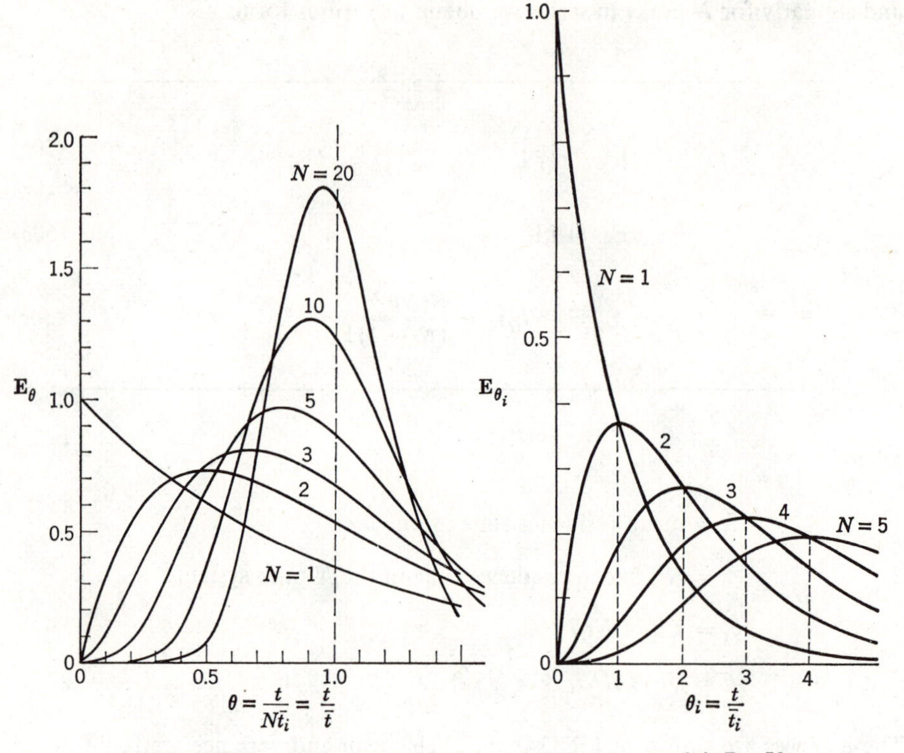

FIGURE 24. RTD curves for the tanks in series model, Eq. 50.

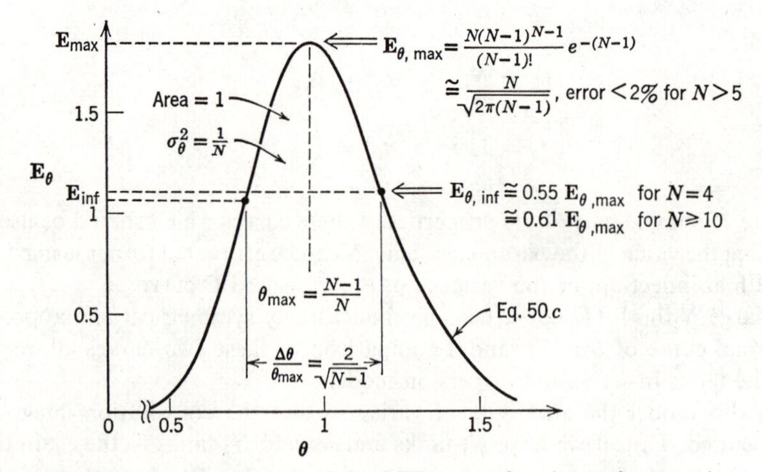

FIGURE 25. Properties of the RTD curve for the tanks in series model.

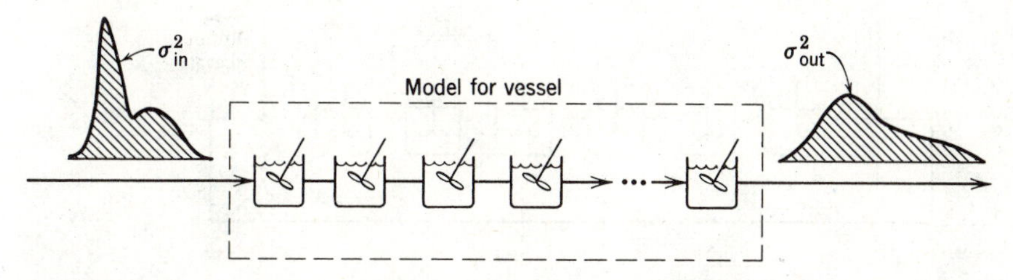

FIGURE 26. For any one-shot tracer input Eq. 53 relates input, output, and number of tanks.

If we introduce any one-shot tracer input into N tanks as shown in Fig. 26 then from Eq. 51

$$\Delta\sigma^2 = \sigma^2_{out} - \sigma_{in}{}^2 = \frac{\bar{t}^2}{N} \tag{53}$$

Conversion for the Tanks-in-Series Model

Chapter 6 treats the conversion in a series of mixed flow reactors and that analysis applies directly to this model. Thus Figs. 6.5 and 6.6 compare the performance of this flow with plug flow.

For large N, thus small deviation from plug flow, the performance expressions for first-order reactions reduce to

$$\frac{V_{N\text{tanks}}}{V_{\text{plug}}} = 1 + \frac{k\tau}{N} \tag{54}$$

and

$$\frac{C_{\text{A},N\text{tanks}}}{C_{\text{A,plug}}} = 1 + \frac{(k\tau)^2}{N} \tag{55}$$

Extensions

Because of the independence of stages (in the sense considered earlier in this chapter) it is easy to evaluate what happens to the **C** curve when tanks are added or subtracted. Thus this model becomes useful in treating recycle flow and closed recirculation systems. Let us briefly look at these applications.

Closed Recirculation System. If we introduce a δ signal into an N stage system, as shown in Fig. 27, the recorder will measure tracer as it flows by the first time, the second time, and so on. In other words it measures tracer which has passed through N tanks, $2N$ tanks, and so on. In fact it measures the superposition of all these signals.

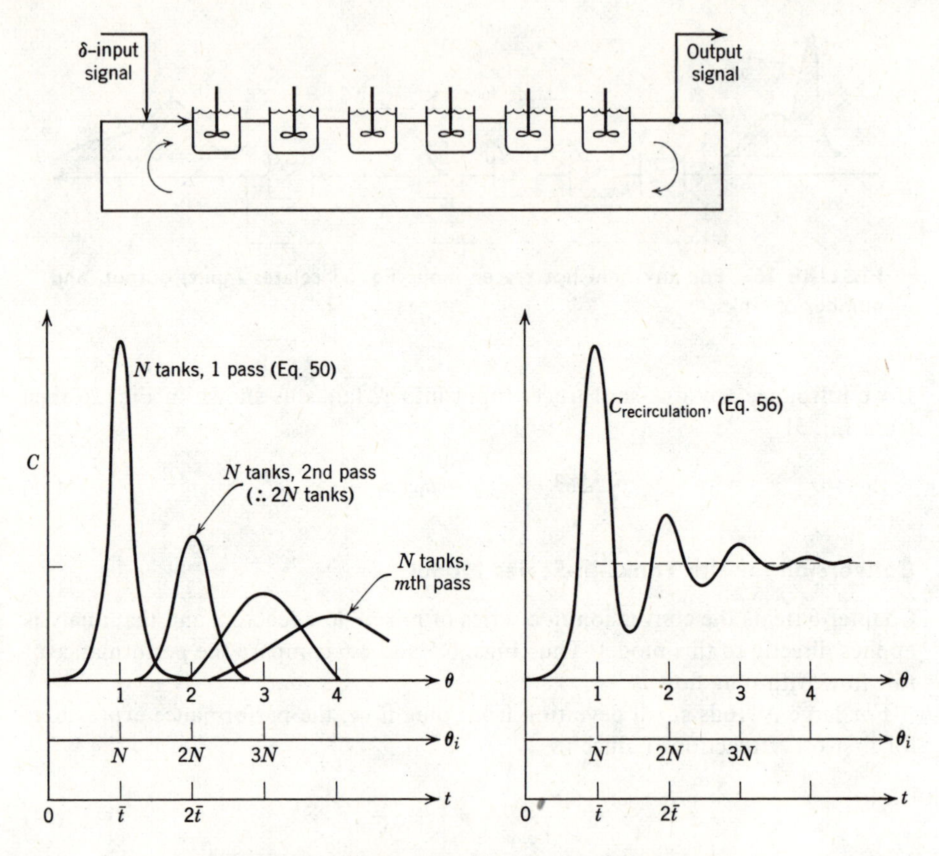

FIGURE 27. Tracer signal in a recirculating system.

To obtain the output signal for these systems simply sum up the contributions from the first, second, and succeeding passes. If m is the number of passes we then have from Eq. 50

$$\bar{t}_i C = e^{-t/\bar{t}_i} \sum_{m=1}^{\infty} \frac{(t/\bar{t}_i)^{mN-1}}{(mN-1)!} \tag{56a}$$

$$C_{\theta_i} = e^{-\theta_i} \sum_{m=1}^{\infty} \frac{\theta_i^{mN-1}}{(mN-1)!} \tag{56b}$$

$$C_{\theta} = Ne^{-N\theta} \sum_{m=1}^{\infty} \frac{(N\theta)^{mN-1}}{(mN-1)!} \tag{56c}$$

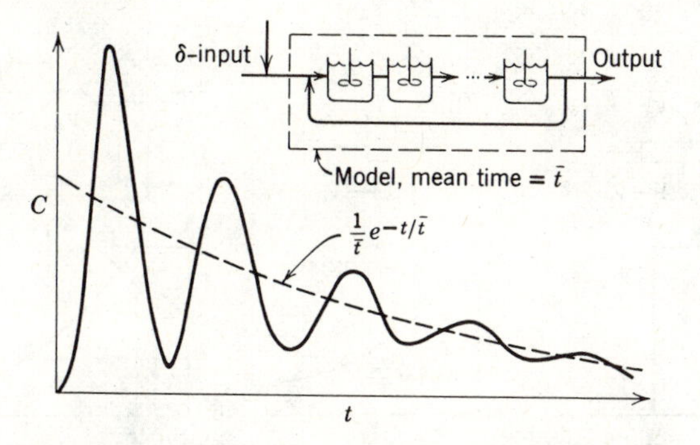

FIGURE 28. Recirculation with slow throughflow.

Figure 27 shows the resulting C curve. As an example of the expanded form of Eq. 56 we have for 5 tanks in series

$$C = \frac{5}{\bar{t}}\, e^{-5t/\bar{t}} \left[\frac{(5t/\bar{t})^4}{4!} + \frac{(5t/\bar{t})^9}{9!} + \cdots \right]$$

$$C_{\theta_i} = e^{-\theta_i} \left(\frac{\theta_i^{\,4}}{4!} + \frac{\theta_i^{\,9}}{9!} + \frac{\theta_i^{\,14}}{14!} + \cdots \right)$$

$$C_\theta = 5e^{-5\theta} \left[\frac{(5\theta)^4}{4!} + \frac{(5\theta)^9}{9!} \cdots \right]$$

where the terms in brackets represent the tracer signal from the first, second, and successive passes.

Recirculation systems can be represented equally well by the dispersion model (see van der Vusse (1962), Voncken *et al.* (1964), and Harrell and Perona (1968)). Which approach one takes simply is a matter of taste, style, and mood.

Recirculation with Throughflow. For relatively rapid recirculation compared to throughflow the system as a whole acts as one large stirred tank, hence the observed tracer signal is simply the superposition of the recirculation pattern and the exponential decay of an ideal stirred tank. This is shown in Fig. 28.

This form of curve is encountered in closed recirculation systems in which tracer is broken down and removed by a first-order process, or in systems using radioactive tracer. Experiments on living organisms give this sort of superposition because tracer is constantly being eliminated by the system.

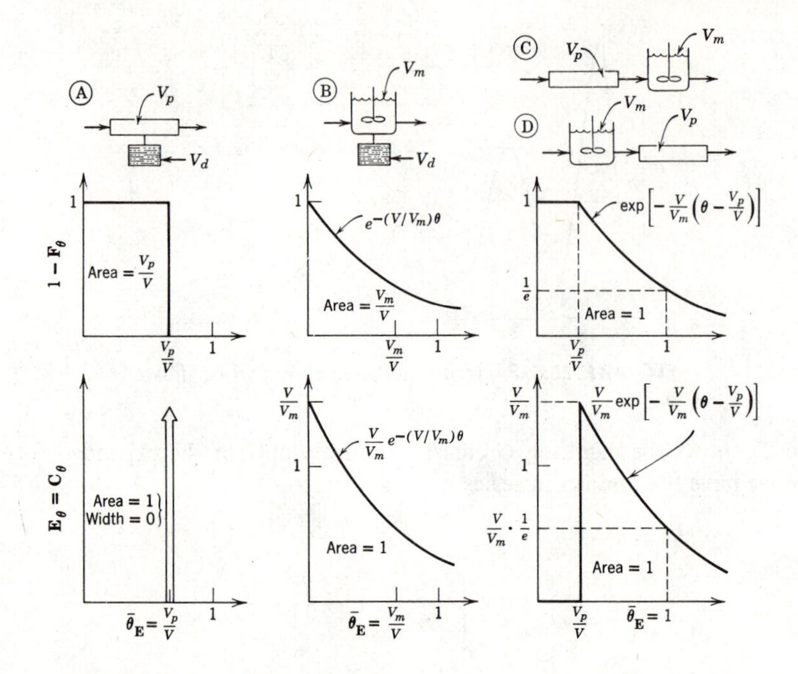

FIGURE 29. Simple flow models and their tracer response curves.

MULTIPARAMETER MODELS

When one-parameter models are unable to account satisfactorily for deviations from the ideals of plug and mixed flow then more complicated models must be attempted. These usually consider the real reactor to consist of different regions (plug, dispersed plug, mixed, deadwater) interconnected in various ways (bypass recycle, or crossflow).

The term deadwater accounts for the portion of fluid which is relatively slow moving, and which, as an idealization, sometimes is taken to be completely stagnant.

Figure 29 represents the simplest of these models. Note how different and distinctive are the shapes of these curves. This property suggests a way of characterizing unknown flows, and this method is in fact used to diagnose pathological flows in vessels. We consider this matter later.

In the discussion to follow we will introduce some of the numerous multiparameter models which have been used to represent real reactors from trickle beds to sparged reactors and fluidized beds.

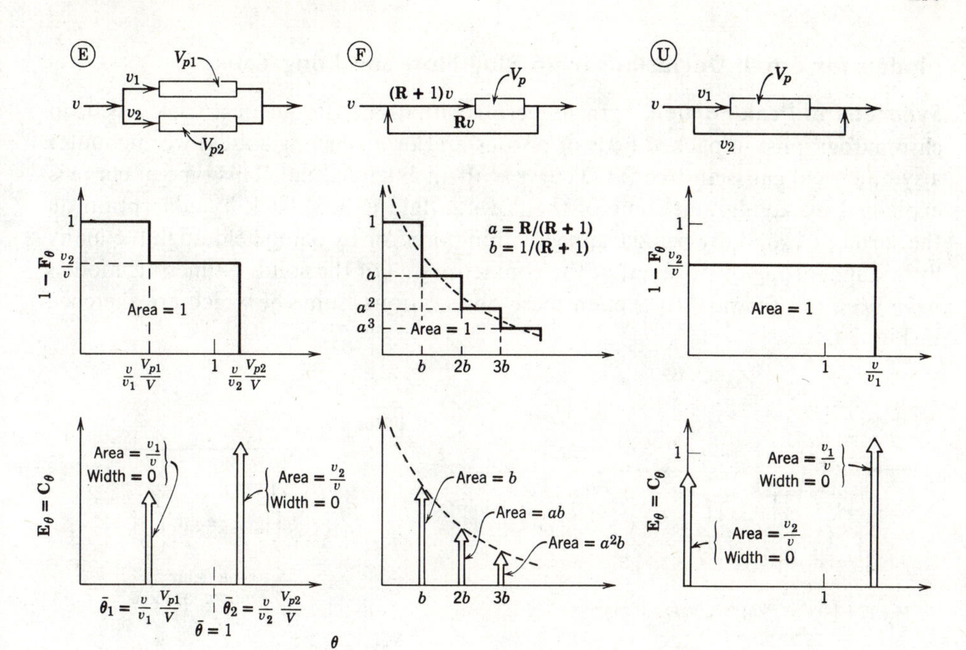

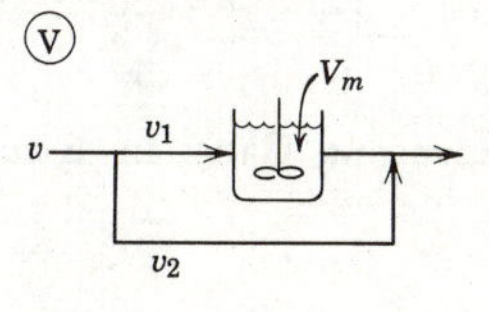

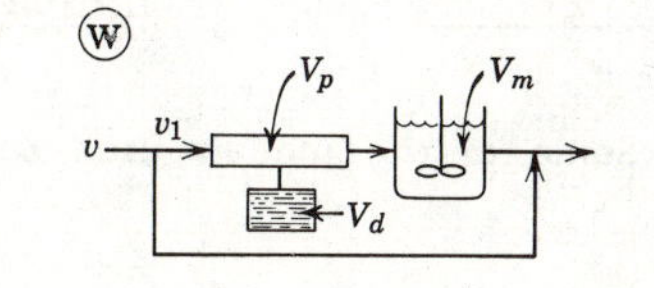

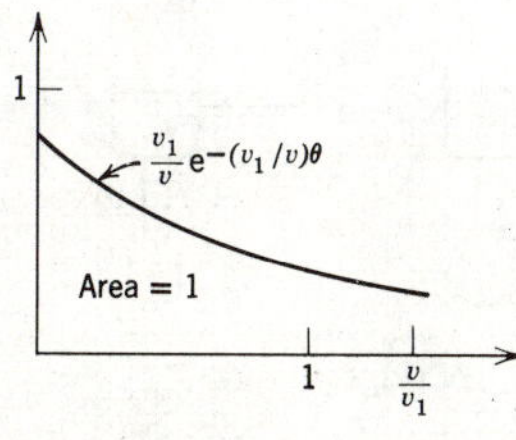

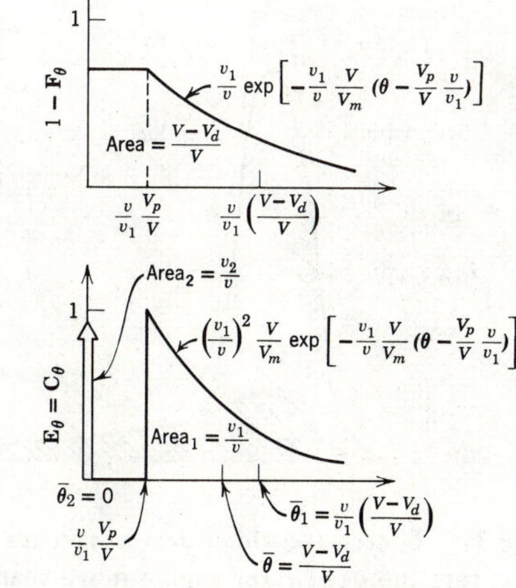

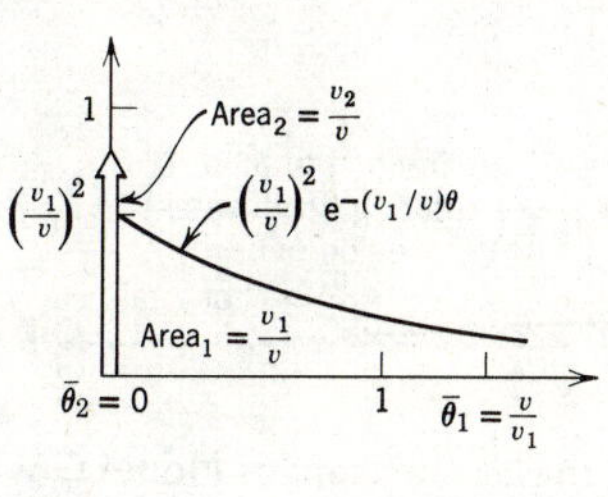

FIGURE 29. *Continued.*

Models for Small Deviations from Plug Flow and Long Tails

Symmetrical Peak Portion. In numerous situations (liquid in a trickle bed, in chromatographs, in packed beds of porous and/or adsorbing solids) we encounter a symmetrical gaussian-like RTD curve with an extended tail. This type of curve is explained by saying that some of the flowing fluid is held back by adsorption on the surface of solid, by being trapped within pores, or by being held up in the many little stagnant regions present at the contact points of the solid. Numerous models have been put forward to explain these observations, some of which are sketched in Fig. 30.

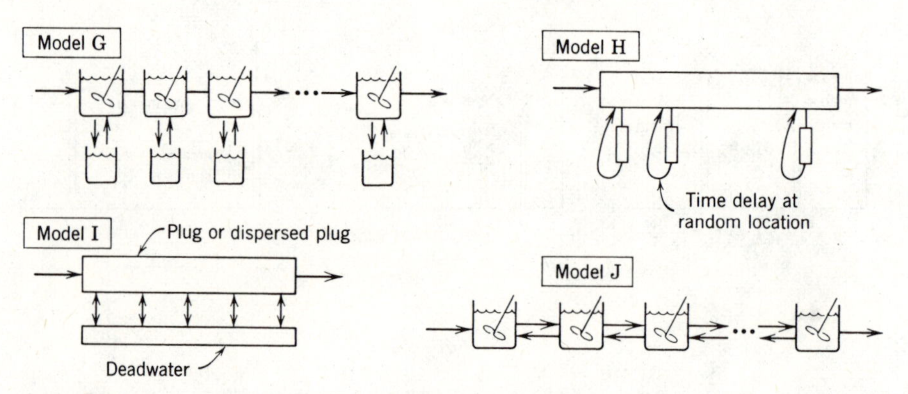

FIGURE 30. Some of the models used to represent Gaussian **E**-curves with long tails.

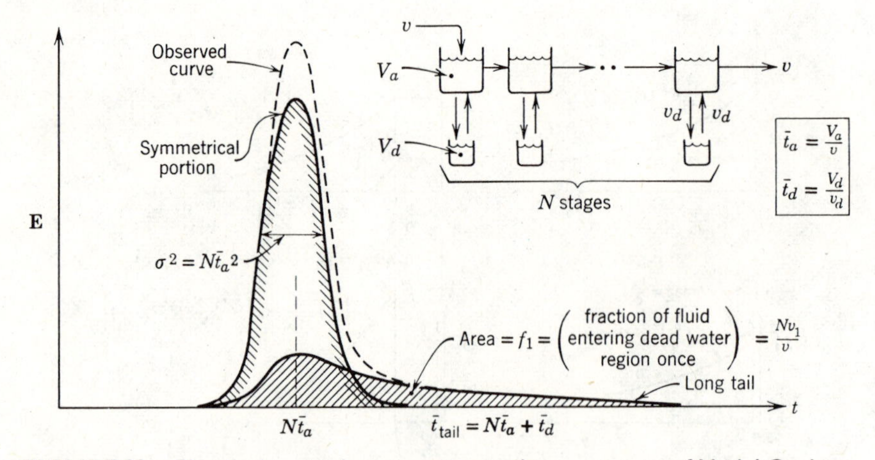

FIGURE 31. Sketch showing how to estimate the parameters of Model G where no fluid enters the deadwater region more than once.

Examination of these models will show that Models H and I are both special cases of Model G for an infinite number of tanks in series. Model J, however, is different from the other three because it alone permits backflow of fluid.

Probably the easiest of these models to treat is Model G, so let us examine it in somewhat more detail. As shown in Fig. 31 the three parameters of this model are

N: the number of stages consisting of active and deadwater regions, in series.

$X = \frac{v_d}{v}$: the crossflow ratio per tank.

$\frac{V_d}{V} = \frac{V_d}{V_a + V_d}$: fraction of volume which is not active.

In terms of these parameters we have for the fluid

$\bar{t}_a = \frac{V_a}{v}$: mean total residence time in an active region.

$\bar{t}_d = \frac{V_d}{v_d}$: mean delay time for each entry into a deadwater region.

In passing through the vessel some fluid is delayed once, some twice, some even more times, by flow into the deadwater regions, so the RTD represents the over-all effect of all these possible paths. Now if f_i is the fraction of fluid which has been delayed i times, and $\mathbf{E}_i$ the RTD for that fluid we can then write

$$\mathbf{E} = f_0\mathbf{E}_0 + f_1\mathbf{E}_1 + \cdots + f_i\mathbf{E}_i + \cdots \tag{57a}$$

Since the fluid must still flow through N active regions no matter how many delays it experiences $\mathbf{E}_i$ is the convolution of N active and i deadwater regions, or

$$\mathbf{E}_i = \mathbf{E}_{N\,\text{active}} * \mathbf{E}_{i\,\text{delays}} \tag{57b}$$

where both $\mathbf{E}_{N\,\text{active}}$ and $\mathbf{E}_{i\,\text{delays}}$ are given by Eq. 50 using the proper mean times $\bar{t}_a$ and $\bar{t}_d$. Equation 57 is thus the general solution for the RTD for Model G.

For relatively small crossflow ($Nv_d < v$) most of the fluid passes through the vessel with no delay, a small fraction is delayed just once, and the fraction delayed more than once is even smaller still and can safely be ignored. Thus the first term of Eq. 57 is large, the second is small, and the succeeding terms can be dropped, giving

$$\mathbf{E} = f_0\mathbf{E}_{N\,\text{active}} + f_1[\mathbf{E}_{N\,\text{active}} * \mathbf{E}_{1\,\text{delay}}] \tag{58}$$

The sketch of Fig. 31 shows that Eq. 58 represents the superposition of a large N-tanks curve and a smaller curve with long tail, and as Levich *et al.* (1967) first showed these can be approximated (for large N) by a gaussian curve and an exponential decay curve.

This analysis provides a method for estimating the values of the parameters to fit the physical system. Thus the location of the gaussian portion estimates the fraction of vessel volume which is active, its width and peak height estimates N, while the exponential tail estimates the crossflow. Figure 31 shows these relationships. As an independent measure L/d_p also estimates N.

For larger crossflow (increased vessel length) the fluid has more chance of being delayed, hence the gaussian (straight through) portion spreads and shrinks while the contributions from multiple delays become increasingly important. The precise solution for this case is difficult to evaluate; however, it suffices to note that the over-all **E** curve becomes increasingly symmetrical, approaching the simple tanks in series model. Chromatographic columns usually closely approximate this limiting behavior.

Distributions with long tails indicate that only a small portion of the fluid is ever delayed by adsorption or transfer into a deadwater region. This may or may not be desirable. Chromatographic columns with such curves are probably too short to behave properly.

With regard to conversion it usually does not pay to try to correct for the tail. Plug flow predictions, or at most the corrections of Eqs. 45 and 46 will be accurate enough. There are situations, however, where the tail of the RTD curve assumes an overriding importance. As an example suppose we have the reaction

$$\text{A(colorless)} \rightarrow \text{R}_{\text{desired}}\text{(colorless)} \xrightarrow[\text{reactor}]{\text{too long in}} \text{S(colored)}$$

Often just a small amount of S will color the product stream sufficiently to make it unacceptable. In such a situation it is important to be able to characterize the tail, to find out what variables affect it, and how to minimize it.

The Real Stirred Tank

For most applications the real stirred tank with sufficient agitation can be taken to approximate the ideal of mixed flow. There are some cases, however, where deviation from this ideal should be considered, for example in large tanks with insufficient agitation and for fast reactions where the time of reaction is short compared to the time for mixing and for achieving uniformity of composition. It is here that mixing models are needed. Not only will these be useful for the real stirred tank, they will have numerous other applications, such as to represent the distribution of chemicals and drugs in animals and man.

There are many ways of treating the deviations from the ideal. We here present two types of analyses depending on whether the time scale of interest is short or long compared to the mean residence time of fluid in the system. We consider these cases in turn, but first, one brief comment on experimentation.

The real stirred tank reactor can be used either as a batch reactor or as a flow reactor, and if the flow pattern is not too different in these two arrangements then tracer experiments in either of them will give the information needed to construct a suitable flow model. Since batch experimentation is often simpler to perform it is used when justified.

Short time Scale Models With tracer experimentation we want to know whether the flow in the stirred tank deviates significantly from the ideal, and when it does we want to know how to treat this deviation. We may also want some measure of the intensity of agitation, or the rate at which an element of fluid is broken up and dispersed within the tank. We take up these two topics in turn.

Imagine a pulse of tracer introduced into a real stirred tank which has no throughflow of fluid, with tracer concentration measured at some representative point within the vessel. Figure 32 shows typical results of such an experiment. Let us examine these tracer curves.

First of all we note that the closed recirculating N tank model of Fig. 27 and Eq. 56 often reasonably fits the curves of Fig. 32 and can therefore be used to model the real stirred tank. With this model the curves of Fig. 32*a*, *b*, and *c* represent flow in approximately 40, 10, and 4 tanks respectively.

Next we note that the periodic output in Fig. 32*a* is certainly due to fluid recirculation. The time between successive peaks characterizes this action, and we call this the *turnover time* t_T. In Fig. 32*a* the turnover time can be found by inspection.

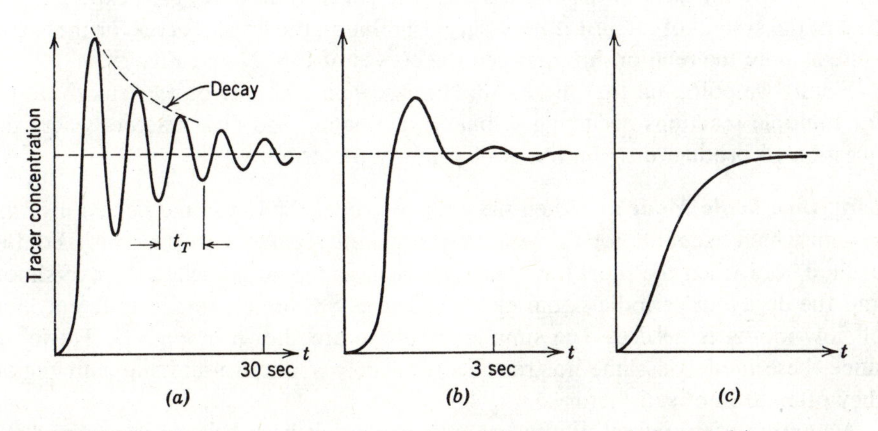

FIGURE 32. Tracer curves for a batch operated stirred tank. Sketch (*a*) represents a rapid turnover with slow breakup of fluid elements.

In Fig. 32*c* this is not possible, and this quantity can only be found from the model after properly matching the experimental curve.

The above findings give the number of tanks N and the turnover time t_T, both of which are dependent on the geometry of the system, energy input, and fluid properties.

Let us proceed to the rate of break-up and dispersal of tracer within the vessel. For the tracer curve of Fig. 32*a* an obvious measure is the rate of decay of the periodic signal, say its half life. However, such a measure is awkward to use for the tracer curves of Figs. 32*b* and *c*. In essence then, this measure is only easy to use when the characteristic decay time is much longer than the turnover time.

To develop a measure useful for any of the tracer curves of Fig. 32 note that since the stirred tank acts as an N compartment system the tracer has only been able to spread uniformly through a single compartment whose volume is V/N in time t_T. Thus we may define a *bulk dispersion coefficient* D_b as follows

$$D_b = \frac{(V/N)^{2/3}}{t_T}, \quad \left[\frac{L^2}{t}\right] \tag{59}$$

Since D_b is determined by N and t_T which in turn are functions of system geometry, power input, and fluid properties, D_b itself is a function of these quantities.

In certain polymerizations whether the product is highly crosslinked or not, granular or gelatinous, its m.w. and physical properties, depends strongly on the mean overall intensity of fluid break-up and dispersion in the vessel. In these cases D_b is likely to be the pertinent parameter to represent this action, to be kept constant or controlled in scale up and design.

So far our discussion has centered on the batch system, and the resulting model can be used directly for the flow system as long as the flow pattern is suspected to be not too different in the two cases. If the patterns do differ then experimentation in the system of interest is necessary. In relating the tracer curves for these two systems note the relationship between the curves of Fig. 27 and Fig. 28.

Finally we point out that these short time scale models are of particular interest for multiple reactions occurring in batch, semibatch, and flow reactors where the question of product distribution is of primary importance.

Long time Scale Models When we wish to model the long time behavior of the system which accounts for the relatively stagnant regions with their long holdup of fluid, and when the short time behavior such as the initial delay, the overshoot, and the decaying periodic signal of Fig. 32 is of no interest then a different kind of flow model is needed. The simplest of these are shown in Fig. 29. However, since these models assume no transfer at all between active and stagnant region they often are not satisfactory.

A number of more realistic two parameter models have been proposed and are used to deal with this situation. Consider those shown in Fig. 33.

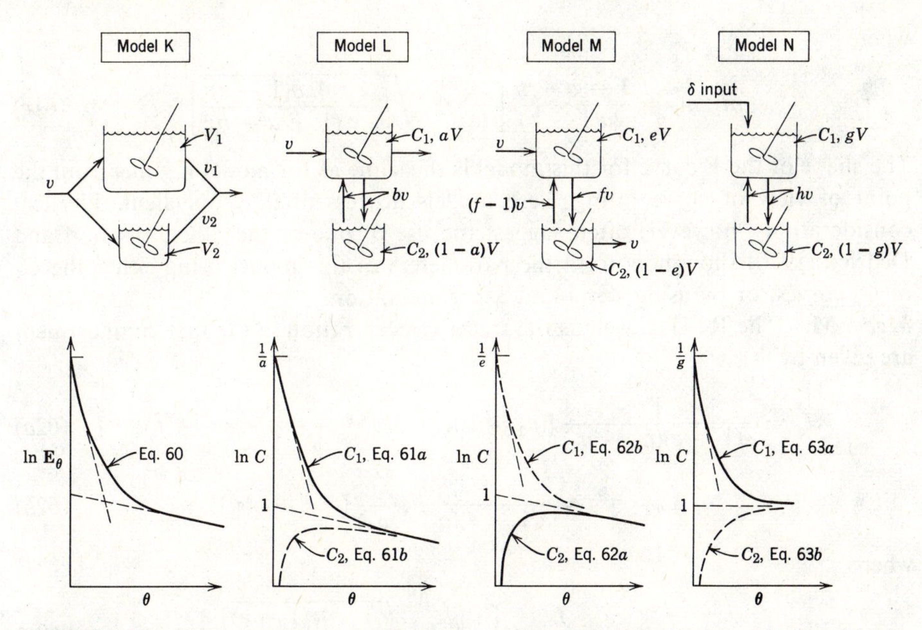

FIGURE 33. Two parameter models for long time-scale behavior of real stirred tanks. Models K, L, M are for flow operations, N for batch operations.

***Model* K.** The RTD for this model is given by

$$\mathbf{E}_\theta = \frac{v_1}{v}\frac{1}{\bar{\theta}_1} e^{-\theta/\bar{\theta}_1} + \frac{v_2}{v}\frac{1}{\bar{\theta}_2} e^{-\theta/\bar{\theta}_2} \tag{60a}$$

where

$$\bar{\theta}_1 = \frac{\bar{t}_1}{\bar{t}} = \frac{V_1/v_1}{V/v} \qquad \text{and} \qquad \bar{\theta}_2 = \frac{\bar{t}_2}{\bar{t}} = \frac{V_2/v_2}{V/v} \tag{60b}$$

This RTD represents the sum of two exponentials. Evaluating the slopes and/or intercepts of the two distinct straight line portions of the log $\mathbf{E}_\theta$ versus θ plot is one way of fitting the parameters of this model. This is the simplest of the models presented in this section.

***Model* L.** The RTD expression and also the tracer concentration in the stagnant compartment of this model are given by

$$\mathbf{E}_\theta = C_1 = \frac{1}{a(1-a)(m_1-m_2)}\left[(m_1 - am_1 + b)e^{m_1\theta} - (m_2 - am_2 + b)e^{m_2\theta}\right] \tag{61a}$$

and

$$C_2 = \frac{b}{a(1-a)(m_1-m_2)}\left[e^{m_1\theta} - e^{m_2\theta}\right] \tag{61b}$$

where

$$m_1, m_2 = \frac{1 - a + b}{2a(1 - a)}\left[-1 \pm \sqrt{1 - \frac{4ab(1 - a)}{(1 - a + b)^2}}\right] \tag{61c}$$

The shape of the **E** curve for this model is the same as for model K, thus from the point of view of curve fitting these models are essentially equivalent. Physical considerations, however, often suggest the use of one or the other. Bischoff and Dedrick (1970) show how to fit the parameters of this model using either the C_1 or C_2 curves, or by using step input experimentation.

***Model* M.** The RTD as well as the tracer concentration in the first compartment are given by

$$C_1 = \frac{1}{e(1 - e)(n_1 - n_2)}[(n_1 - en_1 + f)e^{n_1\theta} - (n_2 - en_2 + f)e^{n_2\theta}] \tag{62a}$$

$$\mathbf{E}_\theta = C_2 = \frac{f}{e(1 - e)(n_1 - n_2)}[e^{n_1\theta} - e^{n_2\theta}] \tag{62b}$$

where

$$n_1, n_2 = \frac{f}{2e(1 - e)}\left[-1 \pm \sqrt{1 - \frac{4e(1 - e)}{f}}\right] \tag{62c}$$

Note that the RTD for this model first rises, then falls, and therefore is quite different from the RTD for models K and L.

***Model* N.** This model really represents the special cases of models L and M for batch operations, or for no throughflow of fluid. The concentrations of tracer in the two compartments resulting from a pulse input into compartment 1 are given by

$$C_1 = 1 + \frac{1 - g}{g}\exp\left[-\frac{t}{g(1 - g)(V/hv)}\right] \tag{63a}$$

and

$$C_2 = 1 - \exp\left[-\frac{t}{g(1 - g)(V/hv)}\right] \tag{63b}$$

The above models are particularly useful for representing the distribution of drugs in mammals. Often the time needed to eliminate a drug from the body is rather long, even in the order of years. On this time scale the fluctuations in drug concentration in the circulating blood stream just after injection damp out quickly and can be ignored. Thus, the blood can be represented by one uniform compartment of the model, the organ or tissue in question by the other compartment.

To account for different kinds of organs and tissues may require the use of a number of compartments, all interchanging material with the blood. Computer solution is usually required for such models.

DIAGNOSING ILLS OF OPERATING EQUIPMENT

The observed **C**-curve can give clues to poor contacting and flow in process equipment. For example, if flow is expected to approximate plug flow then Fig. 34 shows some of the deviations which may occur. Consider these in turn.

Figure 34*a* has the **C**-curve at the correct location and with not too large a spread; hence, there is nothing wrong here. In Fig. 34*b* the early appearance of tracer indicates channelling of fluid and stagnant regions in the reactor. Figure 34*c* indicates that the fluid is recirculating in the reactor. This type of behavior may occur with slow moving fluid in short wide vessels, e.g., liquids in gas-liquid con-

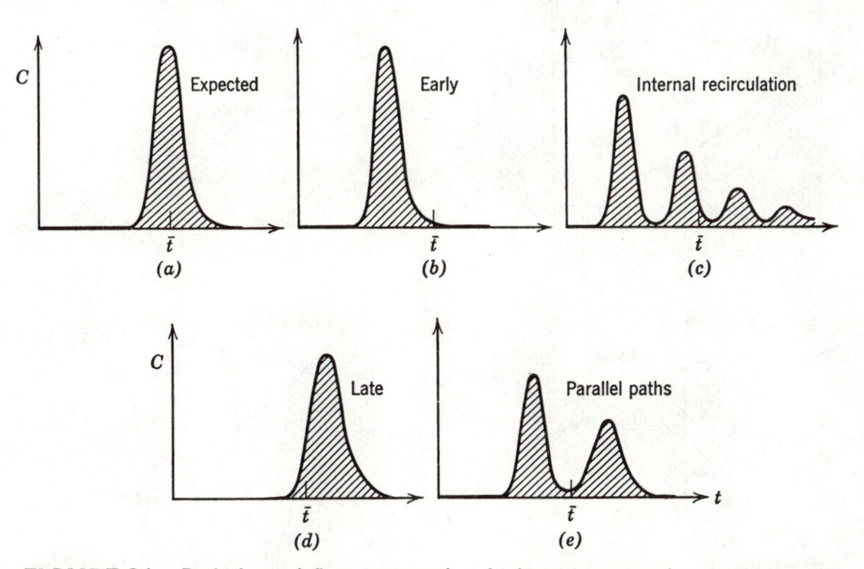

FIGURE 34. Pathological flow in vessels which are supposed to approximate plug flow.

tactors. Figure 34*d* shows tracer appearing later than expected. There are three possible explanations for this: either an error in flow rate measurement, an error in volume available for fluid (has pore voidage or volume of recording leads been accounted for?), or the tracer is not an inert material as it should be, but is adsorbed and held back on the surfaces. Finally Fig. 34*e* shows fluid channelling down two parallel paths. In all the above mentioned cases of poor flow significant improvement should be possible by inserting redistributors and/or by proper baffling.

If mixed flow is expected, then Fig. 35 shows some of the poor patterns which may occur. The first four sketches show the expected, early, recycling, and late flow, and the explanations for these observations are similar to the corresponding plug flow cases. Figure 35*e* shows a shift in time which probably indicates an instrumentation time lag.

Although it is not necessary, sometimes it is helpful to introduce distribution functions related to the RTD, to help diagnose these ills. These functions are the internal age distribution for fluid present in the vessel

$$\bar{t}\mathbf{I} = 1 - \int_0^t \mathbf{E}\, dt \tag{64a}$$

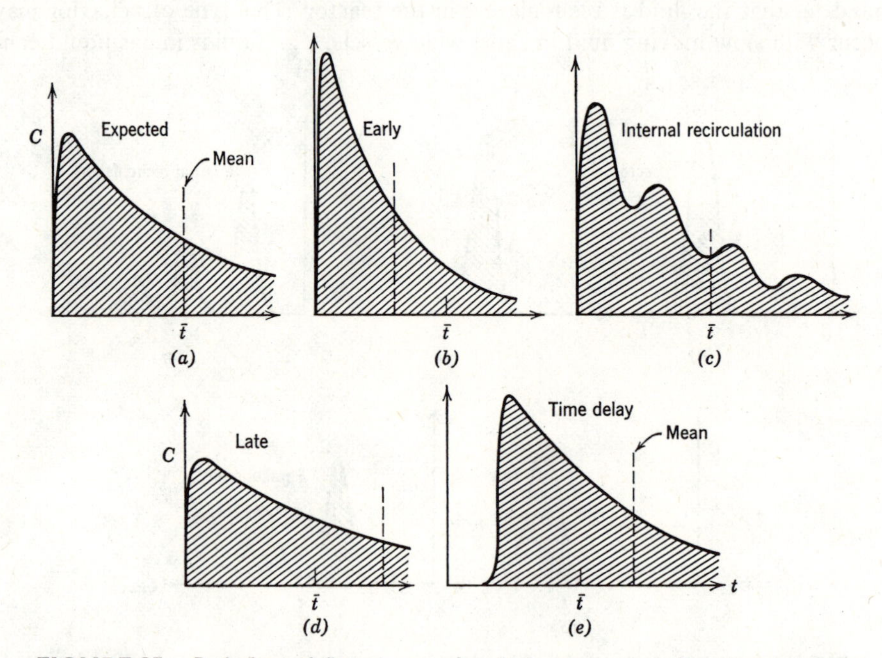

FIGURE 35. Pathological flow in vessels which are supposed to approximate mixed flow. The dotted line represents the observed mean of the tracer curve.

and the intensity function introduced by Naor and Shinnar (1963)

$$\lambda = \frac{\mathbf{E}}{\mathbf{I}} \tag{64b}$$

Himmelblau and Bischoff (1968) show how to use these functions.

In certain operations it is imperative that the flow be made to approach some ideal as closely as possible—usually plug flow. This requirement may cause difficulties, especially for slow flow in large vessels. Continuous large scale chromatographic separations is an important example of such a process. Figure 36 shows some ways of more closely approaching plug flow in equipment.

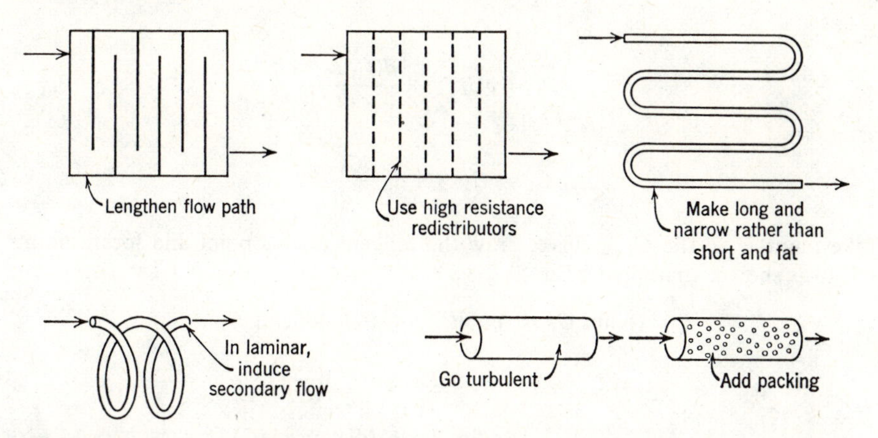

FIGURE 36. Ways of improving the flow, to more closely approach plug flow.

EXAMPLE 7. Correcting a C-curve for an unwanted flow region

Suppose a tracer response is available for vessel 1 followed by an ideal stirred tank (vessel 2) of known size or $\bar{t}_2$. How do we remove the contribution of the stirred tank and find the **C**-curve of vessel 1 alone? Figure E7 sketches the set up.

SOLUTION

We present a simple graphical procedure which can be used as long as the known vessel is an ideal stirred tank.

Suppose that C_{1+2} in Fig. E7 represents the curve for the two vessels in series. Then a material balance for tracer about the ideal stirred tank gives

$$\text{input} = \text{output} + \text{accumulation}$$

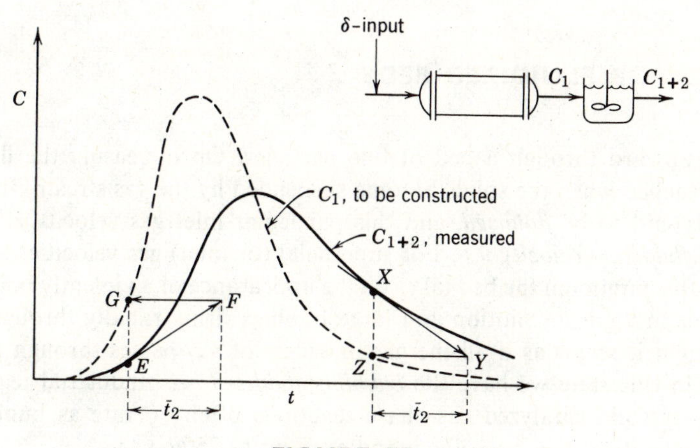

FIGURE E7.

or

$$vC_1 = vC_{1+2} + \frac{d(C_{1+2}V)}{dt}$$

or

$$C_1 = C_{1+2} + \bar{t}_2 \frac{dC_{1+2}}{dt} \tag{i}$$

Take point E on the C_{1+2} curve, draw the tangent at this point and locate points F and G as shown. In applying Eq. (i)

$$\begin{aligned}(\text{point } G) &= (\text{point } E) + (\bar{t}_2)(\text{slope at } E)\\ &= (\text{point } E) + (GF)\left(\frac{AG}{GF}\right)\\ &= (\text{point } E) + (\text{distance } AG)\end{aligned}$$

we verify that point G is on the C_1 curve which we are looking for.

Repeating this procedure for various locations (points X, Y, Z are illustrated) then gives the desired curve, shown as the dashed line.

Comment. Note that the general approach to this problem involves deconvolution, which is difficult. The fact that one of the vessels is a stirred tank is what simplifies matters so greatly.

This procedure can be extended to more than one stirred tank in series with the vessel to be studied. Simply repeat the procedure. It can also be used with little modification with the **F** curve.

The type of problem illustrated here occurs in experimentation when we want to remove end effects. We may also note that in systems where fluid flows at different velocities (say laminar flow) the introduction of tracer in small completely mixed fore and after sections is probably the simplest way to guarantee proper tracer input (proportional to flow velocity) and proper measurement (mixing cup method).

MODELS FOR FLUIDIZED BEDS

Pass gas upward through a bed of fine particles. On increasing the flow rate a point is reached when the solids become suspended by the gas stream. In this state the bed is said to be *fluidized*, and this particular inlet gas velocity is called the *minimum fluidizing velocity* u_{mf}. For superficial (or inlet) gas velocities u_o much in excess of this minimum the bed takes on the appearance of a violently boiling liquid with solids in vigorous motion and large bubbles rising rapidly through the bed. At first sight it seems as if all the gas in excess of u_{mf} passes through the bed as bubbles. In this state we have the *bubbling fluidized bed.* Industrial reactors particularly for solid catalyzed gas-phase reactions often operate as bubbling beds with gas velocities $u_o = 5 \sim 30u_{mf}$, and even up to $250u_{mf}$.

Calculations show that the conversion in bubbling beds may vary from plug flow to well below mixed flow, and the perplexing and embarrassing thing about this is that we cannot reliably estimate or guess what it will be for any new situation. Because of this, scale-up is cautious and uncertain, and preferably left to others.

It was soon recognized that this difficulty stemmed from lack of knowledge of the contacting and flow pattern in the bed: in effect, the bypassing of much of the solids by the rising bubble gas. This led to the realization that adequate prediction of bed behavior had to await a reasonable flow model for the bed.

Since the bubbling bed represents such severe deviations from ideal contacting, not just minor ones as with other single-fluid reactors (packed beds, tubes, etc.), it would be instructive to see how this problem of flow characterization has been attacked. A wide variety of approaches have been tried. We consider these in turn.

Dispersion and Tanks in Series Models. The first attempts at modeling naturally tried the simple one-parameter models; however observed conversion well below mixed flow cannot be accounted for by these models so this approach has been dropped by most workers.

RTD Models. The next class of models relied on the RTD to calculate conversions. But since the rate of catalytic reaction of an element of gas depends on the amount of solid in its vicinity, the effective rate constant is low for bubble gas, high for emulsion gas. Thus any model which simply tries to calculate conversion from the RTD and the fixed rate constant in effect assumes that all elements of gas, both slow and fast moving, spend the same fraction of time in each of the phases. As we will show when we treat the details of gas contacting in fluidized beds this assumption is a shaky one, hence the direct use of the RTD to predict conversions, as developed for linear systems in this chapter, is quite inadequate.

Contact Time Distribution Models. To overcome this difficulty and still use the information given by the RTD, models were proposed which assumed that faster gas stayed mainly in the bubble phase, the slower in the emulsion. Gilliland and Knudsen (1970) used this approach and propose that the effective rate constant depends on the length of stay of the element of gas in the bed, thus

$$\left.\begin{array}{l}\text{short stay means small } k\\ \text{long stay means large } k\end{array}\right\} \quad \text{or } k = k_0 t^m$$

where m is a fitted parameter. Thus combining with Eq. 24 we find for the conversion

$$\frac{\overline{C}_A}{C_{A0}} = \int_0^\infty e^{-kt}\mathbf{E}\,dt = \int_0^\infty e^{-k_0 t^{m+1}}\mathbf{E}\,dt \tag{65}$$

The problem with this approach involves obtaining a meaningful $\mathbf{E}$ function to use in Eq. 65 from a measured $\mathbf{C}$-curve which is both a stochastic quantity and which is obtained at the exit of a bed where considerable backmixing occurs and where an element of tracer may pass the recording point again and again. In fact recent findings suggest that some exit gas from the bed reenters the emulsion and moves back to the entrance of the bed! What does the **C**-curve mean in this situation?

Two-region Models. Recognizing that the bubbling bed consists of two rather distinct zones, the bubble phase and the emulsion phase, experimenters spent much effort in developing models based on this fact. Since such models contain six parameters, see Fig. 37, many simplifications and special cases have been explored (15 to date), and even the complete 6-parameter model has been used.

This approach has been middling successful as a correlating technique to fit any particular set of data; however, it has not been of much use in bringing together data from diverse systems, or for scale-up to new conditions. The difficulty is that we do not know how to assign values to the parameters for new conditions.

Hydrodynamic Flow Models. The discouraging result with the previous approaches lead us reluctantly to the conclusion that we must know more about what goes on in the bed if we hope to develop a reasonable predictive flow model. In particular we must learn more about the behavior of rising gas bubbles since they probably cause much of the difficulty.

Two developments are of particular importance in this regard. The first is Davidson's remarkable theoretical development and experimental verification (see Davidson and Harrison (1963) for details) of the essentials of flow in the vicinity of a rising gas bubble. For bubbles typical of a vigorously bubbling fluidized bed he showed the following:

1. Bubble gas stays with the bubble, recirculating very much like a smoke ring and only penetrating a small distance into the emulsion. This zone of penetration is called the cloud since it envelops the rising bubble.

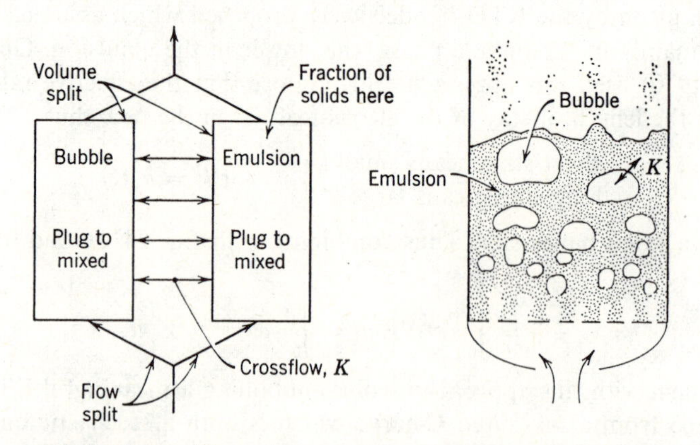

FIGURE 37. Two-phase model to represent the bubbling fluidized bed.

2. All related quantities such as velocity of rise, cloud thickness, recirculation rate, are simple functions of the size of rising bubble.

The surprising and important fact about these findings, shown in Fig. 38, is that bubble gas is much more segregated from the emulsion than ever suspected.

The second important finding was the experimental observation by Rowe and Partridge (1962) that each bubble of gas drags a substantial wake of solids up the bed.

These two developments laid the foundation for a class of hydrodynamic models using bubble size as parameter, with all other quantities derived from bubble size. We will briefly treat the first and simplest of these, the *bubbling bed model* of Kunii and Levenspiel (1968) which assumes:

1. Bubbles are of one size and are evenly distributed in the bed.
2. Flow of gas in the vicinity of rising bubbles follows the Davidson model.
3. Each bubble drags along with it a wake of solids, creating a circulation of solids in the bed, with upflow behind bubbles and downflow in the rest of the emulsion.
4. The emulsion stays at minimum fluidizing conditions; thus the relative velocity of gas and solid remains unchanged.

Based on these assumptions material balances for solids and for the gas give in turn

$$\begin{pmatrix}\text{upflow of solids}\\ \text{with bubble}\end{pmatrix} = \begin{pmatrix}\text{downflow of solids}\\ \text{in emulsion}\end{pmatrix} \tag{66}$$

$$(\text{total throughflow of gas}) = \begin{pmatrix}\text{upflow in}\\ \text{bubble}\end{pmatrix} + \begin{pmatrix}\text{upflow in}\\ \text{emulsion}\end{pmatrix} \tag{67}$$

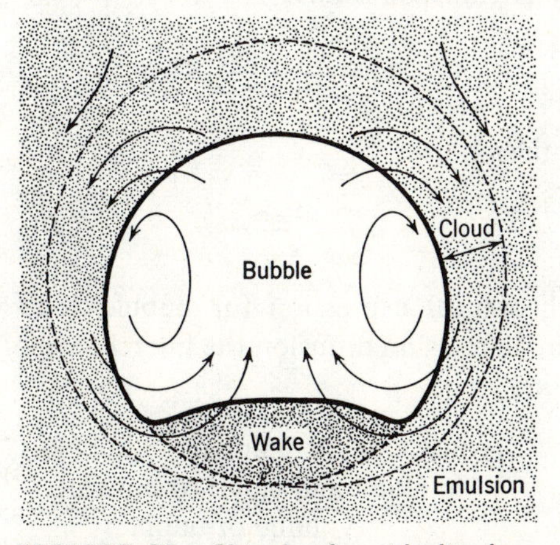

FIGURE 38. Sketch of an idealized gas bubble incorporating Davidson's gas motion and Rowe's wake.

Letting

$$u_{br} = 0.711(gd_b)^{1/2}, \quad \text{rise velocity of a single bubble in a fluidized bed} \tag{68}$$

$$\epsilon_{mf} = \text{void fraction of the bed at minimum fluidizing conditions}$$

$$\alpha = \text{volume of wake/volume of bubble}$$

these material balances give:

Rise velocity of bubbles, clouds, and wakes:

$$u_b = u_o - u_{mf} + u_{br} = u_o - u_{mf} + 0.711(gd_b)^{1/2} \tag{69}$$

Bed fraction in bubbles:

$$\delta = \frac{u_o - (1 - \delta - \alpha\delta)u_{mf}}{u_b} \cong \frac{u_o - u_{mf}}{u_b} \tag{70}$$

Bed fraction in clouds:

$$\beta = \frac{3\delta u_{mf}/\epsilon_{mf}}{u_{br} - u_{mf}/\epsilon_{mf}} \tag{71}$$

Bed fraction in wakes:

$$\alpha\delta \tag{72}$$

Bed fraction in downflowing emulsion including clouds:

$$1 - \delta - \alpha\delta \tag{73}$$

Downflow velocity of emulsion solids:

$$u_s = \frac{\alpha\delta u_b}{1 - \delta - \alpha\delta} \tag{74}$$

Rise velocity of emulsion gas:

$$u_e = \frac{u_{mf}}{\epsilon_{mf}} - u_s \tag{75}$$

Using Davidson's theoretical expression for bubble-cloud circulation and the Higbie theory for cloud-emulsion diffusion the interchange of gas between bubble and cloud is then found to be

$$K_{bc} = \frac{\left(\begin{matrix}\text{volume of gas going from bubble} \\ \text{to cloud and cloud to bubble}\end{matrix} \Big/ \text{sec}\right)}{(\text{volume of bubble})}$$

$$= 4.5\left(\frac{u_{mf}}{d_b}\right) + 5.85\left(\frac{\mathscr{D}^{1/2}g^{1/4}}{d_b^{5/4}}\right) \tag{76}$$

and between cloud and emulsion

$$K_{ce} = \frac{\text{(interchange volume/sec)}}{\text{(volume of bubble)}} \cong 6.78\left(\frac{\epsilon_{mf}\mathscr{D}u_b}{d_b^3}\right)^{1/2} \tag{77}$$

One surprising consequence of this model is its prediction that the gas flow in the emulsion reverses and becomes downward at higher u_o, or at $u_o > 3 \sim 11u_{mf}$. This prediction has been verified in recent years by a number of methods.

The above expressions show that if we know ϵ_{mf}, estimate α, and measure u_{mf} and u_o, then all flow quantities and regional volumes can be determined in terms of one parameter, the bubble size. Figure 39 then represents the model as visualized. The application of this model to chemical conversion is simple and direct, and is treated in Chapter 14.

Various other hydrodynamic models have been proposed recently, using other combinations of assumptions such as:

Changing bubble size with height in the bed
Negligible bubble-cloud resistance
Negligible cloud-emulsion resistance
Nonspherical bubbles.

In all cases the underlying rationale for these hydrodynamic models rests on the observation that beds with identical solids and gas flow rates may develop either large bubbles or small bubbles depending on distributor design, baffle arrangement, etc., thus bubble size must enter as the primary parameter in the model. A

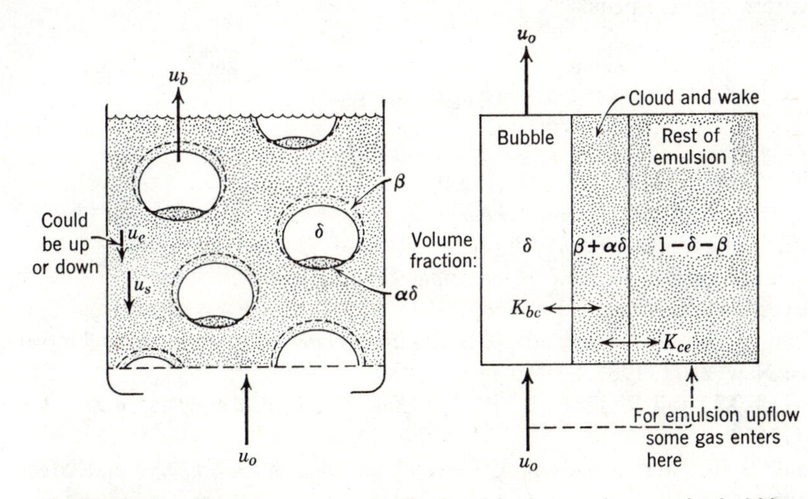

FIGURE 39. Flow pattern in the fluidized bed according to the bubbling bed model.

consequence of this argument is that models which do not allow for different bubble sizes at given imposed bed conditions certainly cannot be adequate.

The power of this class of model should be apparent. For example, even the simplest of these models, the one considered here, gives unexpected predictions (e.g., that most of the gas in the bed may be flowing downward) which are subsequently verified. More important still, this type of model can be tested, it can be shown to be wrong, it can be rejected, because its one parameter, the bubble size, can be compared with observation.

FINAL COMMENTS

Tracer experimentation is the most convenient tool for determining when deviations from ideal flow are serious, and for handling these deviations. It gives conversion predictions which hold rigorously for first-order reactions, but which only hold approximately for other kinetics. Chapter 10 treats these nonlinear kinetics.

In well-operated equipment flow usually is close enough to ideal so that any lowering in performance is a minor consideration. There are exceptions however, particularly when product distribution is a factor. Large deviations from ideality usually indicate some malfunctioning of the equipment such as channelling, bypassing, etc.

In freely bubbling fluidized beds the gas flow normally deviates greatly from the ideal. Unfortunately simple tracer methods are not too useful here. Nevertheless, proper equipment design (baffles and distributor) can greatly reduce the undesirable flow pattern of these beds.

REFERENCES

Aris, R., *Chem. Eng. Sci.*, **9**, 266 (1959).
———, *Proc. Roy. Soc.* (London) **A235**, 67 (1956).
Bischoff, K. B., Ph.D. Thesis, Illinois Institute of Technology, 1961.
———, and Dedrick, R. L., *J. theor. Biol.*, **29**, 63 (1970).
Danckwerts, P. V., *Chem. Eng. Sci.*, **2**, 1 (1953).
Davidson, J. F., and Harrison, D., *Fluidized Particles*, Cambridge University Press, New York, 1963.
Douglas, J. M., and Bischoff, K. B., *Ind. Eng. Chem. Process Design Develop.*, **3**, 130 (1964).
Gilliland, E. R., and Knudsen, C. W., Paper 16d, A.I.Ch.E. Annual Meeting, Chicago, Dec. 1970.
———, Mason, E. A., and Oliver, R. C., *Ind. Eng. Chem.*, **45**, 1777 (1953).

Harrell, Jr., J. E., and Perona, J. J., *Ind. Eng. Chem. Process Design Develop.*, **7**, 464 (1968).
Himmelblau, D. M., and Bischoff, K. B., *Process Analysis and Simulation*, John Wiley & Sons, New York, 1968.
Kunii, D., and Levenspiel, O., *Fluidization Engineering*, John Wiley & Sons, New York, 1969.
——— and Levenspiel, O., *Ind. Eng. Chem. Fundamentals*, **7**, 446 (1968).
Levenspiel, O., *Ind. Eng. Chem.*, **50**, 343 (1958).
——— and Bischoff, K. B., *Ind. Eng. Chem.*, **51**, 1431 (1959); **53**, 313 (1961).
———, Lai, B. W., and Chatlynne, C. Y., *Chem. Eng. Sci.*, **25**, 1611 (1970b).
———, and Smith, W. K., *Chem. Eng. Sci.*, **6**, 227 (1957).
———, and Turner, J. C. R., *Chem. Eng. Sci.*, **25**, 1605 (1970a).
Levich, V. G., Markin, V. S., and Chismadzhev, Y. A., *Chem. Eng. Sci.*, **22**, 1357 (1967).
Naor, P., and Shinnar, R., *Ind. Eng. Chem. Fundamentals*, **2**, 278 (1963).
Pasquon, I., and Dente, M., *J. Catalysis*, **1**, 508 (1962).
Pethö, A., *Chem. Eng. Sci.*, **23**, 807 (1968).
Rowe, P. N., and Partridge, B. A., Proc. Symp. on Interaction between Fluids and Particles, Inst. Chem. Engrs. p. 135, June 1962.
Taylor, G. I., *Proc. Roy. Soc.* (London), **219A**, 186 (1953); **225A**, 473 (1954a).
———, *Proc. Roy. Soc.* (London), **223A**, 446 (1954b).
Tichacek, L. J., *A.I.Ch.E. Journal*, **9**, 394 (1963).
van der Laan, E. Th., *Chem. Eng. Sci.*, **7**, 187 (1958).
van der Vusse, J. G., *Chem. Eng. Sci.*, **17**, 507 (1962).
Voncken, R. M., Holmes, D. B., and den Hartog, H. W., *Chem. Eng. Sci.*, **19**, 209 (1964).
von Rosenberg, D. U., *A.I.Ch.E. Journal*, **2**, 55 (1956).
Wehner, J. F., and Wilhelm, R. H., *Chem. Eng. Sci.*, **6**, 89 (1956).

PROBLEMS

These are loosely grouped as follows:

Problems 1–20: General treatment and one parameter models
Problems 21–35: Devising and fitting multiparameter models

1. A specially designed vessel is to be used as a reactor for a first-order liquid reaction. Since flow in this vessel is suspected to be nonideal, tracer tests are conducted and the following concentration readings represent the response at the vessel outlet to a delta-function tracer input to the vessel inlet. What conversion can we expect in this reactor if conversion in a mixed flow reactor employing the same space time is 82.18%.

Time t, sec	10	20	30	40	50	60	70	80
Tracer concentration (arbitrary reading)	0	3	5	5	4	2	1	0

We suspect that the dispersion and tanks in series models are poor representations of the flow pattern.

2. Fluid flows at a steady rate through 10 tanks in series. A pulse of tracer is introduced into the first tank, the time this tracer leaves the system is measured giving

$$\text{maximum concentration} = 100 \text{ millimol/liter}$$
$$\text{tracer spread} = 1 \text{ min}$$

If 10 more tanks are connected in series with the original 10 tanks what would be

(*a*) The maximum concentration of leaving tracer?
(*b*) The tracer spread?
(*c*) How does the relative spread change with number of tanks?

3. Find the number of tanks in series to represent the close-to-symmetrical **E** curve of Fig. P14.

4. From a pulse input into a vessel the following output signal is obtained

Time, minutes	1	3	5	7	9	11	13	15
Concentration, (arbitrary)	0	0	10	10	10	10	0	0

We want to represent the flow through the vessel with the tanks-in-series model. Using the variance matching procedure determine the number of tanks to use.

5. We are receiving complaints of a large fish kill along the Ohio River, indicating that someone had discharged highly toxic material into the river. Our water monitoring stations at Cincinnati and Portsmouth Ohio (119 miles apart) report that a large slug of phenol is moving down the river and we strongly suspect that this is the cause of the pollution. The slug took 9 hours to pass the Portsmouth monitoring station and its concentration peaked at 8 a.m. Monday. About 24 hours later the slug peaked at Cincinnati, taking 12 hours to pass this monitoring station.

Phenol is produced at a number of locations on the Ohio River, and their distance upriver from Cincinnati are as follows:

Ashland, Ky.—150 miles upstream
Huntington, W.Va.—168
Pomeroy, O.—222
Parkersburg, W.Va.—290
Marietta, O.—303
Wheeling, W.Va.—385
Steubenville, O.—425
Pittsburgh, Pa.—500

What can you say about the probable pollution source?

6. Water is drawn from a lake, flows through a pump and passes down a long pipe in turbulent flow. A slug of tracer enters the intake line at the lake, and is recorded at two locations in the pipe L meters apart. The mean residence time of fluid between recording points is 100 sec, and the spread in the two recorded signals is

$$\sigma_1^2 = 800 \text{ sec}^2$$
$$\sigma_2^2 = 900 \text{ sec}^2$$

What would be the spread of a **C** curve for a section of this pipe, free from end effects and of length $L/5$?

7. It is known that the longer the tube the closer does flow approach plug flow. As an indication of the lengths involved imagine fluid flowing in a tube 1 cm in diameter. If we consider that an RTD similar to 10 tanks in series approximates plug flow find what length of tube gives plug flow under the following flow conditions.

(*a*) A liquid (Sc = 1000) flows at $\text{Re}_t = 10, 10^3, 10^5$.
(*b*) A gas (Sc = 1) flows at $\text{Re}_t = 10, 10^3, 10^5$.

8. Order the following setups with regard to the extent of their deviations from plug flow. In all cases the vessel is 3 meters long and 3 cm in diameter.

(*a*) Empty tube, liquid flowing at $\text{Re}_t = 10{,}000$
(*b*) Empty tube, gas flowing at $\text{Re}_t = 100{,}000$
(*c*) Tube packed with 4 mm spheres, $\epsilon = 0.4$, $\text{Re}_p = 200$

9. At present we are processing a gas stream in a laminar flow tubular reactor. We plan to quadruple the processing rate (at fixed τ) and for this we have two alternatives:

(*a*) quadruple the length of reactor leaving d_t unchanged
(*b*) double the diameter of reactor leaving L unchanged.

Compare the deviation from plug flow of these proposed larger units with that of the present unit. Which scale-up would you recommend?

Data: Assume that the reactors are long enough for the dispersion model to apply, that laminar flow prevails in all these arrangements even though the Reynolds number is rather high.

10. A "closed" vessel has flow for which $D/uL = 0.2$. We wish to represent this vessel by the tanks in series model. What value of N should we select?

11. A 12 m-length of pipe is packed with 1 m of 2-mm material, 9 m of 1 cm-material, and 2 m of 4-mm material. Estimate the variance in output **C** curve for this packed section if the fluid takes 2 min to flow through the section. Assume a constant bed voidage and a constant intensity of dispersion given by $D/ud_p = 2$.

12. Let us introduce a term called the length of a dispersion unit, and let this be the length of vessel which provides the mixing equivalent to one ideal stirred tank.

(*a*) For a vessel long enough so that deviation from plug flow is small, find the length of a dispersion unit.
(*b*) Find the length of a dispersion unit for water flowing in a pipe at Reynolds numbers of 5, 500, and 50,000.
(*c*) Find the length of a dispersion unit for water and air flowing in a packed bed at a particle Reynolds number of 100.
(*d*) What does the answer of part (*c*) suggest for a model for the actual mixing process occurring in a packed bed.

13. *Behavior of short laminar flow reactors.* Consider laminar flow in a tubular reactor which is so short that the dispersion model is not applicable. For this situation prediction of behavior becomes very difficult, however we can still estimate the poorest expected performance of such reactors. As an example take a first-order reaction $A \rightarrow R$ with $X_{A,\text{plug}} = 0.99$. Recalling for laminar flow that the maximum velocity in the centerline of the pipe is twice the average velocity and that this represents the smallest residence time in the reactor estimate the lower bound to the expected conversion.

14. *Behavior of short laminar flow reactors* (*continued*). To sharpen the prediction of the previous problem note that laminar flow with no molecular diffusion (no radial mixing) has a RTD given by

$$\left.\begin{aligned} \mathbf{E}_\theta &= \frac{1}{2\theta^3} \qquad \text{for } \tfrac{1}{2} \leqslant \theta < \infty \\ &= 0 \qquad \text{elsewhere} \end{aligned}\right\} \tag{78}$$

The radial contribution to molecular diffusion greatly narrows this RTD leading to the dispersion model behavior. Figure P.14 shows these two distributions.

Since a wide distribution leads to a large deviation from plug flow the RTD of Eq. 78 represents the poorest expected behavior of the reactor. For the reaction of the previous problem determine the lower bound to the conversion and compare your result with the result of the previous problem.

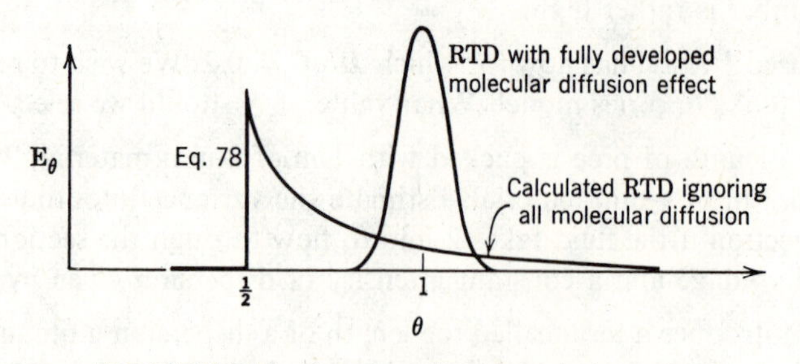

FIGURE P14. Laminar flow in circular tubes.

15. The kinetics of a homogeneous liquid reaction are studied in a flow reactor, and to approximate plug flow the 48 cm long reactor is packed with 5 mm nonporous pellets. If the conversion is 99% for a mean residence time of 1 sec calculate the rate constant for the first order reaction

(*a*) assuming that the liquid passes in plug flow through the reactor,
(*b*) accounting for the deviation of the actual flow from plug flow.
(*c*) What is the error in calculated k if deviation from plug flow is not considered.

Data: Bed voidage $\epsilon = 0.4$
Particle Reynolds number $\mathrm{Re}_p = 200$

16. Tubular reactors for thermal cracking are designed on the assumption of plug flow. On the suspicion that nonideal flow may be an important factor now being ignored, let us make a rough estimate of its role. For this assume isothermal operations in a 2.5-cm-i.d. tubular reactor, using a Reynolds number of 10,000 for flowing fluid. The cracking reaction is approximately first order. If calculations show that 99% decomposition can be obtained in a plug flow reactor 3 m long, how much longer must the reactor be if nonideal flow is taken into account?

17. A reactor has flow characteristics given by the non-normalized **C**-curve in Table P17, and by the shape of this curve we feel that the dispersion or tanks-in-series models should satisfactorily represent flow in the reactor.

TABLE P17

Time	Tracer Concentration	Time	Tracer Concentration
1	9	10	67
2	57	15	47
3	81	20	32
4	90	30	15
5	90	41	7
6	86	52	3
8	77	67	1

(*a*) Find the conversion expected in this reactor, assuming that the dispersion model holds.
(*b*) Find the number of tanks in series which will represent the reactor and the conversion expected, assuming that the tanks-in-series model holds.
(*c*) Find the conversion by direct use of the tracer curve.
(*d*) Comment on the difference in these results, and state which one you think is the most reliable.

Data: The elementary liquid-phase reaction taking place is A + B → products, with a large enough excess of B so that the reaction is essentially first order. In addition, if plug flow existed, conversion would be 99% in the reactor.

18. Repeat the previous problem with the following modifications. The elementary liquid-phase reaction taking place is A + B → products, with equimolar quantities of A and B fed into the reactor. If plug flow existed, conversion would be 99% in the reactor.

19. A pipeline 100 km long will be constructed to transport wine from a wine-producing center to the distribution point. Red and white wine are to flow in turn through this pipeline. Naturally, in switching from one to the other, a region of *vin rosé* is formed. The quantity of *vin rosé* is to be minimized since it is not popular and does not fetch a good price on the market.

(*a*) How does the pipeline size at given Reynolds number affect the quantity of *vin rosé* formed in the switching operation?

(*b*) For fixed volumetric flow rate in the turbulent flow region, what pipe size minimizes the *vin rosé* formed during the switching operation?

(*c*) Assuming that the pipeline is operating at present, what flow rate should we select to minimize the formation of *vin rosé*?

20. Often a pipeline must transport more than one material. These materials are then transported successively, and switching from one to another forms a zone of contamination between the two flowing fluids. Let A refer to the leading fluid and let *B* refer to the following fluid in a 30-cm-i.d. pipe.

(*a*) If the average Reynolds number of the flowing fluids is 10,000, find the 10%–90% contaminated width 10 km downstream from the point of feed.

(*b*) Find the 10%–99% contaminated width at this location (10%–99% contamination means allowing up to 10% of B in A but only allowing 1% of A in B).

(*c*) Find the 10%–99% contaminated width 160 km downstream from the point of feed.

(*d*) For a given flow rate, how does contaminated width vary with length of pipe.

For additional readings see *Petroleum Refiner*, **37**, 191 (March 1958).

21. Strongly radioactive waste fluids are stored in "safe-tanks" which are simply long small diameter (e.g., 20 m by 10 cm) slightly sloping pipes. To avoid sedimentation and development of "hot spots," and also to insure uniformity before sampling the contents, fluid is recirculated in these pipes.

To model the flow in these tanks fluid is recirculated in a closed loop, a pulse of tracer is introduced and the curve of Fig. P21 is recorded. Develop a suitable model for this system and evaluate the parameters.

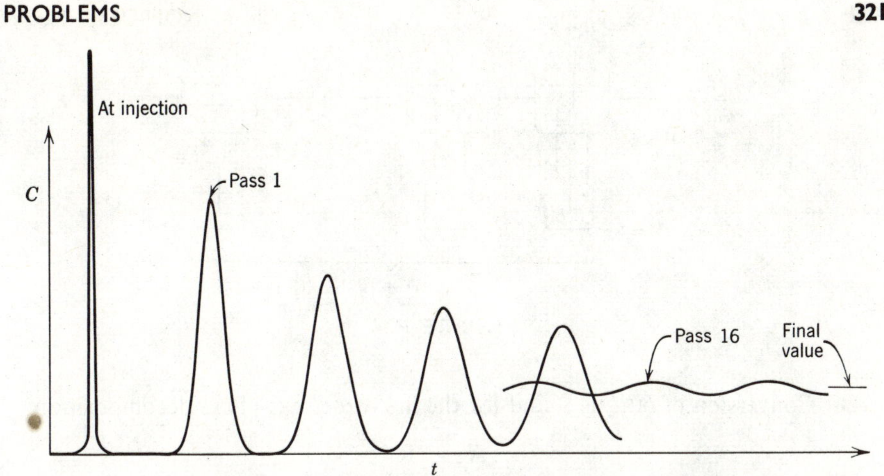

FIGURE P21. RTD for a closed recirculating system.

22. Figures P22*a* and P22*b* show the results of tracer tests on two different vessels using step inputs of tracer (switching from salt water to tap water). In both cases v = 100 liters/min and V = 80 liters. Devise flow models to represent these results.

23. (*a*) Devise a model to represent the flow in a vessel whose dimensionless response to a step input of tracer is given in Fig. P23.

(*b*) To show that sometimes more than one flow model is consistent with a given tracer curve try to develop a second model for this vessel.

24. (*a*) Sketch the main features of the expected **C** curve for the model of Fig. P24.

(*b*) Repeat for a vessel 4 times as long (N = 40).

25. A tracer test is made on a vessel of volume V = 100 liters using a flow rate of water of v = 100 liter/min. The output to a pulse input is shown in Fig. P25. Devise a flow model for this vessel.

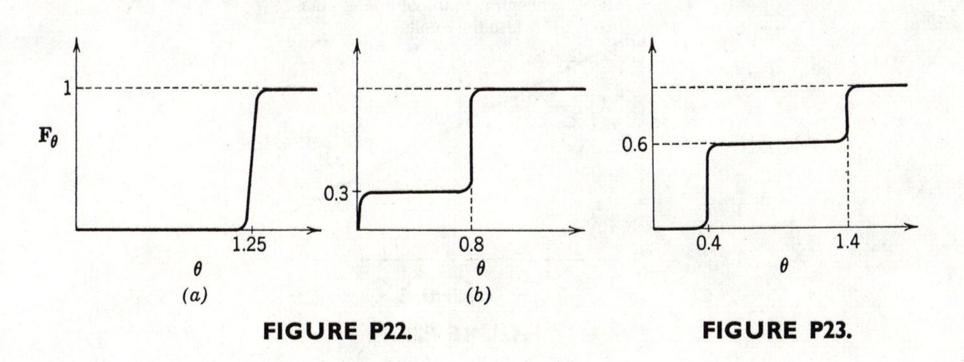

FIGURE P22. **FIGURE P23.**

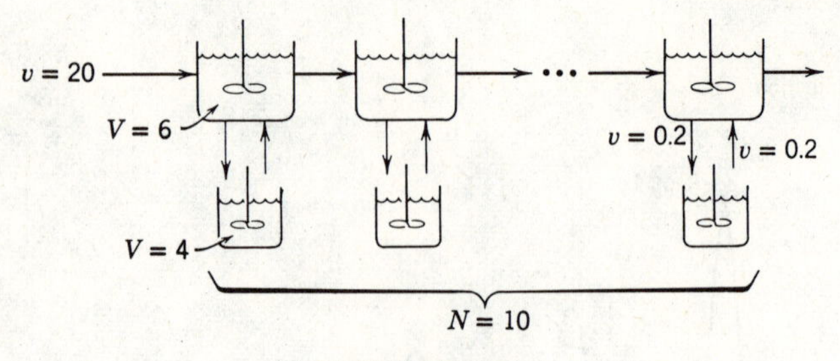

FIGURE P24.

26. Conversion of 60% is found for the first-order gas-phase decomposition

$$A \rightarrow 2R$$

taking place in a flow reactor of volume $V = 100\ \text{cm}^3$, using a flow rate $v_{\text{in}} = 100\ \text{cm}^3/\text{min}$ of feed consisting of 25% A and 75% inerts.

Before we can evaluate the rate constant for the reaction we must know how the fluid flows through the reactor, and for this we make a tracer test as follows. We introduce a δ-input during the reaction and we measure the output tracer concentration at various times. The result is shown in Fig. P25.

(*a*) Develop a flow model for this vessel.
(*b*) Determine the first-order rate constant for the reaction.

27. Gilliland *et al.* (1953) report that the flow of tracer gas through their fluidized bed ($d_t \simeq 7.6$ cm, $L \simeq 2$ m) of catalyst particles is well fitted by the **F** curve of Fig. P27 with a slope of -1.5. Develop a model to represent this flow. Also show in general how the values a and b are related to the parameters of the model and are related to each other.

28. Given an **F** curve for two flow regions in series, one of which is an ideal stirred tank of known size, show how to find the **F** curve for the other region.

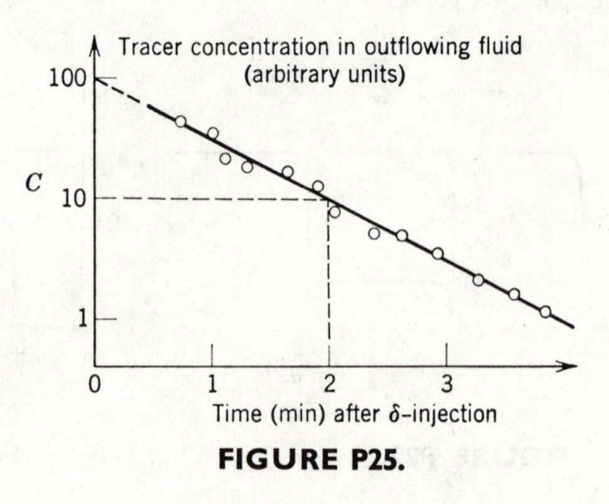

FIGURE P25.

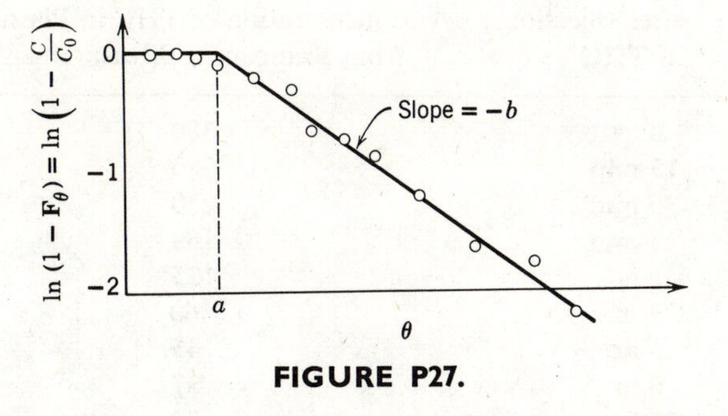

FIGURE P27.

29. Develop flow models to represent the two sets of tracer curves of Fig. P29.

30. Under steady flow conditions and with sufficient agitation a 5-liter experimental reactor produces a polymer with the desired properties. The fluid is rather viscous and a batch experiment with a fluid of similar physical properties gives a tracer curve for this reactor shown in Fig. 32*b*.

In scaling up to semiworks operations in a 500-liter reactor we need the same intensity of agitation, and our first plan is to maintain the same power input per unit volume of fluid. The tracer curve under these conditions is shown in Fig. 32*a*. Is this sufficient agitation or not?

31. We suspect that the stirred tank reactor with the following accurate **C** curve is not behaving ideally. Fit a flow model to this real reactor. The **C** curve was obtained in the 100-liter reactor using a flow rate of 20 liters fluid/min.

θ	0.1	0.4	0.8	1.5	1.8	2.5	3.7	4.4	5.6	7.1
$\mathbf{C}_\theta$	1.4	0.80	0.38	0.12	0.080	0.040	0.020	0.015	0.010	0.006

32. Δ^9-Tetrahydrocannabinol, or THC, is the major active component of marihuana. When ^{14}C-labeled THC was administered intravenously into a human it persisted in the plasma as follows (*Science*, **170**, 1320 (1970)):

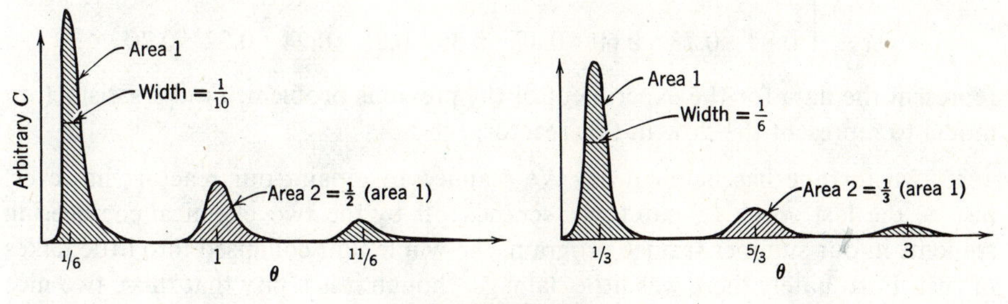

FIGURE P29.

Time after Injection of THC	Concentration of THC in Plasma from Radioactive Measurements
0	380
15 min	280
30 min	220
45 min	160
1 hr	125
2 hr	60
3 hr	45
6 hr	37
24 hr	30
48 hr	21
72 hr	16

(*a*) Fit these data with flow model K of Fig. 33 and evaluate the parameters of this model.

(*b*) As shown by the data, the sharp initial drop in concentration followed by a slow decay suggests that THC is rapidly absorbed by the tissues, then slowly released. If this is so, then model L is a more reasonable physical representation of this phenomenon than model K. Fit model L to the data and evaluate its parameters.

(*c*) Compare for this case, and in general, the parameters of these two models.

33. If the temperature of a reactor is changed the rate constant of the reaction will vary and different conversions will be obtained, all at given space-time. Suppose that conversion experiments of this kind (vary k at constant τ) are run in a particular reactor with a first-order reaction, and a plot of the resulting C_A/C_{A0} versus $k\tau$ data show an initial slope of $-2/3$ followed by an exponential decay to a limiting value of $C_A/C_{A0} = 0.25$ at high $k\tau$. Develop a model to represent the flow in this reactor.

34. Suppose that the following values

$\frac{C_A}{C_{A0}}$	0.5	1	2	4	8	18	38	98	∞
$k\tau$	0.84	0.73	0.60	0.47	0.36	0.28	0.24	0.22	0.20

represent the data for the experiment of the previous problem. Find a satisfactory model to represent the flow in this reactor.

35. Misfortune has befallen us! We planned to repaint our reactor; however, just as the last speck of paint was scraped off by the two chemical engineering students in our summer trainee program, the whole unit collapsed into little flakes of rust. Fortunately there was little damage, though it is a pity that those two nice boys disappeared completely. I always suspected that that stuff was corrosive.

Leaving sentiment aside, I wish to point out that shutdowns are costly, so we don't want this to happen again.

We are not operating on a shoestring any longer and can afford to replace this reactor by a commercial unit for which mixed flow is guaranteed. I am sure, however, that it does not have to be as large as our last unit. Ours may have been somewhat inefficient since we used a salvaged gasoline tank for the reactor and an outboard motor for the mixer, and had the inlet and overflow pipes close to each other. What size of commercial unit should we order so that the conversion will be identical to that in the old unit?

Here is all the information available on the unit, including results of the experiments by those two nice boys made with large bottles of India ink, a flashlight, and a photographer's exposure meter. Perhaps you can figure out what they were up to.

Data:

Reactor: Cylindrical tank about 6 m long with 30 m^3 of usable volume.
Reaction: Elementary second order, equimolar feed, 60% conversion.
Flow rate: 750 liters feed/min.

A large bottle of India ink evenly distributed throughout the reactor gives a concentration reading of 100, and one bottle was used for each experiment.

Time, min	Concentration of Ink at Reactor Outflow
0–20 sec	Rapid fluctuations with jumps up to 5000
10	90
20	55
30	35
40	20
50	10
60	7
80	3
100	1

10

MIXING OF FLUIDS

The problem associated with the mixing of fluids during reaction is important for extremely fast reactions in homogeneous systems, as well as for all heterogeneous systems. This problem has two overlapping aspects: first the *degree of segregation* of the fluid, or whether mixing occurs on the microscopic level (mixing of individual molecules) or the macroscopic level (mixing of clumps, groups, or aggregates of molecules), and second the *earliness of mixing*, or whether fluid mixes early or late as it flows through the vessel.

First consider the degree of segregation of a fluid, and for convenience let us define terms to represent the extremes in segregation of a fluid. For this suppose that liquid A is available in two forms as shown in Fig. 1. In the first form the liquid is as we normally imagine it, with individual molecules free to move about and collide and intermix with all other molecules of the liquid. Let us call such a liquid a *microfluid.* In the second form liquid A is available in a large number of small

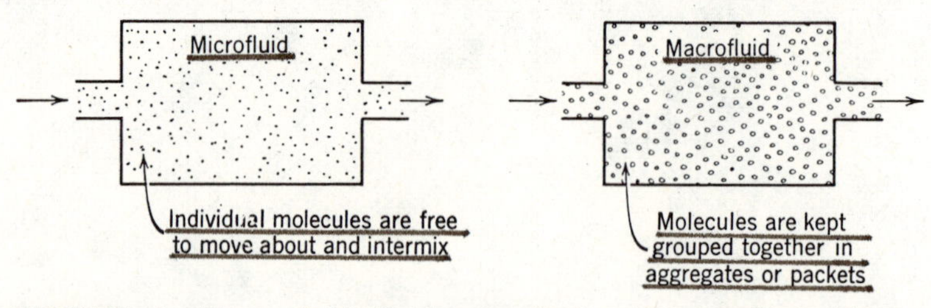

FIGURE 1. Flow of idealized microfluid and macrofluid. Tracer measurements cannot distinguish between them; however, for chemical reactions these fluids may behave differently.

sealed packets, each containing a large number, say about 10^{12} to 10^{18} of molecules. Let us call this type of liquid a *macrofluid.*

A fluid that does not exhibit these extremes in behavior is called a *partially segregated* fluid. A microfluid exhibits no segregation and a macrofluid exhibits complete segregation, but a real fluid exhibits segregation to a lesser or greater extent, depending on its properties and the kind of mixing which is taking place.

Next consider the earliness of mixing. In general numerous flow patterns, some with earlier mixing, others with later mixing, can give the same residence time distribution (RTD) through the vessel, thus this factor also enters the picture.

In essence then, degree of segregation concerns mixing on the molecular scale, while earliness of mixing concerns the gross flow pattern through the vessel.

Since the degree of segregation and the earliness of mixing can influence the performance of the reacting systems, both with respect to product distribution and capacity, we should like to know which form of mixing is advantageous, and which to promote. We do this by considering the micro and macrofluid extremes as well as earliest and latest mixing. This treatment is followed by a brief outline of recent attempts at modeling the intermediate situation of partial segregation.

In our development we first treat systems in which a single fluid is reacting and then systems in which two fluids are contacted and reacted.

SELF-MIXING OF A SINGLE FLUID

Degree of Segregation

The normally accepted state of a liquid or gas is that of a microfluid, and all previous discussions on homogeneous reactions have been based on the assumption. Let us now consider a single reacting macrofluid being processed in turn in batch, plug flow, and mixed flow reactors, and let us see how this state of aggregation can result in behavior different from that of a microfluid.

Batch Reactor. Let the batch reactor be filled with a macrofluid containing reactant A. Since each aggregate or packet of macrofluid acts as its own little batch reactor, conversion is the same in all aggregates and is in fact identical to what would be obtained with a microfluid. Thus for batch operations the degree of segregation does not affect conversion or product distribution

Plug Flow Reactor. Since plug flow can be visualized as a flow of small batch reactors passing in succession through the vessel macro and microfluids act alike. Consequently the degree of segregation does not influence conversion or product distribution in any way.

Mixed Flow Reactor. When a microfluid containing reactant A is treated as in Fig. 2, the reactant concentration everywhere drops to the low value prevailing in

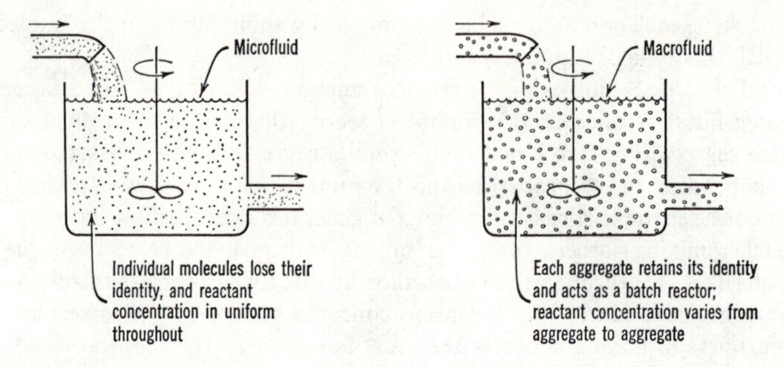

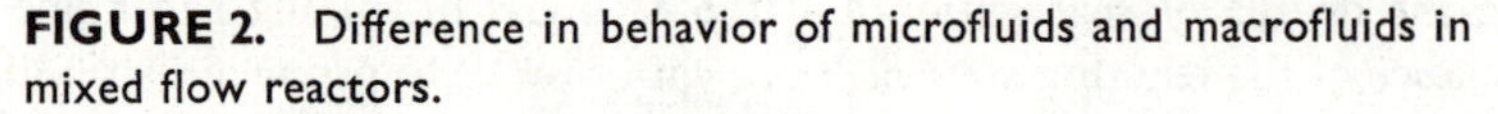
FIGURE 2. Difference in behavior of microfluids and macrofluids in mixed flow reactors.

the reactor. No clump of molecules retains its high initial concentration of A. We may characterize this by saying that each molecule loses its identity and has no determinable past history. In other words, by examining its neighbors we cannot tell whether a molecule is a new-comer or an old-timer in the reactor.

For this system the conversion of reactant is found by the usual methods for homogeneous reactions, or

$$X_A = \frac{(-r_A)V}{F_{A0}} \tag{5.11}$$

or, with no density changes,

$$\frac{C_A}{C_{A0}} = 1 - \frac{(-r_A)\bar{t}}{C_{A0}} \tag{1}$$

where $\bar{t}$ is the mean residence time of fluid in the reactor.

When a macrofluid enters a mixed flow reactor, the reactant concentration in the aggregates does not drop immediately to a low value but decreases in the same way as it would in a batch reactor. Thus a molecule in a macrofluid does not lose its identity, its past history is not unknown, and its age can be estimated by examining its neighboring molecules.

To determine the performance equation for the macrofluid in a mixed reactor imagine first of all the millions upon millions of little equal-size aggregates of macrofluid, batch reactors all, churning about in the vessel. The extent of reaction in each aggregate depends only on its length of stay in the reactor and the kinetics of the reaction. This also holds for any aggregate in the exit stream. Thus the fraction of reactant unconverted in the exit stream is obtained by determining the extent of reaction in all the leaving aggregates. Thus we may write for the exit stream as a whole

$$\begin{pmatrix}\text{fraction of}\\ \text{reactant}\\ \text{unreacted}\end{pmatrix} = \sum_{\substack{\text{all aggregates}\\ \text{in the exit}\\ \text{stream}}} \begin{pmatrix}\text{fraction of reactant}\\ \text{remaining in an}\\ \text{aggregate of age}\\ \text{between } t \text{ and}\\ t + \Delta t\end{pmatrix}\begin{pmatrix}\text{fraction of exit}\\ \text{stream consisting}\\ \text{of aggregates of}\\ \text{age between } t\\ \text{and } t + \Delta t\end{pmatrix} \tag{2}$$

Since the RTD of aggregates in the reactor is given by the exit age distribution function as defined in Chapter 9, Eq. 2 becomes

$$1 - \bar{X}_A = \frac{\bar{C}_A}{C_{A0}} = \int_0^\infty \left(\frac{C_A}{C_{A0}}\right)_{\text{batch}} \mathbf{E}\, dt \tag{3}$$

For mixed reactors the exit age distribution function is known. Thus from Chapter 9 the fraction of the exit stream that has an age between t and $t + dt$ is

$$\mathbf{E}\, dt = \frac{v}{V} e^{-vt/V}\, dt = \frac{e^{-t/\bar{t}}}{\bar{t}}\, dt \tag{4}$$

Replacing Eq. 4 in Eq. 3 gives

$$1 - \bar{X}_A = \frac{\bar{C}_A}{C_{A0}} = \int_0^\infty \left(\frac{C_A}{C_{A0}}\right)_{\text{batch}} \frac{e^{-t/\bar{t}}}{\bar{t}}\, dt \tag{5}$$

This is the general equation for determining conversion of macrofluid in mixed reactors and it may be solved once the kinetics of the reaction is given. Consider various reaction orders.

For a *first-order reaction*, the expression for batch operations, found in Chapter 3,

$$\left(\frac{C_A}{C_{A0}}\right)_{\text{batch}} = e^{-kt} \tag{6}$$

is applicable to the conversion in any single aggregate. On replacing into Eq. 5 we obtain

$$\frac{\bar{C}_A}{C_{A0}} = \frac{1}{\bar{t}} \int_0^\infty e^{-kt} e^{-t/\bar{t}}\, dt$$

which on integration gives the expression for conversion of a macrofluid in a mixed reactor

$$\frac{\bar{C}_A}{C_{A0}} = \frac{1}{1 + k\bar{t}} \tag{7}$$

This equation is identical with that obtained for a microfluid; for example, see Eq. 6.5. Thus we conclude that the degree of segregation has no effect on conversion for first-order reactions.

For a *second-order reaction* of a single reactant in a batch reactor Eq. 3.16 gives

$$\left(\frac{C_A}{C_{A0}}\right)_{\text{batch}} = \frac{1}{1 + C_{A0}kt} \tag{8}$$

On replacing into Eq. 5 we find

$$\frac{\bar{C}_A}{C_{A0}} = \frac{1}{\bar{t}}\int_0^\infty \frac{e^{-t/\bar{t}}}{1 + C_{A0}kt}\,dt$$

and by letting $\alpha = 1/C_{A0}k\bar{t}$ and converting into reduced time units $\theta = t/\bar{t}$, this expression becomes

$$\frac{\bar{C}_A}{C_{A0}} = \alpha e^{\alpha}\int_\alpha^\infty \frac{e^{-(\alpha+\theta)}}{\alpha+\theta}\,d(\alpha+\theta) = \alpha e^{\alpha}\,\text{ei}(\alpha) \tag{9}$$

This is the conversion expression for second-order reaction of a macrofluid in a mixed reactor. The integral, represented by $\text{ei}(\alpha)$ or $-\text{Ei}(-\alpha)$, is called the *exponential integral.* It is a function alone of α, and its value is tabulated in a number of tables of integrals. This equation may be compared with the corresponding expression for microfluids,

$$\frac{C_A}{C_{A0}} = \frac{1}{1 + C_A k\bar{t}} \tag{10}$$

For an *nth-order reaction* the conversion in a batch reactor can be found by the methods of Chapter 3 to be

$$\left(\frac{C_A}{C_{A0}}\right)_{\text{batch}} = [1 + (n-1)C_{A0}^{n-1}kt]^{1/(1-n)} \tag{11}$$

Insertion into Eq. 5 gives the conversion for an nth-order reaction of a macrofluid.

Figures 3 and 4, (and also Fig. 6) illustrate the difference in performance of macrofluids and microfluids in mixed flow reactors and they show clearly that a rise in segregation improves reactor performance for reaction orders greater than unity but lowers performance for reaction orders smaller than unity. Table 1 summarizes the relationships used in preparing these charts. Greenhalgh *et al.* (1959) give an alternate presentation of these charts.

Early and Late Mixing of Fluid

Each flow pattern of fluid through a vessel has associated with it a definite clearly defined residence time distribution (RTD), or exit age distribution function E. The converse is not true, however. Each RTD does not define a specific flow pattern,

TABLE I. Conversion equations for macrofluids and microfluids with $\varepsilon = 0$ in ideal reactors

	Plug Flow	Mixed Flow	
	Microfluid or Macrofluid	Microfluid	Macrofluid
General kinetics	$\tau = -\int_{C_0}^{C} \frac{dC}{-r}$	$\tau = \frac{C_0 - C}{-r}$	$\frac{\overline{C}}{C_0} = \frac{1}{\tau}\int_0^\infty \left(\frac{C}{C_0}\right)_{\text{batch}} e^{-t/\tau}\, dt$
nth-order reaction	$\frac{C}{C_0} = [1 + (n-1)R]^{1/(1-n)}$	$\left(\frac{C}{C_0}\right)^n R + \frac{C}{C_0} - 1 = 0$	$\frac{\overline{C}}{C_0} = \frac{1}{\tau}\int_0^\infty [1 + (n-1)C_0^{n-1}kt]^{1/(1-n)} e^{-t/\tau}\, dt$
$(R = C_0^{n-1}k\tau)$	$R = \frac{1}{n-1}\left[\left(\frac{C}{C_0}\right)^{1-n} - 1\right]$	$R = \left(1 - \frac{C}{C_0}\right)\left(\frac{C_0}{C}\right)^n$	
Zero-order reaction	$\frac{C}{C_0} = 1 - R, \quad R \leqslant 1$	$\frac{C}{C_0} = 1 - R, \quad R \leqslant 1$	$\frac{\overline{C}}{C_0} = 1 - R + Re^{-1/R}$
$\left(R = \frac{k\tau}{C_0}\right)$	$C = 0, \quad R \geqslant 1$	$C = 0, \quad R \geqslant 1$	
First-order reaction	$\frac{C}{C_0} = e^{-R}$	$\frac{C}{C_0} = \frac{1}{1+R}$	$\frac{\overline{C}}{C_0} = \frac{1}{1+R}$
$(R = k\tau)$	$R = \ln\frac{C_0}{C}$	$R = \frac{C_0}{C} - 1$	$R = \frac{C_0}{C} - 1$
Second-order reaction	$\frac{C}{C_0} = \frac{1}{1+R}$	$\frac{C}{C_0} = \frac{-1 + \sqrt{1+4R}}{2R}$	$\frac{\overline{C}}{C_0} = \frac{e^{1/R}}{R}\,\text{ei}\left(\frac{1}{R}\right)$
$(R = C_0 k\tau)$	$R = \frac{C_0}{C} - 1$	$R = \left(\frac{C_0}{C} - 1\right)\frac{C_0}{C}$	

$R = C_0^{n-1}k\tau$, reaction rate group for nth-order reaction, a time or capacity factor.
$\tau = \bar{t}$ since $\varepsilon = 0$ throughout.

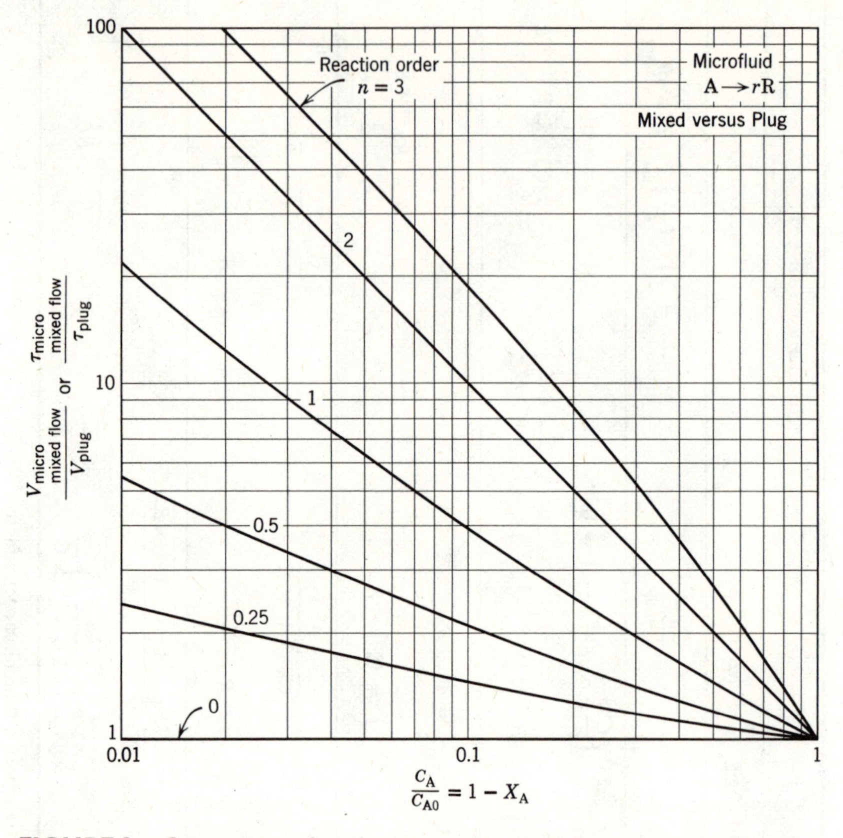

FIGURE 3. Comparison of performance of a mixed flow reactor with a plug flow reactor, both treating a microfluid with nth-order kinetics and $\varepsilon_A = 0$.

hence a number of flow patterns, some with earlier, others with later mixing of fluids may be able to give the same RTD.

Let us evaluate the range of behavior for these different RTD's.

Idealized Pulse RTD. Reflection shows that the only pattern of flow consistent with this RTD is one with no intermixing of fluid of different ages, hence that of plug flow. Consequently it is immaterial whether we have a micro or macrofluid. In addition the question of early or late mixing of fluid is of no concern since there is no mixing of fluid of different ages.

Exponential Decay RTD. The mixed flow reactor can give this RTD. However other flow patterns can also give this RTD; for example, a set of parallel plug flow reactors of proper length, a plug flow reactor with sidestreams, or a combination of these. Figure 5 shows a number of these patterns. Note that in patterns *a* and *b*

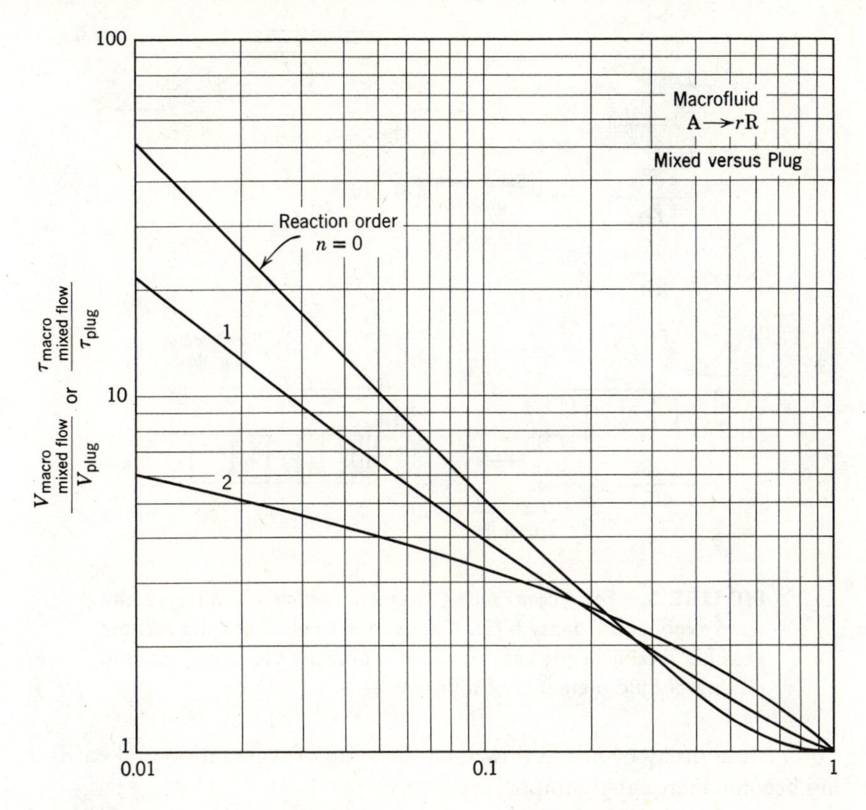

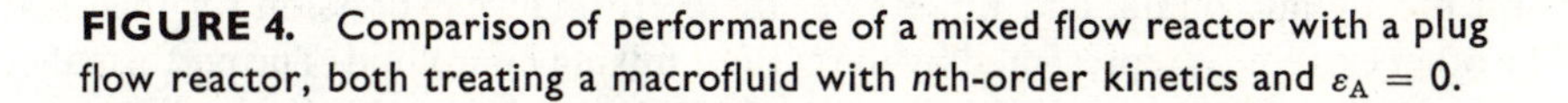

FIGURE 4. Comparison of performance of a mixed flow reactor with a plug flow reactor, both treating a macrofluid with *n*th-order kinetics and $\varepsilon_A = 0$.

entering fluid elements mix immediately with material of different ages, while in patterns *c* and *d* no such mixing occurs.

Now by definition of a macrofluid the mixing on a molecular level occurs late (not at all) for all these flow patterns. On the other hand, for microfluids the earliest possible mixing is obtained with patterns *a* and *b*, but the latest possible mixing is obtained with patterns *c* and *d*. Thus conversion for macrofluids (all cases) and for microfluids with the latest possible mixing will be identical and given by Eq. 5, while conversion for microfluids with earliest possible mixing will be given by the ordinary mixed reactor performance equation, Eq. 1 or 5.13.

Arbitrary RTD. In extending this argument we see that when the RTD is close to that of plug flow then the state of segregation of fluid as well as early or late mixing of fluid has little effect on conversion. However, when the RTD approaches

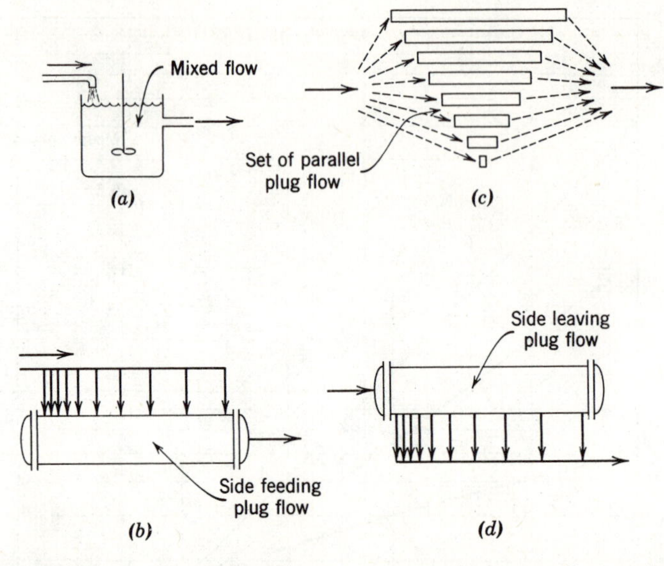

FIGURE 5. Four contracting patterns which can all give the same exponential decay RTD. Cases *a* and *b* represent the earliest possible mixing while cases *c* and *d* represent the latest possible mixing of fluid elements of different ages.

the exponential decay of mixed flow then the state of segregation and earliness of mixing become increasingly important.

For any RTD one extreme in behavior is represented by the macrofluid and the latest mixing microfluid, and Eq. 3 gives the performance expression for this case. The other extreme is represented by the earliest mixing microfluid. The performance expression for this case has been developed by Zwietering (1959) but is difficult to use. Also see Weinstein and Adler (1967) for a general treatment of these extreme cases in terms of the flow patterns of Fig. 5*b* and 5*d*.

Although these extremes give the upper and lower bound to the expected conversion for real vessels, it is usually simpler and preferable to develop a model to reasonably approximate the real vessel and then calculate conversions from this model. This is what actually was done in developing Fig. 9.23 for second-order reaction with axial dispersion.

Summary of Findings for a Single Fluid

1. *Factors affecting the performance of a reactor.* In general we may write

$$\text{Performance:}\quad X_A \text{ or } \varphi\left(\frac{R}{A}\right) = f\left(\begin{matrix}\text{kinetics, RTD, degree of segregation,}\\ \text{earliness of mixing}\end{matrix}\right) \tag{12}$$

2. *Effect of kinetics, or reaction order.* Segregation and earliness of mixing affect the conversion of reactant as follows

$$\text{For } n > 1 \cdots X_{\text{macro}} \text{ and } X_{\text{micro,late}} > X_{\text{micro,early}}$$

For $n < 1$ the inequality is reversed, and for $n = 1$ conversion is unaffected by these factors. This result shows that segregation and late mixing improves conversion for $n > 1$, and decreases conversion for $n < 1$.

3. *Effect of mixing factors for nonfirst-order reactions.* Segregation plays no role in plug flow; however, it increasingly affects the reactor performance as the RTD shifts from plug to mixed flow.

Earliness of mixing plays no role for macrofluids in any flow, and for plug flow of any fluid; however, it increasingly affects performance as the RTD approaches mixed flow and as microfluid behavior is approached.

4. *Effect of conversion level.* At low conversion levels X is insensitive to RTD, earliness of mixing, and segregation. At intermediate conversion levels, the RTD begins to influence X, however earliness and segregation still have little effect. This is the case of Example 1. Finally at high conversion levels all these factors may play important roles.

5. *Effect on product distribution.* Although segregation and earliness of mixing can usually be ignored when treating single reactions, this often is not so with multiple reactions where the effect of these factors on product distribution can be of dominating importance, even at low conversion levels.

As an example consider free-radical polymerization. When an occasional free radical is formed here and there in the reactor it triggers an extremely rapid chain of reactions, often thousands of steps in a fraction of a second. The local reaction rate and conversion can thus be very high. In this situation the immediate surroundings of the reacting and growing molecules, hence the state of segregation of the fluid, can greatly affect the type of polymer formed.

EXAMPLE 1. Effect of segregation and earliness of mixing on conversion

A second-order reaction occurs in a reactor whose RTD is given in Fig. E1. Calculate the conversion for the flow schemes shown in this figure. For simplicity take $C_0 = 1$ $k = 1$, and $\tau = 1$ for each unit.

SOLUTION

Scheme a. Referring to Fig. E1*a*, we have for the mixed reactor

$$\tau = 1 = \frac{C_0 - C_1}{kC^2} = \frac{1 - C_1}{C_1^2}$$

or

$$C_1 = \frac{-1 + \sqrt{1 + 4}}{2} = 0.618$$

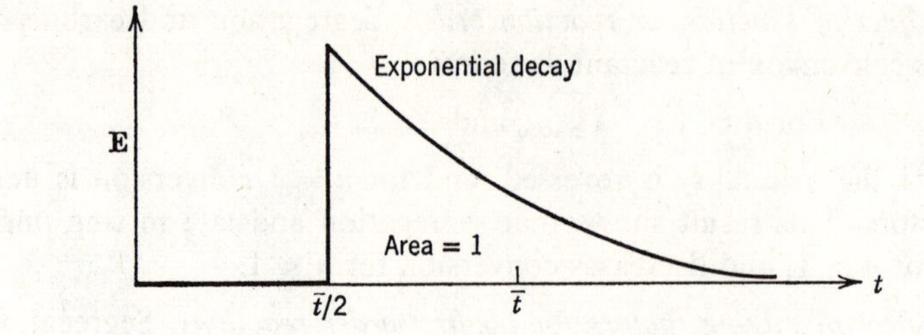

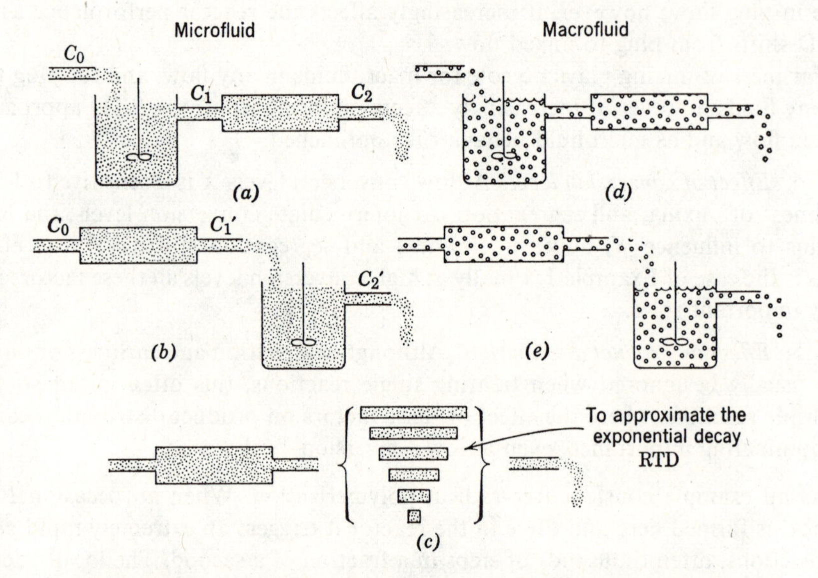

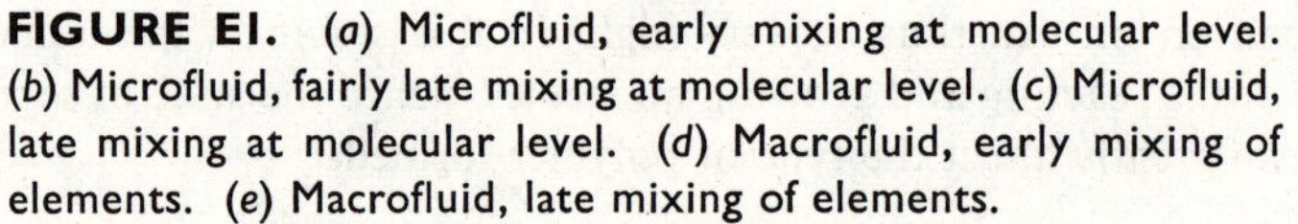

FIGURE E1. (*a*) Microfluid, early mixing at molecular level. (*b*) Microfluid, fairly late mixing at molecular level. (*c*) Microfluid, late mixing at molecular level. (*d*) Macrofluid, early mixing of elements. (*e*) Macrofluid, late mixing of elements.

For the plug flow reactor

$$\tau = 1 = -\int_{C_1}^{C_2} \frac{dC}{kC^2} = \frac{1}{k}\left(\frac{1}{C_2} - \frac{1}{C_1}\right)$$

or

$$\text{Micro-early:} \quad C_2 = \frac{C_1}{C_1 + 1} = \frac{0.618}{1.618} = 0.382$$

Scheme b. Referring to Fig. E1b, we have for the plug flow reactor

$$\tau = 1 = -\int_{C_0}^{C_1'} \frac{dC}{kC^2} = \frac{1}{C_1'} - 1$$

or

$$C_1' = 0.5$$

For the mixed reactor

$$\tau = 1 = \frac{C_1' - C_2'}{kC_2'^2} = \frac{0.5 - C_2'}{C_2'^2}$$

or

$$\text{Micro-fairly late:} \quad C_2' = 0.366$$

Schemes c, d, and e. From Fig. 9.29 the exit age distribution function for the two equal-size reactor system is

$$\begin{aligned} \mathbf{E}_\theta &= 2e^{1-2\theta} \qquad &\text{when } \theta > \tfrac{1}{2} \\ &= 0, &\text{when } \theta < \tfrac{1}{2} \end{aligned}$$

or

$$\begin{aligned} \mathbf{E} &= \frac{2}{\bar{t}}\, e^{1-2t/\bar{t}}, \qquad &\text{when } \frac{t}{\bar{t}} > \tfrac{1}{2} \\ &= 0, &\text{when } \frac{t}{\bar{t}} < \tfrac{1}{2} \end{aligned}$$

Thus Eq. 3 becomes

$$C = \int_{\bar{t}/2}^{\infty} \frac{1}{1 + C_0 kt} \cdot \frac{2}{\bar{t}}\, e^{1-2t/\bar{t}}\, dt$$

With the mean residence time in the two-vessel system $\bar{t} = 2$ min, this becomes

$$C = \int_1^{\infty} \frac{e^{1-t}}{1+t}\, dt$$

and replacing $1 + t$ by x we obtain the exponential integral

$$C = \int_2^{\infty} \frac{e^{2-x}}{x}\, dx = e^2 \int_2^{\infty} \frac{e^{-x}}{x}\, dx = e^2 \mathrm{ei}(2)$$

From a table of integrals we find $\mathrm{ei}(2) = 0.04890$ from which

$$\text{Micro-late, and macro-late or early:} \quad C = 0.362$$

The results of this example confirm the statements made above, that macrofluids and late mixing microfluids give higher conversions than early mixing microfluids for reaction orders greater than unity. The difference is small here because the conversion levels are low; however, this difference becomes more important as conversion approaches unity.

Models for Partial Segregation

Numerous ways have been suggested for treating intermediate amounts of segregation. Briefly some of these are as follows:

Intensity of Segregation. Danckwerts (1958) pictured that in partial segregation molecules of different ages and compositions intermixed and diffused to some extent from aggregate to aggregate. He assumed however that the fluctuations in composition caused by these intruders within an aggregate quickly evened out so that a probe would only measure the mean age or composition of the aggregate. On this basis he defined the intensity of segregation of fluid in a vessel as

$$J = \frac{\text{variation in average ages of aggregates}}{\text{variation in ages of the individual molecules}} = \frac{\sigma^2_{\text{among aggregates}}}{\sigma^2_{\text{molecules}}} = \frac{\sigma^2_{\text{among aggregates}}}{\int_0^\infty (t - \bar{t})^2 \mathbf{E}\, dt} \tag{13}$$

With this definition the upper limit for J is unity for a macrofluid and any RTD. The lower limit is zero for a microfluid in mixed flow. For any other RTD the lower limit for a microfluid lies between zero and one depending on the RTD. Note that with this definition J is not an independent measure of the degree of segregation since it accounts for the RTD as well. Also earliness of mixing must be known before conversions can be determined with this definition.

Instead of ages we could measure compositions of components and thereby extend this measure to the degree of segregation of separate fluid streams. For further readings on this whole approach to partial segregation see Danckwerts (1957, 1958), Zwietering (1959), and Rippin (1967). Unfortunately it has been hard to apply these ideas to real systems.

With a simple and artificial example problem P10 illustrates this definition.

Coalescence Model. Curl (1963) proposed a different measure of partial segregation based on the model of dispersed phase droplets flowing through an ideal stirred tank. Suppose a reaction takes place in these droplets. Then if droplets do not coalesce they will all have different concentrations and this phase will act as a macrofluid. On the other hand if the droplets do coalesce and disperse their compositions will become increasingly alike and this phase will approach microfluid behavior. The coalescence parameter of this model is thus

$$I = \begin{pmatrix} \text{average number of coalescences} \\ \text{experienced by a drop passing} \\ \text{through the vessel} \end{pmatrix} \tag{14}$$

and with this definition

$$I \to 0 \quad \text{represents a macrofluid}$$

$$I \to \infty \quad \text{represents a microfluid}$$

Spielman and Levenspiel (1965) showed how the conversion could easily be obtained from this model by using a modified Monte Carlo procedure. Their curves for partial segregation for zero- and second-order kinetics are shown in Fig. 6.

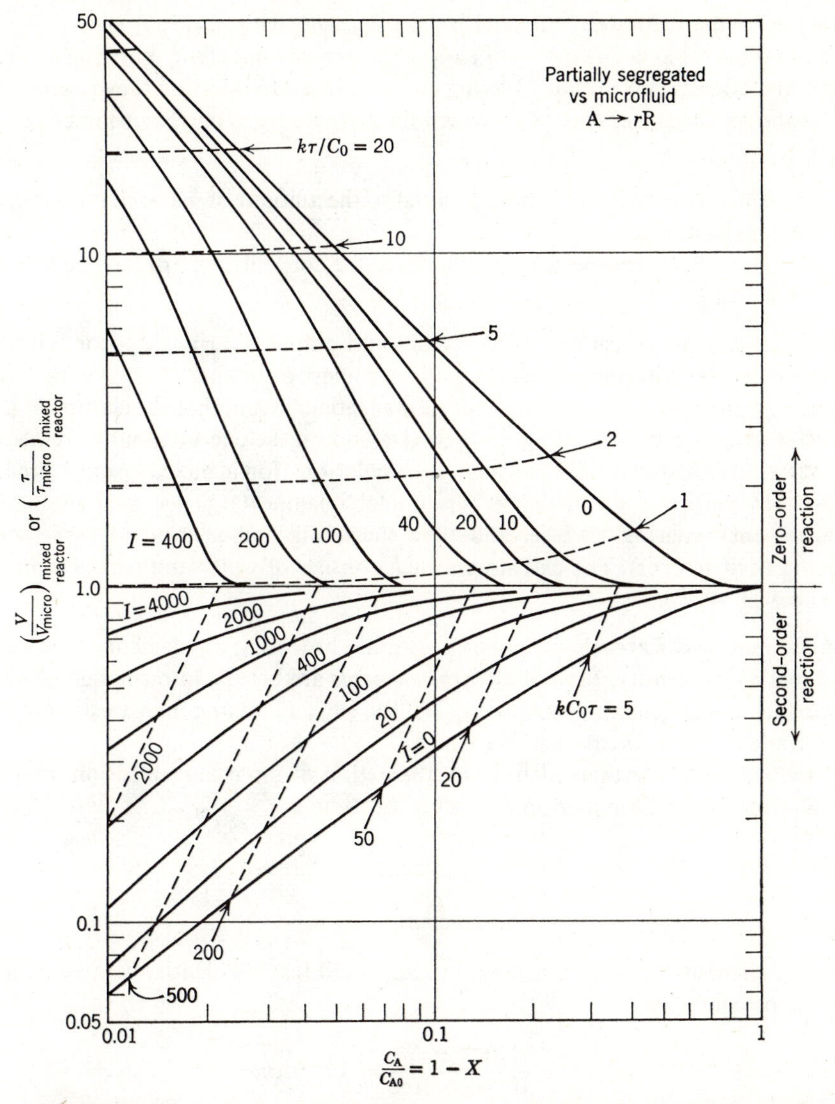

FIGURE 6. Comparison of performance of mixed reactors treating coalescing (partially segregated) fluids and microfluids; for zero- and second-order reactions and $\varepsilon_A = 0$.

They also emphasize the versatility of this procedure, that it can be used for any kinetics, for single and multiple reactions, for multiple feed streams, and for any RTD. Recently Rao and Dunn (1970) used the Monte Carlo approach for reaction of single and separate side by side feed streams in dispersed plug flow.

Two-environment Models. In their model Ng and Rippin (1965) view a reactor with arbitrary RTD as consisting of clumps of unmixed inlet fluid (entering environment) and already mixed fluid (leaving environment). In addition they propose the following possible mechanisms by which fluid moves from one environment to the other.

(*a*) The rate of transfer is proportional to the amount of material remaining in the inlet environment.

(*b*) The rate of transfer is proportional to the concentration difference between environments.

These mechanisms reflect in turn turbulent and diffusional transfer. For arbitrary RTD this model still does not account for earliness of mixing.

Suzuki (1970) proposed a somewhat similar model except that the clumps of inlet fluid shrink at a rate depending on their size and on the energy input provided by the stirrer. With this model conversion calculations for a mixed reactor are not difficult to perform. To visualize this model imagine large ice cubes (entering environment) entering a vessel and then shattering and melting to form water (leaving environment). The exit stream then consists of water and partially melted ice cubes of all sizes.

Applications and Extensions. Let us estimate how long an element of ordinary fluid retains its identity. First of all *large elements* are broken into smaller elements by the turbulence generated by stirrers, baffles, etc. and mixing theory estimates the time needed for this breakup.

Small elements lose their identity by the action of molecular diffusion, and the Einstein random walk equation estimates this time as

$$t = \frac{(\text{size of element})^2}{\begin{pmatrix}\text{diffusion}\\ \text{coefficient}\end{pmatrix}} = \frac{d^2_{\text{element}}}{\mathscr{D}}$$

Thus an element of water 1 micron in size would lose its identity in a very short time, approximately

$$t = \frac{(10^{-4}\ \text{cm})^2}{10^{-5}\ \text{cm}^2/\text{sec}} = 10^{-3}\ \text{sec}$$

while an element of viscous polymer 100 microns in size and 1000 times as viscous as water would retain its identity for a long time, roughly

$$t = \frac{(10^{-2}\ \text{cm})^2}{10^{-8}\ \text{cm}^2/\text{sec}} = 10^4\ \text{sec} \approx 3\ \text{hrs}$$

In general then ordinary fluids behave as microfluids except for very viscous materials and for systems in which very fast reactions are taking place.

The concept of micro- and macrofluid is of particular importance in heterogeneous systems because one of the two phases of such systems usually approximates a macrofluid. For example, the solid phase of fluid-solid systems can be treated exactly as a macrofluid because each particle of solid is a distinct aggregate of molecules. For such systems, then, Eq. 3 with the appropriate kinetic expression is the starting point for design. As another example, the dispersed phase of a liquid-liquid system is partially segregated. This segregation rises as drop coalescence is depressed. The coalescence model of Curl's for partial segregation is directly applicable here.

In the chapters to follow we apply these concepts of micro- and macrofluids to heterogeneous systems of various kinds.

MIXING OF TWO MISCIBLE FLUIDS

Here we consider one topic, the role of the mixing process when two completely miscible reactant fluids A and B are brought together. The mixing of two fluids, when these are immiscible, is a proper subject for study in the chapters on heterogeneous reactions; therefore discussion of such systems will be left for the following chapters.

When two miscible fluids A and B are mixed, we normally assume that they first form a homogeneous mixture which then reacts. However, when the time required for A and B to become homogeneous is not short with respect to the time for reaction to take place, reaction occurs during the mixing process, and the problem of mixing becomes important. Such is the case for very fast reactions or with very viscous reactant fluids.

To help understand what occurs imagine that we have A and B available, each first as a microfluid, and then as a macrofluid. In one beaker mix micro A with

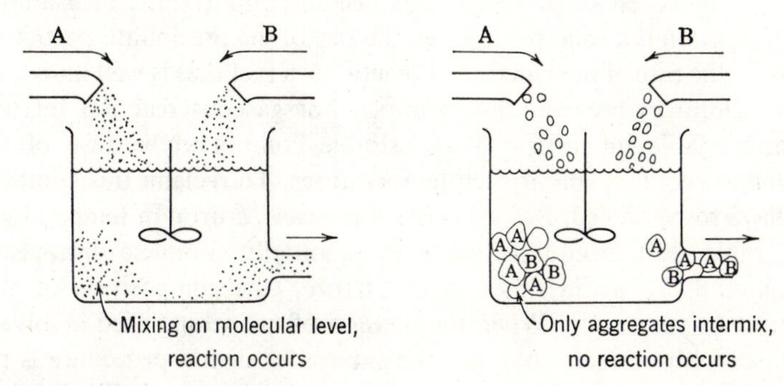

FIGURE 7. Difference in behavior of microfluids and macrofluids in the reaction of A and B.

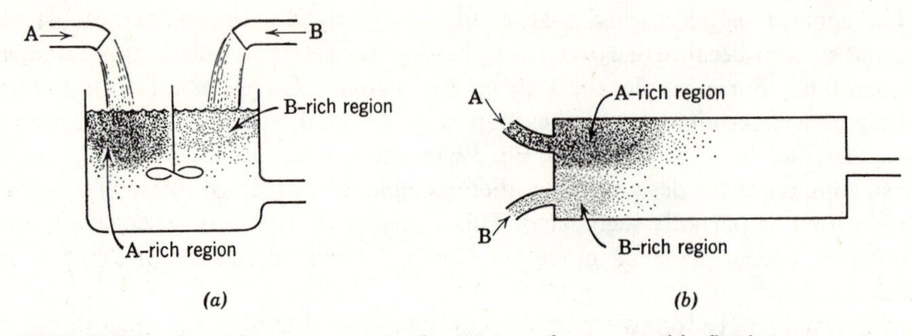

FIGURE 8. Partial segregation in the mixing of two miscible fluids in a reactor.

micro B and in another beaker mix macro A with macro B and let them react. What do we find? Micro A and B behave in the expected manner, and reaction occurs. However, on mixing the macrofluids no reaction takes place because molecules of A cannot contact molecules of B. These two situations are illustrated in Fig. 7. So much for the treatment of the two extremes in behavior.

Now a real system acts as shown in Fig. 8 with regions of A-rich fluid and regions of B-rich fluid.

Though partial segregation requires an increase in reactor size, this is not the only consequence. For example, when reactants are viscous fluids, their mixing in a stirred tank or batch reactor often places layers or "streaks" of one fluid next to the other. As a result reaction occurs at different rates from point to point in the reactor giving a nonuniform product which may be commercially unacceptable. Such is the case in polymerization reactions in which monomer must be intimately mixed with a catalyst. For reactions such as this, proper mixing is of primary importance and often the rate of reaction and product uniformity correlate well with the mixing energy input to the fluid.

For fast reactions the increase in reactor size needed because of segregation is unimportant; however, other side effects become important. For example, if the product of reaction is a solid precipitate, the size of the precipitate particles may be influenced by the rate of intermixing of reactants, a fact that is well known from the analytical laboratory. As another example, hot gaseous reaction mixtures may contain appreciable quantities of a desirable compound because of favorable thermodynamic equilibrium at such temperatures. To reclaim this component the gas may have to be cooled. But, as is often the case, a drop in temperature causes an unfavorable shift in equilibrium with essentially complete disappearance of desired material. To avoid this and to "freeze" the composition of hot gases, cooling must be very rapid. When the method of quenching used involves mixing the hot gases with an inert cold gas, the success of such a procedure is primarily dependent on the rate at which segregation can be destroyed. Finally the length, type, and temperature of a burning flame, the combustion products obtained, the noise levels of jet engines, and the physical properties of polymers as they are affected by the molecular weight distribution of the material are some of the many phenomena or end results of phenomena that are closely influenced by the rate and intimacy of fluid mixing.

Product Distribution in Multiple Reactions

When multiple reactions take place on mixing two reactant fluids and when these reactions proceed to an appreciable extent before homogeneity is attained, segregation is important and can affect product distribution.

Consider the homogeneous-phase competitive consecutive reactions

$$\begin{aligned} A + B &\xrightarrow{k_1} R \\ R + B &\xrightarrow{k_2} S \end{aligned} \tag{15}$$

occurring when A and B are poured into a batch reactor. If the reactions are slow enough so that the contents of the vessel are uniform before reaction takes place, the maximum amount of R formed is governed by the k_2/k_1 ratio. This situation, treated in Chapter 7, is one in which we may assume microfluid behavior. If, however, the fluids are very viscous or if the reactions are fast enough, they will occur in the narrow zones between regions of high A concentration and high B concentration. This is shown in Fig. 9. The zone of high reaction rate will contain a

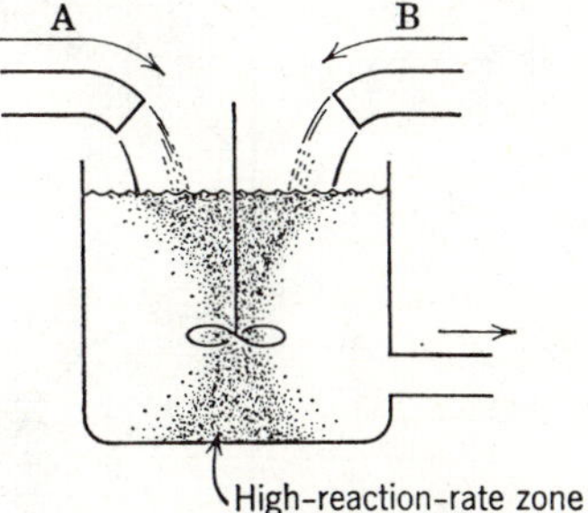

FIGURE 9. When reaction rate is very high, zones of nonhomogeneity exist in a reactor. This condition is detrimental to obtaining high yields of intermediate R from the reactions

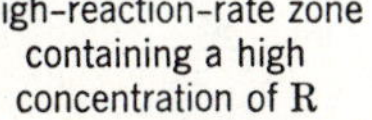

$$\begin{aligned} A + B &\rightarrow R \\ R + B &\rightarrow S \end{aligned}$$

higher concentration of R than the surrounding fluid. But from the qualitative treatment of this reaction in Chapter 7 we know that any nonhomogeneity in A and R will depress formation of R. Thus partial segregation of reactants will depress the formation of intermediate.

For increased reaction rate, the zone of reaction narrows, and in the limit, for an infinitely fast reaction, becomes a boundary surface between the A-rich and B-rich regions. Now R will only be formed at this plane. What will happen to it? Consider a single molecule of R formed at the reaction plane. If it starts its random wanderings (diffusion) into the A zone and never moves back into the B zone, it will not react further. However, if it starts off into the B zone or if at any time during its wanderings it moves through the reaction plane into the B zone, it will be attacked by B to form S. Interestingly enough, from probabilities associated with a betting

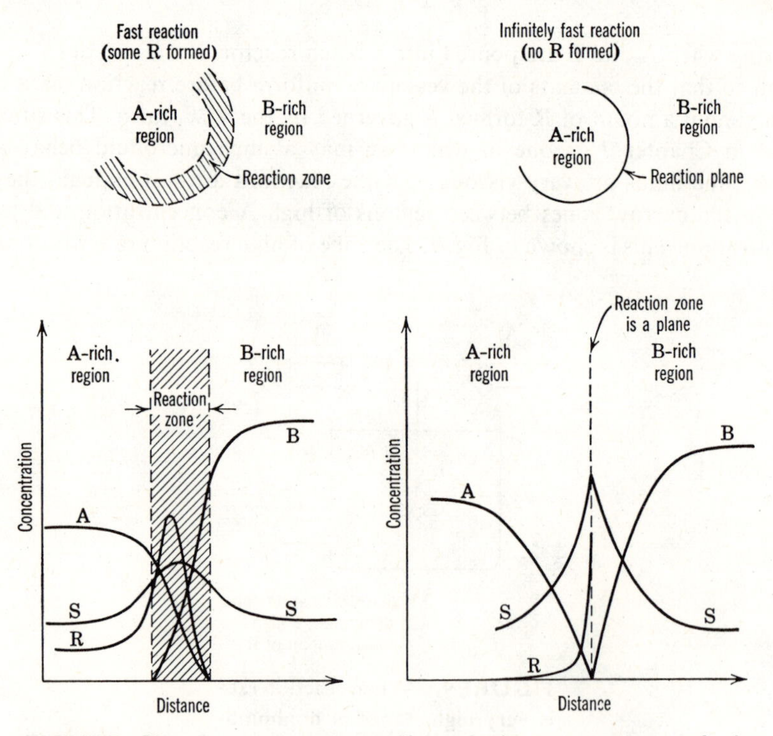

FIGURE 10. Concentration profiles of the components of the reactions

$$A + B \rightarrow R$$
$$R + B \rightarrow S$$

at a representative spot in the reactor between A-rich and B-rich fluid for a very fast and for an infinitely fast reaction.

game treated by Feller (1957), we can show that the odds in favor of a molecule of R escaping from the B zone become smaller and smaller as the number of diffusion steps taken by a molecule gets larger and larger. This finding holds, no matter what pattern of wanderings is chosen for the molecules of R. Thus we conclude that no R is formed. Looked at from the point of view of Chapter 7, an infinitely fast reaction gives a maximum nonhomogeneity of A and R in the mixture, resulting in no R being formed. Figure 10 shows the concentration of materials at a typical reaction interface and illustrates these points.

This behavior of multiple reaction could provide a powerful tool in the study of partial segregation in homogeneous systems. It has been used recently by Paul and Treybal (1971) who simply poured reactant B into a beaker of A and measured the amount of R formed for a very fast reaction of Eq. 15.

These observations on the extremes in behavior serve as a guide to the selection and design of equipment favoring the formation of intermediate when reaction is very fast. The important point is to achieve homogeneity in A and R throughout the reaction mixture before reaction has proceeded to any significant extent. This is done

(*a*) By making the reaction zone as large as possible by vigorous mixing.
(*b*) By dispersing B in A in as fine a form as possible, rather than A in B.
(*c*) By slowing the reaction.

The intermixing and reaction of separate streams, such as shown in Fig. 8*b* is treated by Rao and Dunn (1970), and Rao and Edwards (1971), and the related problem of separate feeds, each with its particular RTD, is treated by Treleaven and Tobgy (1971).

REFERENCES

Curl, R. L., *A.I.Ch.E. Journal*, **9**, 175 (1963).
Danckwerts, P. V., *Chem. Eng. Sci.*, **7**, 116 (1957).
———, *Chem. Eng. Sci.*, **8**, 93 (1958).
Feller, W., *An Introduction to Probability Theory and its Applications*, Vol. I, 2nd ed., John Wiley & Sons, New York, 1957, p. 254.
Greenhalgh, R. E., Johnson, R. L., and Nott, H. D., *Chem. Eng. Progr.*, **55**, No. 2, 44 (1959).
Ng, D. Y. C., and Rippin, D. W. T., *Third Symposium on Chemical Reaction Engineering*, Pergamon Press, Oxford, 1965.
Paul, E. L., and Treybal, R. E., *A.I.Ch.E. Journal*, **17**, 718 (1971).
Rao, D. P., and Dunn, I. J., *Chem. Eng. Sci.*, **25**, 1275 (1970).
———and Edwards, L. L., *Ind. Eng. Chem. Fundamentals*, **10**, 389 (1971).
Rietema, K., *Chem. Eng. Sci.*, **8**, 103 (1958).

Rippin, D. W. T., *Chem. Eng. Sci.*, **22**, 247 (1967).
Spielman, L. A., and Levenspiel, O., *Chem. Eng. Sci.*, **20**, 247 (1965).
Suzuki, M., personal communication, 1970.
Treleaven, C. R. and Tobgy, A. H., *Chem. Eng. Sci.*, **26,** 1259 (1971).
Weinstein, H., and Adler, R. J., *Chem. Eng. Sci.*, **22**, 65 (1967).
Zweitering, Th. N., *Chem. Eng. Sci.*, **11**, 1 (1959).

PROBLEMS

1. For a first-order reaction with $\varepsilon = 0$ show that the performance equations for a macrofluid in plug flow (Eq. 3) and a microfluid in plug flow (Eq. 5.21) are identical.

2. Devise a sketch to illustrate the difference in performance of second-order reactions as a function of RTD, degree of segregation, and earliness of mixing.

3. Derive the expression given in Table 1 for the zero-order reaction of a macrofluid in a mixed flow reactor. As first shown by Rietema (1958) zero-order reactions may approximate the conversion behavior of high concentration droplets reacting with a continuous phase under conditions where mass transfer between phases controls.

4. Find the expression for conversion of a macrofluid in two equal-size mixed reactors for a zero-order reaction. If conversion is 99% for the microfluid, what is it for a macrofluid having the same reaction rate?

5. Repeat the previous problem for a second-order reaction.

6. Dispersed noncoalescing droplets containing reactant A pass through 3 ideal stirred tanks in series. The mean holding time in each tank is 1.5 hr and the rate constant for the first-order decay reaction is 0.1 min. Find the fractional conversion of A in the exit stream from the three reactors.

7. For an infinitely fast reaction

$$A + B \xrightarrow{k_1} R$$

$$R + B \xrightarrow{k_2} S$$

select any model for the diffusion of R from the reaction surface between A-rich and B-rich regions and show what fraction of a given batch of molecules of R formed at the interface still remain unreacted after 5, 10, and 15 diffusion steps. Plot this curve. Does the result seem to confirm the statement, made in this chapter, that in the limit, for an infinitely fast reaction, no intermediate R is obtained.

8. In the presence of a catalyst, reactant A decomposes with first-order kinetics as follows:

$$A \xrightarrow[\text{catalyst}]{k_1} R \xrightarrow[\text{catalyst}]{k_2} S$$

Reactant A and catalyst are introduced separately into a mixed reactor; A forms the dispersed phase while the catalyst is introduced in the continuous phase.

(*a*) Assuming a uniform composition within each droplet of dispersed phase, no movement of A or R into the continuous phase, and identical concentration of catalyst within all the droplets, determine how the reaction proceeds and determine the expected $C_{R,max}$.

(*b*) How does this compare with the setup in which reactant A forms the continuous phase while catalyst forms the dispersed phase.

9. Repeat the previous problem with the following change. The catalyst is replaced by reactant B and the reaction is

$$A + B \xrightarrow{k_1} R$$

$$R + B \xrightarrow{k_2} S$$

with elementary second-order kinetics.

10. Figure P10 represents three different groupings of material in a continuous flow reactor containing just 16 molecules.

(*a*) Determine the intensity of segregation J for each of these cases.

(*b*) What do these three cases represent in terms of the extent of segregation of fluid?

11. Substrate ($C_{A0} = 2$ mol/liter) and enzyme ($C_{E0} = 0.001$ mol/liter) are introduced into a batch reactor. They react, conversion is 90% in 4.1 minutes, and the rate equation which represents this behavior is found to be

$$-r_A = 10^3 \frac{C_{E0}C_A}{1 + C_A}, \quad \frac{\text{mol}}{\text{liter}\cdot\text{min}}$$

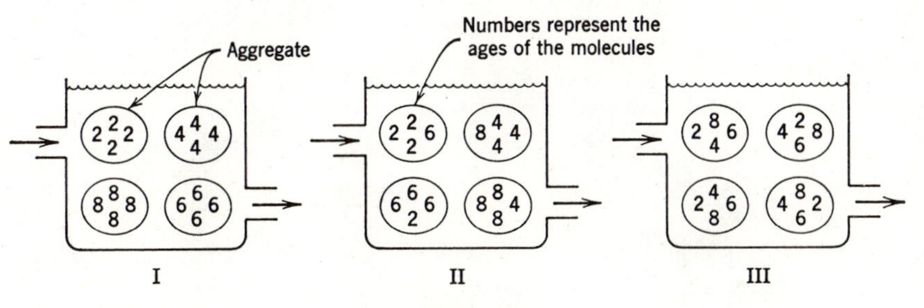

FIGURE P10.

We plan to build and operate a mixed flow reactor using a continuous feed of substrate (C_{A0} = 2 mol/liter) with enzyme (C_{E0} = 0.001 mol/liter).

(*a*) What should be the space-time τ to achieve 90% conversion of substrate? The reactor has been ordered for this τ value, and is presently being assembled. In the meantime, we have been informed of the following disturbing fact. The enzyme is not dispersed on a molecular scale in the solution, as we had assumed, but is present in finely dispersed but distinct flocs or clumps. So, although our calculations were based on microfluid behavior, the enzyme in fact is a macrofluid.

(*b*) On the basis of this new information, what conversion can we expect in the reactor system presently being built?

11

INTRODUCTION TO DESIGN FOR HETEROGENEOUS REACTING SYSTEMS

These final chapters treat the kinetics and design of chemical reactors for heterogeneous systems of various kinds, each chapter considering a different system (see Chapter 1 for discussions of heterogeneous and homogeneous systems). For these systems there are two complicating factors that must be accounted for beyond what is normally considered in homogeneous systems.

1. *The complications of the rate equation.* Since more than one phase is present, the movement of material from phase to phase must be considered in the rate equation. Thus the rate expression in general will incorporate mass transfer terms in addition to the usual chemical kinetics term. These mass transfer terms are different in type and numbers in the different kinds of heterogeneous systems; hence no single rate expression has general application.

2. *The contacting patterns for two-phase systems.* In homogeneous systems we considered two ideal flow patterns of the reacting fluid, plug and mixed flow. In ideal contacting of heterogeneous systems, each fluid may be in plug or mixed flow. Thus many combinations of contacting patterns are possible. On top of this, if one of the phases is discontinuous, as are droplets or solid particles, its macrofluid characteristics will have to be accounted for. Thus each of the many methods of contacting of two phases has associated with it a specific form of performance equation which must be developed for that particular contacting pattern.

This chapter considers some of the aspects of these two factors, rate and contacting pattern, as they affect the treatment of heterogeneous systems of all types.

RATE EQUATION FOR HETEROGENEOUS REACTIONS

In general the rate equation for a heterogeneous reaction accounts for more than one process. This leads us to ask how such processes involving both physical transport and reaction steps can be incorporated into one overall rate expression. The problem of combining rates for different processes is met in conductive heat transfer through layers of different materials, in convective mass transfer from one liquid to another through stagnant boundary films, and also in complex reactions. In all these situations, however, the over-all rate combines processes of the same kind. Let us consider the general problem of combining rates for processes of different kinds.

Let $r_1, r_2, \ldots, r_n$ be the rates of change for the individual processes that are to be accounted for by an overall rate. If the changes take place by parallel paths then the overall rate will be greater than the rate for any of the individual paths. In fact, if the various parallel paths are independent of each other, the overall rate will be simply the sum of all the individual rates, or

$$r_{\text{overall}} = \sum_{i=1}^{n} r_i$$

On the other hand, if the over-all change requires that a number of steps take place in succession, then at steady state all these steps will proceed at the same rate. Thus

$$r_{\text{overall}} = r_1 = r_2 = \cdots = r_n$$

In certain heterogeneous systems, such as fluid–solid noncatalytic reactions, the resistances to reaction can be considered to occur in series. In other systems, such as solid catalyzed reactions, more involved series-parallel relationships exist.

Two points should be mentioned here. First, when rates are to be compared or combined, they should be defined in the same manner. As an example, suppose we wish to combine a mass transfer and a reaction step. Since the rate of mass transfer is defined as the flow of material normal to a unit surface, or

$$Q_{\text{transfer}} = \frac{1}{S}\frac{dN_{\text{A}}}{dt}$$

then the reaction step must similarly be defined, or

$$Q_{\text{reaction}} = r''_{\text{A}} = \frac{1}{S}\frac{dN_{\text{A}}}{dt}$$

Thus the rate of reaction should be based on unit area rather than on unit volume, as was used for homogeneous reactions.

Second, in combining rates we normally do not know the concentration of materials at intermediate positions. Thus we must express the rate in terms of the over-all concentration difference. This is easily done if the rate expressions for all the steps of the process are linear in concentration (first power in concentrations or concentration differences). If the functional relationships are not all linear then the resultant rate expression becomes unwieldy. The following examples illustrate these points, and the comments which follow them suggest ways of bypassing these cumbersome expressions.

EXAMPLE 1. Combining linear rate expressions

The irreversible reaction

$$A(\text{gas}) + B(\text{solid}) \rightarrow R(\text{gas})$$

takes place as shown in Fig. E1. Dilute A diffuses through a stagnant film onto a plane surface consisting of B. There A and B react to yield gaseous product R which diffuses back through the film into the main gas stream. By diffusion the flux of A to the surface is given by

$$Q_g = \frac{1}{S}\frac{dN_A}{dt} = -\mathscr{D}\frac{\Delta C}{\Delta x} = -\frac{\mathscr{D}}{\Delta x}(C_g - C_s) = -k_g(C_g - C_s) \qquad \text{(i)}$$

The reaction is first order with respect to A, and based on unit surface is given by

$$Q_s = \frac{1}{S}\frac{dN_A}{dt} = -k_s C_s \qquad \text{(ii)}$$

where k_s is the reaction rate constant based on unit surface.

At steady state write the overall rate of reaction in terms of k_g, k_s, and the concentration of A in the main gas stream C_g, noting that the concentration of A on the surface C_s cannot be measured and should not appear in the final expression.

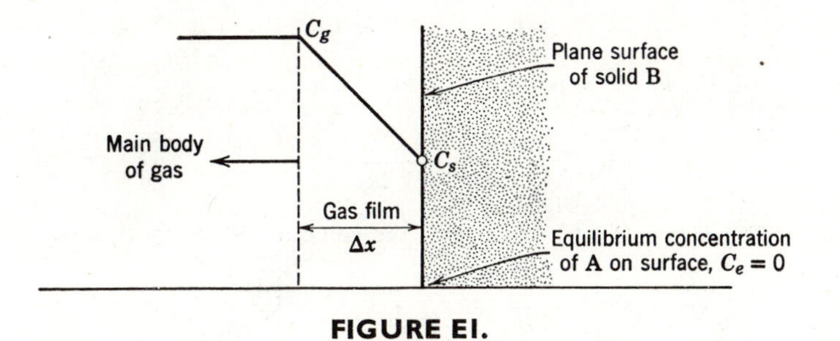

FIGURE E1.

SOLUTION

At steady state the flow rate to the surface is equal to the reaction rate at the surface (processes in series). Thus

$$Q_g = Q_s$$

or from Eqs. (i) and (ii)

$$k_g(C_g - C_s) = k_s C_s$$

Therefore

$$C_s = \frac{k_g}{k_g + k_s} C_g \tag{iii}$$

Replacing Eq. (iii) in either Eq. (i) or Eq. (ii) eliminates the surface concentration which cannot be measured. Thus

$$Q_g = Q_s = \frac{1}{S}\frac{dN_A}{dt} = -\frac{1}{1/k_g + 1/k_s} C_g = -k_{\text{overall}} C_g \tag{iv}$$

Comment. This result shows that $1/k_g$ and $1/k_s$ are additive resistances. It so happens that the addition of resistances to obtain on overall resistance is permissible only when the rate is a linear function of the driving force and when the processes occur in series.

EXAMPLE 2. Combining nonlinear rate expressions

Repeat Example 1 for the case where the reaction step is second order with respect to gas phase reactant A.

SOLUTION

For the mass transfer step:

$$Q_g = k_g(C_g - C_s) \tag{i}$$

For the reaction step:

$$Q_s = k_s C_s^2 \tag{v}$$

At steady state

$$Q_g = Q_s$$

or

$$k_g(C_g - C_s) = k_s C_s^2$$

which is a quadratic in C_s. Solving for C_s we obtain

$$C_s = \frac{-k_g + \sqrt{k_g^2 + 4k_s k_g C_g}}{2k_s} \tag{vi}$$

Eliminating C_s in either Eq. (i) or (ii) with the value found in Eq. (vi), we find

$$Q_g = Q_s = -\frac{k_g}{2k_s}\left(2k_s C_g + k_g - \sqrt{k_g^2 + 4k_s k_g C_g}\right) \tag{vii}$$

Comment. Comparing Eqs. (iv) and (vii), we see clearly that when the individual rate expressions are not all linear then the individual resistances are not additive, neither do they combine in a simple way. To bypass cumbersome expressions such as Eq. (vii) we use one of the following simplifying procedures.

The Concept of the Rate-controlling Step. Since chemical reaction rates vary so widely from reaction to reaction, and also with temperature, very often we find that one or other of the two steps provides the major resistance to the over-all change. In such a case we say that this slow step is rate-controlling and can be considered alone. To illustrate, in this example if the chemical step is rate-controlling then the over-all rate is given by

$$Q_g = Q_s = k_s C_s^2$$

Similarly, if the mass transfer step is rate-controlling then

$$Q_g = Q_s = k_g C_g$$

This procedure leads to the simplest of expressions and therefore is attempted whenever possible. Much of our treatment in the chapters to follow uses this simplification.

Linearization of a Nonlinear Rate. An alternative procedure replaces the nonlinear rate-concentration curve by a linear approximation, and then combines this with the other rate terms, as in Example 1. The way this is done is to expand the nonlinear rate in a Taylor series and then only retain the linear terms (see Problem P3).

Linearization is not as simple or desirable as taking the rate-controlling step, but at times it has to be used.

CONTACTING PATTERNS FOR TWO-PHASE SYSTEMS

There are many ways that two phases can be contacted, and for each the design equation will be unique. If the rate expression is also particular to that heterogeneous system, its peculiarities will be incorporated into the design equation. Thus we may say that the performance equation is tailored to fit the reaction rate and contacting pattern.

With ideal flow of both phases, we have eight principal ways to contact the phases. Figure 1 shows these. Note that no distinction need be made between macrofluids and microfluids when the phase is in plug flow; however, this may have to be done when the material is in mixed flow.

In fluid–solid systems flow patterns (*g*) and (*d*) are the more important, since these in some cases represent fluidized beds and continuous-belt processing of solids. In fluid–fluid systems flow patterns (*g*), (*a*), and (*b*) are of primary interest,

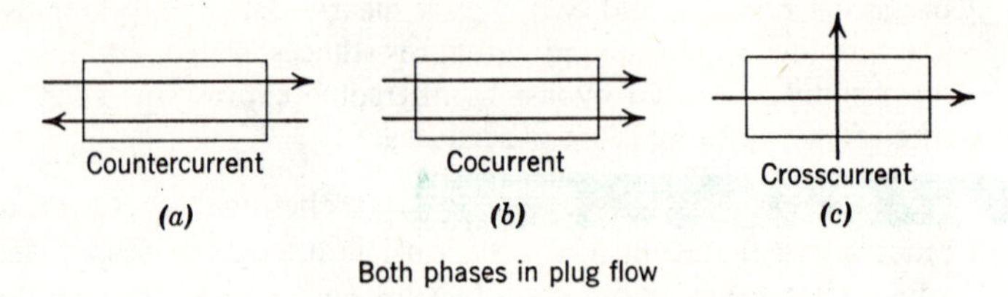

Both phases in plug flow

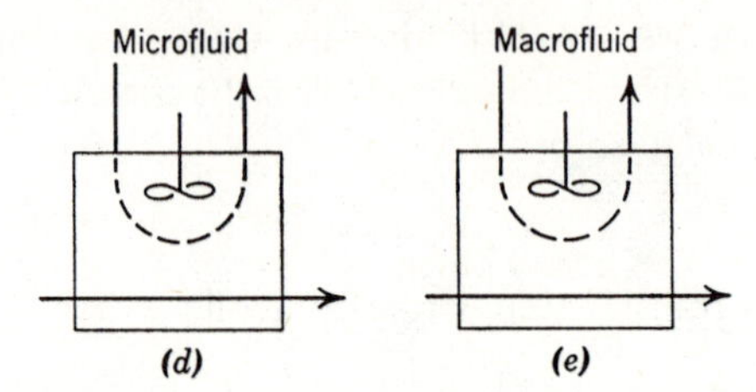

One phase in plug flow, the other in backmix flow

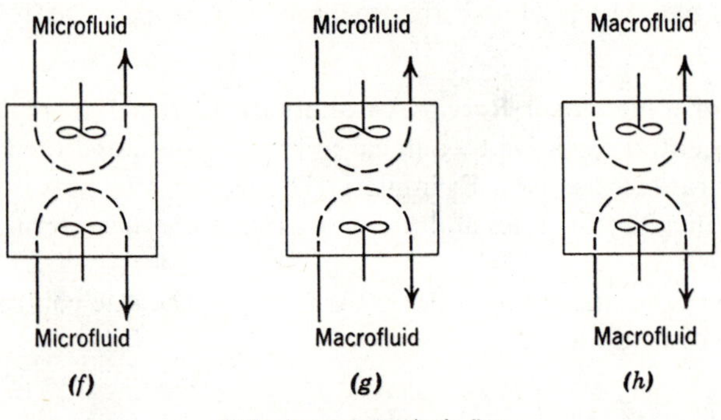

Both phases in backmix flow

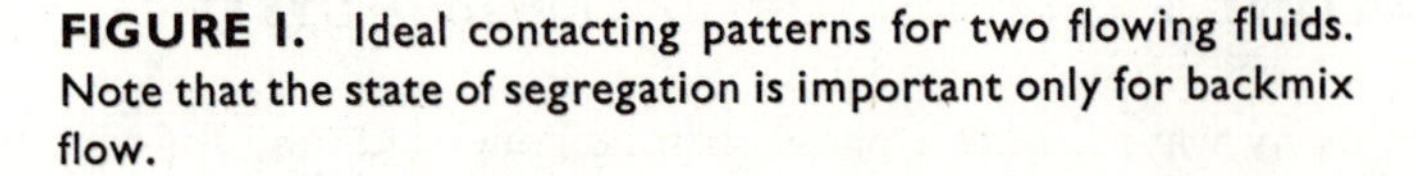

FIGURE 1. Ideal contacting patterns for two flowing fluids. Note that the state of segregation is important only for backmix flow.

since they approximate single mixer-settler, cascade, and tower operations. These are the flow patterns we shall take up in the following chapter.

Design equations for these ideal flow patterns may be developed without too much difficulty. However, when real flow deviates considerably from these, we can do one of two things: we may develop models to mirror actual flow closely, or we may calculate performance with ideal patterns which "bracket" actual flow.

Fortunately, most real reactors for heterogeneous systems can be satisfactorily approximated by one of the eight ideal flow patterns of Fig. 1. This can be seen by comparing these patterns with the sketches of typical reactors for fluid–solid and fluid–fluid systems (Figs. 12.13, 13.6, and 14.17). Notable exceptions are the reactions taking place in fluidized beds.

PROBLEMS

1. Gaseous reactant diffuses through a gas film and reacts on the surface of a solid according to a reversible first order rate,

$$-r'' = k_s(C_s - C_e)$$

Develop an expression for the rate of reaction accounting for both the mass transfer and reaction steps.

2. Repeat Problem 1 if the rate of surface reaction has the form of the Langmuir-Hinshelwood adsorption equation

$$-r'' = \frac{k_1 C_s}{1 + k_2 C_s}$$

3. For the reaction of Example 2

(*a*) linearize the second order rate expression, and

(*b*) with this approximation develop a rate expression accounting for both the mass transfer and reaction steps. Compare this expression with the rigorous expression and see which is simpler to use.

Data: We plan to operate at conditions such that the concentration of gaseous reactant at the surface is somewhere in the vicinity of $C_0 = 2$. The series representation of any function $f(C)$ in the vicinity of C_0 is given by the Taylor expansion,

$$f(C) = f(C_0) + \frac{C - C_0}{1!} f'(C_0) + \frac{(C - C_0)^2}{2!} f''(C_0) + \cdots$$

Use this expression to linearize the rate expression.

4. Linearize the rate expression of Problem 2, then combine with the mass transfer step to obtain an over-all rate. Compare your result with the rigorous expression obtained in Problem 2.

5. In slurry reactors pure reactant gas is bubbled through liquid containing suspended catalyst particles. Let us view these kinetics in terms of the film theory as shown in Fig. P5. Thus to reach the surface of solid the reactant which enters the liquid must diffuse through the liquid film into the main body of liquid, and then through the film surrounding the catalyst particle. At the surface of the particle reactant yields product according to first order kinetics. Derive an expression for the rate of reaction in terms of these resistances.

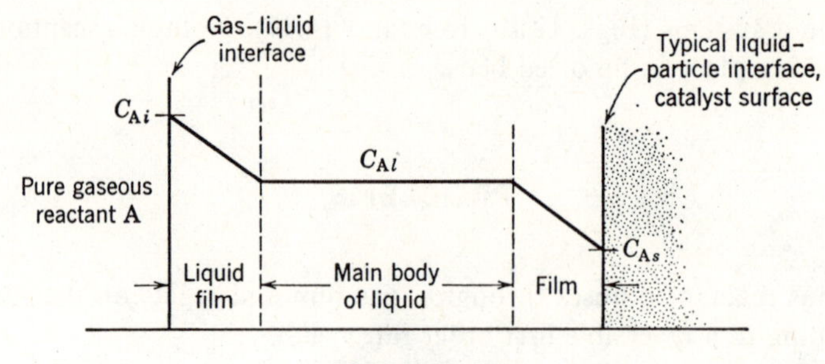

FIGURE P5.

6. Gas containing A contacts and reacts with a semi-infinite slab of solid B as follows

$$A(g) + B(s) \rightarrow R(g) + S(s)$$

As reaction progresses, a sharp reaction plane advances slowly into the solid leaving behind it a layer of product through which gaseous A and R must diffuse. Overall then three resistances act in series, that of the gas film, the ash layer, and reaction, as shown in Fig. P6.

Noting that the rate of thickening of the ash layer is proportional to the rate of reaction at that instant, or

$$\frac{dl}{dt} = M(-r''_A)$$

show that the time to reach any thickness l is the sum of the time required if each resistance acted alone, or

$$t_{\text{actual}} = t_{\text{film alone}} + t_{\text{ash alone}} + t_{\text{reaction alone}}$$

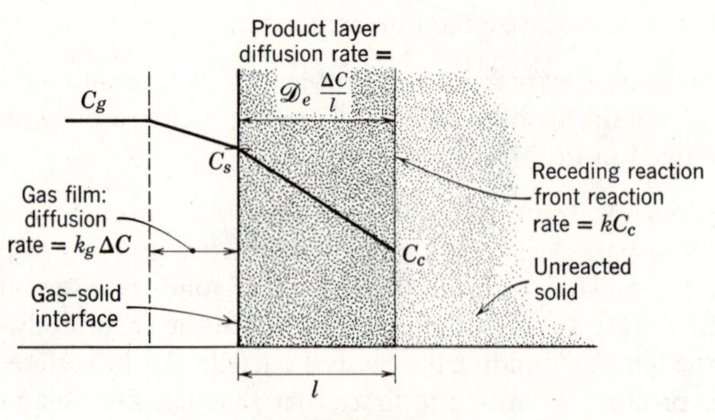

FIGURE P6.

12

FLUID-PARTICLE REACTIONS

This chapter treats the class of heterogeneous reactions in which a gas or liquid contacts a solid, reacts with it, and transforms it into product. Such reactions may be represented by

$$\text{A(fluid)} + b\text{B(solid)} \rightarrow \text{fluid products} \tag{1}$$

$$\rightarrow \text{solid products} \tag{2}$$

$$\rightarrow \text{fluid and solid products} \tag{3}$$

As shown in Fig. 1, solid particles remain unchanged in size during reaction when they contain large amounts of impurities which remain as a nonflaking ash or if they form a firm product material by the reactions of Eq. 2 or Eq. 3. Particles shrink in size during reaction when a flaking ash or product material is formed or when pure B is used in the reaction of Eq. 1.

Fluid–solid reactions are numerous and of great industrial importance. Those in which the solid does not appreciably change in size during reaction are as follows.

1. The roasting (or oxidation) of sulfide ores to yield the metal oxides. For example, in the preparation of zinc oxide the sulfide ore is mined, crushed, separated from the gangue by flotation, and then roasted in a reactor to form hard white zinc oxide particles according to the reaction

$$2ZnS(s) + 3O_2(g) \rightarrow 2ZnO(s) + 2SO_2(g)$$

Similarly, iron pyrites react as follows:

$$4FeS_2(s) + 11O_2(g) \rightarrow 8SO_2(g) + 2Fe_2O_3(s)$$

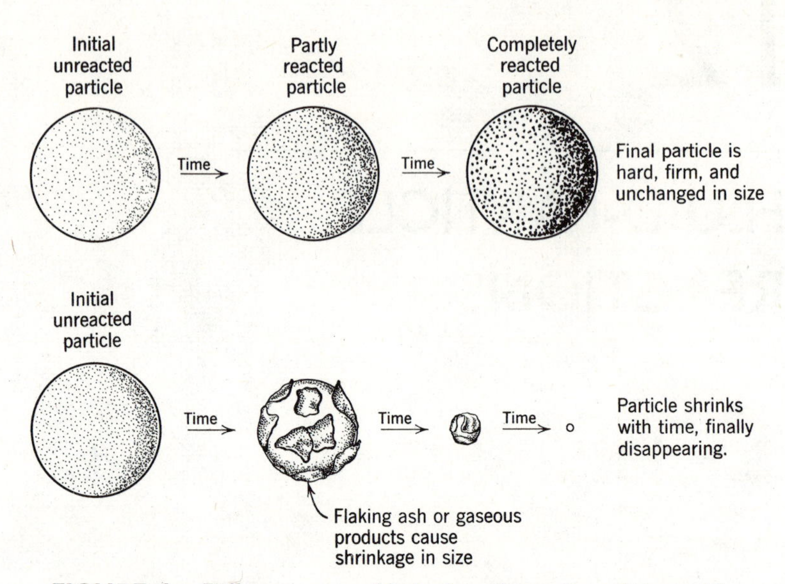

FIGURE 1. Different sorts of behavior of reacting solid particles.

2. The preparation of metals from their oxides by reaction in reducing atmospheres. For example, iron is prepared from crushed and sized magnetite ore in continuous-countercurrent, three-stage, fluidized-bed reactors according to the reaction

$$Fe_3O_4(s) + 4H_2(g) \rightarrow 3Fe(s) + 4H_2O(g)$$

3. The nitrogenation of calcium carbide to produce cyanamide:

$$CaC_2(s) + N_2(g) \rightarrow CaCN_2(s) + C(\text{amorphous})$$

4. The protective surface treatment of solids such as the plating of metals.

The most common examples of fluid–solid reactions in which the size of solid changes are the reactions of carbonaceous materials such as coal briquettes, wood, etc., with low ash content to produce heat or heating fuels. For example, with an insufficient amount of air producer gas is formed by the reactions

$$C(s) + O_2(g) \rightarrow CO_2(g)$$

$$2C(s) + O_2(g) \rightarrow 2CO(g)$$

$$C(s) + CO_2(g) \rightarrow 2CO(g)$$

With steam, water gas is obtained by the reactions

$$C(s) + H_2O(g) \rightarrow CO(g) + H_2(g)$$

$$C(s) + 2H_2O(g) \rightarrow CO_2(g) + 2H_2(g)$$

Other examples of reactions in which solids change in size are as follows.

1. The manufacture of carbon disulfide from the elements:

$$C(s) + 2S(g) \xrightarrow{750\text{–}1000^\circ C} CS_2(g)$$

2. The manufacture of sodium cyanide from sodium amide:

$$NaNH_2(l) + C(s) \xrightarrow{800^\circ C} NaCN(l) + H_2(g)$$

3. The manufacture of sodium thiosulfate from sulfur and sodium sulfite:

$$Na_2SO_3(\text{solution}) + S(s) \rightarrow Na_2S_2O_3(\text{solution})$$

Still other examples are the dissolution reactions, the attack of metal chips by acids, and the rusting of iron.

In Chapter 11 we pointed out that treatment of heterogeneous reaction required the consideration of two factors in addition to those normally encountered in homogeneous reactions: the modification of the kinetic expressions resulting from the mass transfer between phases and the contacting patterns of the reacting phases.

Let us now develop the rate expressions for fluid–solid reactions. These will then be used in design.

SELECTION OF A MODEL

We should clearly understand that every conceptual picture or model for the progress of reaction comes with its mathematical representation, its rate equation. Consequently, if we choose a model we must accept its rate equation, and vice versa. If a model corresponds closely to what really takes place, then its rate expression will closely predict and describe the actual kinetics; if a model widely differs from reality, then its kinetic expressions will be useless. We must remember that the most elegant and high-powered mathematical analysis based on a model which does not match reality is worthless for the engineer who must make design predictions. What we say here about a model holds not only in deriving kinetic expressions but in all areas of engineering.

The requirement for a good engineering model is that it be the closest representation of reality which can be treated without too many mathematical complexities. It is of little use to select a model which very closely mirrors reality but which is so complicated that we cannot do anything with it. Unfortunately, this all too often happens.

For the noncatalytic reaction of particles with surrounding fluid, we consider two simple idealized models, the *progressive-conversion* model and the *unreacted-core* model.

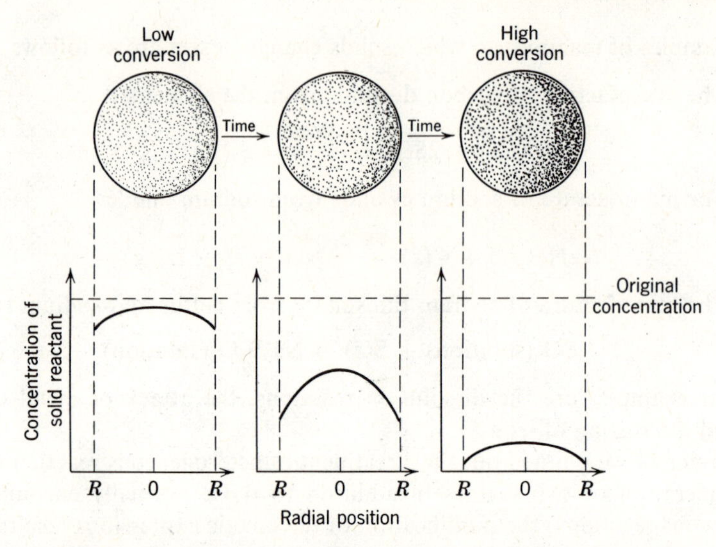

FIGURE 2. According to the progressive-conversion model reaction proceeds continuously throughout the solid particle.

Progressive-conversion Model. Here we visualize that reactant gas enters and reacts throughout the particle at all times, most likely at different rates at different locations within the particle. Thus, solid reactant is converted continuously and progressively throughout the particle as shown in Fig. 2. comportamiento homogeneo

Unreacted-core Model. Here we visualize that reaction occurs first at the outer skin of the particle. The zone of reaction then moves into the solid, and may leave behind completely converted material and inert solid. We refer to these as "ash." Thus, at any time there exists an unreacted core of material which shrinks in size during reaction as shown in Fig. 3.

Comparison of Models with Real Situations. In slicing and examining the cross section of partly reacted solid particles, we usually find unreacted solid material surrounded by a layer of ash. The boundary of this unreacted core may not always be as sharply defined as the model pictures it; nevertheless, evidence from a wide variety of situations indicates that the unreacted-core model approximates real particles more closely in most cases than does the progressive-conversion model. Observations with burning coal, wood, briquettes, and tightly wrapped newspapers also favor the unreacted-core model. For further discussion on these models see page 373.

Since the unreacted-core model seems to reasonably represent reality in a wide variety of situations, we develop its kinetic equations in the following section. In doing this we consider the surrounding fluid to be a gas. However, this is done only for convenience since the analysis applies equally well to liquids.

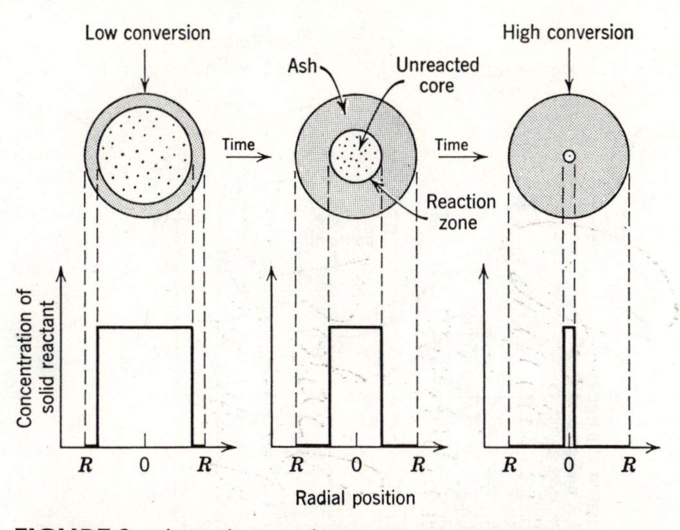

FIGURE 3. According to the unreacted-core model reaction proceeds at a narrow front which moves into the solid particle. Reactant is completely converted as the front passes by.

UNREACTED CORE MODEL FOR SPHERICAL PARTICLES OF UNCHANGING SIZE

This model was first developed by Yagi and Kunii (1955), who visualized five steps occurring in succession during reaction (see Fig. 4).

Step 1. Diffusion of gaseous reactant A through the film surrounding the particle to the surface of the solid.

Step 2. Penetration and diffusion of A through the blanket of ash to the surface of the unreacted core.

Step 3. Reaction of gaseous A with solid at this reaction surface.

Step 4. Diffusion of gaseous products through the ash back to the exterior surface of the solid.

Step 5. Diffusion of gaseous products through the gas film back into the main body of fluid.

At times some of these steps do not exist. For example, if no gaseous products are formed or if the reaction is irreversible, Steps 4 and 5 do not contribute directly to the resistance to reaction. Also, the resistances of the different steps usually vary greatly one from the other; in such cases we may consider that step with the highest resistance to be rate-controlling.

In this treatment we develop the conversion equations for the elementary irreversible reactions (Steps 4 and 5 do not apply) which are represented by Eqs. 1,

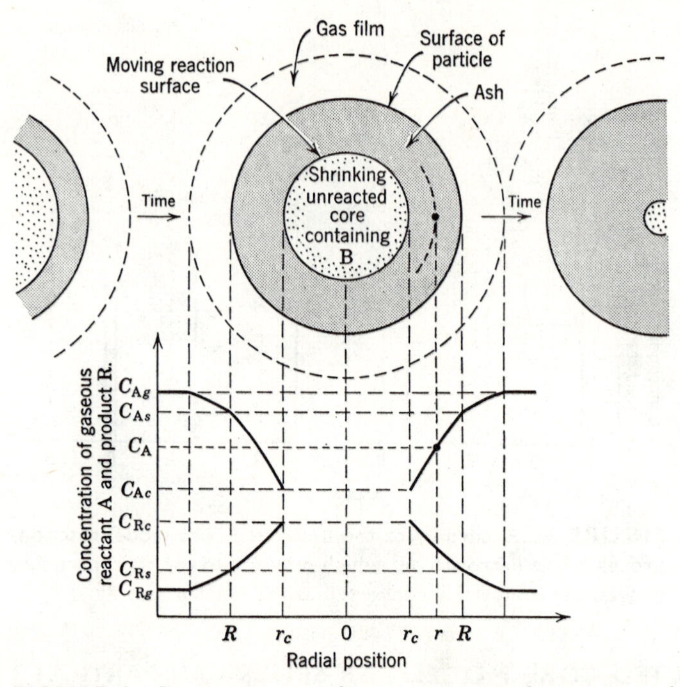

FIGURE 4. Representation of concentrations of reactants and products for the reaction $A(g) + bB(s) \rightleftharpoons rR(g) + sS(s)$ for a particle of unchanging size.

2, or 3. We start with spherical particles in which Steps 1, 2, and 3, in turn, are rate-controlling. We then extend the analysis to nonspherical particles and to situations where the combined effect of these three resistances must be considered together.

Diffusion Through Gas Film Controls

Whenever the resistance of the gas film controls, the concentration profile for gaseous reactant A will be as shown in Fig. 5. From this figure we see that no reactant is present at the surface; hence the concentration driving force, given by $C_{Ag} - C_{As}$, is constant at all times during reaction of the particle. Now since it is convenient to derive the kinetic equations based on available surface, we focus attention on the unchanging exterior surface of a particle S_{ex}. Noting from the stoichiometry of Eqs. 1, 2, and 3 that $dN_B = bdN_A$, we write

$$-\frac{1}{S_{ex}}\frac{dN_B}{dt} = -\frac{1}{4\pi R^2}\frac{dN_B}{dt} = -\frac{b}{4\pi R^2}\frac{dN_A}{dt} = bk_g(C_{Ag} - C_{As}) = bk_g C_{Ag}$$
$$= \text{constant} \tag{4}$$

Nota: $N_B = N_{B0} - X_B N_{B0}$; $dN_B = -N_{B0}\, dX_B$

$N_A = N_{A0} - \frac{X_B}{b} N_{B0}$; $dN_A = -\frac{N_{B0}}{b} dX_B$

$\therefore\ b\,dN_A = dN_B$

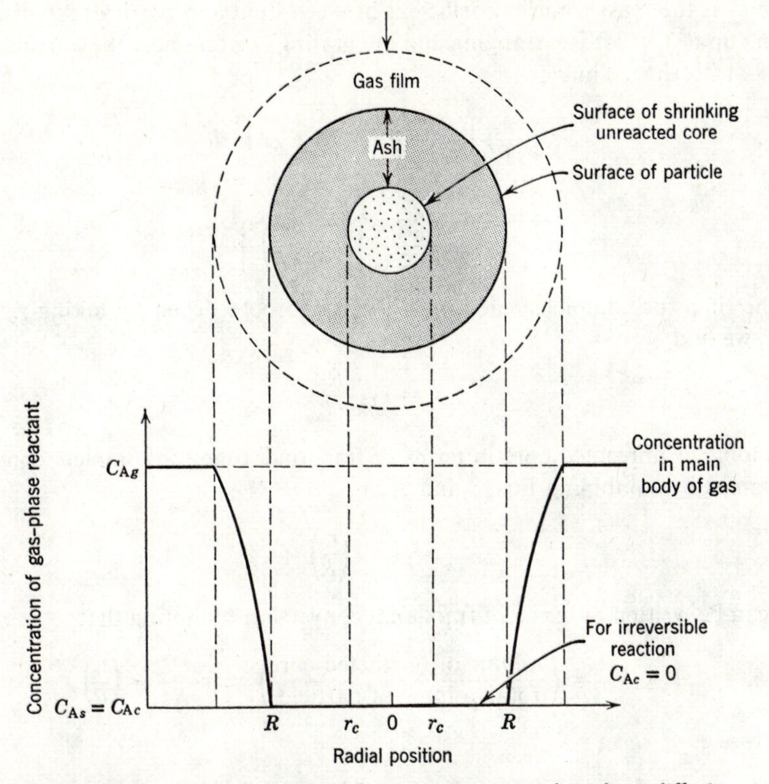

FIGURE 5. Representation of a reacting particle when diffusion through the gas film is the controlling resistance.

If we let ρ_B be the molar density of B in the solid and V be the volume of a particle, the amount of B present in a particle is

$$N_B = \rho_B V = \left(\frac{\text{moles B}}{\text{cm}^3 \text{ solid}}\right)(\text{cm}^3 \text{ solid}) \tag{5}$$

The decrease in volume or radius of unreacted core accompanying the disappearance of dN_B moles of solid reactant or $b dN_A$ moles of fluid reactant is then given by

$$-dN_B = -b\, dN_A = -\rho_B\, dV = -\rho_B\, d(\tfrac{4}{3}\pi r_c^3) = -4\pi\rho_B r_c^2\, dr_c \tag{6}$$

Replacing Eq. 6 in 4 gives the rate of reaction in terms of the shrinking radius of unreacted core, or

$$-\frac{1}{S_{ex}}\frac{dN_B}{dt} = -\frac{\rho_B r_c^2}{R^2}\frac{dr_c}{dt} = bk_g C_{Ag} \tag{7}$$

where k_g is the mass transfer coefficient between fluid and particle; see discussion leading up to Eq. 24. Rearranging and integrating, we find how the unreacted core shrinks with time. Thus

$$-\frac{\rho_B}{R^2}\int_R^{r_c} r_c^2\,dr_c = bk_g C_{Ag}\int_0^t dt$$

or

$$t = \frac{\rho_B R}{3bk_g C_{Ag}}\left[1 - \left(\frac{r_c}{R}\right)^3\right] \tag{8}$$

Let the time for complete reaction of a particle be τ. Then by taking $r_c = 0$ in Eq. 8, we find

$$\tau = \frac{\rho_B R}{3bk_g C_{Ag}} \tag{9}$$

The radius of unreacted core in terms of fractional time for complete conversion is obtained by combining Eqs. 8 and 9, or

$$\frac{t}{\tau} = 1 - \left(\frac{r_c}{R}\right)^3$$

This can be written in terms of fractional conversion by noting that

$$1 - X_B = \left(\frac{\text{volume of unreacted core}}{\text{total volume of particle}}\right) = \frac{\frac{4}{3}\pi r_c^3}{\frac{4}{3}\pi R^3} = \left(\frac{r_c}{R}\right)^3 \tag{10}$$

Therefore

$$\frac{t}{\tau} = 1 - \left(\frac{r_c}{R}\right)^3 = X_B \tag{11}$$

Thus we obtain the relationship between time and radius and conversion, which is shown graphically in Figs. 9 and 10, pp. 374 and 375.

Diffusion through Ash Layer Controls

Figure 6 illustrates the situation in which the resistance to diffusion through the ash controls the rate of reaction. To develop an expression between time and radius, such as Eq. 8 for film resistance, requires a two-step analysis. First we examine a typical partially reacted particle, writing the flux relationships for this condition. Then we apply this relationship for all values of r_c; in other words, we integrate r_c between R and 0.

Consider a partially reacted particle as shown in Fig. 6. Both reactant A and the boundary of the unreacted core move inward toward the center of the particle. But

$$X_B = \frac{\text{moles rxnadas}}{\text{moles iniciales}} = \frac{N_{B0} - N_B}{N_{B0}} = \frac{\rho_B\left(\frac{4}{3}\pi R^3 - \frac{4}{3}\pi r_c^3\right)}{\rho_B\left(\frac{4}{3}\pi R^3\right)} = 1 - \left(\frac{r_c}{R}\right)^3$$

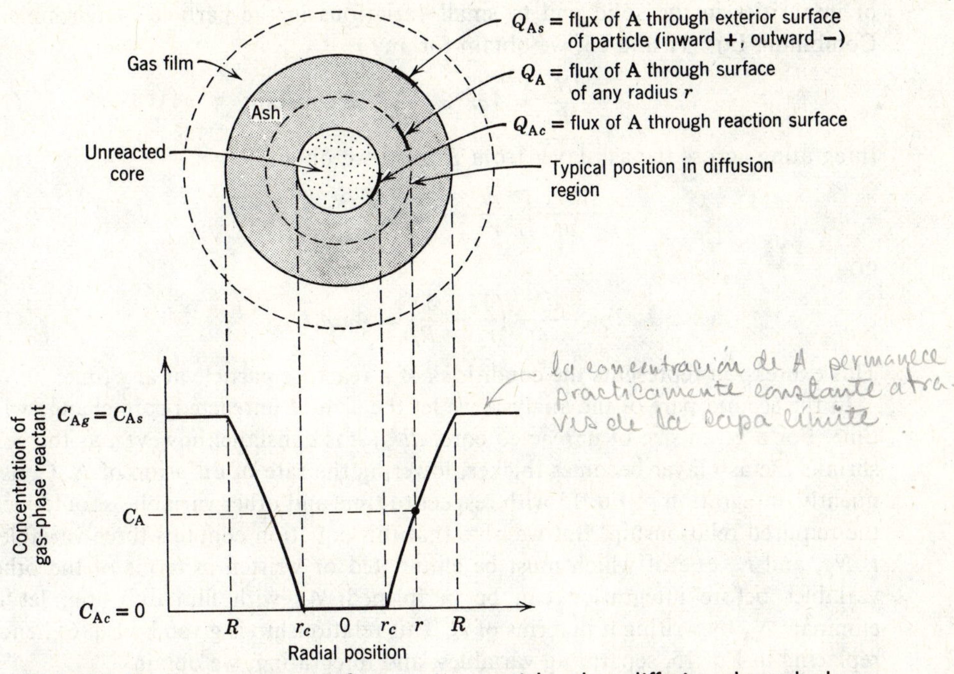

la concentración de A permanece practicamente constante a través de la capa límite

FIGURE 6. Representation of a reacting particle when diffusion through the ash layer is the controlling resistance.

the shrinkage of the unreacted core is slower than the flow rate of A toward the unreacted core by a factor of about 1000, which is roughly the ratio of densities of solid to gas. Because of this it is reasonable for us to assume, in considering the concentration gradient of A in the ash layer at any time, that the unreacted core is stationary. This steady-state assumption allows great simplification in the mathematics which follows. With this assumption the rate of reaction of A at any instant is given by its rate of diffusion to the reaction surface, or

$$-\frac{dN_A}{dt} = 4\pi r^2 Q_A = 4\pi R^2 Q_{As} = 4\pi r_c^2 Q_{Ac} = \text{constant} \tag{12}$$

For convenience, let the flux of A within the ash layer be expressed by Fick's law for equimolar counterdiffusion, though other forms of this diffusion equation will give the same result. Then noting that both Q_A and dC_A/dr are positive, we have

$$Q_A = \mathscr{D}_e \frac{dC_A}{dr} \tag{13}$$

where $\mathscr{D}_e$ is the effective diffusion coefficient of gaseous reactant in the ash layer. Often it is difficult to assign a value beforehand to this quantity because the property of the ash (its sintering qualities, for example) can be very sensitive to small amounts

$-\frac{dN_a}{dt}$ no depende de r

of impurities in the solid and to small variations in the particle's environment. Combining Eqs. 12 and 13, we obtain for any r

$$-\frac{dN_A}{dt} = 4\pi r^2 \mathscr{D}_e \frac{dC_A}{dr} = \text{constant} \tag{14}$$

Integrating across the ash layer from R to r_c, we obtain

$$-\frac{dN_A}{dt} \int_R^{r_c} \frac{dr}{r^2} = 4\pi \mathscr{D}_e \int_{C_{Ag}=C_{As}}^{C_{Ac}=0} dC_A$$

or

$$-\frac{dN_A}{dt}\left(\frac{1}{r_c} - \frac{1}{R}\right) = 4\pi \mathscr{D}_e C_{Ag} \tag{15}$$

This expression represents the conditions of a reacting particle at any time.

In the second part of the analysis we let the size of unreacted core change with time. For a given size of unreacted core, dN_A/dt is constant; however, as the core shrinks the ash layer becomes thicker, lowering the rate of diffusion of A. Consequently, integration of Eq. 15 with respect to time and other variables should yield the required relationship. But we note that this equation contains three variables, t, N_A, and r_c, one of which must be eliminated or written in terms of the other variables before integration can be performed. As with film diffusion, let us eliminate N_A by writing it in terms of r_c. This relationship is given by Eq. 6; hence, replacing in Eq. 15, separating variables, and integrating, we obtain

$$-\rho_B \int_{r_c=R}^{r_c} \left(\frac{1}{r_c} - \frac{1}{R}\right) r_c^2\, dr_c = b\mathscr{D}_e C_{Ag} \int_0^t dt$$

or

$$t = \frac{\rho_B R^2}{6b\mathscr{D}_e C_{Ag}}\left[1 - 3\left(\frac{r_c}{R}\right)^2 + 2\left(\frac{r_c}{R}\right)^3\right] \tag{16}$$

For complete conversion of a particle, $r_c = 0$, and the time required is

$$\tau = \frac{\rho_B R^2}{6b\mathscr{D}_e C_{Ag}} \tag{17}$$

The progression of reaction in terms of the time required for complete conversion is found by dividing Eq. 16 by Eq. 17, or

$$\frac{t}{\tau} = 1 - 3\left(\frac{r_c}{R}\right)^2 + 2\left(\frac{r_c}{R}\right)^3 \tag{18a}$$

which in terms of fractional conversion, as given in Eq. 10, becomes

$$\frac{t}{\tau} = 1 - 3(1 - X_B)^{2/3} + 2(1 - X_B) \tag{18b}$$

These results are presented graphically in Figs. 9 and 10, pp. 374 and 375.

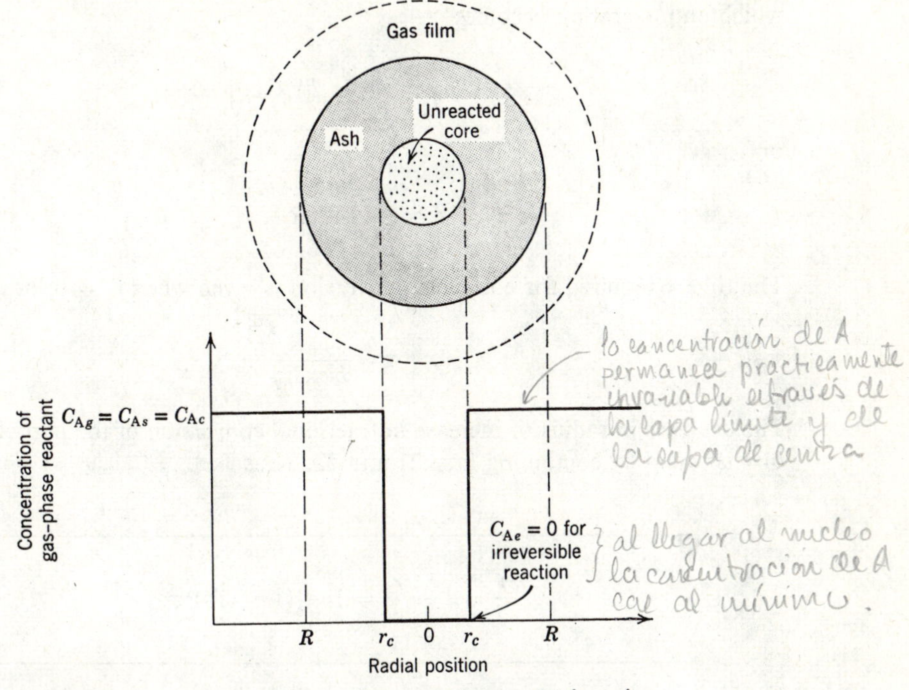

FIGURE 7. Representation of a reacting particle when chemical reaction is the controlling resistance, the reaction being A(g) + bB(s) → products.

Chemical Reaction Controls

Figure 7 illustrates concentration gradients within a particle when chemical reaction controls. Since the progress of the reaction is unaffected by the presence of any ash layer, the quantity of material reacting is proportional to the available surface of unreacted core. Thus, based on unit surface of unreacted core, the rate of reaction for the stoichiometry of Eqs. 1, 2, and 3 is

$$-\frac{1}{4\pi r_c^2}\frac{dN_B}{dt} = -\frac{b}{4\pi r_c^2}\frac{dN_A}{dt} = bk_s C_{Ag} \tag{19}$$

where k_s is the first-order rate constant for the surface reaction. Writing N_B in terms of the shrinking radius, as given in Eq. 6, we obtain

$$-\frac{1}{4\pi r_c^2}\rho_B 4\pi r_c^2 \frac{dr_c}{dt} = -\rho_B \frac{dr_c}{dt} = bk_s C_{Ag} \tag{20}$$

which on integration becomes

$$-\rho_B \int_R^{r_c} dr_c = bk_s C_{Ag} \int_0^t dt$$

or

$$t = \frac{\rho_B}{bk_s C_{Ag}} (R - r_c) \tag{21}$$

The time τ required for complete conversion is given when $r_c = 0$, or

$$\tau = \frac{\rho_B R}{bk_s C_{Ag}} \tag{22}$$

The decrease in radius or increase in fractional conversion of the particle in terms of τ is found by combining Eqs. 21 and 22. Thus

$$\frac{t}{\tau} = 1 - \frac{r_c}{R} = 1 - (1 - X_B)^{1/3} \tag{23}$$

This result is plotted in Figs. 9 and 10, pp. 374 and 375.

RATE OF REACTION FOR SHRINKING SPHERICAL PARTICLES

When no ash forms, as in the burning of pure carbon in air, the reacting particle shrinks during reaction, finally disappearing. This process is illustrated in Fig. 8. For a reaction of this kind we visualize the following three steps occurring in succession.

Step 1. Diffusion of reactant A from the main body of gas through the gas film to the surface of the solid.

Step 2. Reaction on the surface between reactant A and solid.

Step 3. Diffusion of reaction products from the surface of the solid through the gas film back into the main body of gas. Note that the ash layer is absent and does not contribute any resistance.

As with particles of constant size, let us see what rate expressions result when one or the other of the resistances controls.

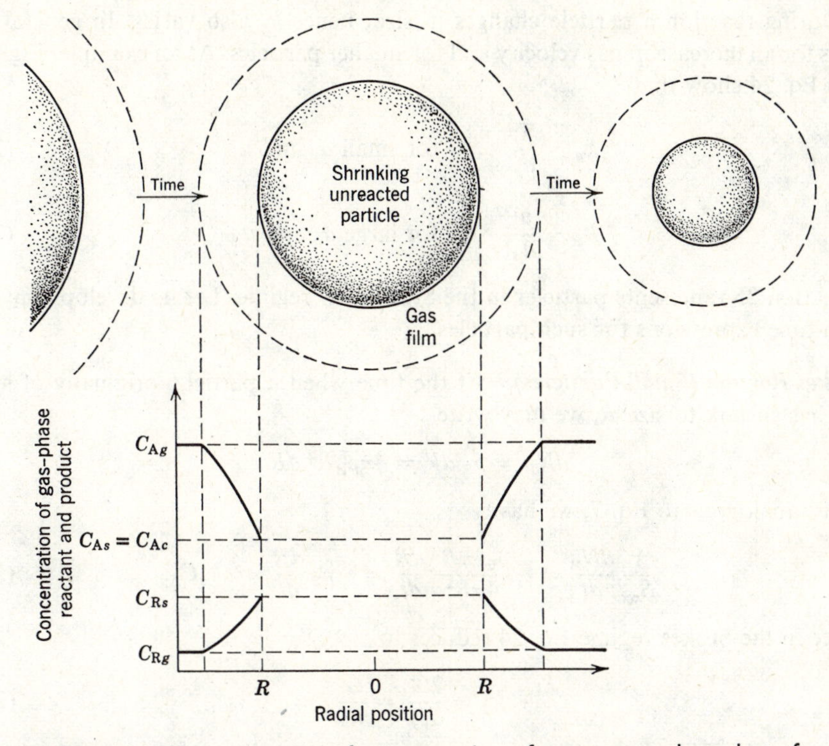

FIGURE 8. Representation of concentration of reactants and products for the reaction $A(g) + bB(s) \rightarrow rR(g)$ between a shrinking solid particle and gas.

Chemical Reaction Controls

When chemical reaction controls, the behavior is identical to that of particles of unchanging size; therefore Fig. 7 and Eq. 21 or 23 will represent the time-conversion behavior of single particles, both shrinking and of constant size.

Gas Film Diffusion Controls

Film resistance at the surface of a particle is dependent on numerous factors, such as the relative velocity between particle and fluid, size of particle, and fluid properties. These have been correlated for various methods of contacting of fluid with solid, such as packed beds, fluidized beds, and solids in free fall. As an example, for mass transfer of a component of mole fraction y in a fluid to free-falling solids Ranz and Marshall (1952) give

$$\frac{k_g d_p y}{\mathscr{D}} = 2 + 0.6(\text{Sc})^{1/3}(\text{Re})^{1/2} = 2 + 0.6\left(\frac{\mu}{\rho\mathscr{D}}\right)^{1/3}\left(\frac{d_p u \rho}{\mu}\right)^{1/2} \tag{24}$$

During reaction a particle changes in size; hence k_g also varies. In general k_g rises for an increase in gas velocity and for smaller particles. As an example, Fig. 12 and Eq. 24 show that

$$k_g \sim \frac{1}{d_p} \qquad \text{for small } d_p \text{ and } u \tag{25}$$

$$k_g \sim \frac{u^{1/2}}{d_p^{\,1/2}} \qquad \text{for large } d_p \text{ and } u \tag{26}$$

Equation 25 represents particles in the Stokes law regime. Let us develop conversion-time expressions for such particles.

Stokes Regime (Small Particles). At the time when a particle, originally of size R_0, has shrunk to size R, we may write

$$dN_B = \rho_B\, dV = 4\pi\rho_B R^2\, dR$$

Thus, analogous to Eq. 7, we have

$$-\frac{1}{S_{ex}}\frac{dN_B}{dt} = -\frac{\rho_B 4\pi R^2}{4\pi R^2}\frac{dR}{dt} = -\rho_B\frac{dR}{dt} = bk_g C_{Ag} \tag{27}$$

Since in the Stokes regime Eq. 24 reduces to

$$k_g = \frac{2\mathscr{D}}{d_p y} = \frac{\mathscr{D}}{Ry} \tag{28}$$

we have on combining and integrating

$$\int_{R_0}^{R} R\, dR = \frac{bC_{Ag}\mathscr{D}}{\rho_B y}\int_0^t dt$$

or

$$t = \frac{\rho_B y R_0^{\,2}}{2bC_{Ag}\mathscr{D}}\left[1 - \left(\frac{R}{R_0}\right)^2\right]$$

The time for complete disappearance of a particle is thus

$$\tau = \frac{\rho_B y R_0^{\,2}}{2bC_{Ag}\mathscr{D}} \tag{29}$$

and on combining we obtain

$$\frac{t}{\tau} = 1 - \left(\frac{R}{R_0}\right)^2 = 1 - (1 - X_B)^{2/3} \tag{30}$$

This relationship of size versus time for shrinking particles in the Stokes regime is shown in Figs. 9 and 10, pp. 374 and 375, and it well represents small burning solid particles and small burning liquid droplets.

Para más información: Analisis de Reactores R. Aris Cap VI

EXTENSIONS

Particles of Different Shape. Conversion-time equations similar to those developed above can be obtained for various shaped particles, and Table 1 summarizes these expressions.

Combination of Resistances. The above conversion-time expressions assume that a single resistance controls throughout reaction of the particle. However, the relative importance of the gas film, ash layer, and reaction steps will vary as conversion progresses. For example, for a constant size particle the gas film resistance remains unchanged, the resistance to reaction increases as the surface of unreacted core decreases, while the ash layer resistance is nonexistent at the start because no ash is present, but becomes progressively more and more important as the ash layer builds up. In general, then, it may not be reasonable to consider that just one step controls throughout reaction.

To account for the simultaneous action of these resistances is straightforward since they act in series and are all linear in concentration. Thus on combining Eqs. 7, 15, and 20 with their individual driving forces and eliminating intermediate concentrations we can show that the time to reach any stage of conversion is the sum of the times needed if each resistance acted alone, or

$$t_{\text{total}} = t_{\text{film alone}} + t_{\text{ash alone}} + t_{\text{reaction alone}} \tag{32a}$$

Similarly, for complete conversion

$$\tau_{\text{total}} = \tau_{\text{film alone}} + \tau_{\text{ash alone}} + \tau_{\text{reaction alone}} \tag{32b}$$

In an alternative approach, the individual resistances can be combined directly to give, at any particular stage of conversion,

$$-\frac{1}{S_{\text{ex}}}\frac{dN_{\text{B}}}{dt} = \frac{bC_{\text{A}}}{\dfrac{1}{k_g} + \dfrac{R(R-r_c)}{r_c\mathscr{D}_e} + \dfrac{R^2}{r_c^2 k_s}} \tag{33a}$$

or

$$-\frac{dr_c}{dt} = \frac{bC_{\text{A}}/\rho_{\text{B}}}{\underset{\text{film}}{\dfrac{r_c^2}{R^2k_g}} + \underset{\text{ash}}{\dfrac{(R-r_c)r_c}{R\mathscr{D}_e}} + \underset{\text{reaction}}{\dfrac{1}{k_s}}} \tag{33b}$$

TABLE I. Conversion—time expressions for various shaped particles, shrinking core model

		Film Diffusion Controls	Ash Diffusion Controls	Reaction Controls
Constant Size Particles	Flat plate	$\frac{t}{\tau} = X_B$	$\frac{t}{\tau} = X_B^2$	$\frac{t}{\tau} = X_B$
	$X_B = 1 - \frac{l}{L}$	$\tau = \frac{\rho_B L}{b k_g C_{Ag}}$	$\tau = \frac{\rho_B L^2}{2b\mathscr{D}_e C_{Ag}}$	$\tau = \frac{\rho_B L}{b k_s C_{Ag}}$
	Cylinder	$\frac{t}{\tau} = X_B$	$\frac{t}{\tau} = X_B + (1 - X_B)\ln(1 - X_B)$	$\frac{t}{\tau} = 1 - (1 - X_B)^{1/2}$
	$X_B = 1 - \left(\frac{r_c}{R}\right)^2$	$\tau = \frac{\rho_B R}{2b k_g C_{Ag}}$	$\tau = \frac{\rho_B R^2}{4b\mathscr{D}_e C_{Ag}}$	$\tau = \frac{\rho_B R}{b k_s C_{Ag}}$
	Sphere	$\frac{t}{\tau} = X_B \quad (11)$	$\frac{t}{\tau} = 1 - 3(1 - X_B)^{2/3} + 2(1 - X_B) \quad (18)$	$\frac{t}{\tau} = 1 - (1 - X_B)^{1/3} \quad (23)$
	$X_B = 1 - \left(\frac{r_c}{R}\right)^3$	$\tau = \frac{\rho_B R}{3b k_g C_{Ag}} \quad (10)$	$\tau = \frac{\rho_B R^2}{6b\mathscr{D}_e C_{Ag}} \quad (17)$	$\tau = \frac{\rho_B R}{b k_s C_{Ag}} \quad (22)$
Shrinking Sphere	Small particle,	$\frac{t}{\tau} = 1 - (1 - X_B)^{2/3} \quad (30)$	Not applicable	$\frac{t}{\tau} = 1 - (1 - X_B)^{1/3}$
	Stokes regime	$\tau = \frac{\rho_B y R_0^2}{2b\mathscr{D} C_{Ag}} \quad (29)$		$\tau = \frac{\rho_B R_0}{b k_s C_{Ag}}$
	Large particle,	$\frac{t}{\tau} = 1 - (1 - X_B)^{1/2} \quad (31)$	Not applicable	$\frac{t}{\tau} = 1 - (1 - X_B)^{1/3}$
	(u = constant)	$\tau = (\text{const})\,\frac{R_0^{3/2}}{C_{Ag}}$		$\tau = \frac{\rho_B R_0}{b k_s C_{Ag}}$

As may be seen, the relative importance of the three individual resistances vary as conversion progresses, or as r_c decreases.

On considering the whole progression from fresh to completely converted constant size particle, we find on the average that the relative roles of these three resistances is given by

$$-\frac{1}{S_{ex}}\frac{dN_A}{dt} = \bar{k}_s C_A = \frac{C_A}{\dfrac{1}{k_g} + \dfrac{R}{2\mathscr{D}_e} + \dfrac{3}{k_s}} \tag{34}$$

For ash-free particles which shrink with reaction, only two resistances, gas film and surface reaction, need be considered. Since these are both based on the changing exterior surface of particles, we may combine them to give at any instant

$$-\frac{1}{S_{ex}}\frac{dN_A}{dt} = \frac{1}{\dfrac{1}{k_g} + \dfrac{1}{k_s}} C_A \tag{35}$$

Various forms of these expressions have been derived by Yagi and Kunii (1955), Shen and Smith (1965), and White and Carberry (1965).

Limitations of the Shrinking Core Model. The assumptions of this model may not match reality precisely. For example, reaction may occur along a diffuse front rather than along a sharp interface between ash and fresh solid, thus giving behavior intermediate between the shrinking core and the continuous reaction models (see p. 360). This problem is considered by Wen (1968), and Ishida *et al.* (1971*a*).

Also, for fast reaction the rate of heat release may be high enough to cause significant temperature gradients within the particles or between particle and the bulk fluid. This problem is treated in detail by Wen and Wang (1970). Also see the discussion on p. 479.

Despite these complications Wen (1968) and Ishida *et al.* (1971*a*, *b*), on the basis of studies of numerous systems, concludes that the shrinking core model is the best simple representation for the majority of reacting gas-solid systems.

There are, however, two broad classes of exceptions to this conclusion. The first comes with the slow reaction of a gas with a very porous solid. Here reaction can occur throughout the solid, and the continuous reaction model may be expected to better fit reality. An example of this is the slow poisoning of a catalyst pellet, a situation treated in Chapter 15.

The second exception occurs when solid is converted by the action of heat, and without needing contact with gas. Baking bread, boiling missionaries, and roasting puppies are mouthwatering examples of such reactions. Here again the continuous reaction model is a better representation of reality. Wen (1968) and Kunii and Levenspiel (1969) treat these kinetics.

DETERMINATION OF THE RATE-CONTROLLING STEP

The kinetics and rate-controlling steps of a fluid–solid reaction are deduced by noting how the progressive conversion of particles is influenced by particle size and operating temperature. This information can be obtained in various ways, depending on the facilities available and the materials at hand. The following observations are a guide to experimentation and to the interpretation of experimental data.

Temperature. The chemical step is usually much more temperature-sensitive than the physical steps; hence, experiments at different temperatures should easily distinguish between ash or film diffusion on the one hand and chemical reaction on the other hand as the controlling step.

Time. Figures 9 and 10 show the progressive conversion of spherical solids when chemical reaction, film diffusion, and ash diffusion in turn control. Results of kinetic runs compared with these predicted curves should indicate the rate-controlling step. Unfortunately, the difference between ash diffusion and chemical

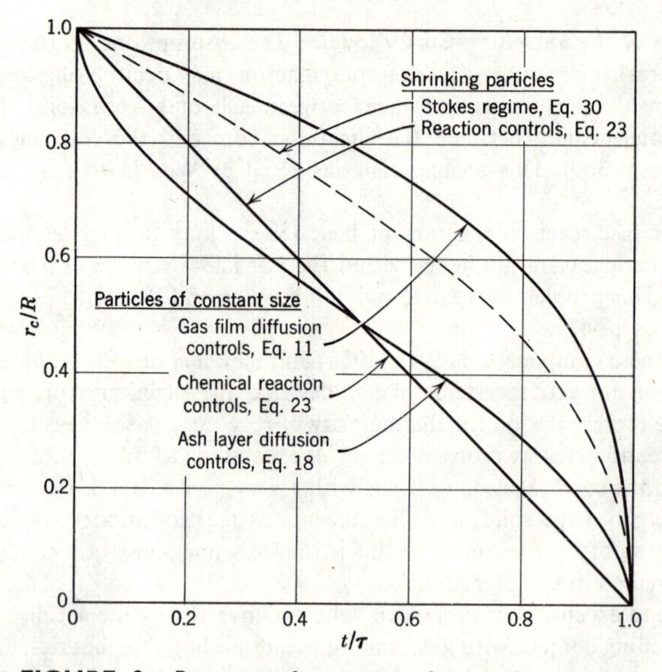

FIGURE 9. Progress of reaction of a single spherical particle with surrounding fluid measured in terms of time for complete reaction.

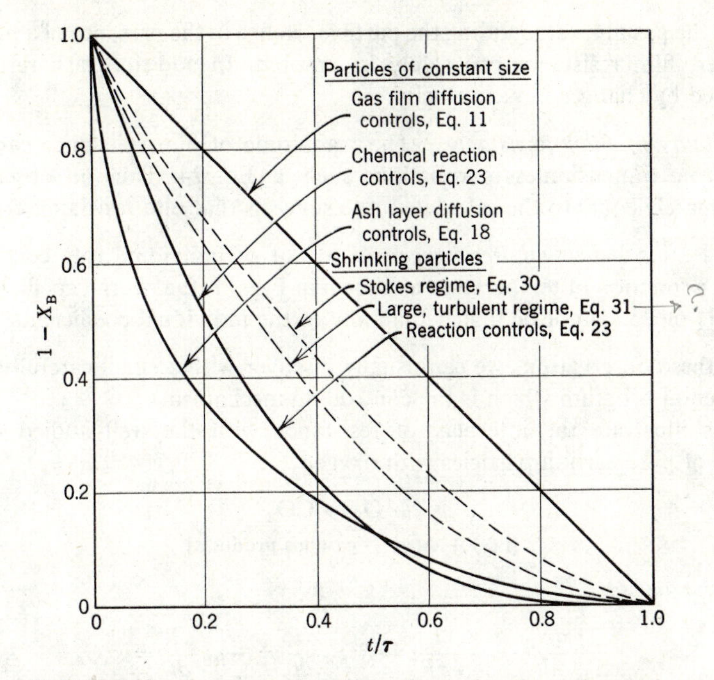

FIGURE 10. Progress of reaction of a single spherical particle with surrounding fluid measured in terms of time for complete conversion.

reaction as controlling steps is not great and may be masked by the scatter in experimental data.

Conversion-time curves analogous to those in Figs. 9 and 10 can be prepared for other solid shapes by using the equations of Table 1.

Particle Size. Equations 16, 21, and 8 with Eq. 24 or 25 show that the time needed to achieve the same fractional conversion for particles of different but unchanging sizes is given by

$t \propto R^{1.5 \text{ to } 2.0}$	for film diffusion controlling (the exponent drops as Reynolds number rises)	(36)
$t \propto R^2$	for ash diffusion controlling	(37)
$t \propto R$	for chemical reaction controlling	(38)

Thus kinetic runs with different sizes of particles can distinguish between reactions in which the chemical and physical steps control.

Ash Versus Film Resistance. When a hard solid ash forms during reaction, the resistance of gas-phase reactant through this ash is usually much greater than through the gas film surrounding the particle. Hence in the presence of a nonflaking ash layer, film resistance can safely be ignored. In addition, ash resistance is unaffected by changes in gas velocity.

Predictability of Film Resistance. The magnitude of film resistance can be estimated from dimensionless correlations such as Eq. 24. Thus an observed rate approximately equal to the calculated rate suggests that film resistance controls.

Overall Versus Individual Resistance. If a plot of individual rate coefficients is made as a function of temperature as shown in Fig. 11, the overall coefficient given by Eq. 34 or 35 cannot be higher than any of the individual coefficients.

With these observations we can usually discover with a small carefully planned experimental program which is the controlling mechanism.

Let us illustrate the interplay of resistance with the well-studied gas–solid reaction of pure carbon particles with oxygen:

$$C + O_2 \rightarrow CO_2$$
$$[B(s) + A(g) \rightarrow \text{gaseous products}]$$

with rate equation

$$-\frac{1}{S_{ex}}\frac{dN_B}{dt} = -\frac{1}{4\pi R^2} 4\pi R^2 \rho_B \frac{dR}{dt} = -\rho_B \frac{dR}{dt} = \bar{k}_s C_A$$

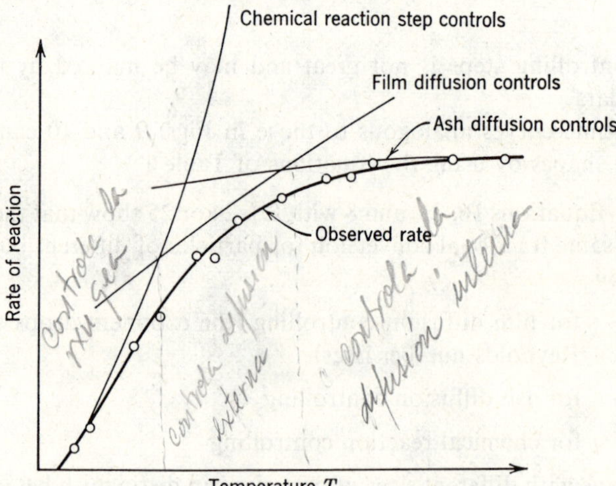

FIGURE 11. Because of the series relationship among the resistances to reaction, the net or observed rate is never higher than for any of the individual steps acting alone.

Since no ash is formed at any time during reaction, we have here a case of kinetics of shrinking particles for which two resistances at most, surface reaction and gas film, may play a role. In terms of these, the overall rate constant at any instant from Eq. 35 is

$$\frac{1}{\bar{k}_s} = \frac{1}{k_s} + \frac{1}{k_g}$$

k_g is given by Eq. 24, while k_s is given by the expression of Parker and Hottel (1936):

$$-\frac{1}{S_{ex}}\frac{dN_B}{dt} = \frac{4.32 \times 10^{11}\, C_{Ag}}{\sqrt{T}} e^{-44{,}000\ \mathrm{cal}/RT} = k_s C_{Ag} \tag{39}$$

where T is in degrees Kelvin and C_{Ag} is in gram moles per liter. Figure 12 shows all this information in convenient graphical form and allows determination of $\bar{k}_s$ for different values of the system variables. Note that when film resistance controls, the reaction is rather temperature insensitive but is dependent on particle size and relative velocity between solid and gas. This is shown by the family of lines, close to parallel and practically horizontal.

In extrapolating to new untried operating conditions, we must know when to be prepared for a change in controlling step and when we may reasonably expect the rate-controlling step not to change. For example, for particles with nonflaking ash a rise in temperature and to a lesser extent an increase in particle size may cause the rate to switch from reaction to ash diffusion controlling. For reactions in which ash is not present, a rise in temperature will again cause a shift from reaction to film resistance controlling.

APPLICATION TO DESIGN

Three factors control the design of a fluid–solid reactor; the reaction kinetics for single particles, the size distribution of solids being treated, and the flow patterns of solids and fluid in the reactor. Where the kinetics are complex and not well known, where the products of reaction form a blanketing fluid phase, where temperature within the system varies greatly from position to position, analysis of the situation becomes difficult and present design is based largely on the experiences gained by many years of operations, innovation, and small changes made on existing reactors. The blast furnace for producing iron is probably the most important industrial example of such a system.

Though some real industrial reactions may never yield to simple analysis, this should not deter us from studying idealized systems. These satisfactorily represent many real systems and in addition may be taken as the starting point for more involved analyses. Here we consider only the greatly simplified idealized systems in

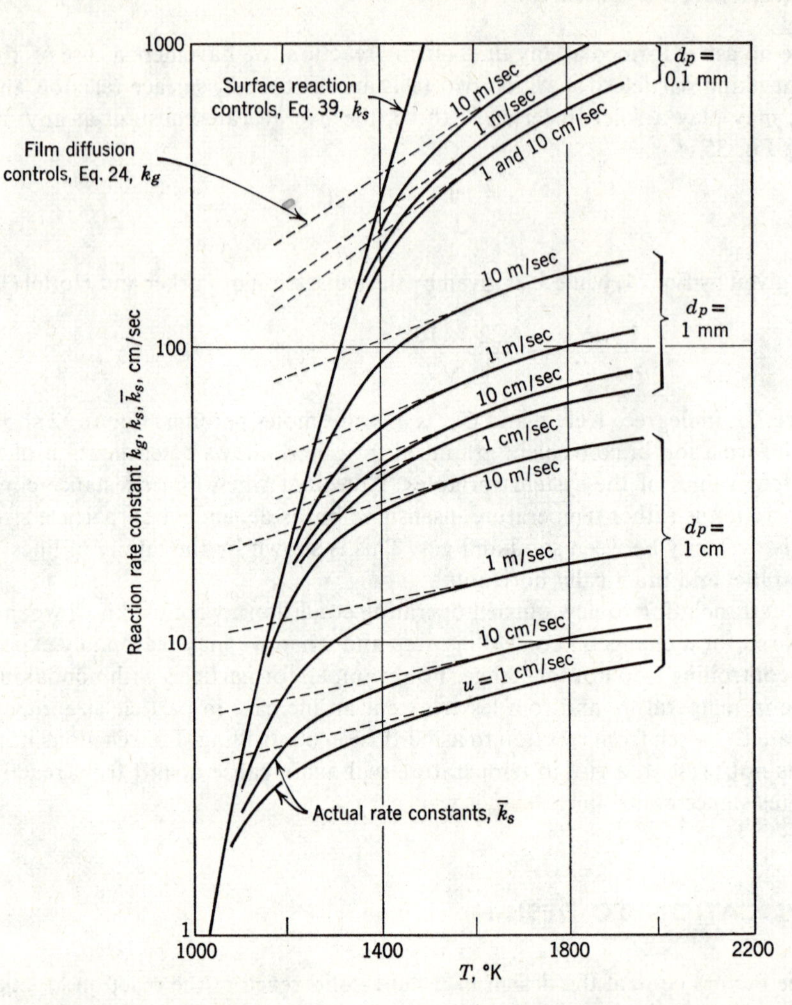

FIGURE 12. Rate of combustion of pure carbon particles, adapted from Yagi and Kunii (1955).

which the reaction kinetics, flow characteristics, and size distribution of solids are known.

Referring to Fig. 13, let us discuss briefly the various types of contacting in gas–solid operations.

Solids and Gas Both in Plug Flow. When solids and gas pass through the reactor in plug flow, their compositions will change during passage. In addition, such operations are usually nonisothermal.

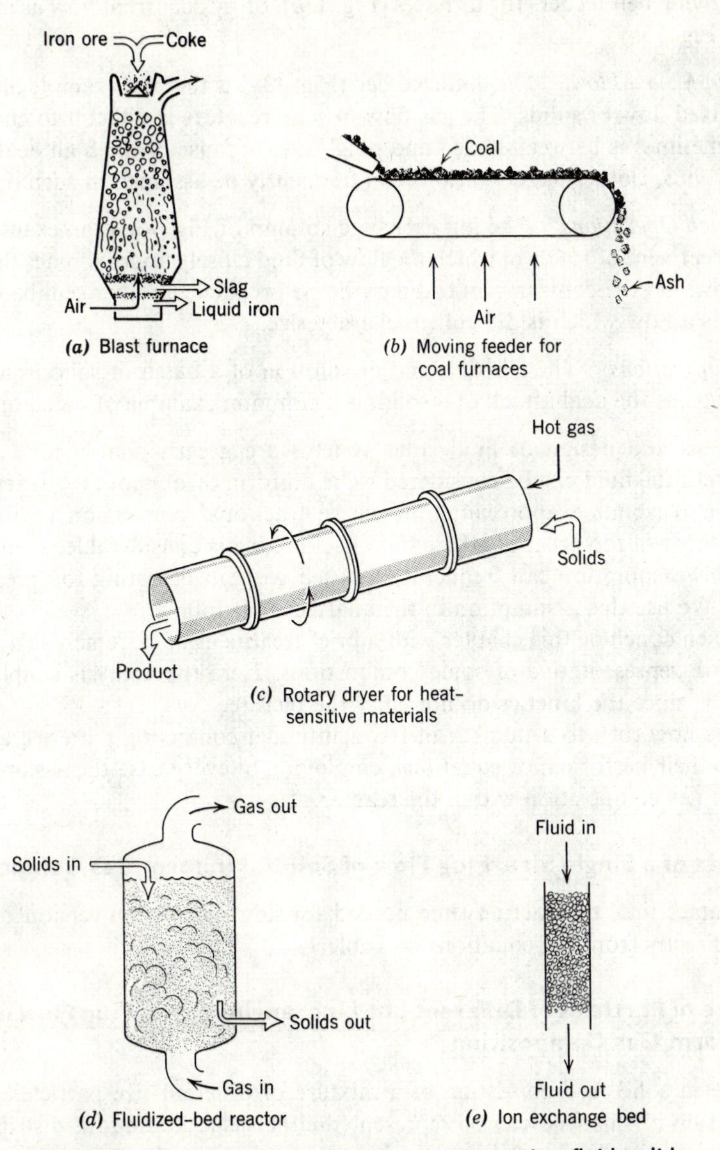

FIGURE 13. Various contacting patterns in fluid–solid reactors: (*a–c*) countercurrent, crosscurrent, and cocurrent plug flow; (*d*) intermediate gas flow, mixed solid flow; (*e*) semibatch operations.

The plug flow contacting of phases may be accomplished in many ways: by countercurrent flow as in blast furnaces and cement kilns (Fig. 13*a*), by cross flow as in moving belt feeders for furnaces (Fig. 13*b*), or by cocurrent flow as in polymer driers (Fig. 13*c*).

Solids in Mixed Flow. The fluidized bed (Fig. 13*d*) is the best example of a reactor with mixed flow of solids. The gas flow in such reactors is difficult to characterize and sometimes is between mixed and plug flow. Because of the high heat capacity of the solids, isothermal conditions can frequently be assumed in such operations.

Semibatch Operations. The ion exchange column of Fig. 13*e* is an example of the batch treatment of solids in which the flow of fluid closely approximates the ideal of plug flow. On the contrary, an ordinary home fireplace, another semibatch operation, has a flow which is difficult to characterize.

Batch Operations. The reaction and dissolution of a batch of solid in a batch of fluid, such as the acid attack of a solid, is a common example of batch operations.

Analysis and design of fluid–solid systems are greatly simplified if the composition of the fluid can be considered to be uniform throughout the reactor. Since this is a reasonable approximation where fractional conversion of fluid-phase reactants is not too great or where fluid backmixing is considerable, as in fluidized beds, this assumption can frequently be used without deviating too greatly from reality. We use this assumption in the analyses that follow.

We then conclude this chapter with a brief treatment of extremely fast reactions which are representative of some combustions. Here the analysis simplifies considerably, since the kinetics do not enter the picture.

Let us now turn to a number of frequently met contacting patterns, and let us develop their performance equations, employing in every case the assumptions of uniform gas composition within the reactor.

Particles of a Single Size, Plug Flow of Solids, Uniform Gas Composition

The contact time or reaction time needed for any specific conversion of solid is found directly from the equations of Table 1.

Mixture of Particles of Different but Unchanging Sizes, Plug Flow of Solids, Uniform Gas Composition

Consider a solid feed consisting of a mixture of different-size particles. The size distribution of this feed can be represented either as a continuous distribution or as a discrete distribution. We use the latter representation because screen analysis, our way of measuring size distributions, gives discrete measurements.

Let F be the quantity of solid being treated in unit time. Since the density of solid may change during reaction, F is defined as the volumetric feed rate of solid in the general case. Where density change of the solid is negligible, F becomes the mass feed rate of solid as well. In addition, let $F(R_i)$ be the quantity of material of size R_i fed to the reactor. If R_m is the largest particle size in the feed, we have for particles of unchanging size

$$F = \sum_{R_i=0}^{Rm} F(R_i), \quad \text{cm}^3/\text{sec or gm/sec}$$

Figure 14 shows the general characteristics of a discrete size distribution.

When in plug flow all solids stay in the reactor for the same length of time t_p. From this and the kinetics for whatever resistance controls the conversion $X_B(R_i)$ for any size of particle R_i can be found. Then the mean conversion $\bar{X}_B$ of the solids leaving the reactor can be obtained by properly summing to find the over-all contribution to conversion of all sizes of particles. Thus

$$\begin{pmatrix}\text{mean value for}\\ \text{the fraction of}\\ \text{B unconverted}\end{pmatrix} = \sum_{\substack{\text{all}\\ \text{sizes}}} \begin{pmatrix}\text{fraction of reactant}\\ \text{B unconverted in}\\ \text{particles of size } R_i\end{pmatrix}\begin{pmatrix}\text{fraction of}\\ \text{feed which is}\\ \text{of size } R_i\end{pmatrix} \tag{40}$$

or in symbols

$$1 - \bar{X}_B = \sum_{R_i=0}^{R_m} [1 - X_B(R_i)] \frac{F(R_i)}{F}, \qquad 0 \leqslant X_B \leqslant 1$$

or

$$1 - \bar{X}_B = \sum_{R(t_p=\tau)}^{R_m} [1 - X_B(R_i)] \frac{F(R_i)}{F} \tag{41}$$

where $R(t_p = \tau)$ is the radius of the largest particle completely converted in the reactor.

These two forms of Eq. 41, actually identical, require some discussion. First of all, we know that the smaller a particle the shorter the time required for complete conversion. Hence some of our feed particles, those smaller than $R(t_p = \tau)$, will be completely reacted. But if we automatically apply our conversion-time equations to these particles we will come up with X_B values greater than unity, which makes no sense physically. Thus in the first form of Eq. 41 X_B must not be allowed to take on values greater than unity. In the second form this condition is incorporated into the lower limit of the summation because particles smaller than $R(t_p = \tau)$ are completely converted and do not contribute to the fraction unconverted, $1 - \bar{X}_B$.

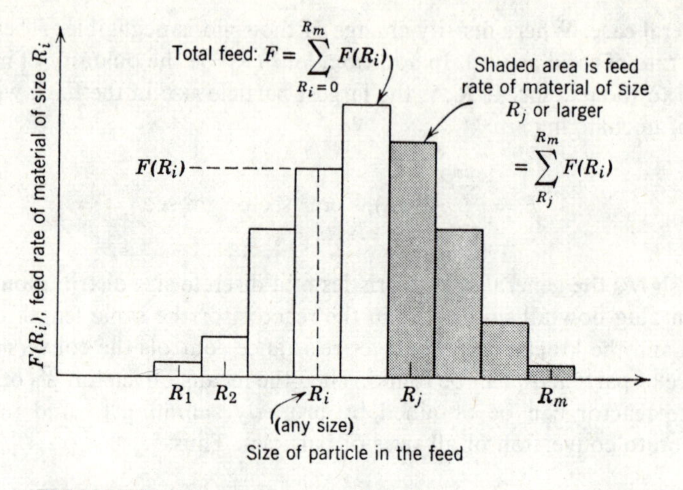

FIGURE 14. Representation of the feed rate of a mixture of particles.

The terms $R(t_p = \tau)$ and $1 - X_B(R_i)$ in Eq. 41 are given by the kinetic expressions of Table 1, and when known allow evaluation of the mean conversion for a mixture feed. The following example illustrates the procedure.

EXAMPLE 1. Conversion of a size mixture in plug flow

A feed consisting

30% of 50-μ-radius particles
40% of 100-μ-radius particles
30% of 200-μ-radius particles

is to be fed continuously in a thin layer onto a moving grate crosscurrent to a flow of reactant gas. For the planned operating conditions the time required for complete conversion is 5, 10, and 20 min for the three sizes of particles. Find the conversion of solids for a residence time of 8 min in the reactor.

SOLUTION

From the statement of the problem we may consider the solids to be in plug flow with $t_p = 8$ min and the gas to be uniform in composition. Hence for a mixed feed Eq. 41 is applicable, or

$$1 - \overline{X}_B = [1 - X_B(50\mu)] \frac{F(50\mu)}{F} + [1 - X_B(100\mu)] \frac{F(100\mu)}{F} + \cdots \qquad \text{(i)}$$

where

$$\frac{F(50\mu)}{F} = 0.30 \qquad \text{and} \qquad \tau(50\mu) = 5 \text{ min}$$

$$\frac{F(100\mu)}{F} = 0.40 \qquad \text{and} \qquad \tau(100\mu) = 10 \text{ min}$$

$$\frac{F(200\mu)}{F} = 0.30 \qquad \text{and} \qquad \tau(200\mu) = 20 \text{ min}$$

Since for the three sizes of particles

$$R_1 : R_2 : R_3 = \tau_1 : \tau_2 : \tau_3$$

we see from Eq. 38 that chemical reaction controls and the conversion-time characteristics for each size are given by Eq. 23 or

$$[1 - X_B(R_i)] = \left(1 - \frac{t_p}{\tau(R_i)}\right)^3$$

Replacing in Eq. (i) we obtain

$$1 - \bar{X}_B = \underbrace{\left(1 - \frac{8 \text{ min}}{10 \text{ min}}\right)^3 (0.4)}_{\text{for } R = 100\mu} + \underbrace{\left(1 - \frac{8}{20}\right)^3 (0.3)}_{\text{for } R = 200\mu}$$

$$= 0.0032 + 0.0648 = 0.068$$

Hence the fraction of solid converted equals 93.2%.

Note that the smallest size of particles is completely converted and does not contribute to the summation of Eq. (i).

Mixed Flow of Particles of a Single Unchanging Size, Uniform Gas Composition

Consider the reactor of Fig. 13*d* with constant flow rates of both solids and gas into and out of the reactor. With the assumption of uniform gas concentration and mixed flow of solids, this model represents a fluidized-bed reactor in which there is no carryover of solids.

The conversion of reactant in a single particle depends on its length of stay in the bed, and for the appropriate controlling resistance is given by Eq. 11, 18, or 23. However, the length of stay is not the same for all particles in the reactor; hence we must calculate a mean conversion $\bar{X}_B$ of material. Recognizing that the solid behaves as a macrofluid, this can be done by the methods of Chapter 10. Thus for the solids leaving the reactor

$$\begin{pmatrix}\text{mean value for}\\ \text{the fraction of}\\ \text{B unconverted}\end{pmatrix} = \sum_{\substack{\text{particles}\\ \text{of all}\\ \text{ages}}} \begin{pmatrix}\text{fraction of reactant}\\ \text{unconverted for}\\ \text{particles staying in}\\ \text{the reactor for time}\\ \text{between } t \text{ and } t + dt\end{pmatrix} \begin{pmatrix}\text{fraction of exit}\\ \text{stream which has}\\ \text{stayed in the}\\ \text{reactor for a time}\\ \text{between } t \text{ and } t + dt\end{pmatrix} \tag{42}$$

or in symbols

$$1 - \bar{X}_B = \int_0^\infty (1 - X_B)\mathbf{E}\, dt, \qquad X_B \leqslant 1$$

Again, when a particle remains in the reactor for a time longer than that required for complete conversion, the calculated conversion becomes greater than 100%. Since this has no physical significance, X_B should remain at unity for particle residence times greater than τ. To guarantee that such particles do not contribute to the fraction unconverted, we modify this equation to read

$$1 - \bar{X}_B = \int_0^\tau (1 - X_B)\mathbf{E}\, dt \tag{43}$$

where $\mathbf{E}$ is the exit age distribution of the solids in the reactor (see Chapter 9).

For mixed flow of solids with mean residence time $\bar{t}$ in the reactor we find from Chapter 9 that

$$\mathbf{E} = \frac{e^{-t/\bar{t}}}{\bar{t}} \tag{44}$$

Thus for mixed flow of the single size of solid which is completely converted in time τ, we obtain

$$1 - \bar{X}_B = \int_0^\tau (1 - X_B)\frac{e^{-t/\bar{t}}}{\bar{t}}\, dt \tag{45}$$

This expression may be integrated for the various controlling resistances.

For *film resistance controlling* Eq. 11 with Eq. 45 yields

$$1 - \bar{X}_B = \int_0^\tau \left(1 - \frac{t}{\tau}\right)\frac{e^{-t/\bar{t}}}{\bar{t}}\, dt \tag{46}$$

which on integration by parts gives

$$\bar{X}_B = \frac{\bar{t}}{\tau}(1 - e^{-\tau/\bar{t}})$$

or in equivalent expanded form, useful for large $\bar{t}/\tau$, (47)

$$1 - \bar{X}_B = \frac{1}{2}\frac{\tau}{\bar{t}} - \frac{1}{3!}\left(\frac{\tau}{\bar{t}}\right)^2 + \frac{1}{4!}\left(\frac{\tau}{\bar{t}}\right)^3 - \cdots$$

For *chemical reaction controlling* Eq. 23 replaced in Eq. 45 gives

$$1 - \bar{X}_B = \int_0^\tau \left(1 - \frac{t}{\tau}\right)^3 \frac{e^{-t/\bar{t}}}{\bar{t}}\, dt \tag{48}$$

Integrating by parts using the recursion formula, found in any table of integrals, we obtain

$$\bar{X}_B = 3\frac{\bar{t}}{\tau} - 6\left(\frac{\bar{t}}{\tau}\right)^2 + 6\left(\frac{\bar{t}}{\tau}\right)^3(1 - e^{-\tau/\bar{t}})$$

or in equivalent form, useful for large $\bar{t}/\tau$, (49)

$$1 - \bar{X}_B = \frac{1}{4}\frac{\tau}{\bar{t}} - \frac{1}{20}\left(\frac{\tau}{\bar{t}}\right)^2 + \frac{1}{120}\left(\frac{\tau}{\bar{t}}\right)^3 - \cdots$$

For *ash resistance controlling* replacement of Eq. 18 in Eq. 45 followed by integration leads to a cumbersome expression which on expansion yields [see Kunii (1958), Yagi and Kunii (1961)]

$$1 - \bar{X}_B = \frac{1}{5}\frac{\tau}{\bar{t}} - \frac{19}{420}\left(\frac{\tau}{\bar{t}}\right)^2 + \frac{41}{4620}\left(\frac{\tau}{\bar{t}}\right)^3 - 0.00149\left(\frac{\tau}{\bar{t}}\right)^4 + \cdots \tag{50}$$

Figures 15 and 16 present these results for solids in mixed flow in convenient graphical form. Figure 16 shows clearly that at high conversion the mixed flow reactor requires a much larger holding time for solids than does a plug flow reactor.

Extension to multistage operations is not difficult. Figures 15 and 16 show these curves for two-stage operations, and Problem P17 treats one of these cases. Kunii and Levenspiel (1969) present expressions for general N stage mixed flow operations.

EXAMPLE 2. Conversion of a single sized feed in a mixed flow reactor

Yagi *et al.* (1951) roasted pyrrhotite (iron sulfide) particles dispersed in asbestos fibers and found that the time for complete conversion was related to particle size as follows:

$$\tau \propto R^{1.5}$$

Particles remained as hard solids during reaction.

A fluidized-bed reactor is planned to convert pyrrhotite ore to the corresponding oxide. The feed is to be uniform in size, τ = 20 min, with mean residence time $\bar{t}$ = 60 min in the reactor. What fraction of original sulfide ore remains unconverted?

SOLUTION

Since a hard product material is formed during reaction, film diffusion can be ruled out as the controlling resistance. For chemical reaction controlling Eq. 38 shows that

$$\tau \propto R$$

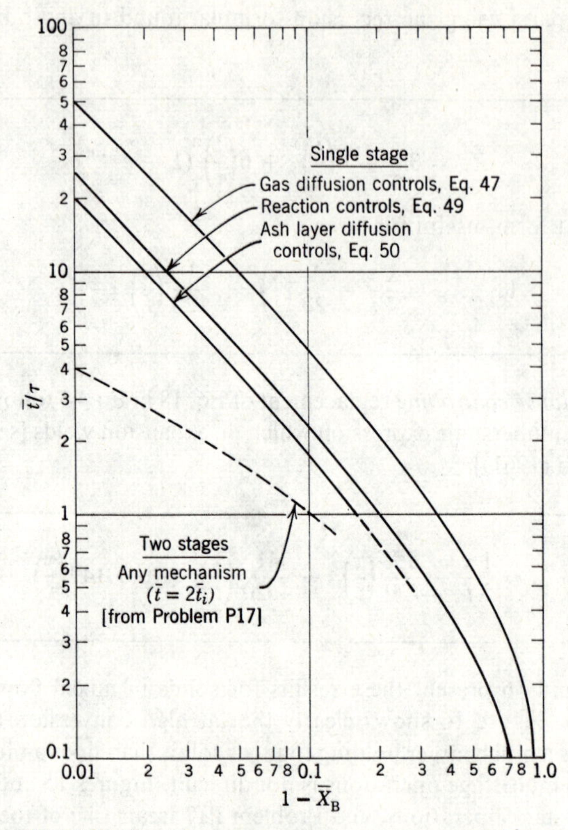

FIGURE 15. Mean conversion versus mean residence time in mixed flow reactors, single size of solid.

whereas for ash layer diffusion controlling Eq. 37 shows that

$$\tau \propto R^2$$

As the experimentally found diameter dependency lies between these two values, it is reasonable to expect that both these mechanisms offer resistance to conversion. Using in turn ash diffusion and chemical reaction as the controlling resistance should then give the upper and lower bound to the conversion expected.

The solids in a fluidized bed approximate mixed flow; hence for chemical reaction controlling Eq. 49, with $\tau/\bar{t} = 20 \text{ min}/60 \text{ min} = \frac{1}{3}$, gives

$$1 - \bar{X}_B = \frac{1}{4}\left(\frac{1}{3}\right) - \frac{1}{20}\left(\frac{1}{3}\right)^2 + \frac{1}{120}\left(\frac{1}{3}\right)^3 - \cdots = 0.078$$

For ash layer diffusion controlling Eq. 50 gives

$$1 - \bar{X}_B = \frac{1}{5}\left(\frac{1}{3}\right) - \frac{19}{420}\left(\frac{1}{3}\right)^2 + \frac{41}{4620}\left(\frac{1}{3}\right)^3 - \cdots = 0.062$$

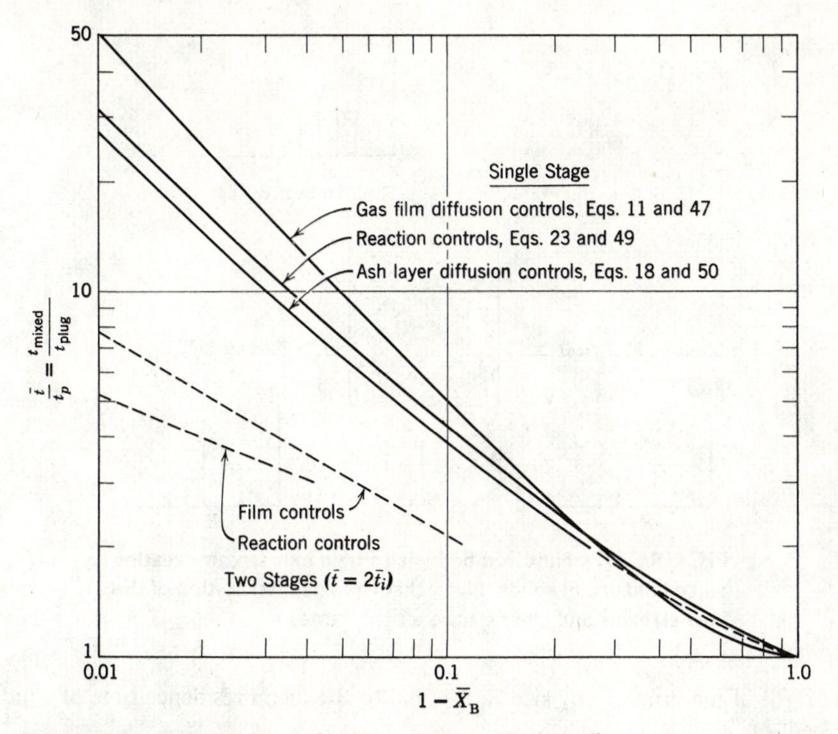

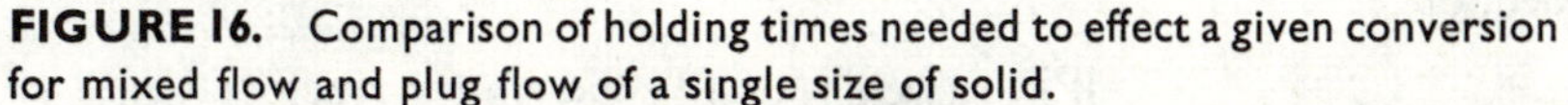

FIGURE 16. Comparison of holding times needed to effect a given conversion for mixed flow and plug flow of a single size of solid.

Hence the fraction of sulfide remaining is between 6.2% and 7.8%, or on averaging

$$1 - \bar{X}_B = 0.07, \qquad \text{or } 7.0\%$$

Mixed Flow of a Size Mixture of Particles of Unchanging Size, Uniform Gas Composition

Often a spectrum of particle sizes is used as feed to a mixed flow reactor. For such a feed and a single exit stream (no carryover) the methods leading to Eqs. 41 and 45, when combined, should yield the required conversion.

Consider the reactor shown in Fig. 17. Since the exit stream is representative of the bed conditions, the size distributions of the bed as well as the feed and exit streams are all alike, or

$$\frac{F(R_i)}{F} = \frac{W(R_i)}{W} \tag{51}$$

where W is the quantity of material in the reactor and where $W(R_i)$ is the quantity of material of any size R_i in the reactor. In addition, for this flow the mean residence

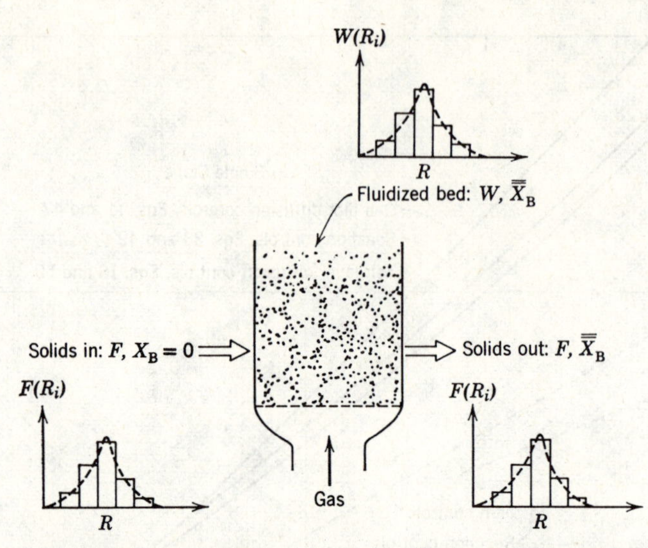

FIGURE 17. Fluidized bed with single exit stream treating a size mixture of solids. Note that the size distribution of the flow streams and the bed are all the same.

time $\bar{t}(R_i)$ of material of any size R_i is equal to the mean residence time of solid in the bed, or

$$\bar{t} = \bar{t}(R_i) = \frac{W}{F} = \frac{\text{(weight of all solids in the reactor)}}{\text{(feed rate of all solids to the reactor)}} \tag{52}$$

Letting $\bar{X}_B(R_i)$ be the mean conversion of particles of size R_i in the bed, we have from Eq. 45

$$1 - \bar{X}_B(R_i) = \int_0^{\tau(R_i)} [1 - X_B(R_i)] \frac{e^{-t/\bar{t}}}{\bar{t}}\, dt \tag{53}$$

However, feed consists of particles of different sizes; hence the over-all mean of B unconverted in all these sizes is

$$\begin{pmatrix}\text{mean value for}\\ \text{fraction of B}\\ \text{unconverted}\end{pmatrix} = \sum_{\substack{\text{all}\\ \text{sizes}}} \begin{pmatrix}\text{fraction uncon-}\\ \text{verted in particles}\\ \text{of size } R_i\end{pmatrix}\begin{pmatrix}\text{fraction of exit or entering}\\ \text{stream consisting of}\\ \text{particles of size } R_i\end{pmatrix} \tag{54}$$

or in symbols

$$1 - \bar{\bar{X}}_B = \sum_{R=0}^{R_m} [1 - \bar{X}_B(R_i)] \frac{F(R_i)}{F}$$

Combining Eqs. 53 and 54 and replacing the former expression with Eqs. 47, 49, or 50 for each size of particle, we obtain in turn, for *film diffusion controlling*,

$$1 - \overline{\overline{X}}_B = \sum^{R_m} \left\{ \frac{1}{2!} \frac{\tau(R_i)}{\bar{t}} - \frac{1}{3!} \left[\frac{\tau(R_i)}{\bar{t}} \right]^2 + \cdots \right\} \frac{F(R_i)}{F} \tag{55}$$

for *chemical reaction controlling*,

$$1 - \overline{\overline{X}}_B = \sum^{R_m} \left\{ \frac{1}{4} \frac{\tau(R_i)}{\bar{t}} - \frac{1}{20} \left[\frac{\tau(R_i)}{\bar{t}} \right]^2 + \cdots \right\} \frac{F(R_i)}{F} \tag{56}$$

for *ash diffusion controlling*,

$$1 - \overline{\overline{X}}_B = \sum^{R_m} \left\{ \frac{1}{5} \frac{\tau(R_i)}{\bar{t}} - \frac{19}{420} \left[\frac{\tau(R_i)}{\bar{t}} \right]^2 + \cdots \right\} \frac{F(R_i)}{F} \tag{57}$$

where $\tau(R_i)$ is the time for complete reaction of particles of size R_i. The following example illustrates the use of these expressions.

EXAMPLE 3. Conversion of a feed mixture in a mixed flow reactor

A feed consisting

30% of 50-μ-radius particles
40% of 100-μ-radius particles
30% of 200-μ-radius particles

is to be reacted in a fluidized-bed steady-state flow reactor constructed from a 4-ft length of 4-in. pipe. The fluidizing gas is the gas-phase reactant, and at the planned operating conditions the time required for complete conversion is 5, 10, and 20 min for the three sizes of feed. Find the conversion of solids in the reactor for a feed rate of 1 kg solids/min if the bed contains 10 kg solids.

Additional information:

The solids are hard and unchanged in size and weight during reaction.

A cyclone separator is used to separate and return to the bed any solids that may be entrained by the gas.

The change in gas-phase composition across the bed is small.

SOLUTION

From the statement of the problem we may treat the fluidized bed as a mixed flow reactor. For a feed mixture Eq. 54 is applicable, and since chemical reaction controls (see Example 1), this equation reduces to Eq. 56 where from the problem statement

$$F = 1000 \text{ gm/min} \qquad W = 10{,}000 \text{ gm} \qquad \bar{t} = \frac{W}{F} = \frac{10{,}000 \text{ gm}}{1000 \text{ gm/min}} = 10 \text{ min}$$

$$F(50\mu) = 300 \text{ gm/min} \quad \text{and} \quad \tau(50\mu) = 5 \text{ min}$$
$$F(100\mu) = 400 \text{ gm/min} \quad \text{and} \quad \tau(100\mu) = 10 \text{ min}$$
$$F(200\mu) = 300 \text{ gm/min} \quad \text{and} \quad \tau(200\mu) = 20 \text{ min}$$

Replacing in Eq. 46 we obtain

$$1 - \overline{\overline{X}}_B = \left[\frac{1}{4}\left(\frac{5\ \text{min}}{10\ \text{min}}\right) - \frac{1}{20}\left(\frac{5}{10}\right)^2 + \cdots\right]\frac{300\ \text{gm/min}}{1000\ \text{gm/min}}$$

for $R = 50\mu$

$$+ \left[\frac{1}{4}\left(\frac{10\ \text{min}}{10\ \text{min}}\right) - \frac{1}{20}\left(\frac{10}{10}\right)^2 + \cdots\right]\frac{400}{1000}$$

for $R = 100\mu$

$$+ \left[\frac{1}{4}\left(\frac{20\ \text{min}}{10\ \text{min}}\right) - \frac{1}{20}\left(\frac{20}{10}\right)^2 + \cdots\right]\frac{300}{1000}$$

for $R = 200\mu$

$$= \left(\frac{1}{8} - \frac{1}{80} + \cdots\right)\frac{3}{10} + \left(\frac{1}{4} - \frac{1}{20} + \frac{1}{120} - \cdots\right)\frac{4}{10}$$

$$+ \left(\frac{1}{2} - \frac{1}{5} + \frac{1}{15} - \frac{2}{110} + \cdots\right)\frac{3}{10}$$

$$= 0.034 + 0.083 + 0.105 = 0.222$$

The mean conversion of solids is then

$$\overline{\overline{X}}_B = 77.8\%$$

Application to a Fluidized Bed with Entrainment of Solid Fines

Carryover of fines may occur in a fluidized bed when the feed consists of a wide size distribution of solids. When this happens we have a reactor as shown in Fig. 18 with one feed stream and two exit streams. Let subscripts 0, 1, 2 refer to the feed, underflow, and carryover streams, respectively. Then by material balance for the entire streams we find

$$F_0 = F_1 + F_2 \tag{58}$$

and for particles of size R_i

$$F_0(R_i) = F_1(R_i) + F_2(R_i) \tag{59}$$

Since mixed flow is assumed, the composition of the underflow stream still represents the composition within the bed, or

$$\frac{F_1(R_i)}{F_1} = \frac{W(R_i)}{W} \tag{60}$$

Now the mean residence time of material of different sizes need not be the same. In fact, since small particles are more likely to be blown out of the bed, intuition

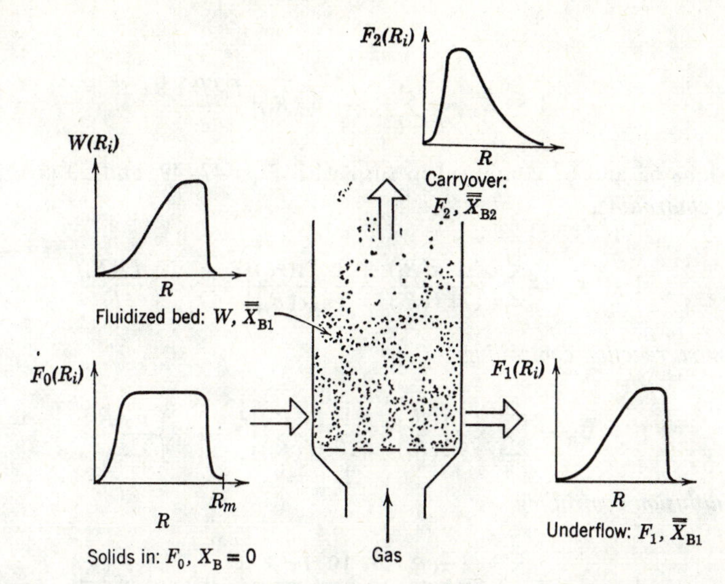

FIGURE 18. Fluidized bed with carryover, showing size distributions of all streams.

suggests that they stay in the bed for a shorter time than do the larger particles. So, for particles of size R_i we find, on combining Eqs. 59 and 60,

$$\bar{t}(R_i) = \frac{(\text{weight of particles of size } R_i \text{ in the bed})}{(\text{flow rate of such particles in or out of the bed})}$$

$$= \frac{W(R_i)}{F_0(R_i)} = \frac{W(R_i)}{F_1(R_i) + F_2(R_i)} = \frac{1}{\dfrac{F_1}{W} + \dfrac{F_2(R_i)}{W(R_i)}} \tag{61}$$

This expression shows that the mean residence time, hence conversion, of particles of any given size is the same in the underflow and carryover stream.

The mean conversion of particles of size R_i from Eq. 43 is then

$$1 - \bar{X}_B(R_i) = \int_0^{\tau(R_i)} [1 - X_B(R_i)] \frac{e^{-t/\bar{t}(R_i)}}{\bar{t}(R_i)}\, dt \tag{62}$$

and for a feed consisting of a size mixture of particles we have, as with Eq. 54,

$$\begin{pmatrix}\text{mean value for}\\ \text{fraction of B}\\ \text{unconverted}\end{pmatrix} = \sum_{\text{all sizes}} \begin{pmatrix}\text{fraction uncon-}\\ \text{verted in particles}\\ \text{of size } R_i\end{pmatrix}\begin{pmatrix}\text{fraction of feed}\\ \text{consisting of particles}\\ \text{of size } R_i\end{pmatrix}$$

or

$$1 - \overline{\overline{X}}_B = \sum^{R_m} [1 - \overline{X}_B(R_i)] \frac{F_0(R_i)}{F_0} \tag{63}$$

Equations 63 and 62 combined in turn with Eqs. 47, 49, and 50 give for *film diffusion controlling*,

$$1 - \overline{\overline{X}}_B = \sum^{R_m} \left\{ \frac{1}{2!} \frac{\tau(R_i)}{\bar{t}(R_i)} - \frac{1}{3!} \left[\frac{\tau(R_i)}{\bar{t}(R_i)} \right]^2 + \cdots \right\} \frac{F_0(R_i)}{F_0} \tag{64}$$

for *chemical reaction controlling*,

$$1 - \overline{\overline{X}}_B = \sum^{R_m} \left\{ \frac{1}{4} \frac{\tau(R_i)}{\bar{t}(R_i)} - \frac{1}{20} \left[\frac{\tau(R_i)}{\bar{t}(R_i)} \right]^2 + \cdots \right\} \frac{F_0(R_i)}{F_0} \tag{65}$$

for *ash diffusion controlling*,

$$1 - \overline{\overline{X}}_B = \sum^{R_m} \left\{ \frac{1}{5} \frac{\tau(R_i)}{\bar{t}(R_i)} - \frac{19}{420} \left[\frac{\tau(R_i)}{\bar{t}(R_i)} \right]^2 + \cdots \right\} \frac{F_0(R_i)}{F_0} \tag{66}$$

Comparison with Eqs. 55 to 57 shows that it is simply the variation of the mean residence time with particle size which distinguishes the conversion in beds with carryover from the ordinary mixed flow in which $\bar{t}(R_i) = \bar{t} =$ constant. For a feed of a single size, $\tau(R_i) = \tau =$ constant, and these expressions reduce even further to Eqs. 47, 49, and 50.

To use these conversion equations we must know $\bar{t}(R_i)$, as yet an unknown quantity, which is dependent on the properties of the two exit streams (see Eq. 61). To find the flow split and properties of these two exit streams requires some independent information on the rate at which particles are blown out of fluidized beds, and this we now consider.

$\bar{t}(R_i)$ from Elutriation Data. Elutriation experiments show that the number of marked particles of a given size blown out of a fluidized bed is proportional to the number of such particles present in the bed, or

$$-\frac{d(\text{number of marked particles})}{dt} = \kappa(\text{number of marked particles in the bed}) \tag{67}$$

where κ, called the elutriation velocity constant, has the units of reciprocal time and is a function of the properties of the system. Yagi and Aochi (1955) and Wen and Hashinger (1960) present generalized relationships for κ which correlate data obtained in a wide variety of physical systems.

Figure 19, illustrating typical elutriation experiments, shows how particle size, gas velocity, and bed depth influence κ. In fluidization of small particles which are

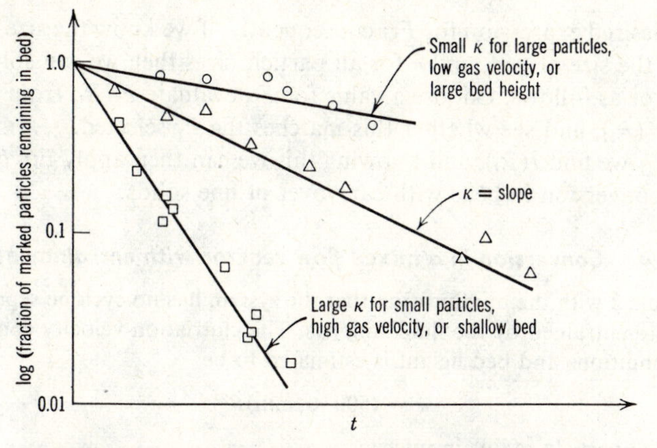

FIGURE 19. Typical results of elutriation experiments a batch of solids showing the dependence of κ on the fluidization variables. Adapted from Yagi and Aochi (1955).

to some extent blown out of the bed, the interrelationship between these variables is approximated by

$$\kappa \sim \frac{(\text{gas velocity})^4}{(\text{bed height})(\text{particle size})^{2 \text{ to } 3}} \tag{68}$$

Equation 68 should not be extrapolated to larger particles which are not blown out of the bed because it predicts a nonzero value for κ, whereas for such particles κ is actually zero.

For particles of size R_i in the bed, and at steady state,

$$\kappa(R_i) = \frac{(\text{rate of carryover of particles of size } R_i)}{(\text{weight of such particles present in the bed})} = \frac{F_2(R_i)}{W(R_i)} \tag{69}$$

With $\kappa(R_i)$ data available from independent experiments, Eq. 61 becomes

$$\bar{t}(R_i) = \frac{W(R_i)}{F_0(R_i)} = \frac{1}{F_1/W + \kappa(R_i)} \tag{70}$$

One last term, F_1, is still to be evaluated before $\bar{t}(R_i)$ can be found. To find F_1 combine Eq. 70 with Eq. 60 and rearrange as follows:

$$F_1(R_i) = \frac{F_0(R_i)}{1 + (W/F_1)\kappa(R_i)} \tag{71}$$

and on summing over all particle sizes

$$F_1 = F_1(R_1) + F_1(R_2) + \cdots + F_1(R_m)$$

$$= \sum^{R_m} \frac{F_0(R_i)}{1 + (W/F_1)\kappa(R_i)} \tag{72}$$

This is the desired expression for F_1; consequently, if we know the size distribution of the feed, the size of bed, and κ for all particle sizes then we can solve for F_1 by trial and error as follows. Choose a value for F_1, evaluate $F_1(R_i)$ from Eq. 71, sum up all the $F_1(R_i)$, and see whether this matches the F_1 selected.

So from F_1 we find $\bar{t}(R_i)$, and knowing this we can then apply Eq. 64, 65, or 66 to find the conversion in beds with carryover of fine solids.

EXAMPLE 4. *Conversion in a mixed flow reactor with entrainment of fines*

Solve Example 3 with the modification that the system has no cyclone separator, therefore solids are entrained by the fluidizing gas. The elutriation velocity constant for the operating conditions and bed height is estimated to be

$$\kappa = (500\ \mu^2/\text{min})R^2$$

where R is the particle radius in microns.

SOLUTION

From Example 3 and solving for $\kappa(R_i)$ we have

$$F_0 = 1000\ \text{gm/min}, \qquad W = 10{,}000\ \text{gm} \qquad \bar{t} = \frac{W}{F_0} = 10\ \text{min}$$

$$\kappa(50\ \mu) = 0.2/\text{min} \qquad \tau(50\ \mu) = 5\ \text{min} \qquad F_0(50\ \mu) = 300\ \text{gm/min}$$

$$\kappa(100\ \mu) = 0.05/\text{min} \qquad \tau(100\ \mu) = 10\ \text{min} \qquad F_0(100\ \mu) = 400\ \text{gm/min}$$

$$\kappa(200\ \mu) = 0.0125/\text{min} \qquad \tau(200\ \mu) = 20\ \text{min} \qquad F_0(200\ \mu) = 300\ \text{gm/min}$$

Figure E4*a* shows all the stream quantities known at this point. The solution procedure then is as follows.

Step 1. Find F_1 by trial and error solution of Eq. 72.
Step 2. Determine $\bar{t}(R_i)$ from Eq. 70.
Step 3. Calculate the conversion from Eq. 65, since chemical reaction controls.

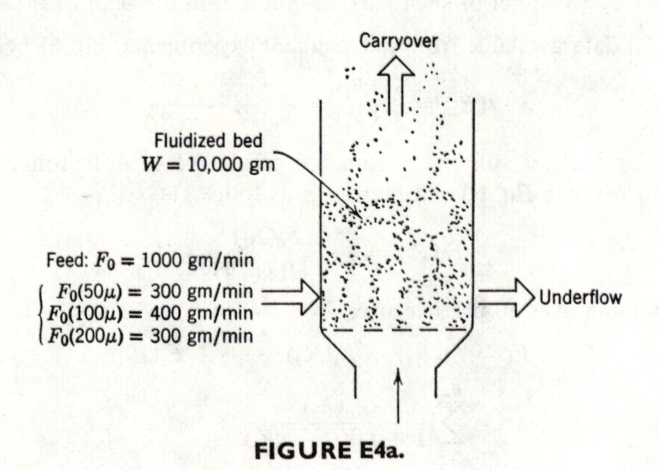

FIGURE E4a.

Step 1. Guess $F_1 = 625$ gm/min; then the value of F_1 calculated by Eq. 72 is

$$F_1 = \underset{R = 50\mu}{\frac{300 \text{ gm/min}}{1 + \dfrac{10{,}000 \text{ gm}}{625 \text{ gm/min}}(0.2/\text{min})}} + \underset{R = 100\mu}{\frac{400}{1 + \dfrac{10{,}000}{125}(0.05)}} + \underset{R = 200\mu}{\frac{300}{1 + \dfrac{10{,}000}{625}(0.0125)}}$$

$$= 71.4 \text{ gm of 50-}\mu \text{ material} + 222.2 + 250$$

$$= 543.6 \text{ gm}, \cdots \text{too low}$$

Guess $F_1 = 400$ gm/min. Then by calculation

$$F_1 = \frac{300}{1 + \dfrac{10{,}000}{400}(0.2)} + \frac{400}{1 + \dfrac{10{,}000}{400}(0.05)} + \frac{300}{1 + \dfrac{10{,}000}{400}(0.0125)}$$

$$= 50 + 177.7 + 228.5$$

$$= 456.2 \text{ gm/min}, \cdots \text{too high}$$

Guess $F_1 = 500$ gm. Then by calculations similar to those just given we find

$$F_1 = 60 + 200 + 240 = 500 \text{ gm}, \cdots \text{check}$$

Thus we have

$$F_1 = 500 \text{ gm/min} \quad \text{and} \quad \left.\begin{aligned} F_1(50\ \mu) &= 60 \text{ gm/min} \\ F_1(100\ \mu) &= 200 \text{ gm/min} \\ F_1(200\ \mu) &= 240 \text{ gm/min} \end{aligned}\right\}$$

Step 2. From Eq. 70 the mean residence times of the various sizes of particles are

$$\bar{t}(50\ \mu) = \frac{1}{F_1/W + \kappa(50\ \mu)} = \frac{1}{\dfrac{500 \text{ gm/min}}{10{,}000 \text{ gm}} + 0.2 \text{ min}} = 4 \text{ min}$$

$$\bar{t}(100\ \mu) = \frac{1}{500/10{,}000 + 0.05} = 10 \text{ min}$$

$$\bar{t}(200\ \mu) = \frac{1}{500/10{,}000 + 0.0125} = 16 \text{ min}$$

Step 3. From Eq. 65 we find the mean overall conversion to be

$$1 - \bar{\bar{X}}_B = \underset{R = 50\mu}{\left[\frac{1}{4}\left(\frac{5 \text{ min}}{4 \text{ min}}\right) - \frac{1}{20}\left(\frac{5}{4}\right)^2 + \frac{1}{120}\left(\frac{5}{4}\right)^3 - \cdots\right]\frac{300 \text{ gm/min}}{1000 \text{ gm/min}}}$$

$$+ \underset{R = 100\mu}{\left[\frac{1}{4}\left(\frac{10 \text{ min}}{10 \text{ min}}\right) - \frac{1}{20}\left(\frac{10}{10}\right)^2 + \cdots\right]\frac{400}{1000}}$$

$$+ \underset{R = 200\mu}{\left[\frac{1}{4}\left(\frac{20 \text{ min}}{16 \text{ min}}\right) - \frac{1}{20}\left(\frac{20}{16}\right)^2 + \cdots\right]\frac{300}{1000}}$$

$$= 0.075 + 0.083 + 0.075 = 0.233$$

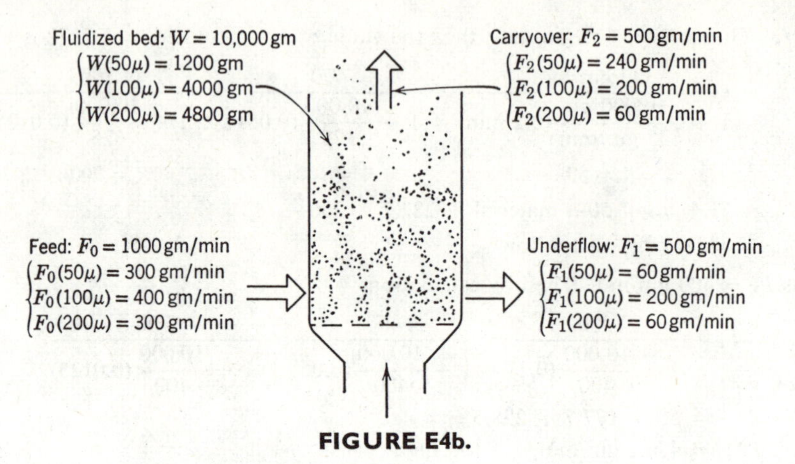

FIGURE E4b.

The mean conversion of solids is then

$$\overline{\overline{X}}_B = 76.7\%$$

The composition of the bed and of the various streams, obtained by material balance, is shown in Fig. E4*b*.

Comparing the solutions for Examples 3 and 4 we see that conversion is not lowered appreciably by the elutriation of fines from the bed. Actually, as shown in the problems at the end of the chapter, conversion can sometimes be increased by allowing solids to be blown out of the bed. This may seem surprising at first, but the reason becomes clear when we realize that the elutriation of fines which are converted in a short time allows the larger sizes of particles to remain in the bed for longer periods, hence increasing their conversion.

The predominance of larger particles in the bed can be seen in Fig. E4*b*.

Instantaneous Reaction

When reaction between gas and solid is fast enough so that any volume element of reactor contains only one or other of the two reactants, but not both, then we may consider reaction to be instantaneous. This extreme is approached in the high temperature combustion of finely divided solids.

In this situation prediction of performance of the reactor is straightforward and is dependent only on the stoichiometry of the reaction. The kinetics do not enter the picture. Let us illustrate this behavior with the following ideal contacting patterns.

Batch Solids. Figure 20 shows two situations, one which represents a packed bed, the other a fluidized bed with no bypassing of gas in the form of large gas bubbles. In both cases the leaving gas is completely converted and remains that way so long as solid reactant is still present in the bed. As soon as the solids are all consumed,

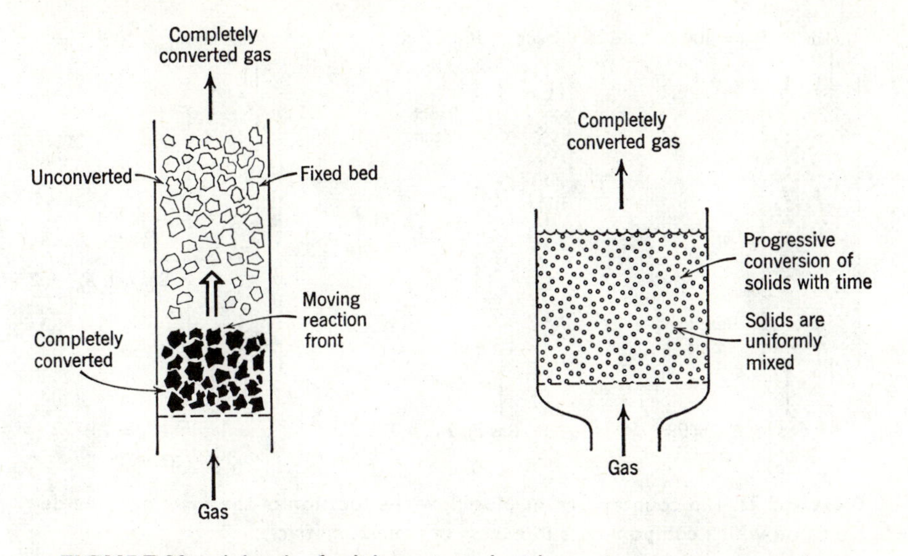

FIGURE 20. A batch of solids contacted with gas; instantaneous reaction.

and this occurs the instant the stoichiometric quantity of gas has been added, then the conversion of gas drops to zero.

Countercurrent Plug Flow of Gas and Solids. Since only one or other reactant can be present at any level in the bed there will be a sharp reaction plane where the reactants meet. This will occur either at one end or the other of the reactor depending on which feed stream is in excess of stoichiometric. Assuming that each 100 moles of solid combine with 100 moles of gas Figs. 21*a* and *b* show what happens when we feed a little less gas than stoichiometric and a little more than stoichiometric.

We may wish reaction to occur in the center of the bed so that both ends can be used as heat exchange regions to heat up reactants. This can be done by matching the gas and solids flow rates; however, this is inherently an unstable system and requires proper control. A second alternative, shown in Fig. 21*c*, introduces a slight excess of gas at the bottom of the bed, and then removes more than this excess at the point where reaction is to occur.

Moving bed reactors for oil recovery from shale is one example of this kind of operation. Another somewhat analogous operation is the multistage counterflow reactor, and the 4- or 5-stage fluidized calciner is a good example of this. In all these operations the efficiency of heat utilization is the main concern.

Cocurrent and Crosscurrent Plug Flow. In cocurrent flow, shown in Fig. 22*a*, all reaction occurs at the feed end, and this represents a poor method of contacting with regard to efficiency of heat utilization and preheating of entering materials.

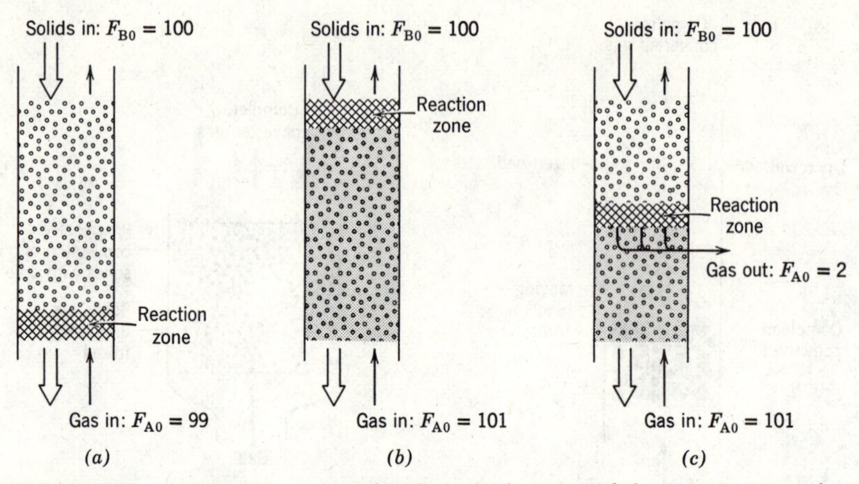

FIGURE 21. In countercurrent plug flow the location of the reaction zone depends on which component is in excess of stoichiometric.

For crosscurrent flow, shown in Fig. 22*b*, there will be a definite reaction plane in the solids whose angle depends solely on the stoichiometry and the relative feed rate of reactants. In practice heat transfer characteristics may somewhat modify the angle of this plane.

Mixed Flow of Solids and Gas. Again in the ideal situation either gas or solid will be completely converted in the reactor depending on which stream is in excess.

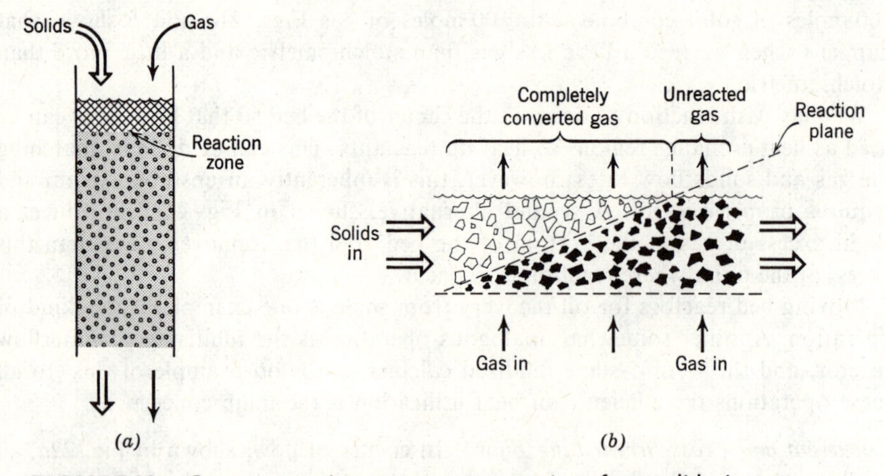

FIGURE 22. Cocurrent and crosscurrent contacting of gas–solids; instantaneous reaction.

The real situation is more complex, however, and to find the distribution of conversion among the particles leaving the bed requires a model for the action within the bed. The most reasonable one assumes that any particle of solid contains so much more reactant than does the surrounding gas that the particle depletes the gas of reactant as it moves through the bed. Thus, on the average, the conversion of solid increases slowly and steadily with time, or

$$X_B \propto (\text{time of stay in bed}) = kt \tag{73}$$

This then gives zero-order macrofluid behavior for the solids.

Incineration of solid wastes in fluidized beds of sand particles in some ways approximate this behavior. Here the emulsion phase containing solid wastes is often starved of oxygen (bubble-emulsion interchange rate in fluidized beds is often rather low), hence solids may well be pyrolized giving off combustible vapors to the vapor space above the bed. In addition, gas bubbles rich in oxygen bypass through the bed to burn the vapors above the bed. The consequence of this is a significant afterburning and temperature rise above the bed. In most beds, however, the residence time of solids is sufficiently long to yield practically complete conversion of solids.

COMMENTS

We have examined a number of contacting patterns for the fluid–solid phases and have presented the design procedure for them. In doing so we have assumed ideal flow for the two phases. For the solids the assumption of mixed flow is usually quite reasonable; however, making a similar assumption for the flow of gas sometimes represents a gross simplification which leads to significant error. For proper design of fluidized units flow patterns for the gas such as given in Chapter 9 should be used. Kunii and Levenspiel (1969) show how to treat this more realistic case.

In addition to these assumptions for flow, we also used a rather simplified model for the kinetics of the reaction, the shrinking core model. We may feel that with these rather severe restrictive assumptions our models have but few applications. This is not so, for models such as those presented here have represented satisfactorily a large number of systems of industrial importance and have been used for their actual design; see Yagi and Kunii (1961) and Kunii (1958). Also, we should note that when reaction is fast enough then the treatment is simplified and becomes independent of the kinetics.

Modification of the methods in this chapter to account for deviation from plug flow in moving bed reactors is treated by Yagi *et al.* (1961) with the dispersion model, and an extension to reactions of particles of changing size is treated in Kunii and Levenspiel (1969).

RELATED READINGS

Kunii, D., and Levenspiel, O., *Fluidization Engineering*, John Wiley & Sons, New York, 1969: on elutriation, Chapter 10; on treatment of solids of changing size, Chapter 11; on heterogeneous gas-solid reactions, Chapter 15.

Wen, C. Y., *Ind. Eng. Chem.*, **60**, 34 (1968).

Shen, J., and Smith, J. M., *Ind. Eng. Chem. Fundamentals*, **4**, 293 (1965).

REFERENCES

Ishida, M., and Wen, C. Y., *Chem. Eng. Sci.*, **26**, 1031 (1971*a*).

Ishida, M., Wen, C. Y., and Shirai, T., *Chem. Eng. Sci.*, **26**, 1043 (1971*b*).

Kunii, D., Ph.D. Thesis, University of Tokyo, 1958.

Kunii, D., and Levenspiel, O., *Fluidization Engineering*, John Wiley & Sons, New York, 1969.

Otake, T., Tone, S., and Oda, S., *Chem. Eng.* (Japan), **31**, 71 (1967).

Parker, A. L., and Hottel, H. C., *Ind. Eng. Chem.*, **28**, 1334 (1936).

Ranz, W. E., and Marshall, W. R., *Chem. Eng. Prog.*, **48**, 173 (1952).

Shen, J., and Smith, J. M., *Ind. Eng. Chem. Fundamentals*, **4**, 293 (1965).

Wen, C. Y., *Ind. Eng. Chem.*, **60**, 34 (1968).

Wen, C. Y., and Hashinger, R. F., *A.I.Ch.E. Journal*, **6**, 220 (1960).

Wen, C. Y., and Wang, S. C., *Ind. Eng. Chem.*, **62**, 30 (1970).

White, D. E., and Carberry, J. J., *Can. J. Chem. Eng.*, **43**, 334 (1965).

Yagi, S., and Aochi, T., Paper presented at the Society of Chemical Engineers (Japan); see Yagi and Kunii (1961).

Yagi, S., and Kunii, D., *5th Symposium (International) on Combustion*. Reinhold, New York, 1955, p. 231; *Chem. Eng.* (Japan), **19**, 500 (1955).

——— and ———, *Chem. Eng. Sci.*, **16**, 364, 372, 380 (1961).

———, ———, Nagahara, K., and Naito, H., *Chem. Eng.* (Japan), **25**, 469 (1961).

PROBLEMS

These are grouped as follows:

Problems 1–8: Kinetics of conversion of solids
Problems 9–19: Simple design
Problems 20–26: Elutriation
Problems 27–30: Design for systems with carryover of solids

Problems 31–36: Miscellaneous including those where the conversion of the gas phase must also be considered.

1. A batch of solids of uniform size is treated by gas in a uniform environment. Solid is converted to give a nonflaking product according to the shrinking core model. Conversion is about $\frac{7}{8}$ for a reaction time of one hour, conversion is complete in two hours. What mechanism is rate controlling?

2. In a shady spot at the end of Brown Street in Lewisburg, Pennsylvania, stands a Civil War memorial—a brass general, a brass cannon which persistent undergraduate legend insists may still fire some day, and a stack of iron cannonballs. At the time this memorial was set up, 1868, the cannonballs were 30 inches in circumference. Today due to weathering, rusting, and the once-a-decade steel wire scrubbing by the DCW, the cannonballs are only 29.75 inches in circumference. Approximately, when will they disappear completely?

3. Calculate the time needed to burn to completion particles of graphite (R_0 = 5 mm, ρ_B = 2.2 gm/cm^3, k_s = 20 cm/sec) in an 8% oxygen stream. For the high gas velocity used assume that film diffusion does not offer any resistance to transfer and reaction. Reaction temperature = 900°C.

4. Particles react with gas of given composition and at given temperature to give a solid product. What can you say about the kinetics of the reaction if the rate of reaction per gram of solid is

(*a*) Proportional to the diameter of particles.
(*b*) Proportional to the square of the particle diameter.
(*c*) Independent of particle size.

5. Two small samples of solids are introduced into a constant environment oven and kept there for one hour. Under these conditions the 4 mm particles are 58% converted, the 2 mm particles are 87.5% converted.

(*a*) Find the rate-controlling mechanism for the conversion of solids.
(*b*) Find the time needed for complete conversion of 1-mm particles in this oven.

6. The reduction of iron ore of density ρ_B = 4.6 gm/cm^3 and size R = 5 mm by hydrogen can be approximated by the unreacted core model. With no water vapor present the stoichiometry of the reaction is

$$4H_2 + Fe_3O_4 \rightarrow 4H_2O + 3Fe$$

with rate approximately proportional to the concentration of hydrogen in the gas stream. The first-order rate constant has been measured by Otake *et al.* (1967) to be

$$k_s = 1.93 \times 10^{+5} e^{-24,000/RT}, \quad \text{cm/sec}$$

(*a*) Taking $\mathscr{D}_e$ = 0.03 cm^2/sec as the average value of the diffusion coefficient for hydrogen penetration of the product layer calculate the time necessary for complete conversion of a particle from oxide to metal at 600°C.

(*b*) Does any particular resistance control? If not, what is the relative importance of the various resistance steps?

7. Spherical particles of zinc blende of size $R = 1$ mm are roasted in an 8% oxygen stream at 900°C and 1 atm. The stoichiometry of the reaction is

$$2ZnS + 3O_2 \rightarrow 2ZnO + 2SO_2$$

Assuming that reaction proceeds by the shrinking core model

(*a*) Calculate the time needed for complete conversion of a particle and the relative resistance of ash layer diffusion during this operation.

(*b*) Repeat for particles of size R = 0.05 mm.

Data: Density of solid, $\rho_B = 4.13$ gm/cm³ $= 0.0425$ mol/cm³
Reaction rate constant, $k_s = 2$ cm/sec
For gases in the ZnO layer, $\mathscr{D}_e = 0.08$ cm²/sec
Note that film resistance can safely be neglected as long as a growing ash layer is present.

8. Let us explore the characteristics of time delay (slow absorption) capsules for taking medicines. In one approach these capsules contain many small particles of slowly dissolving matrix impregnated with active chemical. Sketch the rate of dissolution of active material with time for spherical, cylindrical, and flat plate particles of the same minimum dimension.

9. A 200-ton solid holdup is needed in a single fluidized bed reactor for 99% conversion of particles of unchanging size where ash diffusion controls. What would be the holdup in two fluidized beds in series for identical conversion, feed rate, and gas environment?

10. A large stockpile of coal is burning. Every part of its surface is in flames. In a 24-hr period the linear size of the pile, as measured by its silhouette against the horizon, seems to decrease by about 5%.

(*a*) How should the burning mass decrease in size?

(*b*) When should the fire burn itself out?

(*c*) State the assumptions on which your estimation is based.

11. We happen to have a two-stage fluidized bed to react solids as in Fig. P17. One stage is twice the size of the other and the gas environment will be the same in both stages. To obtain a higher conversion which stage should come first.

12. Solids of unchanging size, $R = 0.3$ mm, are reacted with gas in a steady flow bench scale fluidized reactor with the following result.

$$F_0 = 10 \text{ gm/sec}, \qquad W = 1000 \text{ gm}, \qquad \bar{X}_B = 0.75$$

Also, the conversion is strongly temperature-sensitive suggesting that the reaction step is rate-controlling.

(*a*) Design a commercial sized fluidized bed reactor (find W) to treat 4 metric tons/hr of solid feed of size $R = 0.3$ mm to 98% conversion.

(*b*) How large would a 2-staged fluidized bed be to do this job?

13. Particles of uniform size are 60% converted on the average (shrinking core model with reaction controlling) when flowing through a single fluidized bed.

(*a*) If the reactor is made twice as large but with the same gas environment what would be the conversion of solids?

(*b*) If the output from the first reactor is introduced into a second reactor of same size and same gas environment what would be the conversion?

14. A solid feed consisting of 20 wt % of 1-mm particles and smaller, 30 wt % of 2-mm particles, 50 wt % of 4-mm particles is to be passed through a rotating tubular reactor, somewhat like a cement kiln, where it reacts with gas of uniform composition to give a hard nonfriable solid product. Experiments show that the progress of conversion can reasonably be represented by reaction control for the unreacted core model, and that the time for complete conversion of 4-mm particles is 4 hrs. Find the residence time needed in the tubular reactor for

(*a*) 75% conversion of solids.
(*b*) 95% conversion of solids.
(*c*) 100% conversion of solids.

15. You are a member of the Lavender Hill Philanthropic Society, a worthy organization dedicated to the preservation of important historical monuments.

The latest and by far the boldest and grandest venture ever undertaken by your group is to save Fort Knox from sinking completely out of sight because of the excessive overload on the foundations. The solution to this weighty problem is obvious; eliminate the overload. Preliminary estimates show that this can be done by removing from their vaults 50 tons of long cylindrical 1-in.-diameter gold bars worth about $64,000,000. These will be removed at 8 P.M. on *the* day. Your job is to dispose of them as soon as possible, but certainly before 8 A.M. the next day, when visiting dignitaries of the various constabularies may be expected. After weighing the various alternatives, you hit on the ingenious plan of dumping the bars in the employees' swimming pool, which will be filled for that occasion with aqua regia.

A literature search produces no useful rate data for this reaction, so an experiment is devised with the only sample of gold available, a $\frac{1}{2}$-in.-diameter gold marble. The following results are obtained, using the same fluid as in the pool.

Size of Marble, diameter in inches	0.5	0.4	0.3	0.2	0.1	0
Time, min	0	42	87	130	172	216

(*a*) At what time can the bars be expected to disappear, and can the 8 A.M. deadline be met?

(*b*) Certainly the earlier the bars dissolve the safer the project will be from unforeseen contingencies. With the thought that agitation may speed up the reaction, the project director helpfully suggests that the group's psychologist, and not too reliable member, Harry, with a slight push or prod, may volunteer his services in agitating the pool. Would Harry's services be needed?

Note: Naturally the employees' swimming pool is large enough so that the acid strength is not appreciably lowered during reaction.

16. In a uniform environment 4-mm solid particles are 87.5% converted to product in 5 min. The solids are unchanged in size during reaction, and the chemical reaction step is known to be rate-controlling. What must be the mean residence time of solids to achieve the same mean conversion of reactant in a fluidized-bed reactor operating with the same environment as before, using a feed consisting of equal quantities of 2-mm and 1-mm particles?

17. When the reaction step is rate-controlling for particles of unchanging size

(*a*) Find the expression for the conversion of solids in an *N*-stage fluidized bed where each stage is of the same size and contacts gas of the same composition; see Fig. P.17.

(*b*) For a two-stage reactor show that this expression reduces to

$$1 - \bar{X}_B = 1 - \frac{6}{y} + \frac{18}{y^2} - \frac{24}{y^3} + \left(\frac{6}{y^2} + \frac{24}{y^3}\right)e^{-y}$$

where $y = \tau/\bar{t}_i$, and $\bar{t}_i$ is the holding time of solids per stage.

(*c*) When $\bar{t}_i > \tau$ show that this conversion expression reduces to

$$1 - \bar{X}_B = \frac{y^2}{20}\left(1 - \frac{y}{3} + \frac{y^2}{14} - \frac{y^3}{84} + \cdots\right)$$

18. In a uniform environment 4-mm solid particles are 87.5% converted to product in 5 min. The solids are unchanged in size during reaction, and the ash diffusion step is known to be controlling. What mean conversion is obtainable in a fluidized-bed reactor operating with the same environment but using a feed consisting of equal weights of 2-mm and 1-mm particles. The mean residence time of solids in this reactor is 30 min.

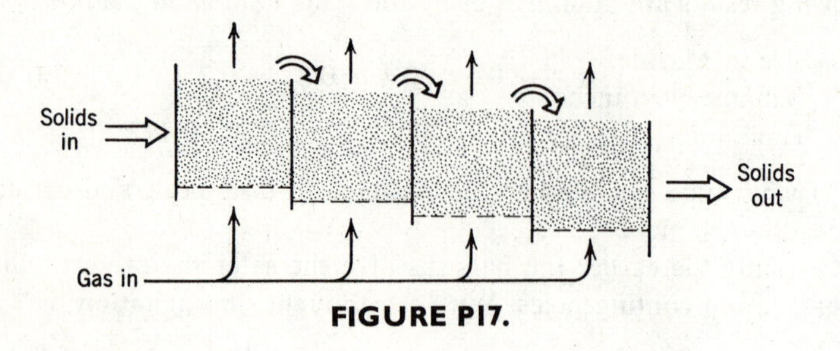

FIGURE P17.

19. A fluidized-bed reactor is planned for the continuous conversion of solid reactant B to solid product R. To find the mean residence time of solids in this flow reactor the following data are obtained in a batch-fluidized unit.

At 1-min intervals, solids are removed from the batch reactor and are analyzed for B and R by an analytical technique which can find the 50% conversion point of B to R. The following results are found.

Size of particles in the batch reactor	4 mm	12 mm
Temperature of the run	550°C	590°C
Time for 50% conversion	15 min	2 hr

What mean residence time is necessary to achieve 98% conversion of B to R if the flow reactor is to be operated at 550°C with a feed of 2-mm particles? The particles are unchanged in size during reaction; therefore it is safe to ignore gas film resistance.

20. (*a*) Particles of a single size are fed to a fluidized bed at a rate of 2 kg/min. The bed contains 60 kg solids. What is the mean residence time of solid if carryover is absent?

(*b*) With the same solid feed rate the gas flow rate is increased with the result that the bed expands, thus containing only 50 kg solids. Another effect is that 0.86 kg solid/min are blown out of the bed. What is the mean residence time of solids under these conditions?

21. A batch of solids (20 kg A, 20 kg B, 60 kg C) are fluidized at high air velocity, and the entrained solids are removed and collected for analysis. After 8 min of operation an analysis of the entrained solids shows 18 kg A, 10 kg B, and no C.

(*a*) Calculate the elutriation constant for these solids.

(*b*) If the entrained solids were collected in a cyclone and immediately returned to the bed what would be entrainment rate of solids under these conditions?

22. Consider the following steady-state operations of a fluidized bed. A mixture of solids A and B (40 kg A/hr and 60 kg B/hr) are fed continuously to a bed which contains 100 kg solids. Because of the high gas velocity used 20 kg solids/hr, all A, are blown out of the bed; the rest of the solids leaves through an overflow pipe. Find the mean residence time in the bed

(*a*) Of the total stream of solids.

(*b*) Of solids A.

(*c*) Of solids B.

23. Steady-state experiments on the elutriation rates of solids are conducted in the simple apparatus consisting of a fluidized bed followed by a cyclone which serves to separate and return to the bed all entrained solids.

In a typical experiment a batch of titanium dioxide composed of 400 gm of uniform 25-μ-radius particles and 600 gm of 50-μ particles is introduced with 5000 gm of larger material, approximately 150-μ-radius, and is fluidized with gas at a fixed velocity. When steady state is achieved, the carryover rate is 22 gm/min consisting of 27.3% of 50-μ particles, the rest being 25-μ particles. As the amount of solid in the recycle system at any instant is small, about 10 gm, it may be neglected in computing the composition of the bed.

Find the elutriation velocity constant as a function of radius of particles (in centimeters), assuming the following type of relationship to hold: $\kappa(R) = \alpha R^{\beta}$. What are the dimensions of α and β?

24. With a uniform feed rate of a single size of particle to a fluidized bed and fixed gas flow rate, 50% of the solids feed is blown out as carryover. With unchanged flow rates of solids and gas but with a doubling of the height of the fluidized bed find how the following change:

(*a*) Mean residence time of solid.
(*b*) Elutriation velocity constant.
(*c*) Fraction of feed blown out as carryover.

Assume unchanged bed density.

25. Glass beads were batch-fluidized in a 10-cm-i.d. column with 21°C 1-atm air at 122 cm/sec, and the quantities of the various sizes of solids blown out in a 2-min interval were determined by Wen and Hashinger (1960) as shown in the following table.

Particle Diameter, microns	Quantity Initially Present, gm	Quantity Removed, gm
70	450	414
98	450	291
146	450	148
277	4050	0

Determine the elutriation velocity constant as a function of particle size for the three small sizes of glass beads.

26. A batch of 60 kg A and 40 kg B is fluidized. The air velocity used is rather high and the solids blown from the bed are trapped in a cyclone and are immediately returned to the bed. At steady state 36 kg A/hr and 8 kg B/hr are blown out and returned.

We plan a continuous operation where this A–B mixture (100 kg/hr) is fed continuously to a bed containing 100 kg solids, the air velocity is kept identical to the batch run, and entrained solids are not returned to the bed. Find the mean residence time of solids A and B in the continuous flow fluidized bed.

27. Solve Example 3 with the following modification: the kinetics of the reaction is ash diffusion controlled with $\tau(R = 100\ \mu) = 10$ min.

28. Solve Example 4 with the following modification: the kinetics of the reaction is ash diffusion controlled with $\tau(R = 100\ \mu) = 10$ min.

29. A pilot plant is to be built to explore the fluidized-bed technique as a means of roasting zinc blende. The reactor is to have an internal diameter of 10 cm and an underflow pipe 18 cm from the bottom of the bed. Feed to the experimental reactor is to be 35.5 gm/min consisting 40% of 10-μ-radius particles and 60% of 40-μ particles. For the optimum gas velocity through the bed, the following bed characteristics can be estimated from values in the literature.

From elutriation data: $\kappa(10\ \mu) = 0.4$/min, $\kappa(40\ \mu) = 0.01$/min.

From bed porosity data: $W = 1100$ gm.

For the temperature selected and ore to be processed, $\tau(10\ \mu) = 2.5$ min, $\tau(40\ \mu) = 10$ min.

(*a*) What conversion of ZnS may be expected in this reactor?

(*b*) The installation of a cyclone to separate and return to the reactor any solids entrained by the fluidized gas is being considered as a possible means of increasing the sulfide conversion. Find the effect on conversion of such a device.

30. A commercial-sized reactor (bed diameter = 1 m, bed height = 2 m) is to be built to treat 15 kg/min of feed of the previous problem. Assuming no change in bed density and gas velocity, find the conversion of sulfide to oxide in such a unit.

31. Consider the following process for converting waste shredded fibers into a useful product. Fibers and fluid are fed continuously into a mixed flow reactor where they react according to the shrinking core model with the reaction step as rate controlling. Develop the performance expression for this operation as a function of the pertinent parameters and ignore elutriation.

32. Consider solids and gas both passing in mixed flow through a reactor, instantaneous reaction, and a feed ratio such that the conversion of solids from the reactor is $\bar{X}_B$.

(*a*) Derive an expression relating $\bar{X}_B$ and the rate constant of the conversion, given by Eq. 73.

(*b*) Sketch the curve for the distribution of conversion among the solid particles and show its main features.

(*c*) What fraction of the particles is completely converted if the mean conversion of solids is $\bar{X}_B = 0.99$?

33. In a gas phase environment, C_{A0}, particles of B are converted to solid product as follows

$$\text{A(gas)} + \text{B(solid)} \rightarrow \text{R(gas)} + \text{S(solid)}$$

Reaction proceeds according to the shrinking core model with reaction control and with time for complete conversion of particles of 1 hr.

A fluidized bed is to be designed to treat 1 ton/hr of solids to 90% conversion using a stoichiometric quantity of A, fed at C_{A0}. Find the weight of solids in the reactor if gas is assumed to be in mixed flow. Note that the gas in the reactor is not at C_{A0}.

34. Repeat problem P33 if the gas is assumed to pass in plug flow through the reactor.

35. Repeat problem P33 if twice the stoichiometric ratio of gas, still at C_{A0}, is fed to the reactor.

36. Repeat problem P33 if twice the stoichiometric ratio of gas, still at C_{A0}, is fed to the reactor, and if the gas is assumed to pass in plug flow through the reactor.

13

FLUID–FLUID REACTIONS

Heterogeneous fluid–fluid reactions are made to take place for one of three reasons. First, the product of reaction may be a desired material. Such reactions are numerous and can be found in practically all areas of the chemical industry where organic syntheses are employed. An example of liquid–liquid reactions is the nitration of organics with a mixture of nitric and sulfuric acids to form materials such as nitroglycerin. The chlorination of liquid benzene and other hydrocarbons with gaseous chlorine is an example of gas–liquid reactions. In the inorganic field we have the manufacture of sodium amide, a solid, from gaseous ammonia and liquid sodium:

$$NH_3(g) + Na(l) \xrightarrow{250°C} NaNH_2(s) + \tfrac{1}{2}H_2$$

Fluid–fluid reactions may also be made to take place to facilitate the removal of an unwanted component from a fluid. Thus the absorption of a solute gas by water may be accelerated by adding a suitable material to the water which will react with the solute being absorbed. Table 1 shows the reagents used for various solute gases.

The third reason for using fluid–fluid systems is to obtain a vastly improved product distribution for homogeneous multiple reactions than is possible by using the single phase alone. We treat this topic briefly at the end of the chapter. Let us return to the first two reasons, both of which concern the reaction of materials originally present in different phases.

The following factors will determine the design method used.

The Over-all Rate Expression. Since materials in the two separate phases must contact each other before reaction can occur, both the mass transfer and the chemical rates will enter the overall rate expression.

TABLE 1. Absorption systems with chemical reaction[a]

Solute Gas	Reagent
CO_2	Carbonates
CO_2	Hydroxides
CO_2	Ethanolamines
CO	Cuprous amine complexes
CO	Cuprous ammonium chloride
SO_2	$Ca(OH)_2$
SO_2	Ozone-H_2O
SO_2	$HCrO_4$
SO_2	KOH
Cl_2	H_2O
Cl_2	$FeCl_2$
H_2S	Ethanolamines
H_2S	$Fe(OH)_3$
SO_3	H_2SO_4
C_2H_4	KOH
C_2H_4	Trialkyl phosphates
Olefins	Cuprous ammonium complexes
NO	$FeSO_4$
NO	$Ca(OH)_2$
NO	H_2SO_4
NO_2	H_2O

[a] Adapted from Teller (1960).

Equilibrium Solubility. The solubility of the reacting components will limit their movement from phase to phase. This factor will certainly influence the form of the rate equation since it will determine whether the reaction takes place in one or both phases.

The Contacting Scheme. In gas–liquid systems semibatch and countercurrent contacting schemes predominate. In liquid–liquid systems mixed flow (mixer-settlers) and batch contacting are used in addition to counter and cocurrent contacting.

Many possible permutations of rate, equilibrium, and contacting pattern can be imagined; however, only some of these are important in the sense that they are widely used on the technical scale. We treat only these cases.

THE RATE EQUATION

For convenience let us call the two phases the gas and liquid phases: let A be the reactant in the gas phase, B the reactant in the liquid phase. If our system consists of two liquids rather than a gas and liquid, we simply let the gas phase of this treatment be the second liquid phase and make the appropriate change in terminology.

Further, let us assume that gaseous A is soluble in the liquid but that B does not enter the gas. Thus A must enter and move into the liquid phases before it can react, and reaction occurs in this phase alone.

Now the overall rate expression for the reaction will have to account for the mass transfer resistance (to bring reactants together) and the resistance of the chemical reaction step. Since the relative magnitude of these two resistances can vary greatly we have a whole spectrum of possibilities, and since each situation requires its own particular analysis, our first problem is to identify these kinetic regimes and to select that one which matches the given physical situation.

Kinetic Regimes for Mass Transfer and Reaction

In terms of the two-film theory of Lewis and Whitman (1924) Fig. 1 shows what may occur as the relative rates of reaction and mass transfer vary from one extreme to the other. Let us start with the infinite fast reaction step.

Case A Instantaneous Reaction with Respect to Mass Transfer. Since an element of liquid can contain either A or B, but not both, reaction will occur at a plane between A-containing and B-containing liquid. Also, since reactants must diffuse to this reaction plane the rate of diffusion of A and B will determine the rate, so that a change in p_A or C_B will move the plane one way or the other.

Case B Instantaneous Reaction; High C_B. For this special case the reaction plane moves to the gas–liquid interface, hence the overall rate will be controlled by the diffusion of A through the gas film. Raising C_B further has no effect on the overall rate.

Case C Fast Reaction; Second-Order Rate. The plane of reaction for case A now spreads into a zone of reaction in which A and B are both present. However, reaction is fast enough so that this reaction zone remains totally within the liquid film. Thus no A enters the main body of liquid to react there.

Case D Fast Reaction; High C_B, hence Pseudo First-order Rate. For the special case where C_B does not drop appreciably within the film, it can be taken to be constant throughout, and the second-order reaction rate (case *C*) simplifies to the more easily solved first-order rate expression.

Cases E and F Intermediate Rate with Respect to Mass Transfer. Here reaction is slow enough for some A to diffuse through the film into the main body of the fluid. Consequently, A reacts both within the film and in the main body of the fluid.

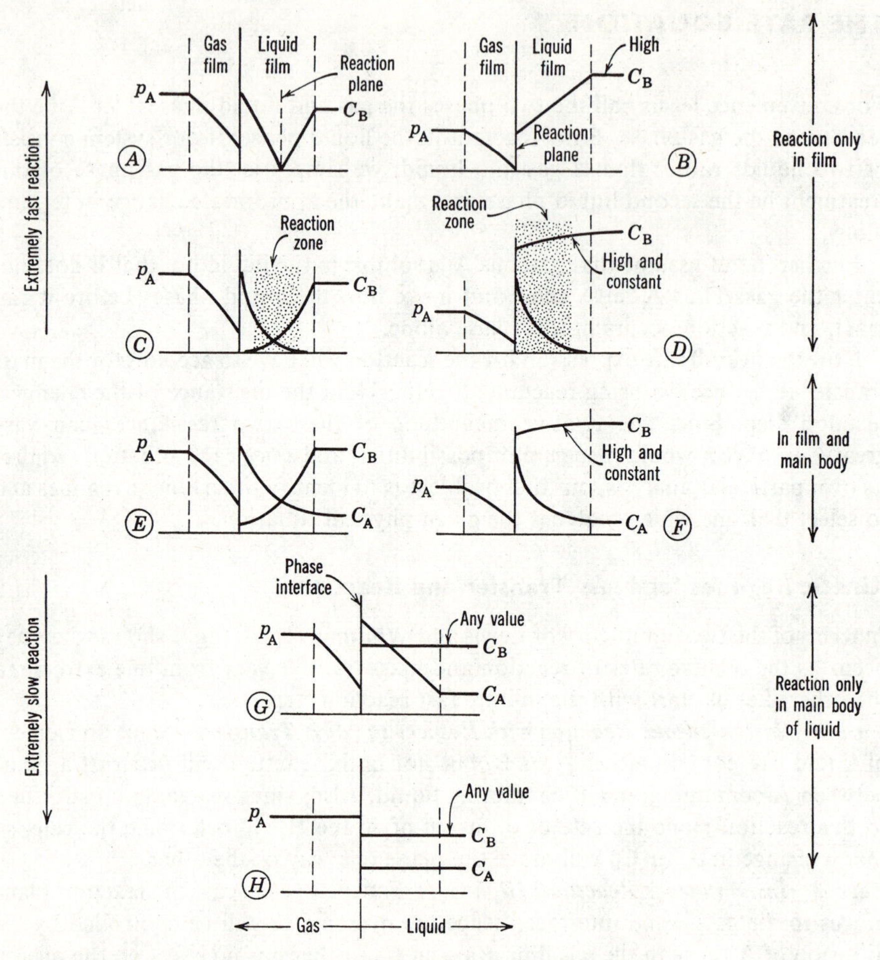

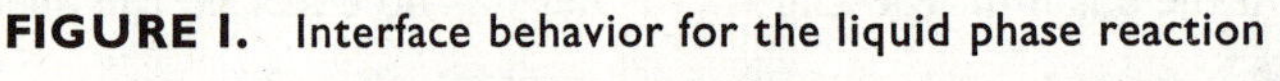

FIGURE I. **Interface behavior for the liquid phase reaction**

A(from gas) + bB(liquid) → products (liquid)

for the complete range of rates of the reaction rate and the mass transfer rate.

Case G Slow Reaction with Respect to Mass Transfer. This represents the somewhat curious case where all reaction occurs in the main body of the liquid; however, the film still provides a resistance to the transfer of A into the main body of liquid.

Case H Infinitely Slow Reaction. Here the mass transfer resistance is negligible, the composition of A and B are uniform in the liquid, and the rate is determined by chemical kinetics alone.

Let us briefly develop the kinetic equations for some of these cases.

Rate Equation for Instantaneous Reaction; Cases *A* and *B*

Consider an infinitely fast reaction of any order:

$$\text{A(from gas)} + b\text{B(liquid)} \rightarrow \text{product} \tag{1}$$

If C_B is not too high we have the situation shown in Fig. 2. At steady state the flow rate of B toward the reaction zone will be b times the flow rate of A toward the reaction zone. Thus the rate of disappearance of A and B are given by

$$-r_A'' = -\frac{r_B''}{b} = \underbrace{k_{Ag}(p_A - p_{Ai})}_{\text{A in gas film}} = \underbrace{k_{Al}(C_{Ai} - 0)\frac{x_0}{x}}_{\text{A in liquid film}} = \underbrace{\frac{k_{Bl}}{b}(C_B - 0)\frac{x_0}{x_0 - x}}_{\text{B in liquid film}} \tag{2}$$

where k_{Ag} and k_{Al}, k_{Bl} are the mass transfer coefficients in gas and liquid phases. The liquid side coefficients are for straight mass transfer without chemical reaction and are therefore based on flow through the whole film of thickness x_0.

At the interface the relationship between p_A and C_A is given by the distribution coefficient, called Henry's law constant for gas–liquid systems. Thus

$$p_{Ai} = H_A C_{Ai} \tag{3}$$

In addition, since the movement of material within the film is visualized to occur by diffusion alone, the transfer coefficients for A and B are related by*

$$\frac{k_{Al}}{k_{Bl}} = \frac{\mathscr{D}_{Al}/x_0}{\mathscr{D}_{Bl}/x_0} = \frac{\mathscr{D}_{Al}}{\mathscr{D}_{Bl}} \tag{4}$$

Eliminating the unknowns x, x_0, p_{Ai}, C_{Ai} in Eqs. 2, 3, and 4, we obtain

$$-r_A'' = -\frac{1}{S}\frac{dN_A}{dt} = \frac{\dfrac{\mathscr{D}_{Bl}}{\mathscr{D}_{Al}}\dfrac{C_B}{b} + \dfrac{p_A}{H_A}}{\dfrac{1}{H_A k_{Ag}} + \dfrac{1}{k_{Al}}} \tag{6}$$

* Alternatives to the film theory are also in use. These models (Higbie (1935), Danckwerts (1950, 1955)) view that the liquid at the interface is continually washed away and replaced by fresh fluid from the main body of the liquid, and that this is the means of mass transport. These unsteady-state surface renewal theories all predict

$$\frac{k_{Al}}{k_{Bl}} = \sqrt{\frac{\mathscr{D}_{Al}}{\mathscr{D}_{Bl}}} \tag{5}$$

as opposed to Eq. 4, above, for the film theory.

With the exception of this one difference, these models, so completely different from a physical standpoint, give essentially identical predictions of steady state behavior. Because of this, and because the film theory is so much easier to develop and use than the other theories, we deal with it exclusively.

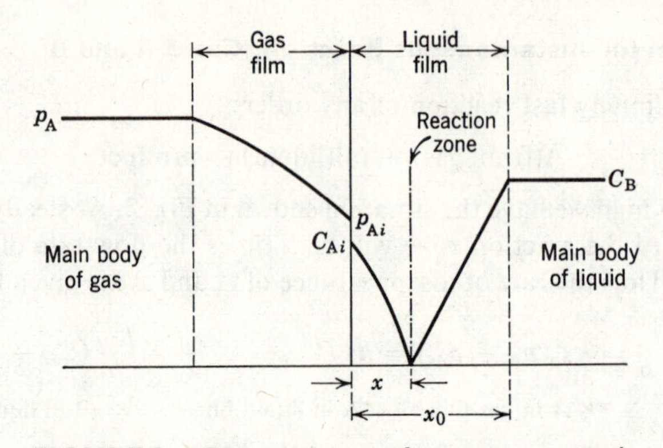

FIGURE 2. Concentrations of reactants as visualized by the two-film theory for an infinitely fast irreversible reaction of any order, A + bB → products.

Special Case of Negligible Gas Phase Resistance. When this happens $k_{Ag} = \infty$ and $p_A = p_{Ai}$, so Eq. 6 reduces to

$$-r''_A = k_{Al}C_{Ai}\left(1 + \frac{\mathscr{D}_{Bl}C_B}{b\mathscr{D}_{Al}C_{Ai}}\right) \tag{7}$$

Comparing with the expression for the maximum rate of straight mass transfer, or

$$-r''_A = k_{Al}C_{Ai} \tag{8}$$

we find that the term in brackets represents the increase in rate of absorption of A resulting from adding reactant B to the liquid. Thus if we define in general

$$E = \begin{pmatrix}\text{enhancement}\\ \text{factor}\end{pmatrix} = \left(\frac{\text{rate with reaction}}{\text{rate for mass transfer alone}}\right) \tag{9}$$

then for the case of infinitely fast reaction rate and no gas phase resistance

$$-r''_A = k_{Al}C_{Ai}E \qquad \text{where } E = 1 + \frac{\mathscr{D}_{Bl}C_B}{b\mathscr{D}_{Al}C_{Ai}} \tag{10}$$

Equations 6 and 10 represent case A of Fig. 1, and can be used directly in design when the quantities $\mathscr{D}_{Bl}/\mathscr{D}_{Al}$, k_{Ag}, k_{Al}, and H_A are known.

Special Case of High C_B. Returning to the general situation shown in Fig. 2, if the concentration of B is raised, or more precisely, if

$$k_{Ag}p_A \leqslant \frac{k_{Bl}}{b}C_B \tag{11}$$

then this condition, combined with Eq. 2, requires that the reaction zone move to and remain at the interface. When this happens, the resistance of the gas phase controls, and the rate is not affected by any further increase in concentration of B. In addition, Eq. 6 simplifies to

$$-r''_A = -\frac{1}{S}\frac{dN_A}{dt} = k_{Ag}p_A \qquad (12)$$

To determine which form of rate expression, Eq. 6 or 12, to use in any specific situation requires knowledge of the concentrations of reactants and of the physical properties of the fluids being used. Nevertheless, we can make a rough order of magnitude estimation as follows. For mass transfer of a constituent through a liquid

$$\mathscr{D}_l \approx 10^{-5}\ \text{cm}^2/\text{sec} \qquad \text{and} \qquad x_0 \approx 10^{-2}\ \text{cm}$$

hence

$$k_l = \frac{\mathscr{D}_l}{x_0} \approx 10^{-3}\ \text{cm/sec} \qquad (13)$$

For a gas phase, on the other hand,

$$\mathscr{D}_g \approx 0.1\ \text{cm}^2/\text{sec} \qquad \text{and} \qquad x_0 \approx 10^{-2}\ \text{cm}$$

hence for driving forces in concentration units

$$k'_g = k_gRT = \frac{\mathscr{D}_g}{x_0} \approx 10\ \text{cm/sec} \qquad (14)$$

Replacing these values in Eq. 11 shows that when

$$\frac{p_A}{RT} = C_{Ag} < \frac{10^{-4}C_{Bl}}{b} \qquad (15)$$

the simple form of rate expression, Eq. 12, may be used. We should bear in mind that Eq. 15 is only an order of magnitude estimate.

Note that the reaction rate constant does not enter into either Eq. 6 or 12, showing that the rate is completely mass transfer controlled. Also, even though the resistance of the liquid film may normally control, when C_B is raised high enough then the resistance always shifts to gas phase controlling.

Examples 1 to 4, p. 431, illustrate the use of these kinetic expressions.

Rate Equations for Fast Reaction; Cases *C* and *D*

For case C with second-order reaction between A and B

$$-r_{A,l} = -\frac{1}{V_l}\frac{dN_A}{dt} = kC_AC_B \tag{16}$$

we may write for the gas and liquid films

$$-r''_A = k_{Ag}(p_A - p_{Ai}) = k_{Al}C_{Ai}E \tag{17}$$

Eliminating C_{Ai} and p_{Ai} with Eq. 3 gives for case C

$$-r''_A = \frac{1}{\underset{\substack{\text{gas}\\\text{film}}}{\dfrac{1}{k_{Ag}}} + \underset{\substack{\text{liquid}\\\text{film}}}{\dfrac{H_A}{k_{Al}E}}}\, p_A \tag{18}$$

where the enhancement factor E is a complex function of k_l, k, b, and C_B/C_{Ai}. A precise analytical expression for E has not yet been found, but an approximate solution ($\pm 10\%$) is given by van Krevelens and Hoftijzer (1948).

For the special case where C_B is high enough to be considered constant the reaction in the liquid becomes pseudo first order, or

$$-r_A = kC_AC_B = (kC_B)C_A = k_1C_A$$

in which case the enhancement factor is a simple expression, as follows:

$$E = \frac{\sqrt{\mathscr{D}_{Al}kC_B}}{k_{Al}} = \frac{\sqrt{\mathscr{D}_{Al}k_1}}{k_{Al}} \tag{19}$$

Eliminating intermediate concentrations in Eq. 17 with Eq. 3 we find the rate for case D to be

$$-r''_A = \frac{1}{\underset{\text{gas film}}{\dfrac{1}{k_{Ag}}} + \underset{\text{liquid film}}{\dfrac{H_A}{\sqrt{\mathscr{D}_{Al}kC_B}}}}\, p_A \tag{20}$$

Note that film thickness does not enter into this expression since reactant A does not penetrate and use the whole film.

Intermediate Rates; Cases *E* and *F*

These cases represent the general situation where reaction occurs both within the film and within the main body of the liquid. Hatta (1929) did the first significant analysis of this problem; however, today the general rate expressions for this regime are still unavailable, while for the special cases the resulting equations are rather complex. The reader may find derivations and discussions of this case in the texts devoted to this subject, such as Sherwood and Pigford (1952), Astarita (1967), and Danckwerts (1970), and also Kramers and Westerterp (1964).

For our purposes it suffices to note that the rate is dependent on both interfacial surface and on volume of liquid. Thus the ratio

$$a_i = \frac{S}{V_l} = \frac{\text{interfacial surface}}{\text{volume of liquid}} \tag{21}$$

becomes a parameter in this regime.

Rate Equation for Slow Reaction; Case *G*

Here the two films and the main body of liquid act as resistances in series; thus we may write for these steps

$$-\frac{1}{S}\frac{dN_A}{dt} = k_{Ag}(p_A - p_{Ai}) = k_{Al}(C_{Ai} - C_A) \tag{22}$$

and

$$-\frac{1}{V_l}\frac{dN_A}{dt} = kC_AC_B \tag{23}$$

Combining and eliminating intermediate concentrations with Eq. 3 gives

$$-\frac{1}{S}\frac{dN_A}{dt} = \frac{1}{\dfrac{1}{k_{Ag}} + \dfrac{H_A}{k_{Al}} + \dfrac{H_Aa_i}{kC_B}}\,p_A$$

or (24)

$$-\frac{1}{V_l}\frac{dN_A}{dt} = \frac{1}{\underset{\text{gas film}}{\dfrac{1}{k_{Ag}a_i}} + \underset{\text{liquid film}}{\dfrac{H_A}{k_{Al}a_i}} + \underset{\text{bulk liquid}}{\dfrac{H_A}{kC_B}}}\,p_A$$

Note that the ratio of surface to volume enters this expression.

Rate Equation for Infinitely Slow Reaction; Case *H*

Here the concentrations of A and B are uniform in the liquid, and the rate is given directly by

$$-r_{A,l} = -\frac{1}{V}\frac{dN_A}{dt} = kC_AC_B \tag{16}$$

Note that the rate is expressed in terms of unit volume of liquid, that the film offers no resistance, hence interfacial area does not enter into the rate expression.

Film Conversion Parameter, *M*

To tell whether reaction is fast or slow we focus on unit surface of gas–liquid interface, we assume that gas phase resistance is negligible, and we define a film conversion parameter

$$M = \frac{\text{maximum possible conversion in film}}{\text{maximum diffusion transport through film}}$$

$$= \frac{kC_{Ai}C_Bx_0}{\dfrac{\mathscr{D}_{Al}}{x_0}\cdot C_{Ai}} = \frac{kC_B\mathscr{D}_{Al}}{{k_{Al}}^2} \tag{25}$$

If $M \gg 1$ all reaction occurs in the film, and surface area is the controlling rate factor. On the other hand, if $M \ll 1$ no reaction occurs in the film, and bulk volume becomes the controlling rate factor. More precisely, it has been found that:

1. If $M > 4$ reaction occurs in the film and we have cases *A, B, C, D*.
2. If $0.0004 < M < 4$ we then have the intermediate cases *E, F, G*.
3. If $M < 0.0004$ we have the infinitely slow reaction of case *H*.

When *M* is large we should pick a contacting device which develops or creates large interfacial areas; energy for agitation is usually an important consideration in these contacting schemes. On the other hand, if *M* is very small all we need is a large volume of liquid. Agitation to create large interfacial areas is of no benefit here.

Table 2 presents typical data for various contacting devices, and from this we see that spray or plate columns should be efficient devices for systems with fast reaction (or large *M*), while bubble contactors should be more efficient for slow reactions (or small *M*).

Clues to the Kinetic Regime from Solubility Data

For reactions which occur in the film the phase distribution coefficient *H* can suggest whether the gas phase resistance is likely to be important or not. To show

TABLE 2. Equipment characteristics for gas–liquid contactors; from Kramers and Westerterp (1964)

Type of Contactor	Interfacial Surface / Volume of Liquid	Interfacial Surface / Volume of Reactor	Volume Fraction of Liquid	Volume of Liquid / Volume of Film
Spray column	~1200 m^2/m^3	~60 m^2/m^3	~0.05	~2–10
Packed column	1200	100	0.08	10–100
Plate column	1000	150	0.15	40–100
Agitated bubble contactor	200	200	0.9	150–800
Bubble contactor	20	20	0.98	4000–10,000

this we write the expression for straight mass transfer of A across the gas and liquid films

$$-\frac{1}{S}\frac{dN_A}{dt} = \frac{1}{\underbrace{\frac{1}{k_{Ag}}}_{\text{gas film}} + \underbrace{\frac{H_A}{k_{Al}}}_{\text{liquid film}}}\Delta p_A \tag{26}$$

Now for slightly soluble gases H_A is large; hence, with all other factors remaining unchanged the above rate equation shows that the liquid film resistance term is large. The reverse holds for highly soluble gases. Thus we see that:

Gas film resistance controls for highly soluble gases.
Liquid film resistance controls for slightly soluble gases.

Since a highly soluble gas is easy to absorb and has its main resistance in the gas phase, we would not need to add a liquid phase reactant B to promote the absorption. On the other hand, a sparingly soluble gas is both difficult to absorb and has its main resistance in the liquid phase; hence it is this system which would benefit greatly by a reaction in the liquid phase.

Thus we conclude that where reaction can significantly aid the absorption of a gas, the gas is only sparingly soluble and the liquid phase resistance is likely to control.

Clues to the Kinetic Regime from Experiment

Ideally, what we need is an experimental setup which is flexible enough to allow the variables to be changed independently and in a known manner, and where we can measure what happens to the rate as a consequence of these changes. Let us

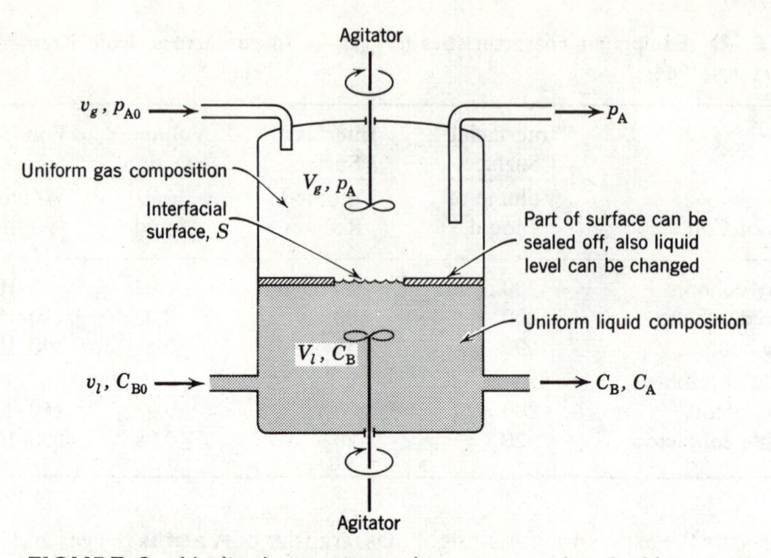

FIGURE 3. Idealized experimental contactor, the double mixed reactor, for studying the kinetics of fluid–fluid systems.

illustrate how such information can be used to establish the kinetic regimes by considering the ideal contactor of Fig. 3 which we will call a double mixed reactor.

The particular features of this device are:

Mixed flow of gas and of liquid, hence uniform composition in the bulk of each phase.

The agitation rate in the phases can be changed independently.

The surface to volume ratio can be varied by changing the interfacial contact area.

The concentration level of reactants can be raised or lowered by changing the feed rate or feed composition.

Because of mixed flow of the phases the rate of absorption and reaction can be found directly by means of a material balance. For example, based on unit volume of liquid we have for component B

$$-r_{B,l} = -\frac{1}{V_l}\frac{dN_B}{dt} = \frac{C_{B0}X_B}{\tau_l} = \frac{v_l(C_{B0} - C_B)}{V_l} \tag{27}$$

or for A, based on volume of liquid,

$$-r_{A,l} = -\frac{1}{V_l}\frac{dN_A}{dt} = -\frac{r_{B,l}}{b} = \frac{v_{\text{inerts}}\pi}{RTV_l}\left[\left(\frac{p_A}{p_{\text{inerts}}}\right)_{\text{in}} - \left(\frac{p_A}{p_{\text{inerts}}}\right)_{\text{out}}\right] \tag{28}$$

Similarly, based on unit surface we have for B

$$-r''_B = -\frac{1}{S}\frac{dN_B}{dt} = -r_{B,l}\cdot\frac{V_l}{S} = \frac{v_l(C_{B0} - C_B)}{S} \tag{29}$$

and for A

$$-r''_A = -\frac{1}{S}\frac{dN_A}{dt} = -\frac{r''_B}{b} = \frac{v_{\text{inerts}}\pi}{RTS}\left[\left(\frac{p_A}{p_{\text{inerts}}}\right)_{\text{in}} - \left(\frac{p_A}{p_{\text{inerts}}}\right)_{\text{out}}\right] \tag{30}$$

Let us illustrate the planning and interpretation of experiments with such a device. Consider the following changes in system variables.

Increase gas phase agitation. If the rate of absorption and reaction rises then the gas phase resistance is important; if it does not rise then the gas phase resistance can be ignored.

Change S, V_l, or both:
(*a*) If dN_A/dt is independent of V_l but is proportional to S then $M > 4$, and we have case A, B, C, or D.
(*b*) If both S and V_l affect dN_A/dt to some extent then $0.0004 < M < 4$, and we have case E, F, or G.
(*c*) If dN_A/dt is proportional to V_l but is independent of S then $M < 0.0004$ and we have case H.

Now that we know which of the three regimes we are in we may try to identify the particular case present by varying the physical factors at our disposal. Problems 7 to 11 at the end of the chapter give practice in this type of analysis.

These methods for setting up rate equations and determining the rate-controlling regimes can be extended to other situations. We illustrate this briefly with the following systems.

Slurry Reaction Kinetics

Here gas which contains reactant A is bubbled through liquid B which contains suspended solid catalyst, and reactant A must reach the catalyst surface to react with B. For such systems the following resistances are viewed to act in series, see Fig. 4.

Step 1. Reactant A must cross the gas film to reach the gas–liquid interface.
Step 2. A must cross the liquid film to reach the main body of liquid.
Step 3. A must cross the liquid film surrounding the catalyst particle to reach the surface.
Step 4. A then reacts on the surface of the catalyst particle with liquid component B.

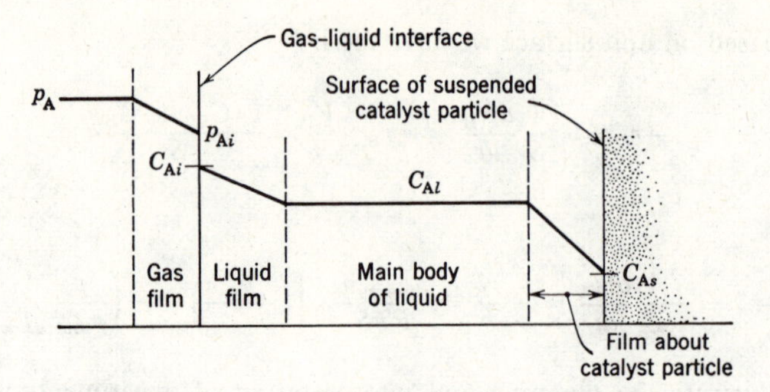

FIGURE 4. Resistances in a slurry reactor for gaseous reactant to reach the surface of suspended solid catalyst.

Let us suppose that B is in large enough excess for the surface reaction to be first order with respect to A with rate constant k_1. Let us also define

$$a_i = \frac{\text{interfacial surface}}{\text{volume of liquid}}$$

and

$$a_s = \frac{\text{surface of suspended catalyst}}{\text{volume of liquid}}$$

Then the rate for these individual steps is given by

$$-r_{A,l} = -\frac{1}{V_l}\frac{dN_A}{dt} = k_{Ag}a_i(p_A - p_{Ai}) = k_{Al}a_i(C_{Ai} - C_{Al})$$
$$= k'_{Al}a_s(C_{Al} - C_{As}) = k_1 a_s C_{As} \tag{31}$$

and on combining we find

$$-r_{A,l} = \frac{1}{\underbrace{\frac{1}{k_{Ag}a_i} + \frac{H_A}{k_{Al}a_i}}_{\text{gas–liquid interface}} + \underbrace{\frac{H_A}{k'_{Al}a_s} + \frac{H_A}{k_1 a_s}}_{\text{catalyst surface}}} p_A \tag{32}$$

It is important to know whether the primary resistance lies at the gas–liquid interface or at the surface of the particles. This is easily found by changing separately gas–liquid interfacial areas and the amount of suspended catalyst in the liquid. This will determine how best to scale up the operation.

The following rearrangement of Eq. 32 is often useful in correlating data in kinetic studies of this system:

$$\frac{p_A}{-r_{A,l}} = \left(\frac{1}{k_{Ag}} + \frac{H_A}{k_{Al}}\right)\frac{1}{a_i} + \left(\frac{H_A}{k'_{Al}} + \frac{H_A}{k_1}\right)\frac{1}{a_s} = C_1\frac{1}{a_i} + C_2\frac{1}{a_s} \tag{33}$$

The hydrogenation of organic compounds represents a typical application of slurry reactors. Since pure hydrogen invariably is used the gas phase resistance just doesn't exist in these operations.

Aerobic Fermentations

Here oxygen from the gas phase must reach the surface of the growing cells which are present in the liquid. With respect to the form of the kinetic equation this situation is very similar to the slurry reactor, and Fig. 5 shows the concentrations and resistances which exist.

The crucial problem in these fermenters is to insure an adequate oxygen supply for proper cellular respiration, and this requires maintaining the concentration of oxygen in the liquid above a particular critical minimum level. For this purpose large gas–liquid interfacial areas are needed. It is also generally found that liquid film resistance controls.

Comments on Rates

The phase interface is normally pictured to be a quiescent layer (film theory), or a surface which is being continually replaced by some sort of eddy mechanism (surface renewal theories). However, recent studies have shown that the presence of impurities in the liquid can completely change the interfacial behavior. The impurities concentrate at the interface and transform it into a vigorously active region with fluid streaming and rapid flows, all due to the complex action of surface tension forces. This is called the Marangoni effect. Often very tiny amounts of impurities will trigger this action, and this generally lowers the mass transfer coefficient.

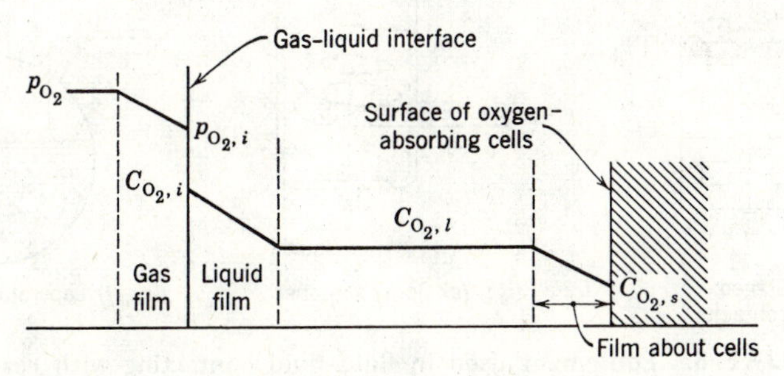

FIGURE 5. Resistances to oxygen absorption by the cells in an aerobic fermenter.

The intrusion of the Marangoni effect and the fact that the values of the system parameters (k_l, k_g, k, H_A, and so on) are often unknown lead us reluctantly to conclude that the best we may hope for in many cases is to find the correct kinetic regime (cases A to H), and then use the simplest equation for this regime. Usually this is sufficient for reasonable scale-up and selection of the favorable contacting devices.

When reaction is used to speed the absorption of a slightly soluble gas (highly soluble gases need no reaction) we certainly want a liquid reactant which gives high rates of reaction. In these cases reaction is usually limited to the interfacial

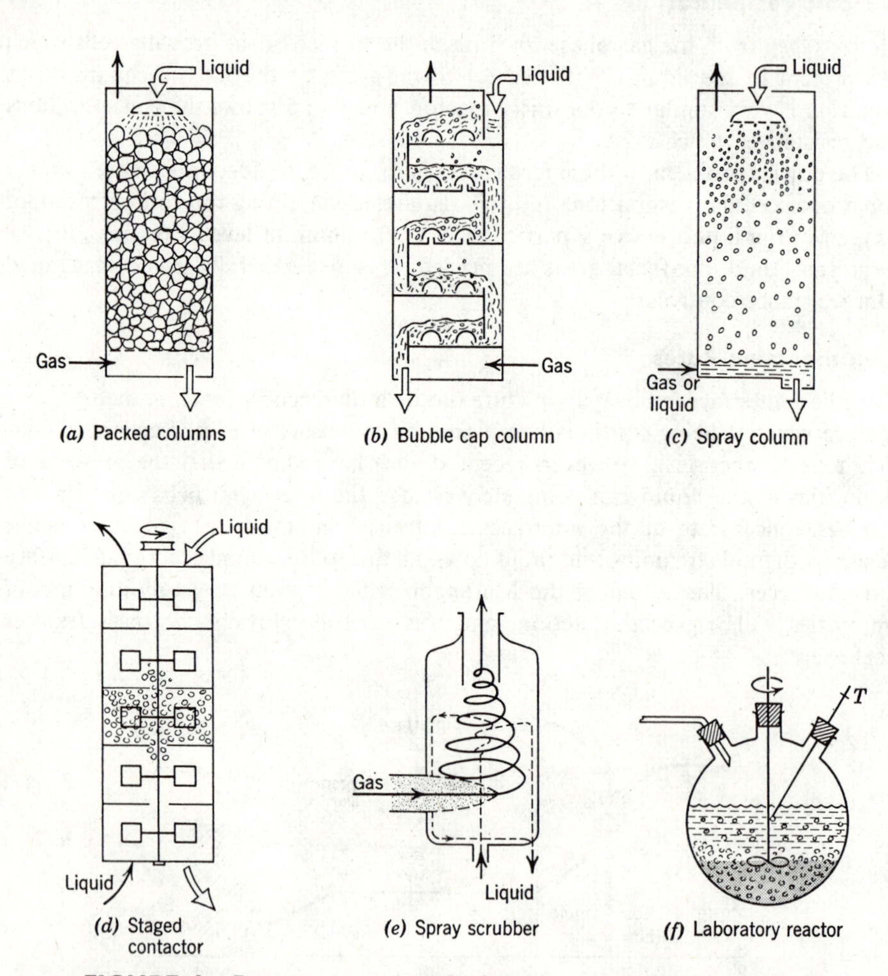

FIGURE 6. Equipment used in fluid–fluid contacting with reaction: (*a–d*) tower operations, (*e–h*) single stage mixed, batch and flow, (*i*) cocurrent multistage mixed, and (*j*) countercurrent multistaged mixed vessels.

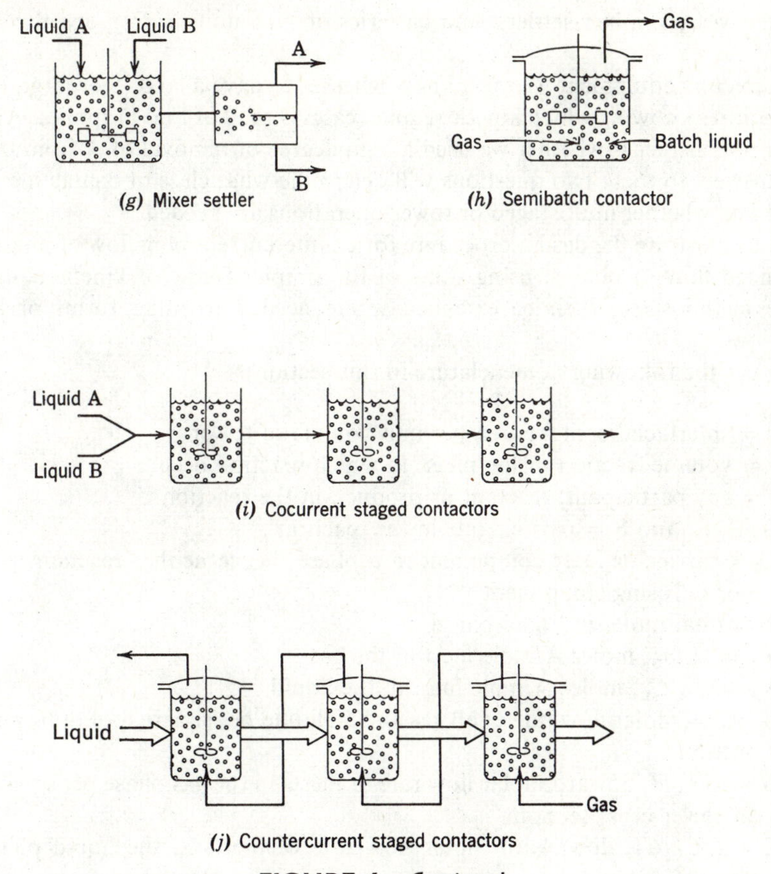

(g) Mixer settler

(h) Semibatch contactor

(i) Cocurrent staged contactors

(j) Countercurrent staged contactors

FIGURE 6. *Continued.*

region; hence cases *A*, *B*, *C*, or *D* usually apply, and towers are used for these reactions on the industrial scale (see Table 2 and related discussion).

When reaction is used to obtain product material, say chlorination of hydrocarbons or hydrogenation of unsaturated hydrocarbons, then we may encounter both fast or slow reactions; hence any of the rate regimes *A* to *H* may be encountered. For slow rates bubble contactors are frequently used on the industrial scale.

APPLICATION TO DESIGN

Figure 6 shows the variety of equipment used for fluid–fluid contacting. Note that many of the contacting devices shown, *a* to *d*, employ flow that can be treated as cocurrent or countercurrent plug flow. Other contacting devices such as single agitated vessels, mixer-settlers, and batteries of such units, *g* to *j*, are also widely used.

In selecting equipment we must know whether to have a large S or large V_l, and this requires knowing which kinetic regime (cases *A* to *H* of Fig. 1) applies. Another factor to consider is whether we need a high degree of removal of A from the gas. The answers to these two questions will determine what class of contacting device to use and whether multistaged or tower operations are needed.

Let us illustrate the design procedure for countercurrent plug flow of phases and for mixed flow of phases, using some of the simpler forms of kinetic equations. These methods can then be extended where needed to other forms of kinetic equations.

We use the following nomenclature in this section:

a = interfacial contact area per unit volume of tower
f = volume fraction of the phase in which reaction occurs
i = any participant, reactant or product, in the reaction
A, B, R, and S = participants in the reaction
U = carrier or inert component in a phase, hence neither reactant, product nor diffusing component
T = total moles in liquid phase
$\mathbf{Y}_A = p_A/p_U$, moles A/mole inert in the gas
$\mathbf{X}_A = C_A/C_U$, moles A/mole inert in the liquid
G', L' = molar flow rate of all gas and all liquid per square foot of tower cross section
$G = G'p_U/\pi$, upward molar flow rate of inerts in the gas phase per square foot of tower cross section
$L = L'C_U/C_T$, downward molar flow rate of inerts in the liquid phase per square foot of tower cross section

With this nomenclature

$$\begin{aligned} \pi &= p_A + p_B + \cdots + p_U \\ C_T &= C_A + C_B + \cdots + C_U \end{aligned} \tag{34}$$

$$d\mathbf{Y}_A = d\left(\frac{p_A}{p_U}\right) = \frac{p_U\,dp_A - p_A\,dp_U}{p_U^2} \tag{35}$$

and

$$d\mathbf{X}_A = d\left(\frac{C_A}{C_U}\right) = \frac{C_U\,dC_A - C_A\,dC_U}{C_U^2} \tag{36}$$

Let us now consider a number of contacting patterns.

Towers for Fast Reaction; Case *A*, *B*, *C*, or *D*

Straight Mass Transfer without Reaction. To determine the tower height, we must in all cases combine the rate expression with a material balance expression. The former is a function of concentration of reactants, and the latter serves to interrelate the changing concentrations of reactants within the tower.

Consider steady-state countercurrent operations. In straight mass transfer without reaction there is only one transferring component; thus a material balance in a differential element of volume shows that

$$\begin{pmatrix}\text{A lost}\\ \text{by gas}\end{pmatrix} = \begin{pmatrix}\text{A gained}\\ \text{by liquid}\end{pmatrix} \tag{37a}$$

or

$$\begin{aligned} G d\mathbf{Y}_A = L d\mathbf{X}_A &= \frac{G\pi\, dp_A}{(\pi - p_A)^2} = \frac{LC_T\, dC_A}{(C_T - C_A)^2} \\ &= d\left(\frac{G'p_A}{\pi}\right) = d\left(\frac{L'C_A}{C_T}\right) = \frac{G'\, dp_A}{\pi - p_A} = \frac{L'\, dC_A}{C_T - C_A} \end{aligned} \tag{37b}$$

For dilute systems $C_A \ll C_T$ and $p_A \ll \pi$; consequently, we have $L \approx L'$ and $G \approx G'$, hence Eq. 37 simplifies to

$$\frac{G}{\pi}\, dp_A = \frac{L}{C_T}\, dC_A \tag{38}$$

At any point in the tower the transfer of A per unit surface of interface is given by

$$-r_A'' = -\frac{1}{S}\frac{dN_A}{dt} = k_{Ag}(p_A - p_{Ai}) = k_{Al}(C_{Ai} - C_A) \tag{39}$$

Combining the material balance, Eq. 37, and the rate expression, Eq. 39, allows us to determine the height of tower. Thus for absorption without reaction

$$\begin{aligned} h &= \frac{G}{a}\int_{\mathbf{Y}_{A1}}^{\mathbf{Y}_{A2}} \frac{d\mathbf{Y}_A}{-r_A''} = G\pi \int_{p_{A1}}^{p_{A2}} \frac{dp_A}{k_{Ag}a(\pi - p_A)^2(p_A - p_{Ai})} \\ &= LC_T \int_{C_{A1}}^{C_{A2}} \frac{dC_A}{k_{Al}a(C_T - C_A)^2(C_{Ai} - C_A)} \\ &= G' \int_{p_{A1}}^{p_{A2}} \frac{dp_A}{k_{Ag}a(\pi - p_A)(p_A - p_{Ai})} \\ &= L' \int_{C_{A1}}^{C_{A2}} \frac{dC_A}{k_{Al}a(C_T - C_A)(C_{Ai} - C_A)} \end{aligned} \tag{40}$$

Again for dilute systems $L \approx L'$ and $G \approx G'$; hence

$$h = \frac{G}{\pi k_{Ag} a} \int_{p_{A1}}^{p_{A2}} \frac{dp_A}{p_A - p_{Ai}} = \frac{L}{C_T k_{Al} a} \int_{C_{A1}}^{C_{A2}} \frac{dC_A}{C_{Ai} - C_A} \tag{41}$$

Figure 7 summarizes the design procedure for countercurrent operations. For cocurrent operations (downward flow of gas) G is replaced by $-G$ throughout.

Mass Transfer with Reaction. For mass transfer with reaction the rate expressions account for the concentrations of both reactants. Thus, in contrast with simple absorption or extraction the material balance must interrelate the concentrations of both these reactants within the tower.

The design procedure is best described by referring to the fairly general type of reaction,

$$A(\text{gas}) + bB(\text{liquid}) \rightarrow \text{products}$$

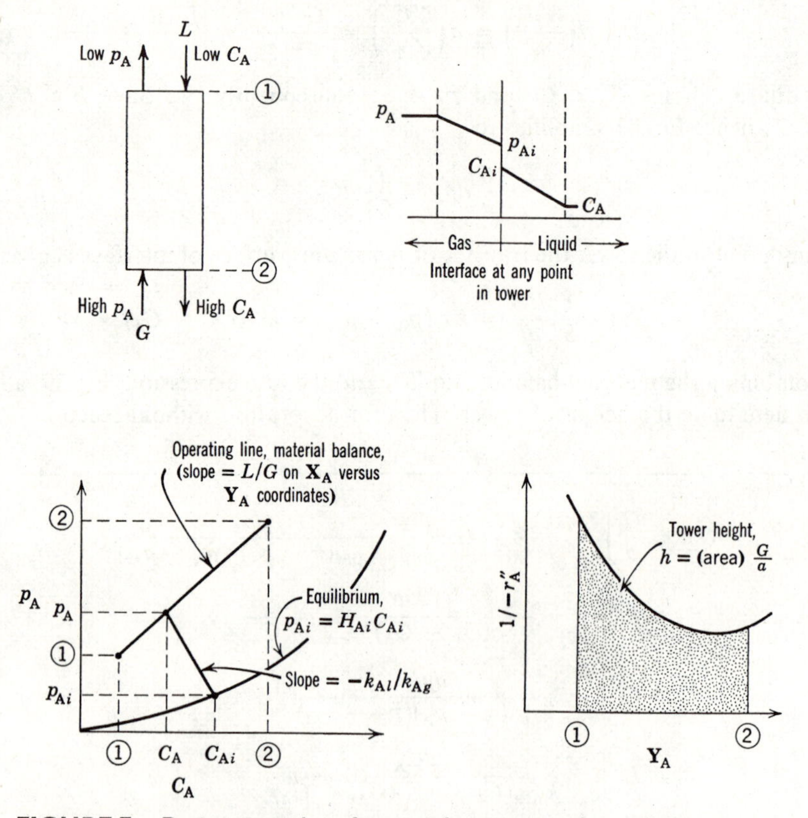

FIGURE 7. Design procedure for straight mass transfer without reaction.

with moderately to infinitely fast rate, the phases being in countercurrent plug flow. Since the reaction is limited to the region close to the phase interface, the two-film theory may be used, and the expression describing the rate of disappearance of reactants is given by Eq. 6 or 12, or 16 and 17, or 20.

The *material balance* for A and B is obtained by noting that of the two reactants, A alone is present in the main body of gas and B alone is present in the main body of liquid. Secondly, for each mole of A reacted b moles of B are consumed. Thus, by referring to Fig. 8, the differential material balance for countercurrent flow is

$$\begin{pmatrix}\text{A lost}\\ \text{by gas}\end{pmatrix} = \frac{1}{b}\begin{pmatrix}\text{B lost}\\ \text{by liquid}\end{pmatrix} \tag{42a}$$

or

$$G\,d\mathbf{Y}_A = -\frac{L\,d\mathbf{X}_B}{b} = G\,d\left(\frac{p_A}{p_U}\right) = -\frac{L}{b}\,d\left(\frac{C_B}{C_U}\right)$$

$$= d\left(\frac{G'p_A}{\pi}\right) = -\frac{1}{b}\,d\left(\frac{L'C_B}{C_T}\right) \tag{42b}$$

For cocurrent operations (downward flow of gas) G is replaced by $-G$ throughout.

The important difference in the material balance with and without reaction is seen by comparing the word equations of Eq. 37 and Eq. 42.

Compositions at any point in the tower are found in terms of the end conditions by integration of Eq. 42; thus

$$G(\mathbf{Y}_A - \mathbf{Y}_{A1}) = -\frac{L(\mathbf{X}_B - \mathbf{X}_{B1})}{b} = G\left(\frac{p_A}{p_U} - \frac{p_{A1}}{p_{U1}}\right) = -\frac{L}{b}\left(\frac{C_B}{C_U} - \frac{C_{B1}}{C_{U1}}\right)$$

$$= \frac{G'p_A}{\pi} - \frac{G_1'p_{A1}}{\pi} = -\frac{1}{b}\left(\frac{L'C_B}{C_T} - \frac{L_1'C_{B1}}{C_{T1}}\right) \tag{43}$$

These expressions give the concentrations of reactants A and B in the phases throughout the tower. Again, for cocurrent flow G is replaced by $-G$ throughout. Figure 8 shows some of the many possible graphical representations of the material balance. In the general case, however, only one representation gives a straight line.

In special cases Eqs. 42 and 43 may be simplified. For example, when all participants in the reaction are dilute $p_U \approx \pi$, $C_U \approx C_T$, and we obtain, for the differential material balance,

$$\frac{G}{\pi}\,dp_A = -\frac{L}{bC_T}\,dC_B \tag{44}$$

and for point conditions

$$\frac{G}{\pi}(p_A - p_{A1}) = -\frac{L}{bC_T}(C_B - C_{B1}) \tag{45}$$

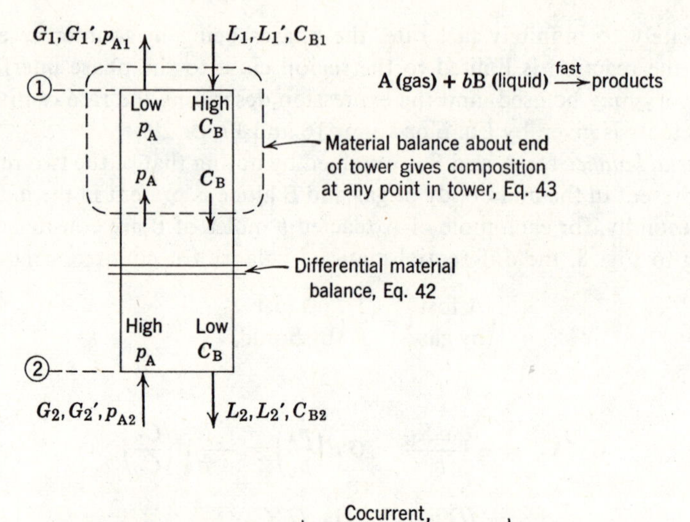

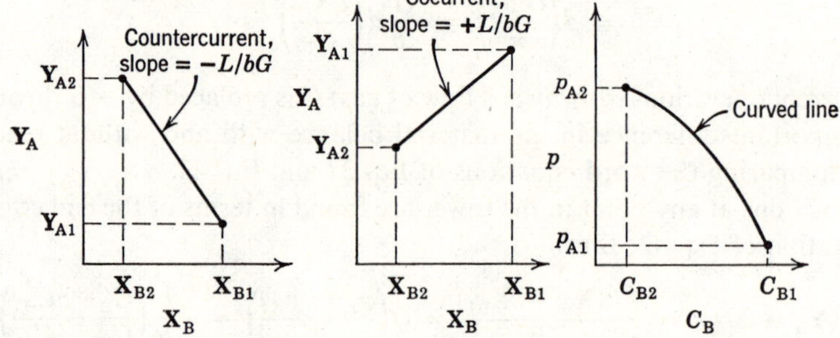

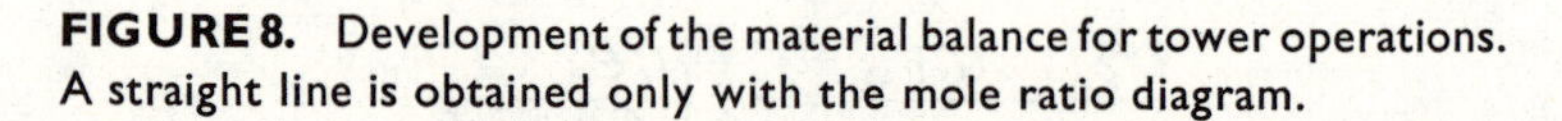

FIGURE 8. Development of the material balance for tower operations. A straight line is obtained only with the mole ratio diagram.

Another special case occurs when no inert or carrier component is present. Thus each phase contains only one pure reactant and whatever products of reaction have accumulated within that phase. For this situation L and G are zero, and the only pertinent forms of the material balance are those involving L' and G'. Example 5 shows one of the many possible forms for the material balance.

The *height of tower* is found by combining the rate equation and material balance for a differential element of tower volume as shown in Fig. 9. Noting that the rate expressions, Eqs. 6, 12, 16, 17, and 20, are all based on unit interfacial area, we have for the disappearance of A

$$G\,d\mathbf{Y}_A = -\frac{L\,d\mathbf{X}_B}{b} = \left(\frac{\text{moles A reacted}}{\text{(interfacial area)(time)}}\right)\left(\frac{\text{interfacial area}}{\text{unit of volume}}\right)\left(\begin{matrix}\text{height of}\\ \text{element}\end{matrix}\right)$$

$$= (-r''_A)a\,dh \tag{46}$$

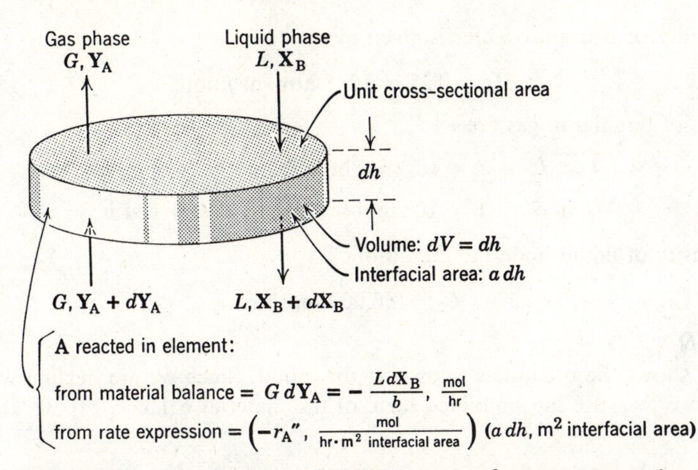

FIGURE 9. Development of design equation for tower operations when reaction is restricted to interfacial region between phases.

Rearranging and integrating, we obtain in terms of A or B

$$h = G\int_{Y_{A1}}^{Y_{A2}} \frac{dY_A}{(-r_A'')a} = \frac{L}{b}\int_{X_{B2}}^{X_{B1}} \frac{dX_B}{(-r_A'')a} \tag{47}$$

where dY_A and dX_B are given by Eqs. 35 and 36. For dilute systems

$$h = \frac{G}{\pi}\int_{p_{A1}}^{p_{A2}} \frac{dp_A}{(-r_A'')a} = \frac{L}{bC_T}\int_{C_{B2}}^{C_{B1}} \frac{dC_B}{(-r_A'')a} \tag{48}$$

Replacing the rate term of Eqs. 47 and 48 by the appropriate expression, either Eq. 6, 12, 16 and 17, or 20 gives the height of tower. Naturally in most cases the integral must be evaluated either graphically or numerically.

The following example illustrates the use of these equations.

EXAMPLE 1. Towers for straight absorption

The concentration of an undesirable impurity A in air is to be reduced from 0.1% to 0.02% by absorption in pure water. Find the height of tower required for countercurrent operations.

Data. For consistency these are all given in units of moles, meters, and hours.

For the packing used

$$k_{Ag}a = 32{,}000 \text{ mol/hr}\cdot\text{m}^3\cdot\text{atm}$$

$$k_{Al}a = 0.1/\text{hr}$$

The solubility of A in pure water is given by

$$H_A = 125 \times 10^{-6} \text{ atm}\cdot\text{m}^3/\text{mol}$$

Flow rates of liquid and gas are

$$L \approx L' = 7 \times 10^5 \text{ mol/hr}\cdot\text{m}^2$$

$$G \approx G' = 1 \times 10^5 \text{ mol/hr}\cdot\text{m}^2 \qquad \text{at } \pi = 1 \text{ atm}$$

Molar density of liquid under all conditions is

$$C_T = 56{,}000 \text{ mol/m}^3$$

SOLUTION

Figure E1 shows the quantities known at this point. Since we are dealing with dilute solutions, we may use the simplified form of the material balance, Eq. 38. Integrating we obtain

$$p_A - p_{A1} = \frac{L}{G}\frac{\pi}{C_T}(C_A - C_{A1})$$

or

$$p_A - 0.0002 = \frac{(7 \times 10^5)(1)}{(1 \times 10^5)(56{,}000)}(C_A - 0)$$

or

$$8000p_A - 1.6 = C_A$$

from which the concentration of A in the leaving liquid is

$$C_{A2} = 8000(0.0010) - 1.6 = 6.4 \text{ mol/m}^3$$

Select a number of partial pressures of A in the tower, determine the corresponding concentration of A in the liquid, calculate the equilibrium partial pressure of A, p_A^*, corresponding to this concentration in the liquid, and then calculate the overall driving force for physical absorption. This procedure is shown in Table E1.

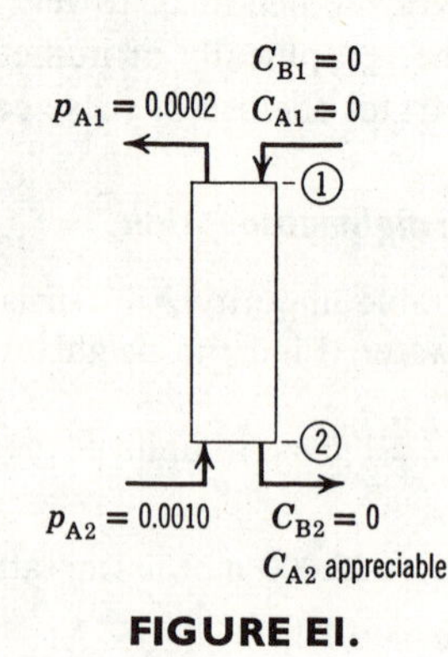

FIGURE E1.

TABLE E1

p_A	C_A	$p_A^* = H_A C_A$	$\Delta p = p_A - p_A^*$
0.0002	0	0	0.0002
0.0006	3.2	0.0004	0.0002
0.0010	6.4	0.0008	0.0002

The overall mass transfer coefficient based on volume of tower is

$$\frac{1}{K_{Ag}a} = \frac{1}{k_{Ag}a} + \frac{H_A}{k_l a} = \frac{1}{32{,}000} + \frac{125 \times 10^{-6}}{0.1} = 0.001283$$

Hence

$$K_{Ag}a = 780 \text{ mol/hr} \cdot \text{m}^3 \cdot \text{atm}$$

The height of tower is then given by Eq. 41, or

$$h = \frac{G}{\pi} \int \frac{dp_A}{(-r_A'')a} = \frac{G}{\pi} \int_{p_{A1}}^{p_{A2}} \frac{dp_A}{K_{Ag}a\, \Delta p_A} = 1 \times 10^5 \int_{0.0002}^{0.0010} \frac{dp_A}{(780)(0.0002)} = 513 \text{ m}$$

Comment. Here the liquid film provides over 95% of the resistance to transfer. Hence we can with little error consider this to be a liquid-film-controlling process. Let us next see how the addition of a reactive liquid affects the performance.

EXAMPLE 2. *Towers for high concentration of liquid reactant; case B*

Replace the unreactive absorbent of Example 1 with a reactive liquid which contains a high concentration of reactant B, $C_{B1} = 800 \text{ mol/m}^3$ or approximately 0.8N.

Data. The reaction

$$A(g) + B(l) \rightarrow \text{products}$$

takes place in the liquid and is extremely rapid.

Assume that the diffusivities of A and B in water are the same. Thus

$$k_{Al} = k_{Bl} = k_l$$

SOLUTION

Figure E2 shows the flow streams. The strategy in solving the problem is as follows.

Step 1. Express the material balance and find C_{B2} in the exit stream.
Step 2. Find which of the two forms of rate equation should be used.
Step 3. Determine the tower height.

Step 1. *Material balance.* For dilute solutions with rapid reaction Eq. 45 gives

$$p_A - p_{A1} = \frac{L\pi}{GbC_T}(C_{B1} - C_B)$$

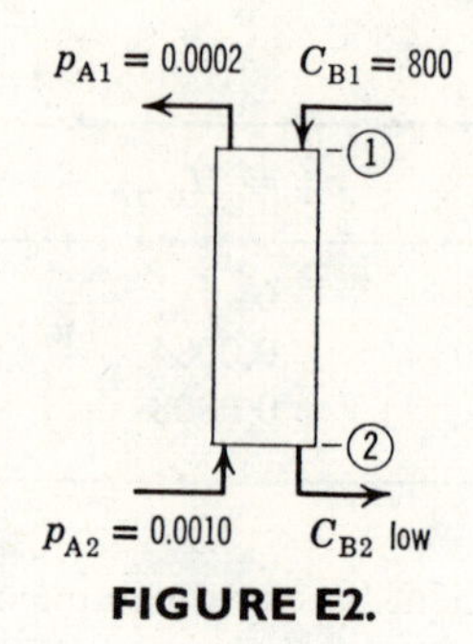

FIGURE E2.

or

$$p_A - 0.0002 = \frac{(7 \times 10^5)(1)}{(1 \times 10^5)(56{,}000)}(800 - C_B)$$

or

$$8000 p_A = 801.6 - C_B$$

hence the concentration of B leaving the tower is

$$C_{B2} = 801.6 - 8000(0.0010) = 793.6 \text{ mol/m}^3$$

Step 2. *Form of rate equation to use.* Check both ends of the tower:

$$\text{at top}\begin{cases} k_{Ag}ap_A = (32{,}000)(0.0002) = 6.4 \text{ mol/hr}\cdot\text{m}^3 \\ k_l a C_B = (0.1)(800) = 80 \text{ mol/hr}\cdot\text{m}^3 \end{cases}$$

$$\text{at bottom}\begin{cases} k_{Ag}ap_A = 32 \\ k_l a C_B = 79.36 \end{cases}$$

At both ends of the tower $k_{Ag}p_A < k_l C_B$; therefore gas-phase resistance controls and Eq. 12 should be used.

Step 3. *Height of tower.* From Eqs. 48 and 12

$$h = \frac{G}{\pi}\int \frac{dp_A}{(-r''_A)a} = \frac{G}{\pi}\int_{p_{A1}}^{p_{A2}} \frac{dp_A}{k_{Ag}ap_A} = 1 \times 10^5 \int_{0.0002}^{0.0010} \frac{dp_A}{32{,}000 p_A} = 5.0 \text{ m}$$

Comment. Even though liquid phase controls in physical absorption, it does not necessarily follow that it should still control when reaction occurs. In fact, we see here that it is the gas phase alone which influences the rate of the overall process. Reaction serves merely to eliminate the resistance of the liquid film. Also note the remarkable improvement in performance; 5 versus 500 meters.

EXAMPLE 3. Towers for low concentration of liquid reactant; case A

Repeat Example 2 using a feed with $C_{B1} = 32$ mol/m³, instead of 800 mol/m³.

SOLUTION

Figure E3 shows the known streams for the tower. Solve by making a material balance, check the form of rate to use, and then apply the performance equation for tower height.

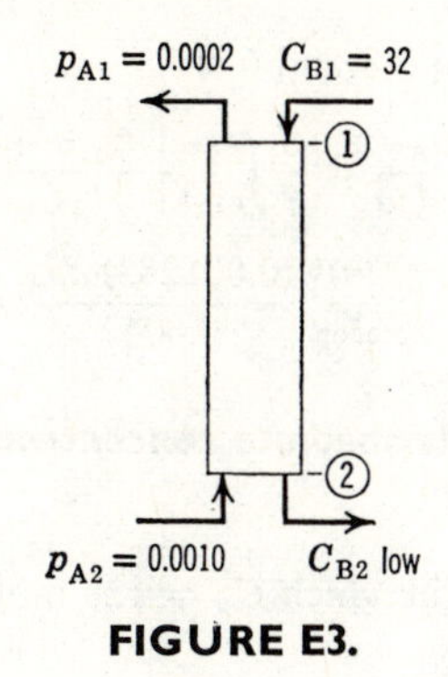

FIGURE E3.

Step 1. *Material balance.* As in Example 2, Eq. 45 gives for any point in the tower

$$8000p_A = 33.6 - C_B$$

hence at the bottom of the tower

$$C_{B2} = 33.6 - (8000)(0.0010) = 25.6 \text{ mol/m}^3$$

Step 2. *Form of rate equation to use.* Check both ends of the tower:

$$\text{at top}\begin{cases} k_{Ag}ap_A = 6.4 \text{ mol/hr}\cdot\text{m}^3 \\ k_l aC_B = 3.2 \text{ mol/hr}\cdot\text{m}^3 \end{cases}$$

$$\text{at bottom}\begin{cases} k_{Ag}ap_A = 32 \\ k_l aC_B = 2.56 \end{cases}$$

At both ends of the tower $k_{Ag}p_A > k_l C_B$; therefore the reaction takes place within the liquid film, and Eq. 6 should be used, or

$$-r''_A = \frac{H_A C_B + p_A}{1/k_{Ag} + H_A/k_l}$$

Step 3. *Height of tower.* At a number of locations evaluate $H_A C_B + p_A$ as shown in Table E3

TABLE E3

p_A	C_B from Material Balance	$H_A C_B$	$p_A + H_A C_B$
0.0002	32.0	0.0040	0.0042
0.0006	28.8	0.0036	0.0042
0.0010	25.6	0.0032	0.0042

Hence the tower height from Eq. 48 and 6 is

$$h = \frac{G}{\pi}\int \frac{dp_A}{(-r''_A)a} = \frac{G}{\pi}\int_{p_{A1}}^{p_{A2}} \frac{1/k_{Ag}a + H_A/k_l a}{H_A C_B + p_A}\, dp_A$$

$$= 1 \times 10^5 \int_{0.0002}^{0.0010} \frac{(0.001283)\, dp_A}{0.0042} = 24.4 \text{ m}$$

EXAMPLE 4. Towers for intermediate concentrations of liquid reactant; cases A and B

Repeat Example 2 using a feed in which $C_B = 128$ mol/m³.

SOLUTION

Refer to Fig. E4*a*, and solve as with the previous examples.

Step 1. *Material Balance.* As with Examples 2 and 3 we have at any point in the tower $8000p_A = 129.6 - C_B$. From this expression we find for the bottom of the tower $C_{B2} = 121.6$.

Step 2. *Form of rate equation to use.* Check both ends of the tower:

$$\text{at top}\begin{cases} k_{Ag}ap_A = 6.4 \text{ mol/hr}\cdot\text{m}^3 \\ k_l aC_B = 12.8 \text{ mol/hr}\cdot\text{m}^3 \end{cases}$$

$$\text{at bottom}\begin{cases} k_{Ag}ap_A = 32 \\ k_l aC_B = 12.16 \end{cases}$$

At the top $k_{Ag}p_A < k_l C_B$; hence Eq. 12 must be used. At the bottom $k_{Ag}p_A > k_l C_B$; hence Eq. 6 must be used.

Let us now find the condition at which the reaction zone just reaches the interface and where the form of rate equation changes. This occurs where

$$k_{Ag}p_A = k_l C_B \qquad \text{or} \qquad 3.2 \times 10^5 p_A = C_B$$

Solving with the material balance we find that the change occurs at $p_A = 0.000395$.

Step 3. *Height of tower.* The rate is found from Table E4. As may be expected, at $p_A = 0.000395$ the calculated rates from Eqs. 6 and 12 are identical.

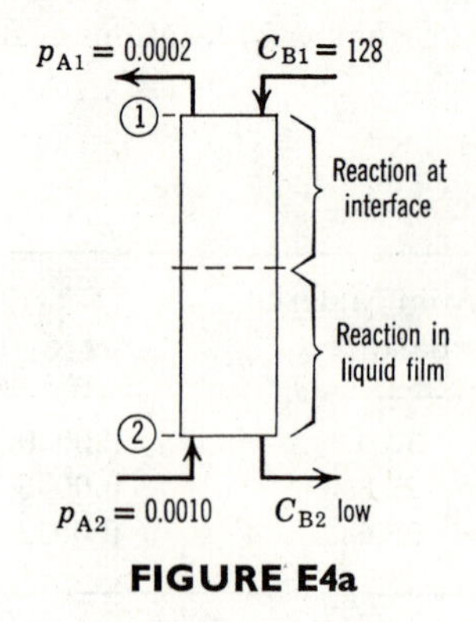

FIGURE E4a

TABLE E4

p_A	C_B	H_AC_B	$p_A + H_AC_B$	$K_{Ag}a$	$(-r''_A)a$	$\frac{1}{(-r''_A)a}$
0.0002				32,000	6.4	0.1563 } Use
0.000395				32,000	12.64	0.0792 } Eq. 12
0.000395	126.4	0.0158	0.0162	780	12.64	
0.0007	124.0	0.0155	0.0162	780	12.64	0.0792 } Use
0.0010	121.6	0.0152	0.0162	780	12.64	0.0792 } Eq. 6

The tower height can be found by graphical integration or analytically as follows:

$$\begin{aligned} h &= h_{\text{upper section}} + h_{\text{lower section}} \\ &= \frac{G}{\pi}\int_{0.0002}^{0.000395} \frac{dp_A}{k_{Ag}ap_A} + \frac{G}{\pi}\int_{0.000395}^{0.0010} \frac{1/k_{gA}a + H_A/k_la}{p_A + H_AC_B}\, dp_A \\ &= \frac{(1 \times 10^5)}{(32{,}000)}\left(\ln\frac{3.95}{2}\right) + (1 \times 10^5)\,\frac{(1/780)(0.000605)}{(0.0162)} \\ &= 2.1 + 4.8 = 6.9 \text{ m} \end{aligned}$$

Figure E4*b* summarizes the method of solution.

Comment. In this example we see that two distinct zones are present. Situations may be encountered where even another zone may be present. For example, if the entering liquid contains insufficient reactant, a point is reached in the tower where all this reactant is consumed. Below this point physical absorption alone takes place in reactant-free liquid. The methods of these examples, when used together, deal in a straightforward manner with this three-zone situation and van Krevelens and Hoftijzer (1948) discuss actual situations where these three distinct zones are present.

Comparing solutions for the four examples shows how reaction increases the effectiveness of the absorption process.

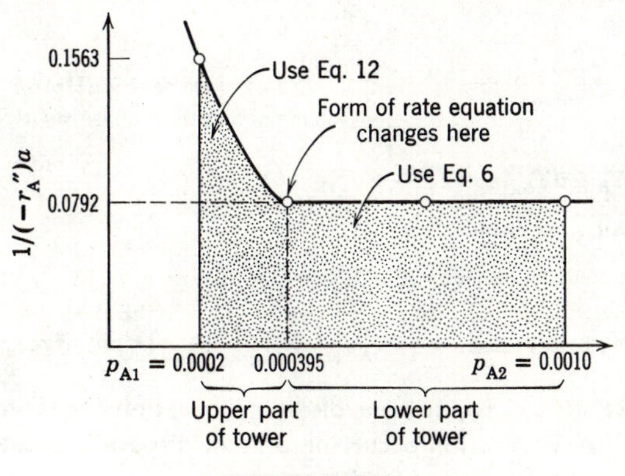

FIGURE E4b.

Towers for Slow Reaction

When reaction is very slow compared to mass transfer, the rate of reaction is best measured in terms of unit volume of reacting phase rather than unit interfacial surface between phases.

Determination of tower height is generally difficult primarily because of the complexities of the material balance, for reactants can be present in appreciable quantities in both phases—flowing up the tower in one phase and flowing down the tower in the other. In general the design methods of extraction (triangular diagrams, Janecke diagrams, and so on) must be used. Despite this difficulty let us illustrate the design equation for the reaction

$$\text{A(gas)} + b\text{B(liquid)} \rightarrow \text{products}$$

with the following restrictions on the material balance; B is insoluble in gas, and the amount of unreacted A in the liquid is small compared to the A in the gas phase. With these restrictions unreacted B flows downward in the liquid, unreacted A flows upward in the gas, and the material balance of Eq. 42 is applicable.

Noting that each mole of A reacting in the liquid is replaced by 1 mole of fresh A from the gas stream, and combining material balance with rate equation as with Eq. 46 we get (see Fig. 10)

$$G\,d\mathbf{Y}_A = -\frac{L\,d\mathbf{X}_B}{b} = \left(\frac{\text{moles of A reacted}}{\text{(volume of liquid)(time)}}\right)\left(\frac{\text{volume of liquid phase}}{\text{total volume}}\right)\left(\begin{matrix}\text{height of}\\ \text{element}\end{matrix}\right)$$

$$= (-r_A) f\,dh$$

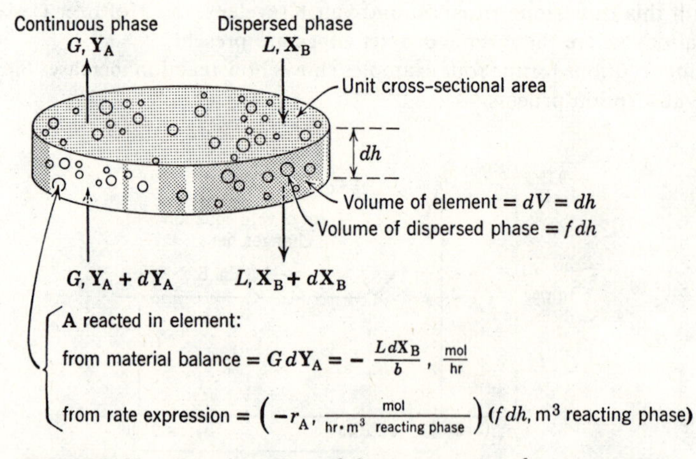

FIGURE 10. Development of design equation for tower operations for slow reaction occurring only in dispersed phase. Rate is based on unit volume of fluid mixture.

Because the rate is based on unit volume of fluid mixture, the volume fraction of reacting phase f appears in Eq. 49 instead of the interfacial area per unit volume as in Eq. 46. On integration we then obtain the desired design equation

$$h = \frac{G}{f}\int_{\mathbf{Y}_{A1}}^{\mathbf{Y}_{A2}} \frac{d\mathbf{Y}_A}{-r_A} = \frac{L}{bf}\int_{\mathbf{X}_{B2}}^{\mathbf{X}_{B1}} \frac{d\mathbf{X}_B}{-r_A} \tag{50}$$

where $d\mathbf{Y}_A$ and $d\mathbf{X}_B$ are given by Eqs. 35 and 36 and $-r_A$ is the appropriate rate expression in volumetric units; for example, Eq. 16.

Equation 50 may take on a variety of special forms when applied to specific physical situations. This is often due to the particular form of the material balance which is applicable.

For example, consider the special case of a spray tower with dilute B present in liquid droplets, A present in gas, and reaction given by Eq. 16. If conditions are such that the concentration of A in the gas is approximately constant throughout the tower Eq. 50 becomes

$$h \approx \frac{L}{bf}\int \frac{d(C_B/C_T)}{kC_AC_B} = \frac{L}{bfkC_AC_T}\int_{C_{B2}}^{C_{B1}} \frac{dC_B}{C_B} = \frac{LH_A}{bfkp_AC_T}\ln\frac{C_{B1}}{C_{B2}} \tag{51}$$

The following example illustrates still another form of the performance equation, different again because of the particular form of material balance which applies.

EXAMPLE 5. Towers for slow reaction; case H

Benzene is to be chlorinated in a tower by countercurrent contacting with a stream of pure gaseous chlorine. The reaction

$$C_6H_6\text{(liquid)} + Cl_2\text{(gas)} \rightarrow C_6H_5Cl\text{(liquid)} + HCl\text{(gas)}$$

is slow, elementary, irreversible, and occurs in the liquid between dissolved chlorine and benzene. With the additional assumptions:

constant molar density of liquid, C_T = constant
constant pressure in gas phase, π = constant
plug flow of both streams
small amount of dissolved and unreacted chlorine in liquid
low solubility of HCl in liquid
H_A constant
the reaction of Cl_2 with C_6H_5Cl to be neglected

derive the expression for the tower height as a function of the variables of the system.

SOLUTION

Let A = chlorine, B = benzene, R = monochlorobenzene, S = hydrogen chloride.

First of all, the material balance cannot be written in terms of L and G because no inerts are present. But we note that for each mole of chlorine used 1 mole HCl is formed and is returned to the gas phase. Similarly, the total molar flow rate of liquid is unchanged because for each mole of benzene reacted a mole of chlorobenzene is formed. Thus the total molar flow rates of gas and liquid remain unchanged, or L' and G' are constant. Thus the material balance of Eq. 42 becomes

$$d\left(\frac{G'p_A}{\pi}\right) = -\frac{1}{b}\,d\left(\frac{L'C_B}{C_T}\right)$$

or

$$\frac{G'}{\pi}\,dp_A = -\frac{L'}{C_T}\,dC_B \tag{i}$$

Combining with the rate expression based on unit volume of reacting or continuous phase

$$-r_A = -r_B = kC_AC_B$$

we obtain

$$\frac{G'}{\pi}\,dp_A = -\frac{L'}{C_T}\,dC_B = kC_AC_Bf\,dh$$

where f is now the volume fraction of liquid or reacting phase. By rearranging and integrating, the height of tower is then

$$h = \frac{G'H_A}{kf\pi}\int_{p_{A1}}^{p_{A2}}\frac{dp_A}{p_AC_B} = \frac{L'H_A}{kfC_T}\int_{C_{B2}}^{C_{B1}}\frac{dC_B}{p_AC_B} \tag{ii}$$

with p_A and C_B related by the material balance. Integrating the differential material balance, Eq. (i), we find at any point in the tower,

$$C_B = C_T - \frac{G'C_T}{L'\pi}(p_A - p_{A1}) \tag{iii}$$

Replacing Eq. (iii) in Eq. (ii) and integrating analytically gives

$$h = \frac{-G'H_A}{fC_Tk[\pi + (G'/L')p_{A1}]}\ln\frac{p_{A1}[\pi - (G'/L')(\pi - p_{A1})]}{\pi^2}$$

or more conveniently in terms of mole fractions

$$h = \frac{-G'H_A}{f\pi C_Tk[1 + (G'/L')y_{A1}]}\ln y_{A1}\left[1 - \frac{G'}{L'}(1 - y_{A1})\right] \tag{52}$$

where

$$y_A = \frac{p_A}{\pi} \quad \text{and} \quad x_A = \frac{C_A}{C_T}$$

Figure E5 shows that the rate is maximum at some intermediate position within the tower, dropping off at either end where the concentration of one or the other of the components is very low.

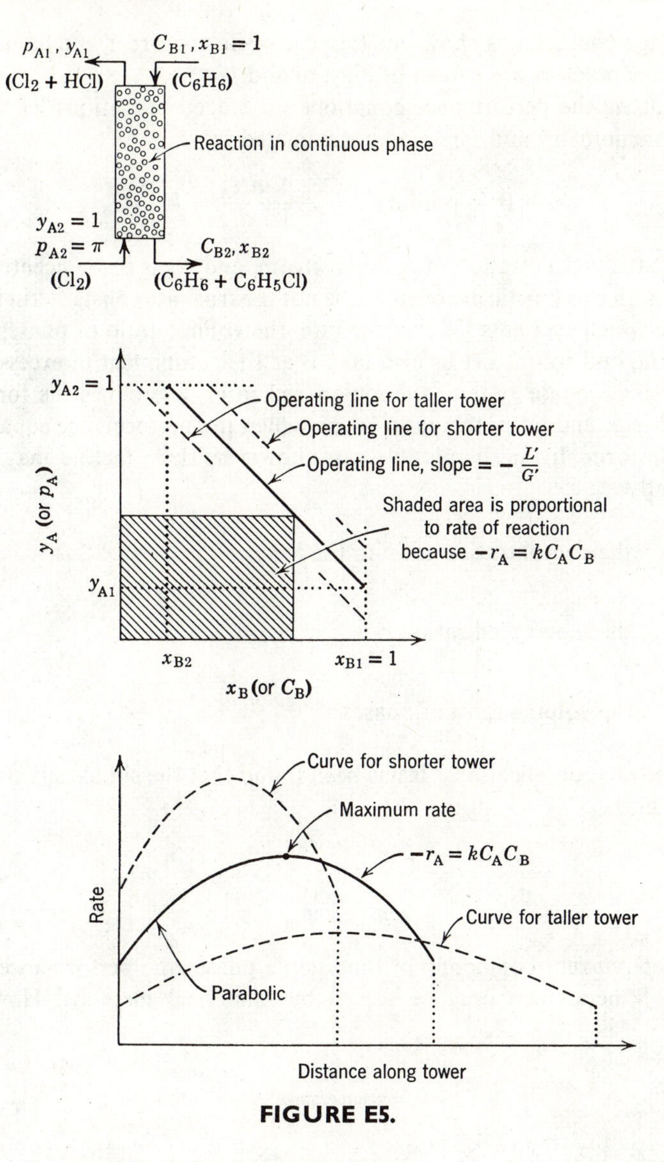

FIGURE E5.

Mixer-settlers (Mixed Flow of Both Phases)

Mixer-settlers are industrial devices for bringing into intimate contact and then separating immiscible fluids. Each of these units operates as an ideal contactor or mixed reactor. For gas–liquid systems where density differences between phases are great, the settler is not needed. Mixer-settlers are frequently used for slow reactions.

A single-stage contactor is shown in Fig. 6*g*, while cocurrent and countercurrent multistaged contactors are shown in Figs. 6*i* and *j*.

In developing the performance equations for staged operations let us suppose that the stoichiometry and rate are approximated by

$$A + B \rightarrow \text{product}, \qquad -\frac{1}{V}\frac{dN_A}{dt} = k_t C_A C_B \tag{53}$$

where C_A is the concentration of A in its stream, and C_B is the concentration of B in its stream. In this kinetic expression k_t is not the true rate constant but is an overall measure which accounts for the true rate, the volume ratio of phases, the state of dispersion, and so on. Let us also take B as the component in excess.

Now in single phase systems the molar feed ratio determines the form of rate equation to use, and when this ratio is unity then the performance equation takes on a simple form. In the immiscible case, however, three factors may be varied independently:

$\frac{C_{B0}}{C_{A0}}$, the concentration ratio in the feed

$\frac{F_{B0}}{F_{A0}}$, the molar feed ratio

$\frac{V_B}{V_A}$, the volume ratio of phases

and this leads to complications. It has been found that the simple situation occurs only when both

$$\bar{t}_A = \bar{t}_B \tag{54}$$

and

$$F_{A0} = F_{B0} \tag{55}$$

For countercurrent contacting of immiscible phases the performance equation for general kinetics can only be solved by numerical methods. However, for

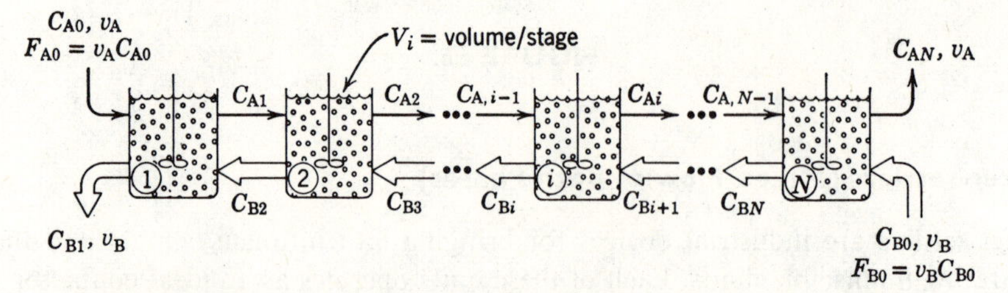

FIGURE 11. Nomenclature for an *N*-stage countercurrent mixed flow contactor.

second-order kinetics (Eq. 53) an analytic solution is possible. Thus for N stages of equal size, as shown in Fig. 11, Ahluwalia and Levenspiel (1968) give

$$V_{\text{total}} = \frac{Nv_A}{C_{B0}k_t L_3}\left[\left(\frac{L_2}{L_1}\right)^{1/N} - 1\right] \tag{56}$$

where

$$L_1 = 1 - \frac{F_{A0}}{F_{B0}} X_A \tag{57}$$

$$L_2 = \frac{1}{1 - X_A} \tag{58}$$

$$L_3 = 1 + \frac{F_{A0}}{F_{B0}}(1 - X_A) \tag{59}$$

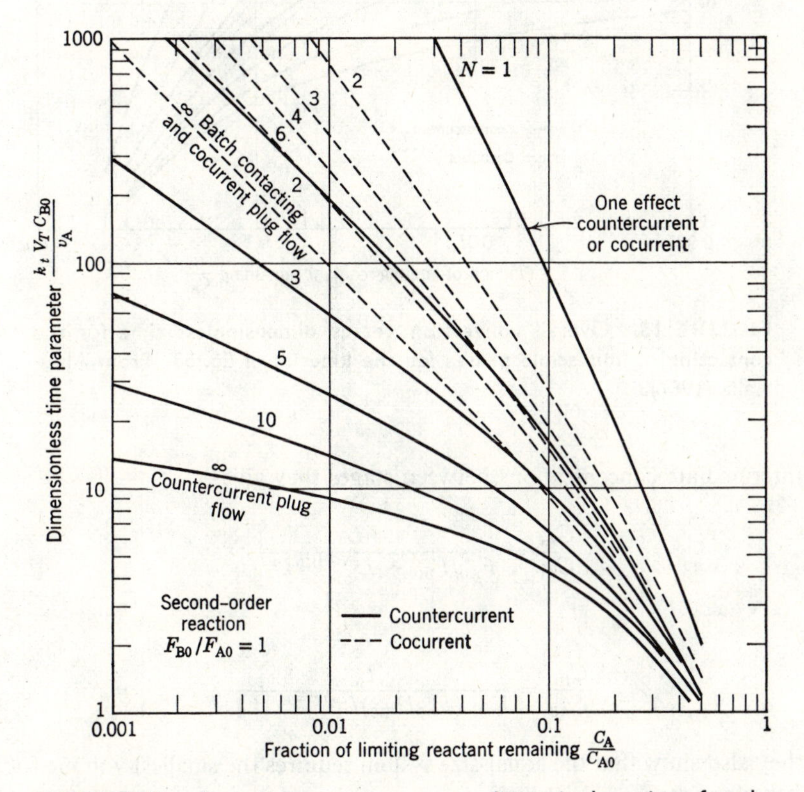

FIGURE 12. Overall conversion versus dimensionless time for the contracting of immiscible phases for the kinetics of Eq. 53. From Ahluwalia (1967).

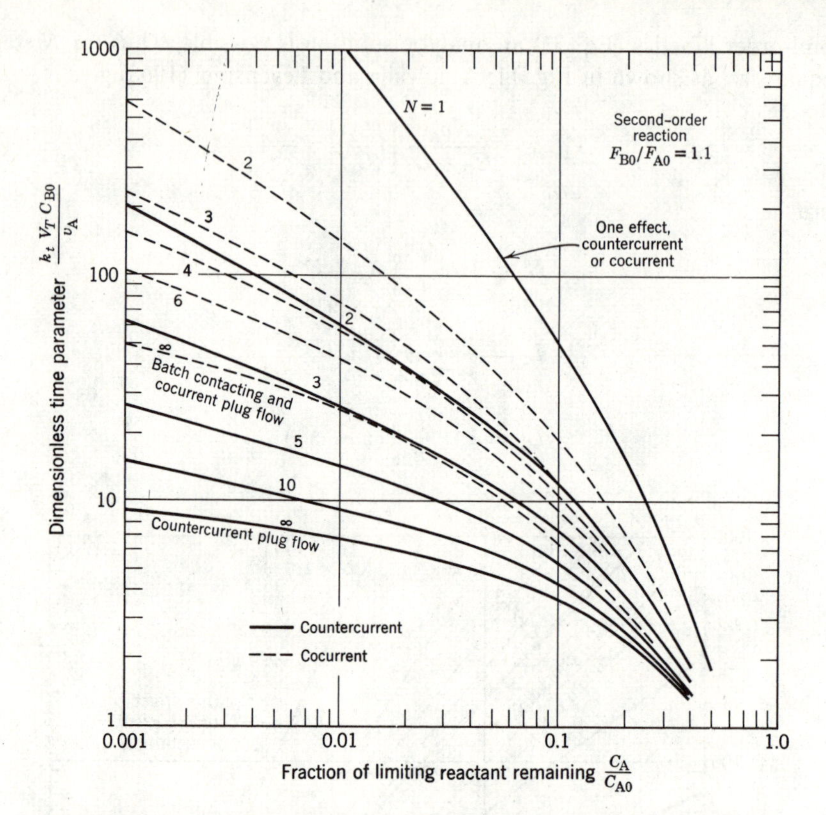

FIGURE 13. Overall conversion versus dimensionless time for the contacting of immiscible phases for the kinetics of Eq. 53. From Ahluwalia (1967).

For intermediate concentrations between stages they give

$$\frac{C_{Ai}}{C_{A0}} = \frac{L_3}{F_{A0}/F_{B0} + L_1^{(N-i)/N} L_2^{i/N}} \tag{60}$$

and

$$\frac{C_{Bi}}{C_{A0}} = \frac{L_3}{1 + (F_{A0}/F_{B0}) L_1^{-(N-i)/N} L_2^{-i/N}} \tag{61}$$

and they also show that the equal-size system requires the smallest volume for that number of stages.

Performance charts for this system have been prepared by Ahluwalia (1967), of which Figs. 12 to 15 are a sample. These charts allow rapid comparison among

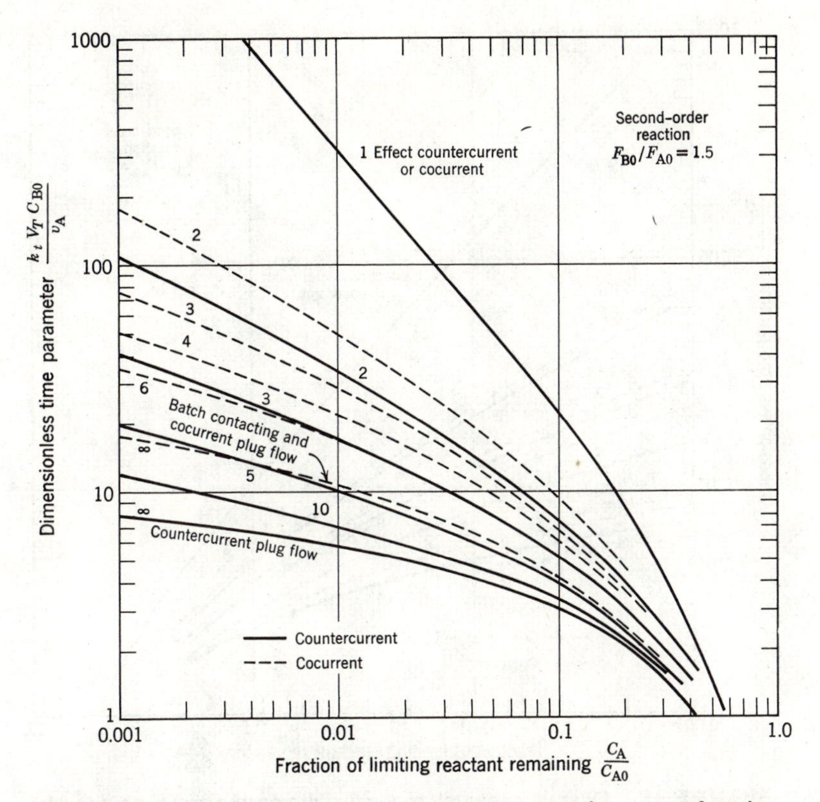

FIGURE 14. Overall conversion versus dimensionless time for the contacting of immiscible phases for the kinetics of Eq. 53. From Ahluwalia (1967).

cocurrent and countercurrent multistage operations, plug flow, single mixer-settlers, and batch reactors, and they generalize the earlier charts of Jenney (1955) which are restricted to the condition of Eq. 54.

One underlying assumption for this whole treatment is that negligible amounts of unreacted A flows from stage to stage with the B stream; the same with unreacted B in the A stream.

EXAMPLE 6. Multistage operations

Reactants A and B are present in separate phases. When these phases are brought into intimate contact in a single mixed reactor, the reaction

$$A + B \rightarrow \text{products}$$

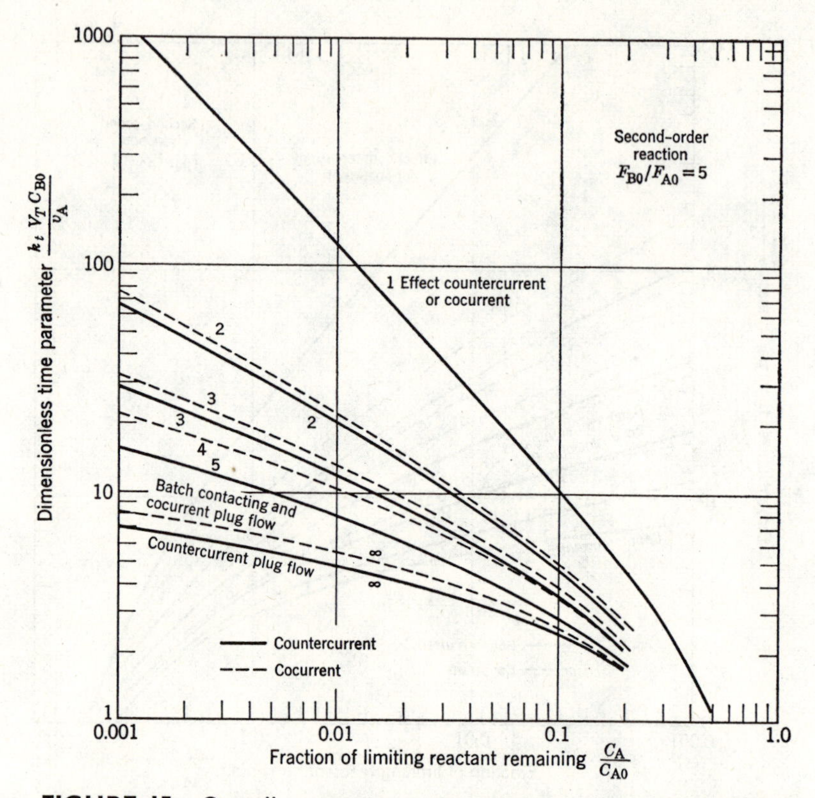

FIGURE 15. Overall conversion versus dimensionless time for the contacting of immiscible phases for the kinetics of Eq. 53. From Ahluwalia (1967).

proceeds slowly with over-all kinetics:

$$-r_A = -\frac{1}{V}\frac{dN_A}{dt} = k_t C_A C_B$$

For equimolar feed the conversion is 95%. With the same feed we plan to raise conversion to 99% by using three mixed reactors in series, each equal in size to the original unit.

(*a*) How much can production be raised by using cocurrent flow of fluids?
(*b*) How much can production be raised by using countercurrent flow of fluids?
(*c*) Find the fractional conversion between the stages of part *b*.
(*d*) Repeat part (*a*) using countercurrent plug flow in a column equal in size to three mixed reactors.

Assume identical mean residence time of the two phases, ignore expansion, and let the volume fraction of the phases remain unchanged.

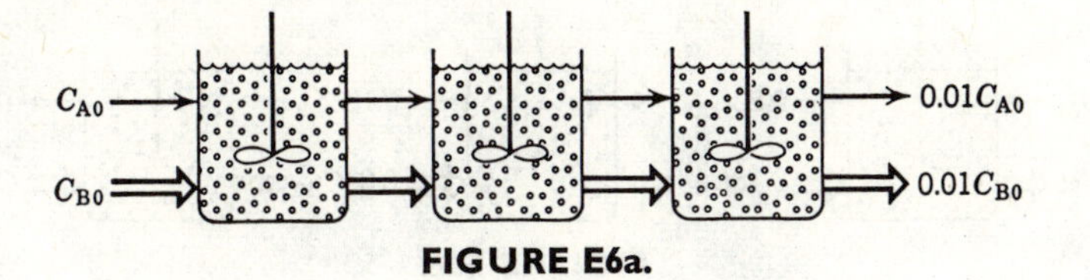

FIGURE E6a.

SOLUTION

(***a***) *Cocurrent, three stages* (see Fig. E6*a*). From Fig. 12 we have for one mixed reactor at 95% conversion

$$\left(\frac{k_t V_T C_{B0}}{v_A}\right)_{N=1} = 380$$

For cocurrent flow in three reactors, 99% conversion, the same figure happens to also give

$$\left(\frac{k_t V_T C_{B0}}{v_A}\right)_{N=3} = 380$$

The holding time is the same in both cases; however, the volume is three times as great in the cocurrent unit. Hence production is three times as high.

We may also note that where $\bar{t}_A = \bar{t}_B$ in each and all stages, all cocurrent flow operations can be handled by the methods and charts of Chapter 6. Thus from Fig. 6.6 we also find

$$k\tau C_{A0} = 380$$

(***b***) *Countercurrent, three stages* (see Fig. E6*b*). From Fig. 12 we again have for 95% conversion

$$\left(\frac{k_t V_T C_{B0}}{v_A}\right)_{N=1} = 380$$

and for three reactors with countercurrent flow of fluids we find from the same figure

$$\left(\frac{k_t V_T C_{B0}}{v_A}\right)_{N=3} = 63$$

Taking ratios

$$\frac{F_{99\%}}{F_{95\%}} = \frac{v_{99}}{v_{95}} = \frac{V_{99}/63}{V_{95}/380} = \frac{3/63}{1/380} = 18$$

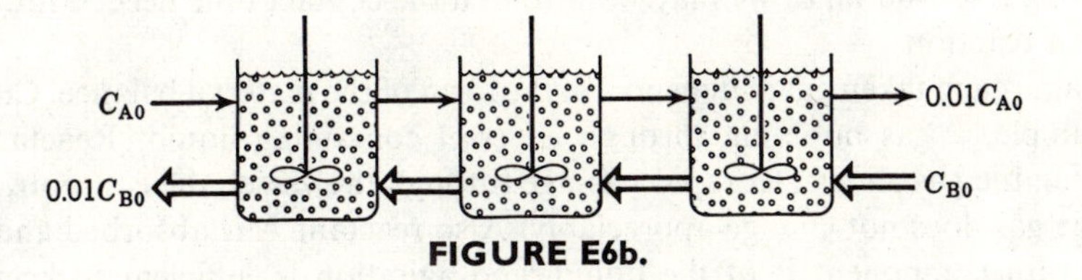

FIGURE E6b.

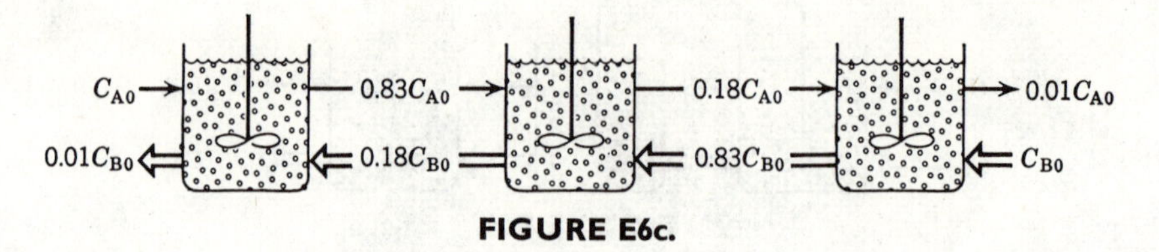

FIGURE E6c.

Hence production is 18 times as high if countercurrent flow is used. This is a sixfold improvement over cocurrent flow.

(*c*) *Intermediate conversions.* First consider the output from stage 1. Equations 57, 58, and 59 give

$$L_1 = 0.01, \qquad L_2 = 100, \qquad L_3 = 1.01$$

and on replacing in Eq. 60 we obtain

$$\frac{C_{A1}}{C_{A0}} = \frac{1.01}{1 + (0.01)^{2/3}(100)^{1/3}} = 0.83$$

By symmetry we find that the fractional conversion for both A and B in the two extreme stages is 17%, so by difference the conversion is 65% in the middle unit. Thus we have the concentrations as shown in Fig. E6*c*.

(*d*) *Countercurrent plug flow.* Figure 12 gives

$$\left(\frac{k_t V_T C_{B0}}{v_A}\right)_{\text{plug}} = 9.2$$

Therefore taking ratios

$$\frac{F_{\text{plug}}}{F_{95\%}} = \frac{v_{\text{plug}}}{v_{95}} = \frac{V_{\text{plug}}/9.2}{V_{95}/380} = \frac{3/9.2}{1/380} = 124$$

Hence production is 124 times as high if countercurrent plug flow is used.

We may also note that the design equation developed in Example 5 can also be used here. Thus for a 1-ft cross section of tower $h = V$ and on letting $G' = L'$ and $y_{A1} = 0.01$ Eq. 52 becomes

$$\frac{Vf\pi C_T k}{G' H_A} = \frac{k_t V_T C_{B0}}{v_A} = -\frac{\ln y_{A1}[1 - (G'/L')(1 - y_{A1})]}{1 + (G'/L')y_{A1}} = \frac{2 \ln 1.01}{1.01} = 9.12$$

which agrees closely with the value obtained from the charts.

Semibatch Contacting Patterns

In semibatch operation where one fluid is continuously passed through a vessel containing a second fluid, we may want to find the contact time needed for a given extent of reaction.

The approach taken again depends on the form of the material balance. Consider, for example, a gas bubbling through a vessel containing liquid. Reactant A is present in the gas and, in its rapid passage through the vessel, the concentration of A in the gas does not change appreciably. Also reactant A is absorbed and reacts slowly with component B of the liquid, and agitation is sufficient to keep compositions throughout the liquid uniform. With the passage of time the concentration of B will fall but the concentration of A will remain unchanged. If the kinetics are first order with respect to both A and B, we then have

$$-r_B = -\frac{1}{V_l}\frac{dN_B}{dt} = kC_AC_B$$

Rearranging and integrating, noting that C_A is constant, we obtain

$$-\int_{C_{B0}}^{C_B} \frac{dC_B}{C_B} = kC_A \int_0^t dt$$

or

$$t = \frac{1}{kC_A} \ln \frac{C_{B0}}{C_B}$$

or

$$\frac{C_B}{C_{B0}} = e^{-C_A kt} = e^{-p_A kt/H_A} \tag{62}$$

Equation 51 and the equation just given are essentially identical expressions, one applied to plug flow, the other to batch contacting of a liquid with a uniform gas environment.

REACTIVE DISTILLATION AND EXTRACTIVE REACTIONS

In this chapter we have so far considered two reasons for using fluid–fluid contacting: either as an aid to a separation process or as a means of forming a desired product. We now consider a third reason, the improvement of the product distribution of homogeneous reactions.

To illustrate the ideas involved suppose we have a homogeneous catalyzed reaction of A to R to S, with R as the desired product. If R is soluble in a second immiscible phase we may want deliberately to introduce that phase into the reactor, extract R as it forms, and thereby obtain an improved product distribution. Schematically, then,

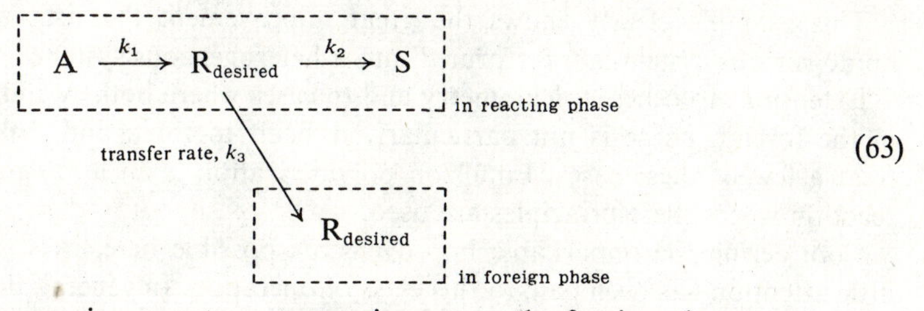

(63)

Figure 16 shows various reactor setups using gas as the foreign phase.

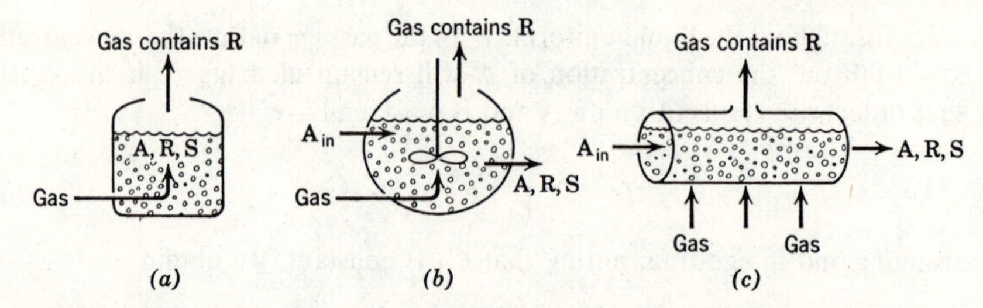

FIGURE 16. Extractive reaction with liquid in (*a*) batch, (*b*) mixed flow, (*c*) plug flow. Reaction A → R → S occurs in the liquid, gas is the foreign phase.

Now it is not too hard to show (see Problem P29) that when $k_3 > k_1$ then the formation of R is maximized by letting the reaction go to completion, in which case the ratio of R (in foreign phase) to S produced is given by k_3/k_2. Since k_1 and k_2 are chemical steps while k_3 is a physical step it is easy to change their relative magnitudes. Lowering k_1 and k_2 with respect to k_3 thereby gives a much improved product distribution.

As an example consider the homogeneous liquid reaction with $k_1 = 1$ and $k_2 = 4$ taking place in a batch reactor. From Chapter 7 we then find

$$\frac{C_{R,max}}{C_{A0}} = 16\%$$

However, if gas is bubbled through the reactor to sweep away the R formed, and if the gas–liquid transfer rate of R on a volumetric basis is $k_3 = 6$, then since $k_3 > k_1$,

$$\frac{C_{R,max}}{C_{A0}} = \frac{k_3}{k_2 + k_3} = \frac{6}{4 + 6} = 60\%$$

To improve the product distribution even further, if the temperature of the reactor is lowered so that $k_1 = 0.1$, $k_2 = 0.4$, while k_3 remains unchanged, then

$$\frac{C_{R,max}}{C_{A0}} = \frac{6}{6 + 0.4} = 94\%$$

This example clearly shows the great improvement that is possible when a homogeneous system is transformed into a heterogeneous system.

Extensions to other stoichiometry and to cases where both A and R are soluble in the foreign phase is not particularly difficult to treat, and Problems P30–32 treat a few of these cases. Emulsion polymerization is an industrially important reaction where these principles are used.

Considering the remarkable improvements possible here, it is surprising how little attention has been paid to this class of operation. In general, then, contacting a homogeneous reacting system with a foreign phase should be considered whenever poor product distribution is encountered with multiple reactions. In such a situation the phase equilibrium relationships (solubilities or vapor pressures) should be examined to determine whether such an operation is likely to be beneficial.

When the two phases are gas and liquid then the operation is called a *reactive distillation*, when both are liquid it is called an *extractive reaction.*

COMMENT

This chapter has only touched lightly on this complex subject of two-fluid contacting. Various extensions of this presentation are possible. One large area deals with other kinetic forms, for single and simultaneous reactions, also the analysis of multiple reactions such as fermentations (see Aiba *et al.* (1965)) and polymerizations. These represent difficult but extremely important subjects.

Another extension extends the design procedures to systems where significant amounts of unreacted A and B are present in both phases, to systems where reaction occurs in both phases, and to slow reactions where mass transfer resistance intrudes (case *G* kinetics). Trambouze *et al.* (1961*a,b*) have made significant advances here.

Finally, the numerous aspects of extractive reactions need to be thoroughly explored and systematized. This has still to be done.

REFERENCES

Ahluwalia, M. S., M.S. Thesis, Illinois Institute of Technology, Chicago, 1967.

———, and Levenspiel, O., *Can. J. Chem. Eng.*, **46**, 443 (1968).

Aiba, S., Humphrey, A. E., and Millis, N. F., *Biochemical Engineering*, Academic Press, New York, 1965.

Astarita, G., *Mass Transfer with Chemical Reaction*, Elsevier, Amsterdam, 1967.

Danckwerts, P. V., *Gas-Liquid Reactions*, McGraw-Hill, New York, 1970.

Danckwerts, P. V., *Trans. Faraday Soc.*, **46**, 300 (1950); *A.I.Ch.E. Journal*, **1**, 456 (1955).

Hatta, S., *Technol. Repts. Tôhoku Univ.*, **10**, 119 (1932), from Sherwood and Pigford (1952).

Higbie, R., *Trans. A.I.Ch.E.*, **31**, 365 (1935).

Jenney, T. M., *Chem. Eng.*, **62**, 198 (Dec. 1955).

Kramers, H., and Westerterp, K. R., *Elements of Chemical Reactor Design and Operations*, Netherlands University Press, Amsterdam, 1963.

Lewis, W. K., and Whitman, W. G., *Ind. Eng. Chem.*, **16**, 1215 (1924).

Sherwood, T. K., and Holloway, F. A. L., *Trans. A. I. Ch. E.*, **36**, 21 (1940).
———, and Pigford, R. L., *Absorption and Extraction*, 2nd ed., McGraw-Hill, New York, 1952.
Teller, A. J., *Chem. Eng.*, **67**, 111 (July 11, 1960).
Trambouze, P., *Chem. Eng. Sci.*, **14**, 116 (1961*a*); with Trambouze, M. T., and Piret, E. L., *A.I.Ch.E. Journal*, **7**, 138 (1961*b*).
van Krevelens, D. W., with Hoftijzer, P., and van Hooren, C. J., *Rec. Trav. Chim.*, **67**, 563, 587, 133 (1948); *Chem. Eng. Sci.*, **2**, 145 (1953).

PROBLEMS

These are grouped as follows:

Problems 1–11 deal with gas–liquid kinetics.
Problems 12–18 deal with tower design.
Problems 19–28 deal with multistaged and batch reactors.

1. CO_2 is to be removed from air by countercurrent contact with water at 25°C.
(*a*) What are the relative resistances of gas and liquid films for this operation?
(*b*) What simplest form of rate equation would you use for tower design?
(*c*) For this removal operation would you expect reaction with absorption to be helpful? Why?

From the literature we have for CO_2 between air and water

$$k_g a = 80 \text{ mol/hr} \cdot \text{liter} \cdot \text{atm}$$
$$k_l a = 25/\text{hr}$$
$$H = 30 \text{ atm} \cdot \text{liter/mol}$$

2. We plan to use an NaOH solution to hasten the removal of CO_2 from air at 25°C (see data of previous problem).
(*a*) What form of rate equation should we use when $p_{CO_2} = 0.01$ atm and the solution is 2N in NaOH?
(*b*) How much can absorption be speeded compared to physical absorption with pure water?

Assume that the reaction is instantaneous and is represented by

$$CO_2 + 2OH^- = H_2O + CO_3^{--}$$

3. Repeat the previous problem for $p_{CO_2} = 0.2$ atm and for a solution which is 0.2N NaOH.

4. Hydrogen sulfide (0.1%) in a carrier gas at 20 atm is to be absorbed at 20°C by a solution containing 0.25 mol/liter methanolamine (MEA). Find the form of rate equation which applies at these conditions and find how much faster is the rate over straight physical absorption in pure water.

Data: H_2S and MEA react as follows

$$H_2S + RNH_2 = HS^- + RNH_3^+$$

and since this is an acid base neutralization we can regard it as irreversible and instantaneous. Also from Danckwerts (1970) we find

$$k_{Al}a = 0.030/\text{sec}$$
$$k_{Ag}a = 6 \times 10^{-5}\ \text{mol/cm}^3\cdot\text{sec}\cdot\text{atm}$$
$$\mathscr{D}_{Al} = 1.5 \times 10^{-5}\ \text{cm}^2/\text{sec}$$
$$\mathscr{D}_{Bl} = 10^{-5}\ \text{cm}^2/\text{sec}$$
$$H_A = 0.115\ \text{atm}\cdot\text{liter/mol—for } H_2S \text{ in water}$$

5. Consider a highly water-soluble gas such as ammonia for which at about 10°C

$$H \approx 0.01\ \text{atm}\cdot\text{liter/mol}$$

Also consider the sparingly soluble gases such as carbon monoxide, oxygen, hydrogen, methane, ethane, nitric oxide, and nitrogen for which

$$H \approx 1000\ \text{atm}\cdot\text{liter/mol}$$

For straight physical absorption of these gases in water, assuming no reaction:

(*a*) What are the relative resistances of the gas and liquid films?
(*b*) Which resistance if any controls the absorption process?
(*c*) What form of rate equation should be used for design in these two cases?
(*d*) How does the solubility of the slightly soluble gas affect its rate of absorption in water?
(*e*) In which case (slightly or highly soluble gas) would chemical reaction be more helpful in speeding the process and why?

For this problem, use the rough order of magnitude estimation of k_l and k_g given in this chapter.

6. Consider the absorption of a base A by water in a packed column. At a location where gas is being absorbed by pure water, the overall rate based on unit volume of tower may be expressed as

$$-r_A = (-r_A'')a = -\frac{1}{V}\frac{dN_A}{dt} = K_{Ag}ap_A$$

where $K_{Ag}a$ is the overall coefficient based on unit volume of tower.

(*a*) Now suppose that an acid B is added to the water to aid the absorption. Assuming instantaneous reaction, show how $K_{Ag}a$ should vary with acid strength. Show this on a plot of $K_{Ag}a$ versus acid strength. Also show how this plot should allow estimation of the individual mass transfer coefficients for physical absorption.

(*b*) Sherwood and Holloway (1940) present the following data for the absorption of ammonia in acid solution of various strengths at 25°C.

$K_{Ag}a$, mol/hr·liter·atm	300	310	335	350	380	370
Acid normality	0.4	1.0	1.5	2.0	2.8	4.2

From this data find the gas film contribution (in percent) to the overall mass transfer resistance in physical absorption of ammonia in air.

7. For the absorption of A in reactive B preliminary experiments in a double mixed reactor using one volume ratio, interfacial area, and agitation rate shows that the rate

$$-r''_A \propto p_A C_B$$

(*a*) What kinetic regimes do these results suggest?
(*b*) If more than one regime is likely what further experiments would distinguish between them?

8. Repeat the previous problem with one change: the rate is found to be well represented by

$$-r''_A \propto \frac{p_A C_B}{\text{const.} + C_B}$$

9. The double mixed reactor of Fig. 3 is used to study the kinetics of absorption of A in a reactive solution. The reaction is rather fast and of stoichiometry

$$\text{A(gas)} + \text{B(liquid)} \rightarrow \text{product}$$

In a particular run at 15.5°C and 1 atm, and with a gas-liquid interfacial area of 100 cm^2, the following results were obtained:

Entering gas: $p_A = 0.5$ atm, $v = 30$ cm^3/min
Leaving gas: $v = 15.79$ cm^3/min
Liquid in: $C_B = 0.20$ mol/liter, $v = 5$ cm^3/min

For this run calculate
(*a*) C_B leaving the reactor
(*b*) The rate of absorption
(*c*) The rate of reaction

10. The absorption of gaseous A in an aqueous solution of reactive B proceeds with stoichiometry

$$\text{A(gas)} + \text{B(liquid)} \rightarrow \text{products}$$

The diffusivities of A and B in water are equal and the phase distribution coefficient for A is found to be

$$H_A = 2.5 \text{ atm}\cdot\text{liter/mol}$$

A number of runs are made in the double mixed reactor of Fig. 3 which has an interfacial area of 100 cm². The feeds and flows are adjusted so that the conditions in the reactor and calculated rates are as follows:

Run	p_A, atm	C_B, mol/cm³	$-r''_A$, mol/sec·cm²
1	0.05	10×10^{-6}	15×10^{-6}
2	0.02	2×10^{-6}	5×10^{-6}
3	0.10	4×10^{-6}	22×10^{-6}
4	0.01	4×10^{-6}	4×10^{-6}

What can you say about the kinetics and rate regime for this absorption and reaction?

11. The double mixed reactor of Fig. 3 is used to study the kinetics of a slurry reaction in which liquid containing suspended catalyst particles and a high concentration of reactant B is contacted with pure gaseous reactant A at 1 atm. The following results are obtained when the liquid volume is 200 cm³.

Run	Gas-Liquid Interfacial Area, cm²	$\dfrac{\text{gm cat}}{\text{cm}^3 \text{ liquid}}$	Rate Based On Interfacial Area, $-r''_A$
1	10	0.1	10
2	10	0.5	25
3	50	0.1	2.5
4	50	0.5	10

Knowing that the chemical step on the catalyst surface is rapid develop an expression for rate of reaction based on unit volume of slurry and indicate which resistance, if any, controls.

12. Consider the infinitely fast reaction of Examples 1 to 4. If the acid strength of the absorbing liquid is raised, a point is reached above which further increases will not affect the over-all rate of mass transfer and the tower height. Find this particular concentration.

13. Repeat Examples 1 and 2 using cocurrent flow of fluids in the tower.

14. Repeat Examples 3 and 4 using cocurrent flow of fluids in the tower.

15. Repeat Example 2 using a dilute acid of concentration

$$C_B = 3.2 \text{ mol/m}^3$$

16. An impurity A in a gas is to be reduced from 1% to 2 ppm by countercurrent contact with liquid containing a reactant of concentration $C_B = 3.2$ mol/m³

$$k_{Ag}a = 32{,}000 \text{ mol/hr}\cdot\text{m}^3\cdot\text{atm}$$
$$k_{Al}a = k_{Bl}a = 0.5/\text{hr}$$
$$L = 7 \times 10^5 \text{ mol/hr}\cdot\text{m}^2$$
$$G = 1 \times 10^5 \text{ mol/hr}\cdot\text{m}^2$$
$$H_A = 1.125 \times 10^{-3} \text{ atm}\cdot\text{m}^3/\text{mol}$$
$$C_T = 56{,}000 \text{ mol/m}^3$$

The reaction A + B → product is rapid.

(*a*) Find the height of tower needed.

(*b*) What recommendations do you have (about the concentration of liquid-phase reactant) that may help improve the process?

(*c*) What incoming concentration of B would give the minimum height of tower? What is this height?

17. The H_2S content of a gas is to be reduced from 1% down to 1 ppm by contact in a packed column with an aqueous solution containing 0.25 mol/liter of methanolamine(MEA). Determine a reasonable L/G to use and the height of tower needed.

Data: H_2S and MEA react as follows:

$$H_2S + RNH_2 \rightarrow HS^- + RNH_3^+$$

and since this is an acid-base neutralization we can regard it as irreversible and instantaneous.

As a reasonable gas flow rate take

$$G = 3 \times 10^{-3} \text{ mol/cm}^2\cdot\text{sec}$$

Also from Danckwerts (1970) take

$$k_{Al}a = 0.03 \text{ sec}$$
$$k_{Ag}a = 6 \times 10^{-5} \text{ mol/cm}^3\cdot\text{sec}\cdot\text{atm}$$
$$\mathscr{D}_{Al} = 1.5 \times 10^{-5} \text{ cm}^2/\text{sec}$$
$$\mathscr{D}_{Bl} = 10^{-5} \text{ cm}^2/\text{sec}$$
$$H_A = 0.115 \text{ liter}\cdot\text{atm/mol—for } H_2S \text{ in water}$$

18. Suppose that 50% conversion of benzene to monochlorobenzene is attained in countercurrent plug flow operations using pure chlorine and pure benzene feed, $L'/G' = 1$ (see Example 5). How much can conversion be increased if the height of tower is doubled? Assume that no side reactions occur.

19. The reaction of A which enters in an aqueous stream ($C_{A0} = 1.5$ mol/liter, $v_A = 2$ liters/min) with B which enters in an organic ($C_{B0} = 1.0$ mol/liter, $v_B = 4.5$ liters/min) takes place in a single 100-liter mixed flow reactor. 20% of the vessel volume is taken up by the aqueous phase while the rest is organic, and under these conditions the rate is slow and given by

$$-\frac{1}{V}\frac{dN_A}{dt} = \left(0.4\,\frac{\text{liter}}{\text{mol}\cdot\text{min}}\right)C_A C_B$$

(*a*) Find the fractional conversion of A and B in the reactor from the charts in this chapter.
(*b*) Calculate the conversions from the performance equation for mixed flow reactors and verify the solution found in part *a*.

20. Repeat problem 19 with the modification that the ratio of phases is reversed, thus giving 80% aqueous and 20% organic. Also assume that reaction occurs in the organic phase alone.

21. If the 100-liter reactor of the previous problem is replaced by two 50-liter mixed reactors determine the composition of exit streams
(*a*) For countercurrent flow
(*b*) For cocurrent flow

22. For countercurrent contacting of immiscible A and B in two equal sized mixed reactors the overall conversion of A is 90%, the overall conversion of B is 60%. What is the conversion between stages?

23. Immiscible A and B are introduced into a 10-liter batch reactor and after 1 hr conversion of A is found to be 98%. Determine the value of the rate constant in the rate expression

$$A + B \rightarrow R \qquad -\frac{1}{V}\frac{dN_A}{dt} = k_t C_A C_B$$

For phase A: $C_{A0} = 1$ mol/liter, $V_A = 6$ liters
For phase B: $C_{B0} = 2$ mol/liter, $V_B = 4$ liters

24. Immiscible A and B are to be reacted in a 3-stage countercurrent mixed reactor system. The feed streams are

For A: $C_{A0} = 2$ mol/liter, $v_A = 100$ liters/hr
For B: $C_{B0} = 2.5$ mol/liter, $v_B = 120$ liters/hr

For the expected volume ratio of phases the rate of reaction is given by

$$-\frac{1}{V}\frac{dN_A}{dt} = \left(2.76\,\frac{\text{liters}}{\text{mol}\cdot\text{hr}}\right)C_A C_B$$

How large must the stages be for 95% conversion of reactant A?

25. Reactants A and B are present in separate phases. For equimolar feed of A and B countercurrent in three reactors, 99% conversion of A is now attained.
(*a*) A fourth reactor is to be added to the three operating at present. For the same conversion how much can production be raised?

(*b*) Find the conversion occurring in each unit, and with a sketch show the compositions of fluids leaving each unit.

26. Consider the three-reactor systems of the previous problem.

(*a*) If the volumetric flow ratio remains unchanged but C_{B0} is raised 50% so that 1.5 mol B/mol A are fed to the unit, how should this affect the amount of A that can be treated?

(*b*) With a sketch show the compositions of fluids leaving each unit.

27. Two fluids are contacted and reacted in a mixer-settler. The continuous phase consists of pure reactant A which is but slightly soluble in phase B′ containing reactant B. The reaction is slow and is confined to the dispersed phase as follows:

$$A + B \rightarrow \text{products}; \qquad -\frac{1}{V_{\text{dispersed}}}\frac{dN_A}{dt} = kC_AC_B$$

Pure A is continually recycled; products of the reaction remain in the dispersed phase.

(*a*) Assuming mixed flow and negligible holding time in the settler, derive an expression for the concentration of B in the stream leaving the mixer-settler in terms of $\bar{t}_B$, k, C_{B0}, and the unchanged concentration of A within the dispersed phase.

(*b*) Find the rate constant for the reaction (moles, liters, minutes) from the following experimental information:

Feed rate of phase B′ = 0.2 cm^3/sec
Fraction of dispersed phase in the reactor = 24%
Volume of fluid in the reactor = 1500 cm^3
Concentration of A in dispersed phase = 2.7×10^{-5} mol/liter
Concentration of B in entering dispersed phase = 0.02 mol/liter
Concentration of B in leaving dispersed phase = 0.0125 mol/liter

28. For a slow reaction between A and B initially present in different phases with second-order kinetics of Eq. 53, is countercurrent or cocurrent plug flow more efficient? Verify this conclusion by comparing reactor volumes for 99% conversion of limiting reactant using (*a*) equimolar feed, (*b*) feed consisting of 1.5 mol B/mol A, and (*c*) a large excess of B.

29. Suppose the successive liquid phase reaction of A to R to S occurs in a batch (or plug flow) reactor, and suppose that R can be swept away by bubbling a gas through the liquid (see Fig. 16*a*). Thus with linear kinetics throughout, we have according to Eq. 63

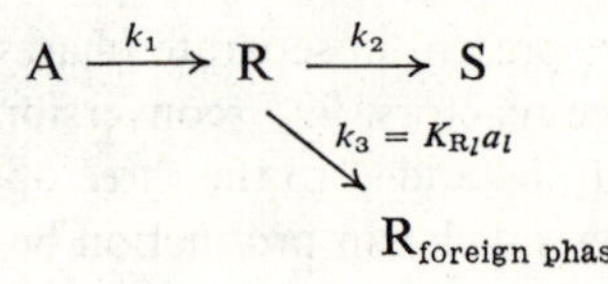

(*a*) Derive the expression for the time when the amount of R is maximized.

(*b*) From this expression show that whenever $k_3 > k_1$ then we should always react to completion to produce the largest amount of R possible. Note that under these conditions all the R formed ends up in the gas phase.

(*c*) Find the fractional yield of R in this case.

30. Repeat problem P29 for a bubble contactor reactor through which liquid passes in mixed flow (see Fig. 16*b*).

31. Given the elementary homogeneous series-parallel reactions

$$A \xrightarrow[k_1 = 1 \text{ liter/mol}\cdot\text{min}]{+B} R_{\text{desired}} \xrightarrow[k_2 = 8 \text{ liters/mol}\cdot\text{min}]{+B} S$$

occuring in a batch reactor where $C_{A0} = 1$ mol/liter, and $C_{B0} = 10$ mol/liter.

(*a*) Find $C_{R\max}/C_{A0}$ in the batch reactor.

Suppose that R is the only reaction component with a significant vapor pressure. Also when gas is vigorously bubbled through the reactor the gas-liquid transfer coefficient for R is

$$K_{Rl}a_l = 80 \text{ min}^{-1}$$

(*b*) Find $C_{R,\max}/C_{A0}$ from the reactor under these conditions.

(*c*) If C_{B0} is lowered from 10 to 1 mol/liter how does this change $C_{R\max}/C_{A0}$ in part (*b*).

(*d*) If the temperature is lowered so that $k_1 = 0.3$ and $k_2 = 1$ while $K_{Rl}a_l$ remains unchanged how does this change $C_{R,\max}/C_{A0}$ in part (*b*).

14

SOLID-CATALYZED REACTIONS

With many reactions the rates are affected by materials which are neither reactants nor products. Such materials, called catalysts, may slow down reactions, in which case they are called negative catalysts, or they may speed up reactions, in which case they are called positive catalysts. Catalysts may be solids or fluids. Design for reactions with fluid catalysts is straightforward and is considered in the chapters on fluid systems. In this and the next chapter we treat reactions with solid catalysts. This chapter deals with the catalyst whose effectiveness does not change with time or with use, while the next chapter extends this treatment to the catalyst whose effectiveness does change with time or with use.

Solid catalyzed reactions usually involve the high-energy rupture or high-energy synthesis of materials, and these reactions play an important role in many industrial processes, such as the production of methanol, ammonia, sulfuric acid, and various petrochemicals. Consider a narrow fraction of natural petroleum. Since this consists of a mixture of many compounds, primarily hydrocarbons, its treatment under extreme conditions will cause a variety of changes to occur simultaneously, producing a spectrum of compounds, some desirable, others undesirable. Although a catalyst can easily speed the rate of reactions a thousandfold or a millionfold, still, when a variety of reactions are encountered, the most important characteristic of a catalyst is its *selectivity*. By this we mean that it only changes the rates of certain reactions, often a single reaction, leaving the rest unaffected. Thus, in the presence of an appropriate catalyst, products containing predominantly the materials desired can be obtained from a given feed.

The following are some general observations.

1. The selection of a catalyst to promote a reaction is not well understood;

therefore in practice extensive trial and error may be needed to produce a satisfactory catalyst.

2. Duplication of the chemical constitution of a good catalyst is no guarantee that the solid produced will have any catalytic activity.

3. This observation suggests that it is the physical or crystalline structure which somehow imparts catalytic activity to a material. This view is strengthened by the fact that heating a catalyst above a certain critical temperature may cause it to lose its activity, often permanently. Thus present research on catalysts is strongly centered on the surface structure of solids.

4. To explain the action of catalysts it is thought that reactant molecules are somehow changed, energized, or affected to form intermediates in the regions close to the catalyst surface. Various theories have been proposed to explain the details of this action. In one theory the intermediate is viewed as an association of a reactant molecule with a region of the surface; in other words the molecules are somehow attached to the surface. In another theory molecules are thought to move down into the atmosphere close to the surface and be under the influence of surface forces. In this view the molecules are still mobile but are nevertheless modified. In still a third theory it is thought that an active complex, a free radical, is formed at the surface of the catalyst. This free radical then moves back into the main gas stream, triggering a chain of reactions with fresh molecules before being finally destroyed. In contrast with the first two theories, which consider the reaction to occur in the vicinity of the surface, this theory views the catalyst surface simply as a generator of free radicals, with the reaction occurring in the main body of the gas.

5. In terms of the transition-state theory the catalyst reduces the potential energy barrier over which the reactants must pass to form products. From Chapter 2 we know that this in turn increases the rate of reaction. This lowering in energy barrier is shown in Fig. 1.

6. Though a catalyst may speed up a reaction, it never determines the equilibrium or end point of a reaction. This is governed by thermodynamics alone. Thus with or without a catalyst the equilibrium constant for the reaction is always the same.

7. Since the solid surface is responsible for catalytic activity, a large readily accessible surface in easily handled materials is desirable. By a variety of methods active surface areas the size of football fields can be obtained per cubic centimeter of catalyst.

Though there are many problems related to solid catalysts, we consider only those which are related to the development of kinetic rate equations needed in design. We simply assume that we have a catalyst available to promote a specific reaction. We wish to evaluate the kinetic behavior of reactants in the presence of this material and then use this information for design.

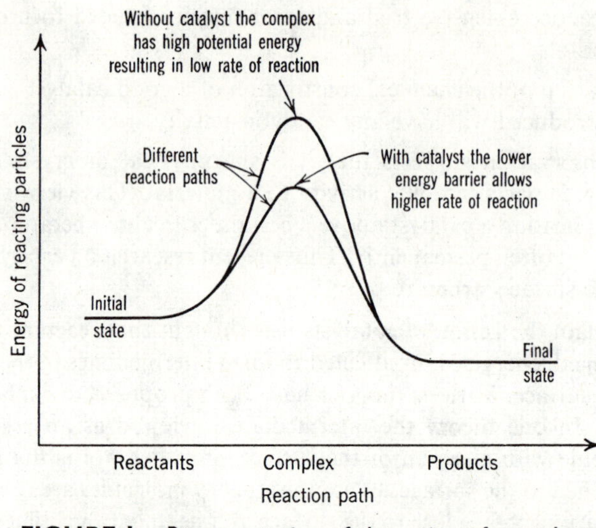

FIGURE 1. Representation of the action of a catalyst.

Thus again, as with noncatalytic heterogeneous reactions, we follow a two-step procedure: to determine the kinetics of the reaction and to use this information for design.

THE RATE EQUATION

In solid-catalyzed reactions, somehow the presence of catalyst surface in the proximity of reactive gas molecules promotes reaction. Thus with porous catalyst pellets reaction occurs at all the gas–solid interfaces both at the outside boundaries and within the pellets themselves. For such systems the most reasonable representation of reality pictures reaction occurring to a lesser or greater extent throughout the catalyst pellets. This is in contrast to the shrinking unreacted-core model with its definite moving reaction front which was used to represent noncatalyzed gas–solid reactions of Chapter 12.

In developing rate expressions for catalytic reactions we must account for the various processes that may cause resistance to reaction. For a single porous catalyst particle we may visualize these as follows.

Gas film resistance. Reactants diffuse from the main body of the fluid to the exterior surface of the catalyst.

Pore diffusion resistance. Because the interior of the pellet contains so much more area than the exterior, most of the reaction takes place within the particle itself. Therefore reactants move through the pores into the pellet.

Surface phenomenon resistance. At some point in their wanderings reactant molecules become associated with the surface of the catalyst. They react to give products which are then released to the fluid phase within the pores.

Pore diffusion resistance for products. Products then diffuse out of the pellet.

Gas film resistance for products. Products then move from the mouth of catalyst pores into the main gas stream.

Resistance to heat flow. For fast reactions accompanied by large heat release or absorption the flow of heat into or out of the reaction zone may not be fast enough to keep the pellet isothermal. If this happens the pellet will cool or heat, strongly affecting the rate. Thus the heat transfer resistance across the gas film or within the pellet could influence the rate of reaction.

We treat these heat effects later. At first we only consider the mass transfer resistances. Again, as with the heterogeneous systems of Chapters 12 and 13, not all these mass transfer terms need to be considered at any one time. Frequently the last two steps may be ignored or may be incorporated in the first two terms. For example, where there is no change in the number of moles during reaction the outward diffusion of product can be taken into account simply by considering equimolar counterdiffusion of reactant into the pores rather than diffusion of reactant through a stagnant fluid.

Consider a single idealized catalyst pore, with its three resistances as shown in Fig. 2. Since these resistances do not all occur in series or in parallel they cannot be combined by the simple methods of Chapter 11. From Fig. 2 we see that the film and surface resistance steps do act in series with each other, and that it is the pore diffusion step which adds the complications since it is not related in a simple way to the other steps. Consequently, as we presently show, film and surface reaction resistances can be treated separately in turn, whereas the pore diffusion resistance can never be treated independently.

Let us now consider the various forms that the rate equation takes when one or the other of the resistances steps is large or dominates.

Film Resistance Controls

When the gas film resistance is much greater than the other resistances, then the rate of reaction is limited by the movement of reactant to the surface as given by the mass transfer coefficient k_g between gas and solid. The rate based on unit exterior surface of particle S_{ex} is thus

$$-\frac{1}{S_{ex}}\frac{dN_A}{dt} = k_g(C_{Ag} - C_{Ae}) \tag{1}$$

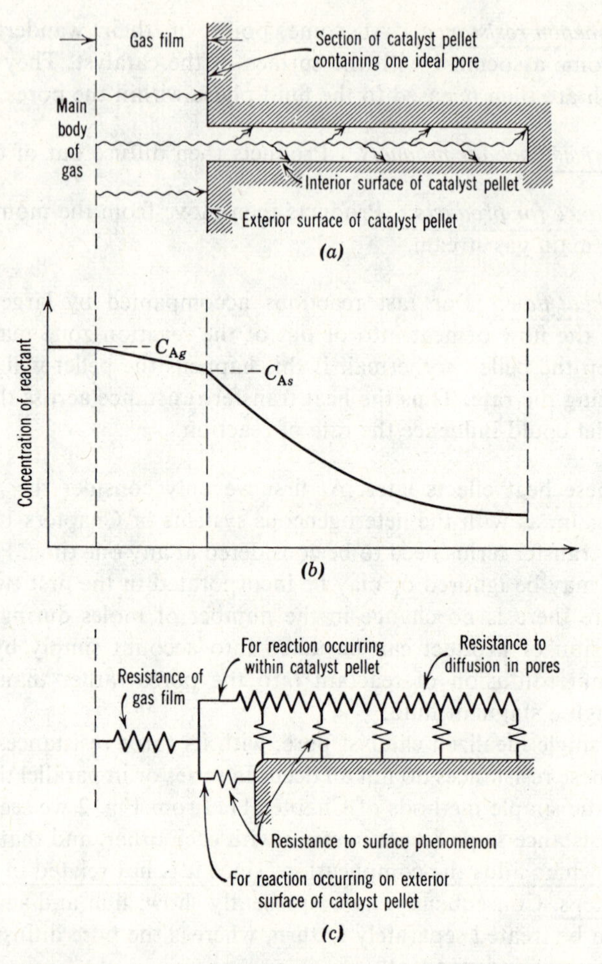

FIGURE 2. Continuous-reaction model for porous catalysts: (*a*) sketch of a catalyst pore, (*b*) concentration of reactants in the vicinity of a pore, (*c*) electrical analog of a pore.

where C_{Ag} is the concentration of reactant A in the gas stream and C_{Ae} is its equilibrium concentration on the surface.

Values for the mass transfer coefficient for various contacting schemes are given in the literature, usually as empirical or semiempirical dimensionless correlations such as Eq. 12.24. Because film resistance is likely to control in very high temperature operations, the observed over-all reaction rate will be the same for both catalytic and noncatalytic gas–solid reactions in this regime.

Surface Phenomenon Controls

Because of the great industrial importance of catalytic reactions, considerable effort has been spent in developing theories from which kinetic equations can rationally be developed. The most useful for our purposes supposes that the reaction takes place on an active site on the surface of the catalyst. Thus three steps are viewed to occur successively at the surface.

Step 1. A molecule is adsorbed onto the surface and is attached to an active site.

Step 2. It then reacts either with another molecule on an adjacent site (dual-site mechanism), with one coming from the main gas stream (single-site mechanism), or it simply decomposes while on the site (single-site mechanism).

Step 3. Products are desorbed from the surface, which then frees the site.

In addition, all species of molecules, free reactants, and free products as well as site-attached reactants, intermediates, and products taking part in these three processes are assumed to be in equilibrium.

Rate expressions derived from various postulated mechanisms are all of the form:

$$\text{rate of reaction} = \frac{(\text{kinetic term})(\text{driving force or displacement from equilibrium})}{(\text{resistance term})} \quad (2)$$

For example, for the reaction

$$A + B \rightleftarrows R + S, \qquad K$$

occurring in the presence of inert carrier material U, the rate expression when adsorption of A controls is

$$-r''_A = \frac{k(p_A - p_R p_S / K p_B)}{1 + K_A p_R p_S / K p_B + K_B p_B + K_R p_R + K_S p_S + K_U p_U}$$

When reaction between adjacent site-attached molecules of A and B controls, the rate expression is

$$-r''_A = \frac{k(p_A p_B - p_R p_S / K)}{(1 + K_A p_A + K_B p_B + K_R p_R + K_S p_S + K_U p_U)^2}$$

whereas for desorption of R controlling it becomes

$$-r''_A = \frac{k(p_A p_B / p_S - p_R / K)}{1 + K_A p_A + K_B p_B + K K_R p_A p_B / p_S + K_S p_S + K_U p_U}$$

Each detailed mechanism of reaction with its controlling factor has its corresponding rate equation, involving anywhere from three to seven arbitrary constants, the K values. For reasons to be made clear, we do not intend to use equations such as these. Consequently, we do not go into their derivations. These are given by Hougen and Watson (1947), Corrigan (1954, 1955), Walas (1959), and elsewhere.

Now, in terms of the contact time or space time, most catalytic conversion data can be fitted adequately by relatively simple first- or nth-order rate expressions;

see Prater and Lago (1956). Since this is so, why should we concern ourselves with selecting one of a host of rather complicated rate expressions suggested by theoretical mechanisms, why not select the simplest empirical rate expression which satisfactorily fits the data?

The following discussion summarizes the arguments for and against the use of simple empirical kinetic equations. → EXPRESIONES DE RATA GLOBAL

Truth and Predictability. The strongest argument in favor of searching for the actual mechanism is that if we find one which we think represents what truly occurs, extrapolation to new and more favorable operating conditions is much more safely done. This is a powerful argument. Other arguments, such as augmenting knowledge of the mechanism of catalysis with the final goal of producing better catalysts in the future, do not concern a design engineer who has a specific catalyst at hand.

Problems of Finding the Mechanism. To prove that we have such a mechanism we must show that the family of curves representing the rate equation type of the favored mechanism fits the data so much better than the other families that the others can be rejected. With the large number of parameters (three to seven) that can be chosen arbitrarily for each rate-controlling mechanism, a very extensive experimental program is required, using very precise and reproducible data, which in itself is quite a problem. We should bear in mind that it is not good enough to select the mechanism that well fits, or even best fits the data. Differences in fit may be so slight as to be explainable entirely in terms of experimental error. In statistical terms these differences may not be "significant." Unfortunately, if a number of alternative mechanisms fit the data equally well, we must recognize that the equation selected can only be considered to be one of good fit, not one that represents reality. With this admitted, there is no reason why we should not use the simplest and easiest-to-handle equation of satisfactory fit. In fact, unless there are good positive reasons for using the more complicated of two equations, we should always select the simpler of the two if both fit the data equally well. The statistical analyses and comments by Chou (1958) on the codimer example in Hougen and Watson (1947) in which 18 mechanisms were examined illustrate the difficulty in finding the correct mechanism from kinetic data, and show that even in the most carefully conducted programs of experimentation the magnitude of the experimental error will very likely mask any of the differences predicted by the various mechanisms.

Thus it is hardly ever possible to determine with reasonable confidence which is the correct mechanism.

Dangers in Extrapolation. Nevertheless, let us suppose that we have found the correct mechanism. Extrapolating to unexplored regions is still dangerous because other resistances may become important, in which case the form of the over-all rate equation changes.

Problems of Combining Resistances. Again let us suppose that we have found the correct mechanism and resultant rate equation for the surface phenomenon. Combining this step with any of the other resistance steps, such as pore or film diffusion, becomes rather impractical. When this has to be done, it is best to replace the multi-constant rate equation by an equivalent first-order expression, which can then be combined with other reaction steps to yield an overall rate expression.

Summary. From this discussion we conclude that it is good enough to use the simplest available correlating rate expression which satisfactorily represents the data.

For additional comments questioning the validity of the active-site approach, suggesting forms of kinetic equations to be used in reactor design, and suggesting what is the real utility of the active site theory, see the opposing points of view presented by Weller (1956) and Boudart (1956).

Form of Rate Equation to be Used when Surface Phenomenon Controls. For design purposes we usually can fit the data satisfactorily by a first-order irreversible or reversible rate equation

$$-r_A = kC_A \qquad \text{or} \qquad -r_A = k(C_A - C_{Ae})$$

by an nth-order irreversible rate equation

$$-r_A = kC_A{}^n$$

or by simplified expressions suggested by the active-site theory

$$-r_A = \frac{kC_A}{1 + k_1C_A} \qquad \text{or} \qquad -r_A = \frac{k(C_A - C_{Ae})}{1 + k_1C_A}$$

and

$$-r_A = \frac{kC_A}{(1 + k_1C_A)^2} \qquad \text{or} \qquad -r_A = \frac{k(C_A - C_{Ae})}{(1 + k_1C_A)^2}$$

and similar equations when more than one reactant is involved. Rate expressions such as these will be used to represent situations in which surface phenomena control.

Qualitative Predictions from Active-site Theory. For design the real value of the active-site theory is that it gives a qualitative idea of what may be expected to happen on extrapolating to new operating conditions. Imagine molecules adsorbing, reacting, and desorbing from the surface. From findings in adsorption we know that a rise in pressure results in an increase in amount of material adsorbed. Hence if adsorption is rate controlling, an increase in reactant concentration will result in an increase in rate of reaction.

Suppose desorption controls. Being an equilibrium process between site-attached and free product molecules, desorption is unaffected by an increase in concentration of reactants. Hence we get no increase in rate with a rise in reactant concentration.

When the chemical reaction controls, we visualize that all the active sites are actively in use. Many cases may be considered here. For example, consider a single-site decomposition of reactant A. Increasing the concentration of reactant in the atmosphere above the surface will not speed up this reaction because the surface is assumed to be already saturated with A. Again consider the dual-site reaction: A + B → products. This mechanism visualizes that a molecule of A on a site reacts by attacking a molecule of B on a neighboring site. Now, if A is in excess on the surface, the reaction rate is mainly a function of the concentration of B on the surface. Increasing C_B or decreasing C_A will both allow more B molecules on the surface and therefore increase the rate of reaction. Increasing C_A will simply swamp the surface with A, crowding out B and further slowing down the rate of reaction. Figure 3 summarizes these conclusions. Other cases can similarly be considered.

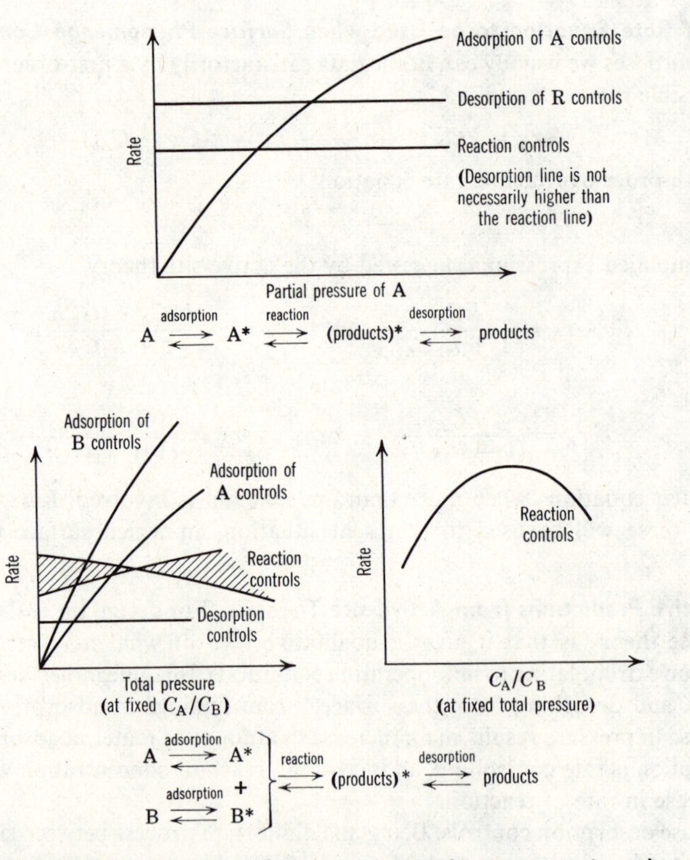

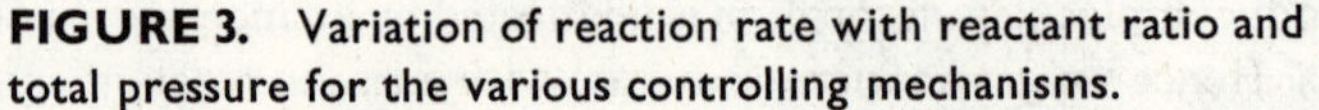

FIGURE 3. Variation of reaction rate with reactant ratio and total pressure for the various controlling mechanisms.

Effect of Change in Operating Pressures. The active-site theory allows us to predict what may happen to the rate when we extrapolate to higher or lower operating pressures. In general, at very low pressures with adsorption controlling we have essentially a first-order reaction. At higher pressures the surface will become increasingly saturated with molecules. If these do not react rapidly enough, surface reaction becomes controlling, causing the rate to level off or even drop. If the reactants on the surface react rapidly but the products do not desorb rapidly enough, the surface will become saturated with product molecules and desorption will control, in which case the reaction rate will again level off.

The effect of pressure on rate of reaction as predicted by the active-site theory is given in Figs. 4 and 5 and shows that at low pressure all reactions are approximated by a first-order rate. At higher pressures the rates level off, becoming zero order, or may even drop. A more detailed presentation using initial rate data is given by Yang and Hougen (1950).

Pore Diffusion Resistance Important

Single Cylindrical Pore, First-order Reaction. Consider at first a single cylindrical pore, with reactant A diffusing into the pore, a first-order reaction

$$A \rightarrow \text{product} \qquad \text{and} \qquad -\frac{1}{S}\frac{dN_A}{dt} = k_s C_A$$

taking place at the walls of the pore, and product diffusing out of the pore, as shown in Fig. 6. This simple model will later be extended.

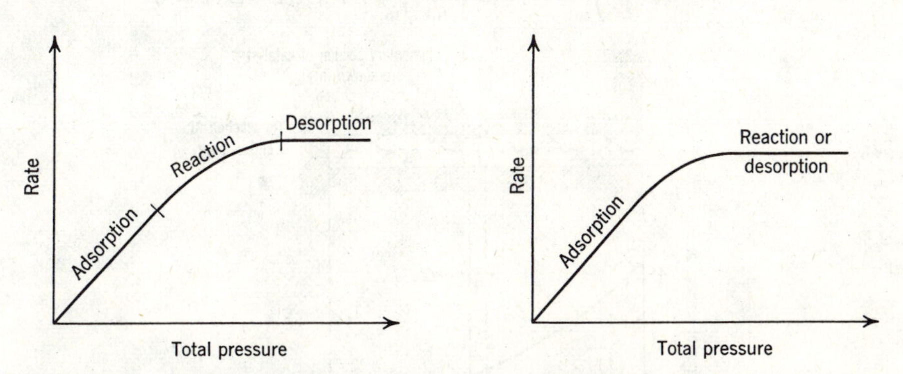

FIGURE 4. Change in controlling mechanism as a function of reactant concentration or total pressure for reaction of two reactants with single-site mechanism,

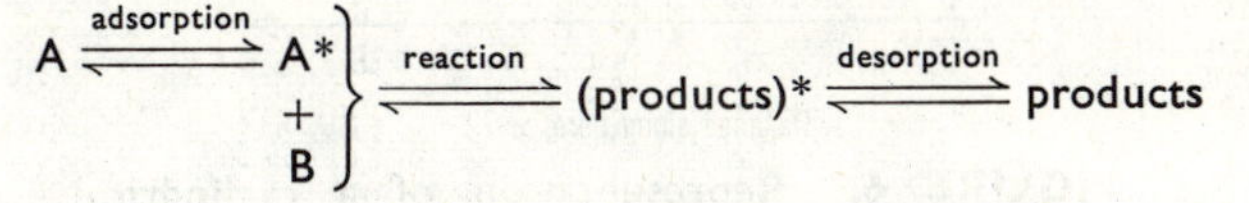

$$\left.\begin{matrix} A \underset{}{\overset{\text{adsorption}}{\rightleftharpoons}} A^* \\ + \\ B \end{matrix}\right\} \overset{\text{reaction}}{\rightleftharpoons} (\text{products})^* \overset{\text{desorption}}{\rightleftharpoons} \text{products}$$

and with fixed reactant ratio, or for reaction of single reactants.

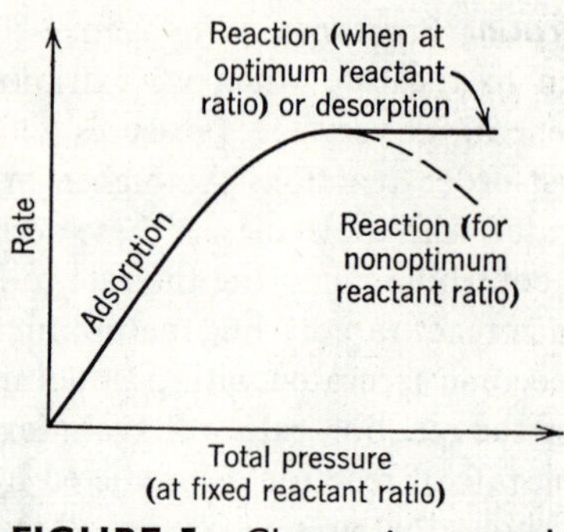

FIGURE 5. Change in controlling mechanism for the dual-site mechanism:

$$\left.\begin{array}{l} A \xrightleftharpoons{\text{adsorption}} A^* \\ \qquad\quad + \\ B \xrightleftharpoons{\text{adsorption}} B^* \end{array}\right\} \xrightleftharpoons{\text{reaction}} R^* \xrightleftharpoons{\text{desorption}} R$$

The flow of materials into and out of any section of pore is shown in detail in Fig. 7. At steady state a material balance for reactant A for this elementary section gives

$$\text{output} - \text{input} + \text{disappearance by reaction} = 0 \tag{4.1}$$

or with the quantities shown in Fig. 7,

$$-\pi r^2 \mathscr{D}\left(\frac{dC_A}{dx}\right)_{out} + \pi r^2 \mathscr{D}\left(\frac{dC_A}{dx}\right)_{in} + k_s C_A(2\pi r\,\Delta x) = 0$$

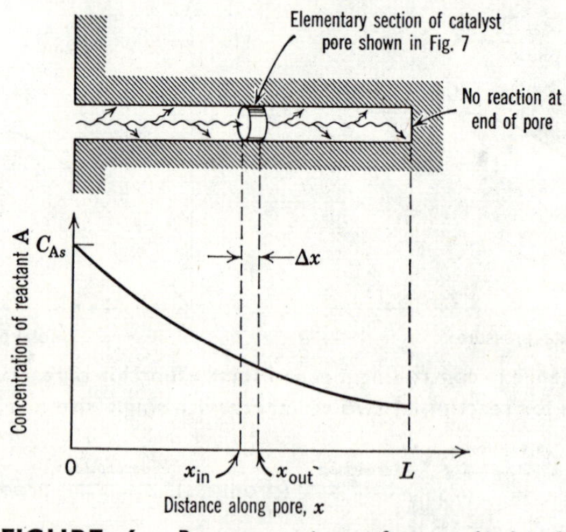

FIGURE 6. Representation of a cyclindrical catalyst pore.

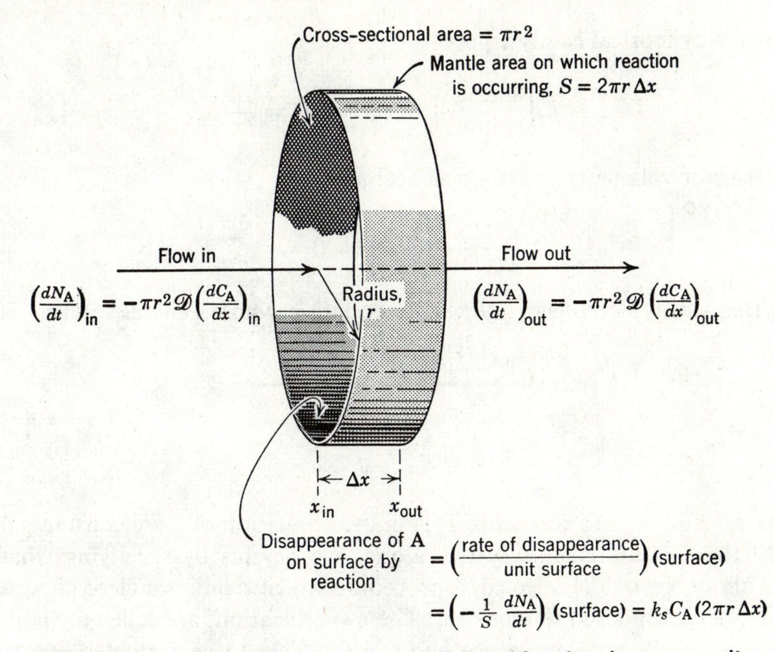

FIGURE 7. Setting up the material balance for the elementary slice of catalyst pore.

Rearranging

$$\frac{\left(\frac{dC_A}{dx}\right)_{out} - \left(\frac{dC_A}{dx}\right)_{in}}{\Delta x} - \frac{2k_s}{\mathscr{D}r} C_A = 0$$

and taking the limit as Δx approaches zero (see the equation above Eq. 9.41a), we obtain

$$\frac{d^2C_A}{dx^2} - \frac{2k_s}{\mathscr{D}r} C_A = 0 \tag{3}$$

Note that the first-order chemical reaction is expressed in terms of unit surface area of catalyst pore; hence k_s has units of length per time. In general, the interrelation between rate constants on different bases is given by

$$\left(k, \frac{1}{\text{hr}}\right)\left(\begin{matrix}\text{volume,}\\ \text{m}^3\end{matrix}\right) = \left(k_m, \frac{\text{m}^3}{\text{hr}\cdot\text{kg}}\right)\left(\begin{matrix}\text{mass of}\\ \text{catalyst, kg}\end{matrix}\right) = \left(k_s, \frac{\text{m}}{\text{hr}}\right)\left(\begin{matrix}\text{surface of}\\ \text{catalyst, m}^2\end{matrix}\right)$$

or

$$kV = k_m W = k_s S \tag{4}$$

Hence for the cylindrical catalyst pore

$$k = k_s\left(\frac{\text{surface}}{\text{volume}}\right) = k_s\left(\frac{2\pi rL}{\pi r^2 L}\right) = \frac{2k_s}{r} \tag{5}$$

Thus in terms of volumetric units Eq. 3 becomes

$$\frac{d^2C_A}{dx^2} - \frac{k}{\mathscr{D}}C_A = 0 \tag{6}$$

This is a frequently met linear differential equation whose general solution is

$$C_A = M_1e^{mx} + M_2e^{-mx} \tag{7}$$

where

$$m = \sqrt{\frac{k}{\mathscr{D}}} = \sqrt{\frac{2k_s}{\mathscr{D}r}}$$

and where M_1 and M_2 are constants. It is in the evaluation of these constants that we restrict the solution to this system alone. We do this by specifying what is particular about the model selected, a procedure which requires a clear picture of what the model is supposed to represent. These specifications are called the boundary conditions of the problem. Since two constants are to be evaluated, we must find and specify two boundary conditions. Examining the physical limits of the conceptual pore, we find that the following statements can always be made. First, at the pore entrance

$$C_A = C_{As}, \qquad \text{at } x = 0 \tag{8a}$$

Second, because there is no flux or movement of material through the interior end of the pore

$$\frac{dC_A}{dx} = 0, \qquad \text{at } x = L \tag{8b}$$

With the appropriate mathematical manipulations of Eqs. 7 and 8 we then obtain

$$M_1 = \frac{C_{As}e^{-mL}}{e^{mL} + e^{-mL}} \qquad M_2 = \frac{C_{As}e^{mL}}{e^{mL} + e^{-mL}} \tag{9}$$

Hence the concentration gradient of reactant within the pore is

$$\frac{C_A}{C_{As}} = \frac{e^{m(L-x)} + e^{-m(L-x)}}{e^{mL} + e^{-mL}} = \frac{\cosh m(L-x)}{\cosh mL} \tag{10}$$

This progressive drop in concentration on moving into the pore is shown in Fig. 8, and this is seen to be dependent on the dimensionless quantity mL, called the *Thiele modulus.*

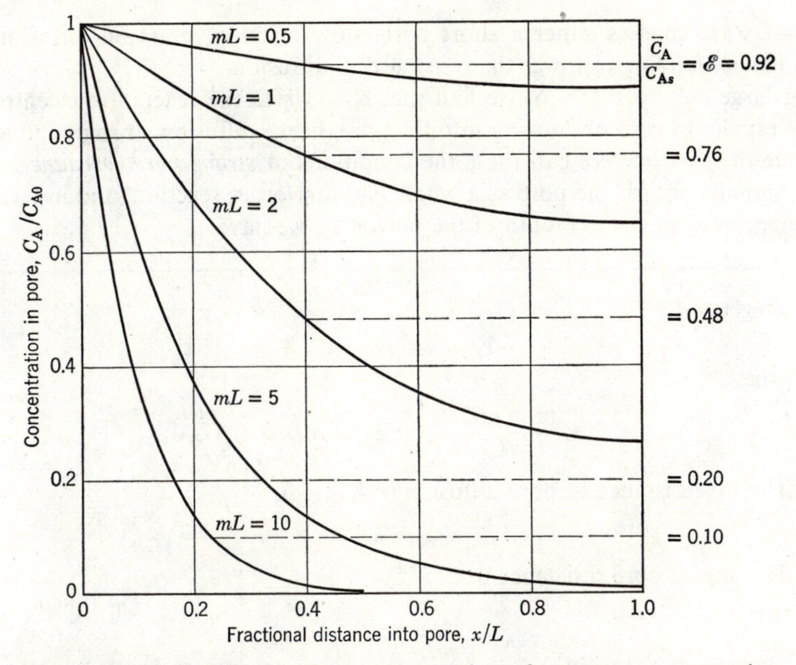

FIGURE 8. Distribution and average value of reactant concentration within a catalyst pore as a function of the parameter *mL*.

To measure how much the reaction rate is lowered because of the resistance to pore diffusion define the quantity $\mathscr{E}$ called the *effectiveness factor* as follows:

$$\text{Effectiveness factor, } \mathscr{E} = \frac{\text{(actual reaction rate within pore)}}{\text{(rate if not slowed by pore diffusion)}}$$

$$= \frac{\bar{r}_{A,\text{ with diffusion}}}{r_{A,\text{ without diffusion}}} \tag{11}$$

In particular, for first-order reactions $\mathscr{E} = \bar{C}_A/C_{As}$ because the rate is proportional to the concentration. Evaluating the average rate in the pore from Eq. 10 gives the relationship

$$\mathscr{E}_{\text{first order}} = \frac{\bar{C}_A}{C_{As}} = \frac{\tanh mL}{mL} \tag{12}$$

which is shown by the solid line in Fig. 9. With this figure we can tell whether pore diffusion modifies the rate of reaction, and inspection shows that this depends on whether *mL* is large or small.

For small *mL*, or $mL < 0.5$, we see that $\mathscr{E} \cong 1$, the concentration of reactant does not drop appreciably within the pore; thus pore diffusion offers negligible resistance to reaction. This can also be verified by noting that a small value for

$\mathscr{E}$ TAMBIEN SE CONOCE por η
$mL = \Phi$ = modulo de Thiele

ver en Smith J.M. gráfico de η vs. Φ_s para mayor claridad. (ver aquí p. 475)

$mL = L\sqrt{k/\mathscr{D}}$ means either a short pore, slow reaction, or rapid diffusion, all three factors tending to lower the resistance to diffusion.

For large mL, or $mL > 5$, we find that $\mathscr{E} = 1/mL$, the reactant concentration drops rapidly to zero on moving into the pore, hence diffusion strongly influences the rate of reaction. We call these the conditions of *strong pore resistance*.

To summarize: for the pore as a whole and first-order reaction, and in terms of the concentration at the mouth of the pore, C_{As}, we have:

In general

$$-r_A = k\bar{C}_A = kC_{As}\mathscr{E} \tag{13}$$

where

$$\mathscr{E} = \frac{\tanh mL}{mL} \qquad \text{and} \qquad mL = L\sqrt{\frac{k}{\mathscr{D}}}$$

For no resistance to pore diffusion ($mL < 0.5$)

$$-r_A = kC_{As} \tag{14}$$

For strong pore resistance ($mL > 5$)

$$-r_A = \frac{kC_{As}}{mL} = \frac{(k\mathscr{D})^{1/2}}{L} C_{As} \tag{15}$$

From the discussion related to Fig. 2 we found that resistance to pore diffusion does not act in series with surface reaction resistance and hence cannot be treated independent of it. Equation 13 seems at first sight to indicate that pore diffusion can be accounted for by a separate, multiplicative correction-type term $\mathscr{E}$. This is true; however, this factor $\mathscr{E}$ involves not only a diffusion term but also a surface reaction term in the form of the rate constant. Thus pore diffusion can never become controlling in the sense that it alone will determine the overall rate of reaction.

We have gone into considerable detail to develop the expressions for the simple case of diffusion into a single cylindrical pore for a first-order irreversible reaction because it shows how the nonadditive pore resistance effect is to be treated. Relaxing the various restrictions of this treatment leads to useful extensions and generalizations. These follow, mostly without proof or derivation, but with pertinent references.

Mixture of Particles of Various Shapes and Sizes. For a catalyst bed consisting of a mixture of particles of various shapes and sizes Aris (1957) showed that the correct mean effectiveness factor is

$$\bar{\mathscr{E}} = \mathscr{E}_1 f_1 + \mathscr{E}_2 f_2 + \cdots \tag{16}$$

where $f_1, f_2, \ldots$ are the volume fractions of particles of sizes 1, 2, ... in the mixture.

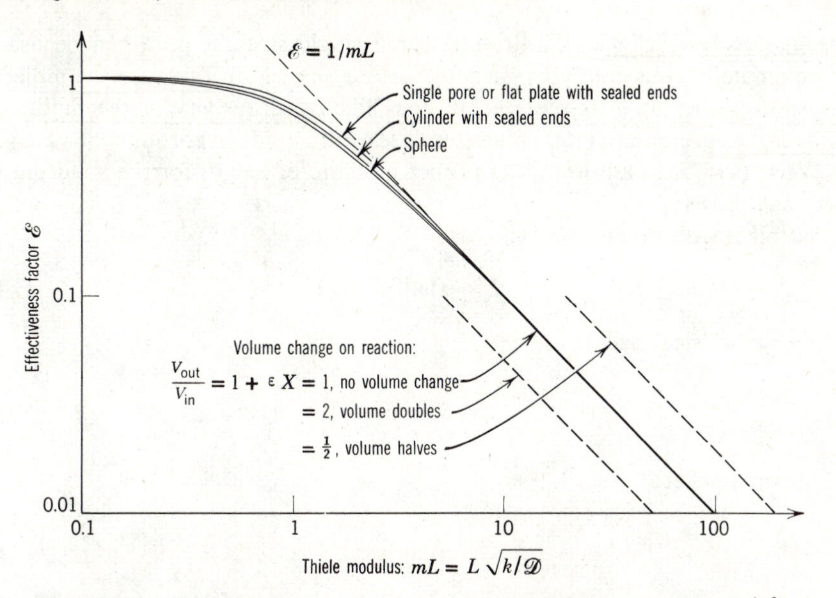

FIGURE 9. The effectiveness factor as a function of the parameter *mL* for various catalyst shapes and for volume change during reaction. Prepared from Aris (1957) and Thiele (1939).

Molar Volume Change. With decrease in fluid density (expansion) during reaction the increased outflow of molecules from the pores makes it harder for reactants to diffuse into the pore, hence lowering $\mathscr{E}$. On the other hand, volumetric contraction results in a net molar flow into the pore, hence increasing $\mathscr{E}$. For a first-order reaction Thiele (1939) found that this flow simply shifted the $\mathscr{E}$ versus mL curve as shown in Fig. 9.

Arbitrary Reaction Kinetics. If the Thiele modulus is generalized as follows

$$mL = \frac{(-r_{As})L}{\left[2\mathscr{D}\int_{C_{Ae}}^{C_{As}}(-r_A)\,dC_A\right]^{1/2}} \tag{17}$$

then the $\mathscr{E}$ versus mL curves for all forms of rate equation closely follow the curve for the first-order reaction. This generalized modulus becomes

for first-order reversible reactions

$$mL = L\sqrt{\frac{k}{\mathscr{D}X_{Ae}}} \tag{18}$$

for nth-order irreversible reactions

$$mL = L\sqrt{\frac{(n+1)kC_{As}^{n-1}}{2\mathscr{D}}} \tag{19}$$

Porous Catalyst Pellets. The results for a single straight pore can be used to approximate porous catalyst pellets of various shapes. In this case the molecular diffusion coefficient $\mathscr{D}$ is replaced by the effective diffusivity of the fluid in the solids $\mathscr{D}_e$. Representative $\mathscr{D}_e$ values for gases and liquids in porous solids are given by Weisz (1959). In addition, the proper measure of length for the solid must be used, as follows:

Flat plate with no end effect:

$$L = \text{(half width)} \tag{20}$$

Long cylindrical pellet:

$$L = \frac{R}{2} \tag{21}$$

Spherical pellet:

$$L = \frac{R}{3} \tag{22}$$

Arbitrary shaped pellet:

$$L = \frac{\text{(volume of pellet)}}{\text{(exterior surface available for reactant penetration and diffusion)}} \tag{23}$$

The precise $\mathscr{E}$ versus mL curves for these various geometries have been calculated by Aris (1957) and they all conveniently coincide with the curve for the single pore shown in Fig. 9, except in the intermediate region. However, even here the maximum variation in $\mathscr{E}$ among the curves is never more than 18%.

With porous catalyst pellets the rate of reaction can be defined usefully in one of many ways, as will be shown in Eqs. 34 and 35; however, when testing for strong pore diffusion (see next section) we must use a rate based on unit volume of catalyst pellet, $-r_A'''$.

Testing a Catalyst for Strong Pore Resistance Effects. Suppose we have an observed rate measurement for given catalyst pellets in a known environment. How can we tell from this whether pore resistance is or is not intruding to slow the rate of reaction? To answer we start by assuming that we at least know the form of rate equation, or reaction order, even though the rate constant may be unknown, and that the resistance to film diffusion is negligible, hence $C_{As} = C_{Ag}$.

First-order reactions. Expressing Eq. 13 in terms of unit volume of catalyst pellet and eliminating the unknown rate constant k gives in general

$$\frac{(-r_A''')L^2}{\mathscr{D}_e C_{Ag}} = \mathscr{E}(mL)^2 \tag{24}$$

With negligible resistance to pore diffusion Fig. 9 shows that $mL < 1$ and $\mathscr{E} = 1$. In this regime Eq. 24 then becomes:

For no pore resistance

$$\frac{(-r_A''')_{obs}L^2}{\mathscr{D}_e C_{Ag}} < 1 \tag{25}$$

On the other hand, in the regime of strong resistance to pore diffusion $mL > 1$ and $\mathscr{E} = 1/mL$; thus Eq. 24 becomes:

For strong pore resistance

$$\frac{(-r_A''')_{obs}L^2}{\mathscr{D}_e C_{Ag}} > 1 \tag{26}$$

This simple criterion was first presented by Weisz and Prater (1954) and requires only the observed rate, the bulk stream concentration, and catalyst measurables.

Arbitrary reaction type. If the reaction rate is of the form $-r_A''' = k \cdot f(C)$ and the generalized Thiele modulus of Eq. 17 is inserted into Eq. 24, then the criteria developed above become

$$\frac{(-r_A''')_{obs}L^2 f(C_{Ag})}{2\mathscr{D}_e \int_{C_{Ae}}^{C_{Ag}} f(C_A)\, dC_A} \left.\begin{array}{l} \cdots < 1 \text{ for negligible pore resistance} \\ \cdots > 1 \text{ for strong pore resistance} \end{array}\right\} \tag{27}$$

These expressions and pertinent related literature are given by Bischoff (1967).

Heat Effects During Reaction

When reaction is so fast that the heat released (or absorbed) in the pellet cannot be removed rapidly enough to keep the pellet close to the temperature of the fluid, then nonisothermal effects intrude. In such a situation two different kinds of temperature effects may be encountered.

Within-particle ΔT. There may be a temperature variation within the pellet.

Film ΔT. The whole pellet may be hotter (or colder) than the surrounding fluid.

For exothermic reactions, heat is released and the particles are hotter than the surrounding fluid, hence the nonisothermal rate is always higher than the isothermal rate as measured by the bulk stream conditions. Similarly, for endothermic reactions the nonisothermal rate is lower than the isothermal rate because the particle is cooler than the surrounding fluid.

Thus our first conclusion: if the harmful effects of thermal shock, or sintering of the catalyst surface, or drop in selectivity, do not occur with hot particles, then we would encourage nonisothermal behavior in exothermic reactions. On the other hand, we would like to depress such behavior for endothermic reactions.

We next ask which form of nonisothermal effect, if any, may be present. The following simple calculations tell.

For *film* ΔT we equate the rate of heat removal through the film with the rate of heat generation by reaction within the pellet. Thus

$$Q_{\text{generated}} = (V_{\text{pellet}})(-r'''_{\text{A,obs}})(-\Delta H_r)$$

$$Q_{\text{removed}} = hS_{\text{pellet}}(T_g - T_s)$$

and on combining we find

$$\Delta T_{\text{film}} = (T_g - T_s) = \frac{L(-r'''_{\text{A.obs}})(-\Delta H_r)}{h} \tag{28}$$

where L is the characteristic length of the pellet.

For *within-particle* ΔT the simple analysis by Prater (1958) for any particle geometry and any kinetics gives the desired expression. Since the temperature and concentration within the particle are represented by the same form of differential equation (Laplace equation) Prater showed that the T and C_{A} distributions must have the same shape; thus at any point in the pellet

$$-k_{\text{eff}}\frac{dT}{dx} = \mathscr{D}_e \frac{dC_{\text{A}}}{dx}(-\Delta H_r)$$

and for the pellet as a whole

$$\Delta T_{\text{particle}} = (T_{\text{center}} - T_s) = \frac{\mathscr{D}_e(C_{\text{As}} - C_{\text{A.center}})(-\Delta H_r)}{k_{\text{eff}}} \tag{29}$$

where k_{eff} is the effective thermal conductivity within the pellet.

For temperature gradients within particles alone the corresponding nonisothermal effective factor curves have been calculated by Carberry (1961), Weisz and Hicks (1962), and others. (See Bischoff (1967) for references.) Figure 10 illustrates

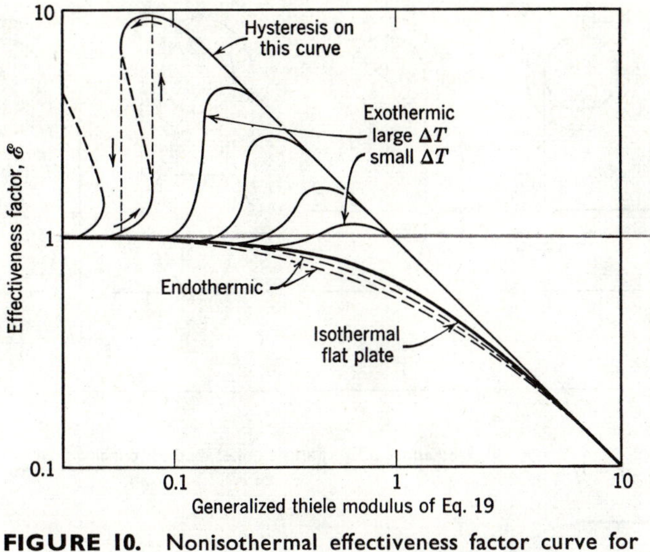

FIGURE 10. Nonisothermal effectiveness factor curve for temperaure variation within the particle. Adapted from Bischoff (1967).

these curves in dimensionless form, and shows that the shape is very similar to the isothermal curve of Fig. 9 with the following exception. For exothermic reactions only, where pore resistance just begins to intrude, the effectiveness factor can become greater than unity. This finding is not unexpected in light of the above discussion.

However, for gas–solid systems Hutchings and Carberry (1966) and McGreavy and coworkers (1969, 1970) show that if reaction is fast enough to introduce nonisothermal effects, then the temperature gradient occurs *primarily across the gas film, not within the particle*. Thus we may expect to find a significant film ΔT, before any within-particle ΔT becomes evident.

Let us illustrate the sequence of events which occur with exothermic reactions as heat effects become progressively more severe. This sequence may be viewed to result from raising the reaction rate, or from raising the temperature of the ambient fluid T_g. Figures 11 and 12 illustrate this sequence, and in all cases the effectiveness factor refers to the rate for a particle bathed uniformly in reactant fluid at temperature T_g. Thus the effectiveness factor accounts for both reactant depletion within the particle and nonisothermal effects.

1. For very slow reaction the concentration of materials is uniform throughout the pellet and the heat generated is removed rapidly enough to keep the pellet at the temperature of the gas.

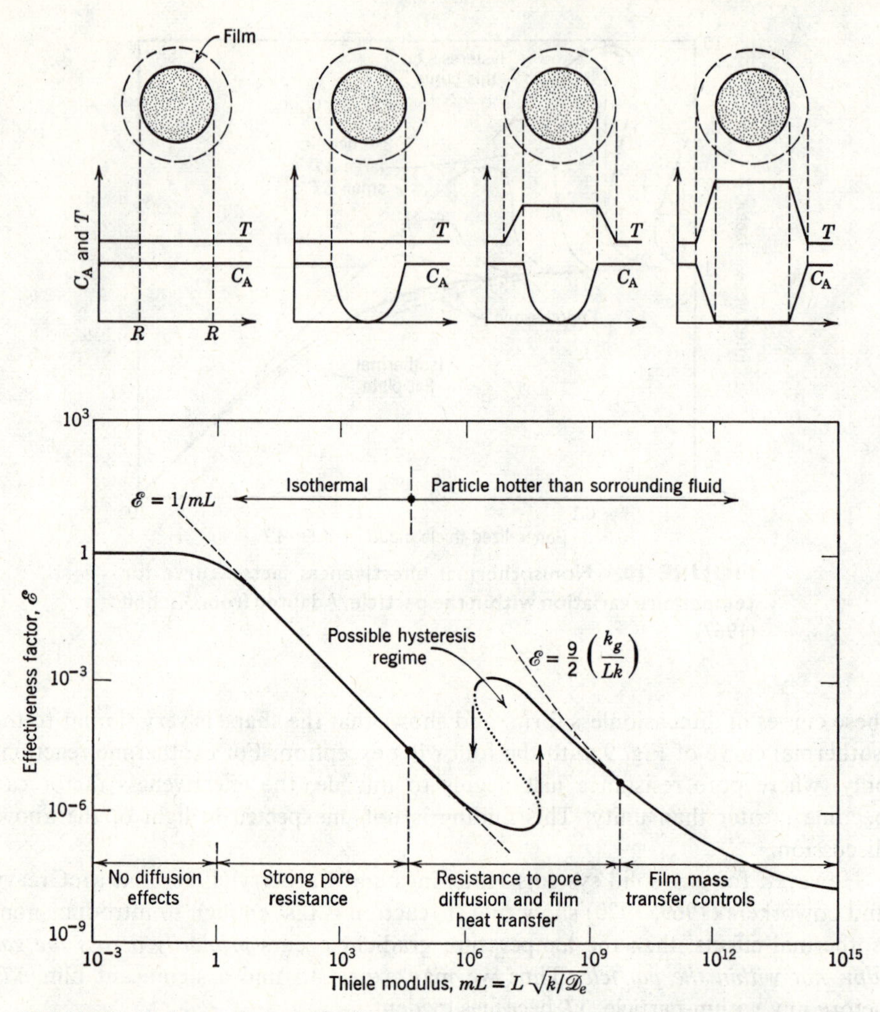

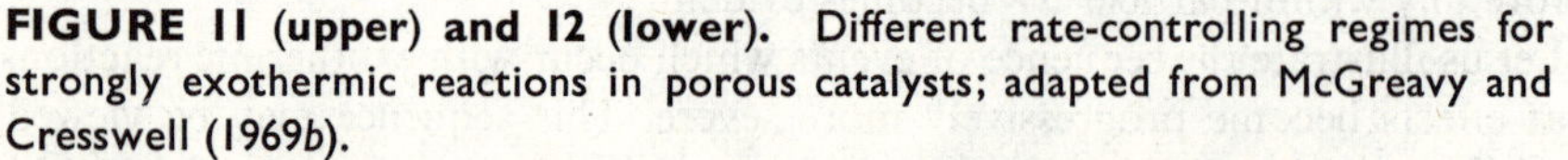
FIGURE 11 (upper) and 12 (lower). Different rate-controlling regimes for strongly exothermic reactions in porous catalysts; adapted from McGreavy and Cresswell (1969*b*).

2. For an increased reaction rate pore resistance becomes the first non-uniformity to intrude. The temperature still remains close to uniform, hence the treatment of the isothermal effectiveness factor, presented in the previous section, falls in this range.

3. For even higher reaction rates the particle, still uniform in temperature, becomes progressively hotter than the surrounding gas. Pore diffusion effects become more pronounced and most of the reaction occurs in a thin shell close to

the catalyst surface. In this range of conditions the heat generation rate may well outstrip the heat removal rate, yielding autothermal behavior (see Chapter 8) with an accompanying temperature jump and hysteresis effects.

4. Finally, for extremely high rates the particle is hot enough so that all reactants are consumed as they reach its exterior surface. In this regime the supply of reactants becomes the slow step, and mass transfer through the gas film will control the rate of reaction.

Although the usual $\mathscr{E}$ versus mL plot of Fig. 12 clearly shows this sequence, McGreavy and Thornton (1970*b*) recommend using a somewhat different plot, that of $\mathscr{E}$ versus T_g. They find this representation to be particularly convenient for design calculations on nonisothermal packed bed reactors.

For endothermic reactions the situation is somewhat simpler. A rise in rate first introduces isothermal diffusion. Higher rates will then cause a cooling of the particle which slows the rate. Thus some stable maximum rate of reaction will be reached.

Combination of Resistances for Isothermal Particles

By the methods outlined in the illustrative example of Chapter 11, the over-all effect of film resistance, pore resistance, and first-order reaction can readily be combined since these are linear processes. Thus, for the gas film we have

$$-\frac{1}{S_{ex}}\frac{dN_A}{dt} = k_g(C_{Ag} - C_{As})$$

whereas for diffusion and reaction within the interior of the particle we have, based on particle volume, V_p, and neglecting the slight contribution to reaction by the exterior surface of the particle,

$$-\frac{1}{V_p}\frac{dN_A}{dt} = kC_{As}\mathscr{E} \tag{13}$$

Combining these two expressions, eliminating the unknown surface concentration, gives

$$-\frac{1}{S_{ex}}\frac{dN_A}{dt} = \frac{1}{S_{ex}/k\mathscr{E}V_p + 1/k_g}C_{Ag}$$

or

$$-\frac{1}{V_p}\frac{dN_A}{dt} = \frac{1}{1/k\mathscr{E} + V_p/k_gS_{ex}}C_{Ag} \tag{30}$$

which for spherical particles of radius R becomes

$$-\frac{1}{S_{ex}}\frac{dN_A}{dt} = \frac{1}{3/k\mathscr{E}R + 1/k_g} C_{Ag}$$

or (31)

$$-\frac{1}{V_p}\frac{dN_A}{dt} = \frac{1}{1/k\mathscr{E} + R/3k_g} C_{Ag}$$

Note that in Eqs. 30 and 31 k is the first-order rate constant based on unit volume of particle, whereas k_g is the mass transfer coefficient based on unit exterior surface of particle. With rates defined on other bases, care must be taken to make the appropriate conversion of units.

Exterior and Interior Surface. Until now, because of its relatively small surface area, we have assumed the contribution to reaction by the exterior surface of a catalyst particle to be negligible. But when reaction is so rapid that the reactant has little chance to penetrate the catalyst particle, this assumption leads to error, and the over-all rate must account for reaction both in the interior and on the outside of the particle. Referring to Fig. 2, we find the over-all effect of all these resistances to be

$$-\frac{1}{S_{ex}}\frac{dN_A}{dt} = \frac{1}{\dfrac{1}{k_g} + \dfrac{1}{k\mathscr{E}(V_p/S_{ex})[S_{in}/(S_{ex} + S_{in})] + k_s}} C_{Ag}$$

or (32)

$$-\frac{1}{V_p}\frac{dN_A}{dt} = \frac{1}{\dfrac{V_p}{k_g S_{ex}} + \dfrac{1}{k\mathscr{E}[S_{in}/(S_{in} + S_{ex})] + k_s(S_{ex}/V_p)}} C_{Ag}$$

Resistance of: gas film; reaction in interior; reaction on exterior

where

$$k_s(S_{in} + S_{ex}) = kV_p$$

Now when

$$\mathscr{E}\frac{S_{in}}{S_{ex}} \ll 1$$

reaction on the exterior surface of the catalyst predominates over reaction within the pores, and Eq. 32 reduces to

$$-\frac{1}{S_{ex}}\frac{dN_A}{dt} = \frac{1}{1/k_g + 1/k_s} C_{Ag} \tag{33}$$

Equation 33 is the expression for reaction on nonporous catalyst particles.

Equation 32 also reduces, under appropriate limiting conditions, to all the previously derived first-order, isothermal expressions with individual resistances controlling or important. By comparing the relative magnitude of the conductance terms in this expression we should be able to estimate which resistance to reaction controls.

EXPERIMENTAL METHODS FOR FINDING RATES

Any type of reactor with known contacting pattern may be used to explore the kinetics of catalytic reactions. Since only one fluid phase is present in these reactions, the rates can be found as with homogeneous reactions. The only special precaution to observe is to make sure that the performance equation used is dimensionally correct and that its terms are carefully and precisely defined. The reason for this precaution is the wide variety of bases (void or bulk or pellet volume, surface area, or mass of catalyst) that may be used to express rates of reaction. To illustrate, for batch constant-volume systems Eq. 5.3 in its various forms becomes

$$\underline{\frac{t}{C_{A0}}} = \int \frac{dX_A}{-r_A} = \underline{\frac{V}{W}\int \frac{dX_A}{-r'_A}} = \frac{V}{S}\int \frac{dX_A}{-r''_A} = \frac{V}{V_p}\int \frac{dX_A}{-r'''_A} = \frac{V}{V_r}\int \frac{dX_A}{-r''''_A} \tag{34}$$

For steady-state plug flow systems Eq. 5.17 in its various forms becomes

$$\underline{F_{A0}\, dX_A} = -r_A\, dV = \underline{-r'_A\, dW} = -r''_A\, dS = -r'''_A\, dV_p = -r''''_A\, dV_r \tag{35}$$

For mixed or recycle flow the corresponding expressions are used.

For convenience the W, $-r'_A$ system of units indicated by the underlined terms of Eqs. 34 and 35 will be used hereafter in this chapter. However, we could just as well have used any other consistent pair of measures.

The experimental strategy in studying catalytic kinetics usually involves measuring the extent of conversion of gas passing in steady flow through a batch of solids. Any flow pattern can be used, as long as the pattern selected is known; otherwise the kinetics cannot be found. A batch reactor can also be used. In turn we discuss the following experimental devices:

Differential (flow) reactor
Integral (plug flow) reactor
Mixed reactor
Batch reactor

Differential Reactor. We have a differential flow reactor when *we choose* to consider the rate to be constant at all points within the reactor. Since rates are concentration-dependent this assumption is usually reasonable only for small conversions or for shallow small reactors. But this is not necessarily so, e.g., for slow reactions where the reactor can be large, or for zero-order kinetics where the composition change can be large.

For each run in a differential reactor the plug flow performance equation becomes

$$\frac{W}{F_{A0}} = \int_{X_{A,in}}^{X_{A,out}} \frac{dX_A}{-r'_A} = \frac{1}{(-r'_A)_{ave}} \int_{X_{A,in}}^{X_{A,out}} dX_A = \frac{X_{A,out} - X_{A,in}}{(-r'_A)_{ave}}$$

from which the average rate for each run is found to be

$$(-r'_A)_{ave} = \frac{F_{A0}(X_{A,out} - X_{A,in})}{W} = \frac{F_{A,in} - F_{A,out}}{W} \tag{36}$$

Thus each run gives directly a value for the rate at the average concentration in the reactor, and a series of runs gives a set of rate-concentration data which can then be analyzed for a rate equation.

The suggested procedure is as follows.

1. Make a series of kinetic runs using different $C_{A,in}$.
2. Select the highest $C_{A,in}$ as the basis for calculating F_{A0} and conversions, and call this concentration C_{A0}.
3. For each run determine F_{A0}, W, $X_{A,in}$, $X_{A,out}$, and $C_{A,ave}$.
4. From Eq. 36 calculate the rate for each run.
5. We now have a series of rate-concentration data.

Applying the differential method of analysis of Chapter 3, a rate equation can be found from this information.

Example 2 illustrates this procedure.

Integral Reactor. When the variation in reaction rate within a reactor is so large that *we choose* to account for these variations in the method of analysis, then we have an integral reactor. Since rates are concentration-dependent, such large variations in rate may be expected to occur when the composition of reactant fluid changes significantly in passing through the reactor. We may follow one of two procedures in searching for a rate equation.

Integral analysis. Here a specific mechanism with its corresponding rate equation is put to the test by integrating the rate equation for the reactor flow conditions. The procedure is as follows.

1. Make a series of runs in a packed bed with fixed feed C_{A0} but with varying W and/or F_{A0} in such a manner that a wide range of W/F_{A0} and $X_{A,out}$ are obtained.
2. Select a rate equation to be tested, and with it integrate the plug flow performance equation to give

$$\frac{W}{F_{A0}} = \int_0^{X_A} \frac{dX_A}{-r'_A} \qquad \text{similar to (5.17)}$$

3. For each experimental run evaluate numerically the left- and the right-hand side of this equation.

4. Plot one against the other and test for linearity.

Equations 5.20 and 5.23 are the integrated forms of Eq. 5.17 for simple kinetic equations, and Example 3*a* illustrates this procedure.

Differential analysis. Integral analysis provides a straightforward rapid procedure for testing some of the simpler rate expressions. However, the integrated forms of these expressions become unwieldy with more complicated rate expressions. In these situations, the differential method of analysis which evaluates the rate directly by differentiating the appropriate curve becomes more convenient. The procedure is closely analogous to the differential method described in Chapter 3. So, by rearranging Eq. 5.16 we obtain the expression which allows us to find the rates of reaction in integral reactors. Thus we have

$$-r'_A = \frac{dX_A}{dW/F_{A0}} = \frac{dX_A}{d(W/F_{A0})} \tag{37}$$

The suggested procedure is as follows.

1. Make a series of runs in the packed bed using a fixed feed C_{A0} but varying F_{A0} and/or W such that a wide range of W/F_{A0} and X_A are obtained.

2. Plot $X_{A,out}$ versus W/F_{A0} for all runs.

3. Fit the best curve to the $X_{A,out}$ versus W/F_{A0} data, making it pass through the origin.

Equation 37 shows that the rate of reaction at any value of X_A is simply the slope of this curve; hence at a number of X_A values find the slope of this curve (or rate of reaction) as well as the corresponding concentration of reactant C_A.

We now have a series of rates versus concentrations which can be analyzed by the methods of Chapter 3 to obtain a rate equation. Example 3*b* illustrates this procedure.

Mixed Reactor. A mixed reactor requires a uniform composition of fluid throughout, and although it may seem difficult at first thought to approach this ideal with gas–solid systems (except for differential contacting), such contacting is in fact practical. One simple experimental device which closely approaches this ideal has been devised by Carberry (1964). It is called the *basket-type mixed reactor* and it is illustrated in Fig. 13. References to design variations and uses of basket reactors are given by Carberry (1969). Another device for approaching mixed flow is the recycle reactor, considered in the next section.

For the mixed reactor the performance equation becomes

$$\frac{W}{F_{A0}} = \frac{X_{A,out}}{-r'_{A,out}}$$

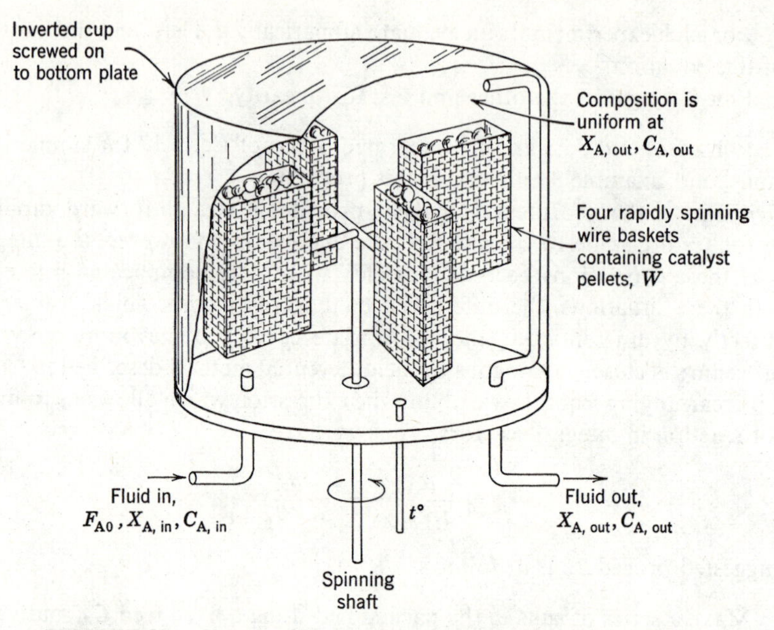

FIGURE 13. Sketch of basket-type experimental mixed reactor.

from which the rate is

$$-r'_{A,out} = \frac{F_{A0}X_{A,out}}{W} \tag{38}$$

Thus each run gives directly a value for the rate at the composition of the exit fluid.

Examples 5.1, 5.2, and 5.3 show how to treat such data.

Recycle Reactor. As with integral analysis of an integral reactor, when we use a recycle reactor we must put a specific kinetic equation to the test. The procedure requires inserting the kinetic equation into the performance equation for recycle reactors

$$\frac{W}{F_{A0}} = (\mathbf{R} + 1) \int_{(\mathbf{R}/\mathbf{R}+1)X_{Af}}^{X_{Af}} \frac{dX_A}{-r'_A} \tag{6.21}$$

and integrating. Then a plot of the left- versus right-hand side of the equation tests for linearity. Figure 14 sketches an experimental recycle reactor.

For a large enough recycle ratio mixed flow is approached, in which case the methods of the mixed reactor (direct evaluation of rate from each run) can be used. Thus a high recycle ratio provides a way of approximating mixed flow with what is essentially a plug flow device. Perkins and Rase (1958), and Livbjerg and Villadsen (1971) present more details about this system.

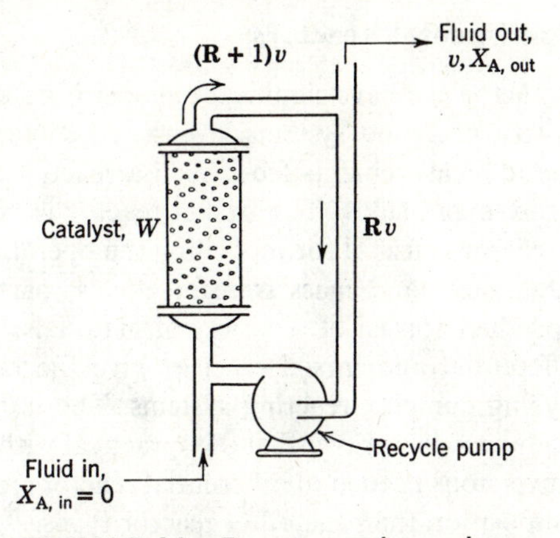

FIGURE 14. Experimental recycle reactor. When the recycle ratio is large enough mixed flow is closely approximated.

Batch Reactor. Figure 15 sketches the main features of an experimental reactor which uses a batch of catalyst and a batch of fluid. In this system we follow the changing composition with time and interpret the results with Eq. 34. The procedure is analogous with the homogeneous batch reactor. To insure meaningful results the composition of fluid must be uniform throughout the system at any instant. This requires that the conversion per pass through the catalyst be small.

A recycle reactor with no through flow becomes a batch reactor. This type of batch reactor was used by Butt *et al.* (1962).

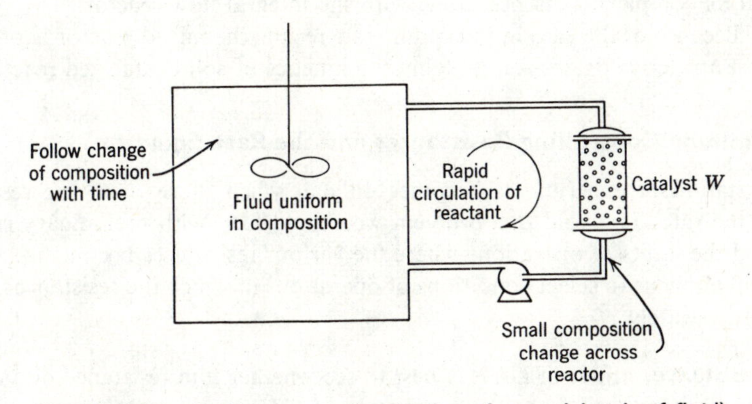

FIGURE 15. Batch reactor (batch of catalyst and batch of fluid) for catalytic reactions.

Comparison of Experimental Reactors

1. The integral reactor can have significant temperature variations from point to point, especially with gas–solid systems, even with cooling at the walls. This could well make kinetic measurements from such a reactor completely worthless when searching for rate expressions. The basket reactor is best in this respect.

2. The integral reactor is useful for modelling the operations of larger packed bed units with all their heat and mass transfer effects, particularly for systems where the feed and product consist of a variety of materials.

3. Since the differential and mixed reactors give the rate directly they are more useful in analyzing complex reacting systems. The test for anything but a simple kinetic form can become awkward and impractical with the integral reactor.

4. The small conversions needed in differential reactors require more accurate measurements of composition than the other reactor types.

5. The recycle reactor with small **R** shares many of the disadvantages of the integral reactor but with large recycle it acts as a mixed reactor and shares its advantages. Actually, to minimize heat effects the catalyst need not be all at one location, but can be distributed throughout the recycle loop.

6. In exploring the physical factors of heat and mass transfer, the integral reactor most closely models the larger fixed bed; however, the basket, recycle, and batch reactors are more suited for finding the limits for such effects, for avoiding the regime where these effects intrude, and for studying the kinetics of the reaction unhindered by these phenomena.

7. The batch reactor, like the integral reactor, gives cumulative effects, thus is useful for following the progress of multiple reactions. In batch reactors it is easier to study reactions free from heat and mass transfer resistances (simply increase the circulation rate), and it is also simple to slow down the progress of reactions (use a larger batch of fluid, or less catalyst); however, direct modelling of the packed bed with all its complexities is best done with the integral flow reactor.

8. Because of the ease in interpreting its results the mixed reactor is probably the most attractive device for studying the kinetics of solid catalyzed reactions.

Determining Controlling Resistances and the Rate Equation

Interpretation of experiments becomes difficult when more than one resistance affects the rate. To avoid this problem we should like, with preliminary runs, to first find the limits of operations where the various resistances become important. This will allow us to select conditions of operations in which the resistances can be studied separately.

Film Resistance. First of all, it is best to see whether film resistance of any kind (for mass or heat transfer) need be considered. This can be done in a number of ways.

1. Experiments can be devised to see whether the conversion changes at different gas velocities but at identical space-time. This is done by using different amounts of catalyst in integral or differential reactors for identical values for space-time, by changing the spinning rate in basket reactors, or by changing the circulation rate in recycle or batch reactors.

2. If data are available we can calculate whether film resistance to heat transfer is important by the estimate of Eq. 28, and whether film resistance to mass transport is important by comparing the observed first-order rate constant based on the volume of particle with the mass transfer coefficient for that type of flow, such as Eq. 12.24; thus compare

$$k_{obs}V_p \text{ versus } k_g S_{ex} \tag{39}$$

If the two terms are of the same order of magnitude we may suspect that the gas film resistance affects the rate. On the other hand, if $k_{obs}V_p$ is much smaller than $k_g S_{ex}$ we may ignore the resistance to mass transport through the film. Example 1 illustrates this type of calculation.

3. As mentioned in the section on temperature effects, if the rate of reaction rises, then the first form of film resistance likely to intrude on the rate is that for heat transfer; that for mass transfer only intrudes at much higher rates. This fact is clearly shown by the typical numerical values of Example 1.

Nonisothermal Effects. We may expect temperature gradients to occur either across the gas film or within the particle. However, the previous discussion indicates that for gas–solid systems the most likely effect to intrude on the rate will be the temperature gradient across the gas film. Consequently, if experiment shows that gas film resistance is absent then we may expect the particle to be at the temperature of its surrounding fluid, hence isothermal conditions may be assumed to prevail. Again see Example 1.

Pore Resistance. It is best to study pore resistance under conditions where film resistance is absent, or where $C_{As} = C_{Ag}$. This will also insure that nonisothermal effects will not intrude.

Let us assume that we are in this regime. Then pore resistance is accounted for by an effectiveness factor in the rate equation. Hence, based on unit mass of catalyst, we have

$$-r'_A = -\frac{1}{W}\frac{dN_A}{dt} = kC_{Ag}\mathscr{E}$$

where $\mathscr{E}$ is a function of mL as given by Fig. 9 or Eq. 13.

Consider a bed of catalyst particles of size R_1, another of size R_2. In the operating region where the resistance to pore diffusion is negligible $\mathscr{E} = 1$, hence for the two beds we have

$$\frac{-r'_{A1}}{-r'_{A2}} = \frac{kC_{Ag}\mathscr{E}_1}{kC_{Ag}\mathscr{E}_2} = \frac{\mathscr{E}_1}{\mathscr{E}_2} = 1 \tag{40}$$

Thus the rate of reaction based on unit mass of bed is the same for the two beds and is independent of the particle size.

In the region of strong pore resistance $\mathscr{E} = 1/mL$, hence for the two sizes of particles we have

$$\frac{-r'_{A1}}{-r'_{A2}} = \frac{\mathscr{E}_1}{\mathscr{E}_2} = \frac{mL_2}{mL_1} = \frac{R_2}{R_1} \tag{41}$$

Thus the rate of reaction varies inversely with particle size.

How do we find the transition between these two regions? If the original runs with different-size particles showed identical rates, thus negligible resistance to pore diffusion, then further runs should be made under conditions in which diffusional resistance becomes increasingly important relative to the resistance to chemical reaction. In other words, we should make runs with progressively larger-size particles or at higher temperature levels until the rates begin to differ in the two beds. On the other hand, if we originally were in the region of strong diffusional resistance, succeeding runs should be made with smaller particles or at lower temperatures. Thus the $\mathscr{E}$ versus mL curve can be found and fitted.

nth-order reactions behave in an unexpected way in the region of strong pore resistance. Combining the nth-order rate with the generalized modulus of Eq. 19 we obtain

$$-r'_A = kC_{Ag}{}^n\mathscr{E} = kC_{Ag}{}^n \cdot \frac{1}{mL} = kC_{Ag}{}^n \cdot \frac{1}{L}\sqrt{\frac{2\mathscr{D}_e}{(n+1)kC_{Ag}^{n-1}}}$$

$$= \left(\frac{2}{n+1} \cdot \frac{k\mathscr{D}_e}{L^2}\right)^{1/2} C_{Ag}^{(n+1)/2} \tag{42}$$

Thus an nth-order reaction behaves like a reaction of order $(n + 1)/2$ or

0 order becomes $\frac{1}{2}$ order
1st order remains 1st order
2nd order becomes 1.5 order
3rd order becomes 2nd order

In addition the temperature dependency of reactions is affected by strong pore resistance. From Eq. 42 the observed rate constant for nth-order reactions is

$$k_{obs} = \left(\frac{2}{n+1} \cdot \frac{k\mathscr{D}_e}{L^2}\right)^{1/2}$$

Taking logs and differentiating with respect to temperature and noting that both the reaction rate and to a lesser extent the diffusional process are temperature-dependent gives

$$\frac{d(\ln k_{obs})}{dT} = \frac{1}{2}\left[\frac{d(\ln k)}{dT} + \frac{d(\ln \mathscr{D}_e)}{dT}\right] \tag{43}$$

With Arrhenius temperature dependencies for both reaction and diffusion* we have

$$k = k_0 e^{-E_{true}/RT} \quad \text{and} \quad \mathscr{D}_e = \mathscr{D}_{e0} e^{-E_{diff}/RT}$$

and replacing in Eq. 43 gives

$$E_{obs} = \frac{E_{true} + E_{diff}}{2} \tag{44}$$

Since the activation energy for gas-phase reactions is normally rather high, say 20 ~ 60 kcal, while that for diffusion is small (about 1 kcal at room temperature or 4 kcal at 1000°C), we can write approximately

$$E_{obs} \cong \frac{E_{true}}{2} \tag{45}$$

These results show that the observed activation energy for reactions influenced by strong pore resistance is approximately one-half the true activation energy.

To sum up: the existence of strong resistance to pore diffusion can be determined:

1. By calculation if $\mathscr{D}_e$ is known (see Eqs. 25, 26, and 27).
2. By comparing the rate for different pellet sizes.
3. By noting the drop in activation energy of the reaction with rise in temperature, coupled with a possible change in reaction order.

Chemical Reaction. Kinetic equations based on the various active-site models may be obtained by the methods outlined by Yang and Hougen (1950) or Corrigan (1955). However, since these require an extensive research program to explore, and must be replaced by the corresponding linear rate expressions anyway if diffusional resistances are to be considered, we consider only the simpler empirical forms of rate equations to represent the kinetics when surface phenomena control. With film and pore diffusion absent, the rate of surface reaction can then be found in a straightforward manner.

PRODUCT DISTRIBUTION IN MULTIPLE REACTIONS

More often than not, solid-catalyzed reactions are multiple reactions; reactions occur side by side, and the products of these decompose further. Of the variety of products formed, usually only one is desired, and it is the yield of this material

* One can *fit* an Arrhenius-type temperature dependency to any process as long as the temperature range is not too large. For the temperature ranges considered in reactions this type of fit for the diffusional process causes no difficulties.

which is to be maximized. In cases such as these the question of product distribution is of primary importance.

Now, the general rules for maximizing a given product in homogeneous reactions apply equally well to solid-catalyzed reactions. These rules were developed in Chapter 7 and they accounted for the type of flow in the reactor. Briefly, then, for reactions in parallel the key to optimizing yields is to maintain the proper high or low concentration levels of reactants within the reactor, whereas for reactions in series the key is to avoid the mixing of fluids of different compositions.

In catalytic reactions diffusional resistance may cause the fluid in the pellet interior to differ in composition from the surrounding fluid. This may result in a product distribution different from what would be observed were this nonhomogeneity absent. Here we consider the difference between the true (at the reacting surface of the catalyst) and observed (material entering the main gas stream from the catalyst) product distribution. Knowing the reasons for this difference will suggest how to control the operating conditions so as to obtain the most favorable over-all yield of desired product.

In terms of fractional yields we may say that the over-all fractional yield Φ is found from the observed instantaneous fractional yields φ_{obs}, whereas the latter is related to the true instantaneous fractional yield φ_{true}, based on what happens at the interior surfaces of the catalyst. Thus

$$\varphi_{true} \xrightarrow[\text{reaction kinetics}]{\text{catalyst properties}} \varphi_{obs} \xrightarrow[\text{Chapter 7}]{\text{methods of}} \Phi$$

Here we examine in what way strong pore diffusion modifies the true instantaneous fractional yield for various types of reactions; however, we leave to Chapter 7 the calculation of the overall fractional yield in reactors with their particular flow patterns of fluid. In addition, we will not consider film resistance to mass transfer since this effect is unlikely to influence the rate.

Decomposition of a Single Reactant by Two Paths

Consider the parallel-path decomposition

$$A \begin{array}{l} \overset{k_1}{\nearrow} \text{R(desired)}, \qquad r_R = k_1 C_A^{a_1} \\ \underset{k_2}{\searrow} \text{S(unwanted)}, \qquad r_S = k_2 C_A^{a_2} \end{array} \tag{46}$$

which has an instantaneous fractional yield at any element of catalyst surface

$$\varphi_{true}\left(\frac{R}{R+S}\right) = \frac{r_R}{r_R + r_S} = \frac{1}{1 + (k_2/k_1)C_A^{a_2 - a_1}} \tag{47a}$$

or for first-order reactions

$$\varphi_{\text{true}} = \frac{1}{1 + (k_2/k_1)} \tag{47b}$$

Strong resistance to pore diffusion. Under these conditions we have

$$r_R = k_1 C_{Ag}{}^{a_1} \cdot \mathscr{E}_1 = k_1 C_{Ag}{}^{a_1} \cdot \frac{1}{mL}$$

and with Eq. 42

$$r_R \cong k_1 C_{Ag}{}^{a_1} \cdot \frac{1}{L} \left[\frac{4\mathscr{D}}{(a_1 + a_2 + 2)(k_1 + k_2) C_{Ag}^{a_1 - 1}} \right]^{1/2}$$

Using a similar expression for r_s and replacing both of these into the defining equation for φ gives

$$\varphi_{\text{obs}} \cong \frac{1}{1 + (k_2/k_1) C_{Ag}^{(a_2 - a_1)/2}} \tag{48a}$$

and for equal order or for first-order reactions

$$\varphi_{\text{obs}} = \frac{1}{1 + (k_2/k_1)} \tag{48b}$$

This result is expected since the rules in Chapter 7 suggest that the product distribution for competing reactions of same order should be unaffected by changing concentration of A in the pores.

Side-by-Side Decompositions of Two Reactants

Consider a feed consisting of two components, both of which react when in contact with a solid catalyst, as follows:

$$\begin{aligned} &A \xrightarrow{k_1} R \text{ (desired)}, \qquad r_R = k_1 C_A{}^a \\ &B \xrightarrow{k_2} S \text{ (unwanted)}, \qquad r_S = k_2 C_B{}^b \end{aligned} \tag{49}$$

The instantaneous fractional yield of desired product is then

$$\varphi_{\text{true}}\left(\frac{R}{R + S}\right) = \frac{1}{1 + (k_2 C_B{}^b / k_1 C_A{}^a)} \tag{50a}$$

or for first-order reactions

$$\varphi_{\text{true}} = \frac{1}{1 + (k_2 C_B / k_1 C_A)} \tag{50b}$$

Strong resistance to pore diffusion. In this regime the observed rates of reaction, as given by Eq. 42, are

$$r_R = k_1 C_{Ag}{}^a \cdot \mathscr{E}_A = \left(\frac{2}{a+1} \cdot \frac{k_1 \mathscr{D}_B}{L^2}\right)^{1/2} C_{Ag}^{(a+1)/2}$$

$$r_S = k_2 C_{Bg}{}^b \cdot \mathscr{E}_B = \left(\frac{2}{b+1} \cdot \frac{k_2 \mathscr{D}_B}{L^2}\right)^{1/2} C_{Bg}^{(b+1)/2}$$

hence the observed fractional yield is

$$\varphi_{\text{obs}} = \frac{1}{1 + \left(\dfrac{a+1}{b+1} \cdot \dfrac{k_2 \mathscr{D}_B C_{Bg}^{b+1}}{k_1 \mathscr{D}_A C_{Ag}^{a+1}}\right)^{1/2}} \tag{51a}$$

For first-order reactions this becomes simply

$$\varphi_{\text{obs}} = \frac{1}{1 + \left(\dfrac{k_2 \mathscr{D}_B}{k_1 \mathscr{D}_A}\right)^{1/2} \dfrac{C_{Bg}}{C_{Ag}}} \cong \frac{1}{1 + \left(\dfrac{k_2}{k_1}\right)^{1/2} \dfrac{C_{Bg}}{C_{Ag}}} \tag{51b}$$

Reactions in Series

As characteristic of reactions in which the desired product can decompose further, consider the successive first-order decompositions

$$A \xrightarrow{k_1} R \xrightarrow{k_2} S$$

When surface reaction is the rate-controlling step, C_A does not drop in the interior of catalyst particles, true rates are observed thus

$$\varphi_{\text{obs}} = \varphi_{\text{true}} \quad \text{or} \quad \left(\frac{k_2}{k_1}\right)_{\text{obs}} = \left(\frac{k_2}{k_1}\right)_{\text{true}} \tag{52}$$

Strong resistance to pore diffusion. An analysis similar to that on page 470 using the appropriate kinetic rate expressions gives the concentration ratio of materials in the main gas stream (or pore mouths) at any point in the reactor. Thus the differential expression (see Wheeler (1951) for details, and compare with Eq. 7.32) is

$$\frac{dC_{Rg}}{dC_{Ag}} = -\frac{1}{1+\gamma} + \gamma \frac{C_{Rg}}{C_{Ag}}, \qquad \gamma = \left(\frac{k_2}{k_1}\right)^{1/2} \tag{53}$$

For mixed flow integration with $C_{R0} = 0$ gives

$$C_{Rg} = \frac{1}{1+\gamma} \cdot \frac{C_{Ag}(C_{A0} - C_{Ag})}{C_{Ag} + \gamma(C_{A0} - C_{Ag})} \tag{54}$$

For plug flow integration with $C_{R0} = 0$ gives

$$\frac{C_{Rg}}{C_{A0}} = \frac{1}{1+\gamma} \cdot \frac{1}{1-\gamma}\left[\left(\frac{C_{Ag}}{C_{A0}}\right)^{\gamma} - \frac{C_{Ag}}{C_{A0}}\right] \tag{55}$$

Comparing Eqs. 54 and 55 with the corresponding expressions for no resistance in pores, Eqs. 7.36 and 7.33, shows that here the distribution of A and R is given by a reaction having the square root of the true k ratio, with the added modification that C_{Rg} is divided by $1 + \gamma$. The maximum yield of R is likewise affected. Thus for plug flow Eq. 3.52 is modified to give

$$\frac{C_{Rg,\max}}{C_{A0}} = \frac{\gamma^{\gamma/1-\gamma}}{1+\gamma}, \qquad \gamma = \left(\frac{k_2}{k_1}\right)^{1/2} \tag{56}$$

and for mixed flow Eq. 7.23 is modified to give

$$\frac{C_{Rg,\max}}{C_{A0}} = \frac{1}{(1+\gamma)(\gamma^{1/2}+1)^2} \tag{57}$$

Table 1 shows that the yield of R is about halved with strong resistance to diffusion in the pores.

TABLE 1. The role of diffusion in pores for first-order reactions in series

	$C_{Rg,\max}/C_{A0}$ for Plug Flow			$C_{Rg,\max}/C_{A0}$ for Mixed Flow		
k_2/k_1	No Resistance	Strong Resistance	Percent Decrease	No Resistance	Strong Resistance	Percent Decrease
$\frac{1}{64}$	0.936	0.650	30.6	0.790	0.486	38.5
$\frac{1}{16}$	0.831	0.504	39.3	0.640	0.356	44.5
$\frac{1}{4}$	0.630	0.333	47.6	0.444	0.229	48.5
1	0.368	0.184	50	0.250	0.125	50
4	0.157	0.083	47.2	0.111	0.057	48.5
16	0.051	0.031	38.2	0.040	0.022	44.5

For more on the whole subject of the shift in product distribution caused by diffusional effects, see Wheeler (1951).

Extensions to Real Catalysts

So far we have considered catalyst pellets having only one size of pore. Real catalysts, however, have pores of various sizes. A good example of this are the pellets prepared by compressing a porous powder. Here there are large openings between the powder and small pores within each powder particle. As a first approximation we may represent this structure by two pore sizes as shown in Fig. 16.

If we define the degree of branching of a porous structure by α where

$\alpha = 0$ represents a nonporous particle
$\alpha = 1$ represents a particle with one size of pore
$\alpha = 2$ represents a particle with two pore sizes, etc.

then every real porous pellet can be characterized by some value of α.

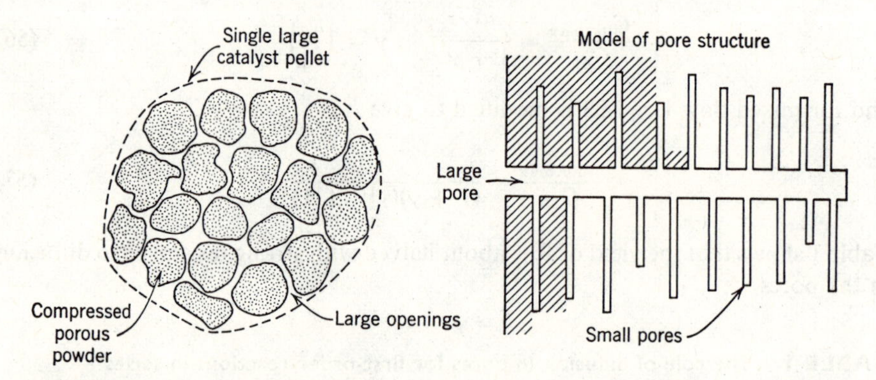

FIGURE 16. Porous structure with two sizes of pores as a model for a pellet of compressed porous powder.

Now for strong pore diffusion in one size of pore we already know that the observed order of reaction, activation energy, and k ratio for multiple reactions will differ from the true value. Thus

$$\text{for } \alpha = 1 \quad \begin{cases} E_{\text{obs}} = \frac{1}{2}E_{\text{diff}} + \frac{1}{2}E \\ n_{\text{obs}} = 1 + \dfrac{n-1}{2} \\ \left(\dfrac{k_2}{k_1}\right)_{\text{obs}} = \left(\dfrac{k_2}{k_1}\right)^{1/2} \cdots \text{for side by side reactions} \end{cases} \tag{58}$$

Carberry (1962*a*, 1962*b*), Tartarelli (1968), and others have extended this type of analysis to other values of α and to reversible reactions. Thus for two sizes of pores, where reaction occurs primarily in the smaller pores (hence much more area), while both sizes of pores offer strong pore diffusional resistance, we find

$$\text{for } \alpha = 2 \quad \begin{cases} E_{\text{obs}} = \frac{3}{4}E_{\text{diff}} + \frac{1}{4}E \\ n_{\text{obs}} = 1 + \dfrac{n-1}{4} \\ \left(\dfrac{k_2}{k_1}\right)_{\text{obs}} = \left(\dfrac{k_2}{k_1}\right)^{1/4} \cdots \text{for side by side reactions} \end{cases} \tag{59}$$

More generally for an arbitrary porous structure

$$\text{for any } \alpha \quad \begin{cases} E_{\text{obs}} = \left(1 - \dfrac{1}{2^\alpha}\right)E_{\text{diff}} + \dfrac{1}{2^\alpha}E \\ n_{\text{obs}} = 1 + \dfrac{n-1}{2^\alpha} \\ \left(\dfrac{k_2}{k_1}\right)_{\text{obs}} = \left(\dfrac{k_2}{k_1}\right)^{1/2^\alpha} \cdots \text{for side by side reactions} \end{cases} \tag{60}$$

These findings show that for large α, diffusion plays an increasingly important role, the observed activation energy decreases to that of diffusion, and the reaction order approaches unity. So, for a given porous structure with unknown α, the only reliable estimate of the true k ratio would be from experiments under conditions where pore diffusion is unimportant. On the other hand, finding the ratio of k values under both strong and negligible pore resistance should yield the value of α. This in turn should shed light on the pore structure geometry of the catalyst.

EXAMPLE 1. Search for the rate-controlling mechanism

An experimental rate measurement on the decomposition of A is made with a particular catalyst (see pertinent data listed below).

(*a*) Is it likely that film resistance to mass transfer influences the rate?

(*b*) Could this run have been made in the regime of strong pore diffusion?

(*c*) Would you expect temperature variations within the pellet or across the gas film?

Data. For the spherical particle:

$d_p = 2.4$ mm or $L = R/3 = 0.4$ mm $= 4 \times 10^{-4}$ m cat

$\mathscr{D}_e = 5 \times 10^{-5}$ m³/hr·m cat, (effective mass diffusivity)

$k_{\text{eff}} = 0.4$ kcal/hr·m cat·°K, (effective thermal conductivity)

For the gas film surrounding the pellet, (from correlations in the literature):

$h = 40$ kcal/hr·m² cat·°K, (heat transfer coefficient)

$k_g = 300$ m³/hr·m² cat, (mass transfer coefficient)

For the reaction:

$$\Delta H_r = -40 \text{ kcal/mol A (exothermic)}$$
$$C_{Ag} = 20 \text{ mol/m}^3 \text{ (at 1 atm and 336°C)}$$
$$-r'''_{A,obs} = 10^5 \text{ mol/hr}\cdot\text{m}^3 \text{ cat}$$

Assume that the reaction is first order.

SOLUTION

(*a*) *Film mass transfer.* From Eq. 39, and introducing numerical values, we obtain

$$\frac{\text{observed rate}}{\text{rate if film resistance controls}} = \frac{k_{obs}V_p}{k_g S_{ex}} = \frac{(-r'''_{A,obs}/C_{Ag})(\pi d_p{}^3/6)}{k_g(\pi d_p{}^2)} = \frac{-r'''_{A,obs}}{C_{Ag}k_g}\cdot\frac{d_p}{6}$$

$$= \frac{10^5 \text{ mol/hr}\cdot\text{m}^3 \text{ cat}}{(20 \text{ mol/m}^3)(300 \text{ m}^3/\text{hr}\cdot\text{m}^2 \text{ cat})}\cdot\frac{2.4 \times 10^{-3} \text{ m cat}}{6}$$

$$= \frac{1}{150}$$

The observed rate is very much lower than the limiting film mass transfer rate. Thus the resistance to film mass transfer certainly should not influence the rate of reaction.
(*b*) *Strong pore diffusion.* Equations 25 and 26 test for strong pore diffusion. Thus

$$\frac{(-r'''_A)_{obs}L^2}{\mathscr{D}_e C_{Ag}} = \frac{(10^5 \text{ mol/hr}\cdot\text{m}^3 \text{ cat})(4 \times 10^{-4} \text{ m cat})^2}{(5 \times 10^{-5} \text{ m}^3/\text{hr}\cdot\text{m cat})(20 \text{ mol/m}^3)} = 16$$

This quantity is greater than unity hence pore diffusion is influencing and slowing the rate of reaction.
(*c*) *Nonisothermal operations.* The estimate for the upper limit to temperature variations is given by Eqs. 28 and 29. Thus within the pellet

$$\Delta T_{max,pellet} = \frac{\mathscr{D}_e(C_{Ag} - 0)(-\Delta H_r)}{k_{eff}}$$

$$= \frac{(5 \times 10^{-5} \text{ m}^3/\text{hr}\cdot\text{m cat})(20 \text{ mol/m}^3)(40 \text{ kcal/mol})}{(0.4 \text{ kcal/hr}\cdot\text{m cat}\cdot°\text{K})}$$

$$= 0.1°\text{C}$$

Across the gas film

$$\Delta T_{max,film} = \frac{L(-r'''_{A,obs})(-\Delta H_r)}{h}$$

$$= \frac{(4 \times 10^{-4} \text{ m})(10^5 \text{ mol/hr}\cdot\text{m}^3)(40 \text{ kcal/mol})}{(40 \text{ kcal/hr}\cdot\text{m}^2\cdot°\text{K})}$$

$$= 40°\text{C}$$

These estimates show that the pellet is close to uniform in temperature, but could well be hotter than the surrounding fluid.

The findings of this example use coefficients close to those observed in real gas-solid systems, and the conclusions verify the discussions of this chapter.

EXAMPLE 2. The rate equation from a differential reactor

The catalytic reaction

$$A \rightarrow 4R$$

is run at 3.2 atm and 117°C in a plug flow reactor which contains 0.01 kg of catalyst and uses a feed consisting of the partially converted product of 20 liter/hr of pure unreacted A. The results are as follows:

Run	1	2	3	4
$C_{A,in}$, mol/liter	0.100	0.080	0.060	0.040
$C_{A,out}$, mol/liter	0.084	0.070	0.055	0.038

Find a rate equation to represent this reaction.

SOLUTION

Since the maximum variation about the mean concentration is 8% (run 1), we may consider this to be a differential reactor and we may apply Eq. 36 to find the reaction rate.

Basing conversion for all runs on pure A at 3.2 atm and 117°C, we have

$$C_{A0} = \frac{N_{A0}}{V} = \frac{p_{A0}}{RT} = \frac{3.2\ \text{atm}}{(0.082\ \text{liter}\cdot\text{atm/mol}\cdot{}^\circ\text{K})(390^\circ\text{K})} = 0.1\ \frac{\text{mol}}{\text{liter}}$$

and

$$F_{A0} = C_{A0}v = \left(0.1\ \frac{\text{mol A}}{\text{liter}}\right)\left(20\ \frac{\text{liter}}{\text{hr}}\right) = 2\ \frac{\text{mol}}{\text{hr}}$$

Because the density changes during reaction, concentrations and conversions are related by

$$\frac{C_A}{C_{A0}} = \frac{1 - X_A}{1 + \varepsilon_A X_A} \qquad \text{or} \qquad X_A = \frac{1 - C_A/C_{A0}}{1 + \varepsilon_A(C_A/C_{A0})}$$

where $\varepsilon_A = 3$ for the basis selected (pure A).

Table E2 shows the details of the calculations. Plotting $-r_A'$ versus C_A as shown in Fig. E2 gives a straight line through the origin, indicating a first-order decomposition.

TABLE E2

$\frac{C_{A,in}}{C_{A0}}$	$\frac{C_{A,out}}{C_{A0}}$	$C_{A,av}$ mol/liter	$X_{A,in} = \dfrac{1 - \frac{C_{A,in}}{C_{A0}}}{1 + \varepsilon_A \frac{C_{A,in}}{C_{A0}}}$	$X_{A,out} = \dfrac{1 - \frac{C_{A,out}}{C_{A0}}}{1 + \varepsilon_A \frac{C_{A,out}}{C_{A0}}}$	$\Delta X_A = X_{A,out} - X_{A,in}$	$-r_A' = \dfrac{\Delta X_A}{W/F_{A0}}$
1	0.84	0.092	$\frac{1-1}{1+3} = 0$	$\frac{1 - 0.84}{1 + 3(0.84)} = 0.0455$	0.0455	$\frac{0.0455}{0.01/2} = 9.1$
0.8	0.70	0.075	0.0588	0.0968	0.0380	7.6
0.6	0.55	0.0575	0.1429	0.1698	0.0269	5.4
0.4	0.38	0.039	0.2727	0.2897	0.0170	3.4

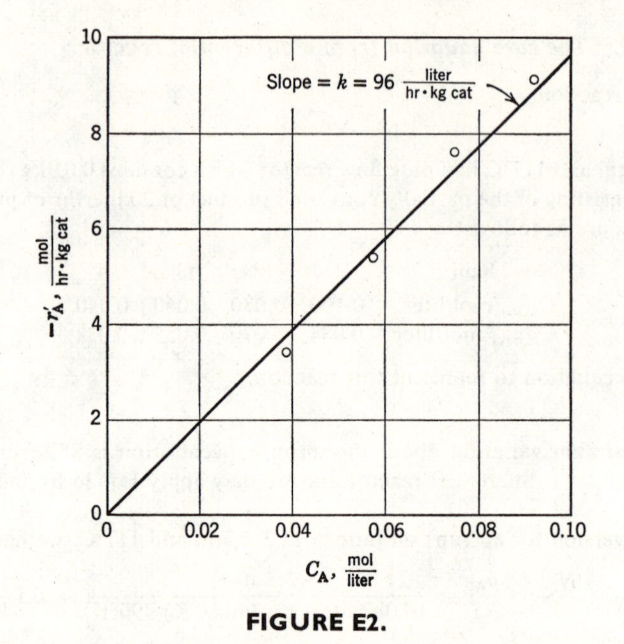

FIGURE E2.

The rate in terms of moles A reacted/hr·kg cat is then found from this figure to be

$$-r'_A = -\frac{1}{W}\frac{dN_A}{dt} = \left(96\ \frac{\text{liters}}{\text{hr}\cdot\text{kg cat}}\right)\left(C_A,\ \frac{\text{mol}}{\text{liter}}\right)$$

EXAMPLE 3. The rate equation from an integral reactor

The catalytic reaction

$$A \rightarrow 4R$$

is studied in a plug flow reactor using various amounts of catalyst and 20 liter/hr of pure A feed at 3.2 atm and 117°C. The concentrations of A in the effluent stream is recorded for the various runs as follows.

Runs	1	2	3	4	5
Catalyst used, kg	0.020	0.040	0.080	0.120	0.160
$C_{A,out}$, mol/liter	0.074	0.060	0.044	0.035	0.029

(*a*) Find the rate equation for this reaction, using the integral method of analysis
(*b*) Repeat part (*a*), using the differential method of analysis.

SOLUTION

(*a*) *Integral Analysis.* From Example 2 we have for all experimental runs

$$C_{A0} = 0.1 \text{ mol/liter}$$
$$F_{A0} = 2 \text{ mol/hr}$$
$$\varepsilon_A = 3$$

Since the concentration varies significantly during the runs, the experimental reactor should be considered to be an integral reactor.

As a first guess try a first-order rate expression. If this will not fit the data we will then try some other simple rate form until we do get a fit. Showing all dimensions and units the plug flow performance equation is

$$\frac{W,\ \text{kg cat}}{F_{A0},\ \dfrac{\text{mol A}}{\text{hr}}} = \int_{X_A=0}^{X_A} \frac{dX_A,\ \dfrac{\text{mol A reacted}}{\text{mol A fed}}}{-r'_A,\ \dfrac{\text{mol A reacted}}{\text{hr}\cdot\text{kg cat}}}$$

With first-order kinetics, keeping consistent units, this becomes

$$\frac{W}{F_{A0}} = \int_0^{X_A} \frac{dX_A}{kC_A} = \frac{1}{kC_{A0}} \int_0^{X_A} \frac{1+\varepsilon_A X_A}{1-X_A}\, dX_A$$

Using Eq. 5.22 to evaluate the integral we find

$$k\frac{C_{A0}W}{F_{A0}} = (1+\varepsilon_A)\ln\frac{1}{1-X_A} - \varepsilon_A X_A$$

and with ε_A, C_{A0}, and F_{A0} replaced by numerical values this becomes

$$\left(4\ln\frac{1}{1-X_A} - 3X_A\right) = k\left(\frac{W}{20}\right)$$

The two terms in parentheses should be proportional to each other with k as the constant of proportionality. Evaluating these terms in Table E3*a* for the data points and plotting as in Fig. E3*a*, we see that there is no reason to suspect that we do not have a linear

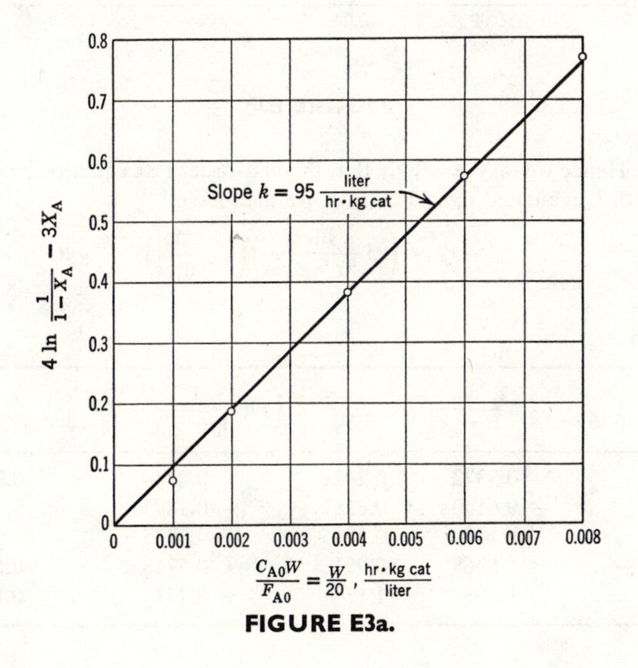

FIGURE E3a.

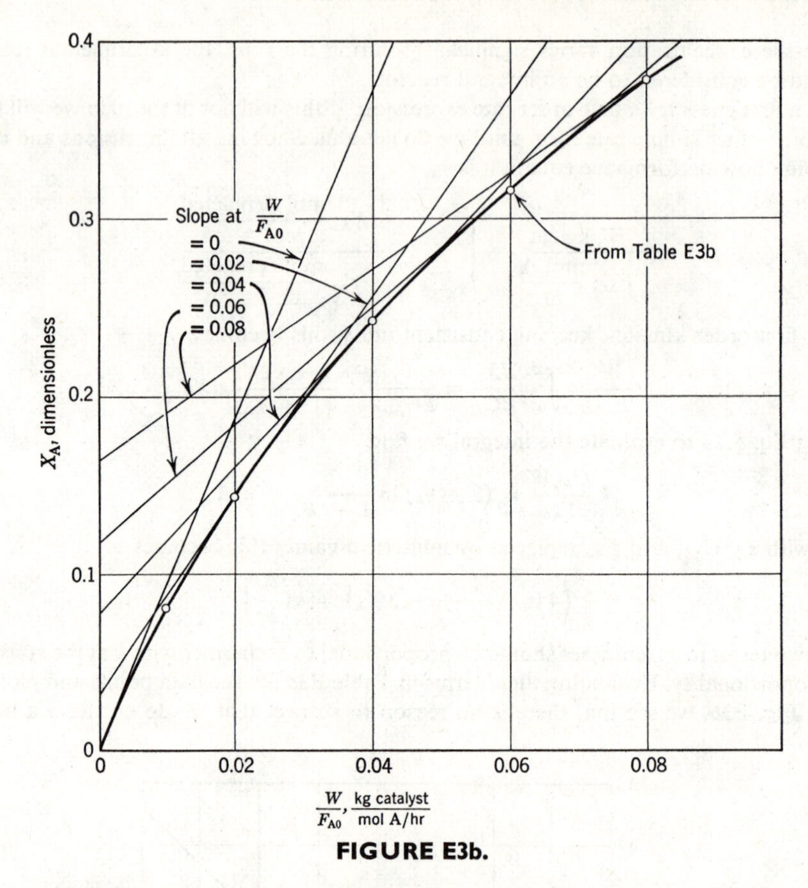

FIGURE E3b.

relationship. Hence we may conclude that the first-order rate equation satisfactorily fits the data. With k evaluated from Fig. E3*a*, we then have

$$-r'_A = \left(95 \frac{\text{liters}}{\text{hr}\cdot\text{kg cat}}\right)\left(C_A, \frac{\text{mol}}{\text{liter}}\right)$$

TABLE E3a

X_A (from Table E3*b*)	$4 \ln \frac{1}{1 - X_A}$	$3X_A$	$\left(4 \ln \frac{1}{1 - X_A} - 3X_A\right)$	W, kg	$\frac{W}{20}$
0.0808	0.3372	0.2424	0.0748	0.02	0.001
0.1429	0.6160	0.4287	0.1873	0.04	0.002
0.2415	1.1080	0.7245	0.3835	0.08	0.004
0.317	1.5268	0.951	0.5758	0.12	0.006
0.379	1.908	1.137	0.771	0.16	0.008

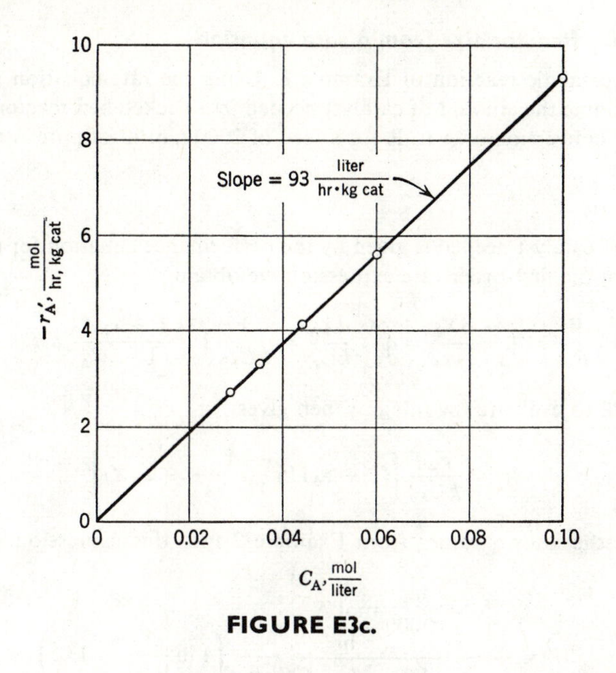

FIGURE E3c.

(*b*) *Differential analysis.* Equation 37 shows that the rate of reaction is given by the slope of the X_A versus W/F_{A0} curve. The tabulation (Table E3*b*) based on the measured slopes in Fig. E3*b* shows how the rate of reaction is found at various C_A. The linear relation between $-r_A$ and C_A in Fig. E3*c* then gives for the rate equation:

$$-r'_A = \left(93 \frac{\text{liters}}{\text{hr}\cdot\text{kg cat}}\right)\left(C_A, \frac{\text{mol}}{\text{liter}}\right)$$

TABLE E3b

W	$\frac{W}{F_{A0}}$	$\frac{C_{A,\text{out}}}{C_{A0}}$	$X_A = \frac{1 - \frac{C_A}{C_{A0}}}{1 + \varepsilon_A \frac{C_A}{C_{A0}}}$	$-r'_A = \frac{dX_A}{d\left(\frac{W}{F_{A0}}\right)}$ (from Fig. E3*b*)
0	0	1	0	$\frac{0.4}{0.043} = 9.3$
0.02	0.01	0.74	0.0808	—
0.04	0.02	0.60	0.1429	5.62
0.08	0.04	0.44	0.2415	4.13
0.12	0.06	0.35	0.317	3.34
0.16	0.08	0.29	0.379	2.715

EXAMPLE 4. Reactor size from a rate equation

Consider the catalytic reaction of Example 2. Using the rate equation found for this reaction determine the amount of catalyst needed in a packed bed reactor (assume plug flow) for 35% conversion of A to R for a feed of 2000 mol/hr of pure A at 3.2 atm and 117°C.

SOLUTION

The amount of catalyst needed is given by the performance equation for plug flow, and on introducing the first-order rate expression we obtain

$$\frac{W}{F_{A0}} = \int_0^{X_A} \frac{dX_A}{-r'_A} = \int_0^{X_A} \frac{dX_A}{kC_A} = \frac{1}{kC_{A0}} \int_0^{X_A} \frac{1 + \varepsilon_A X_A}{1 - X_A}\, dX_A$$

Using Eq. 5.22 to evaluate the integral then gives

$$W = \frac{F_{A0}}{kC_{A0}} \left[(1 + \varepsilon_A) \ln \frac{1}{1 - X_A} - \varepsilon_A X_A \right]$$

Replacing all the known values from Example 2 into this expression gives the final result, or

$$W = \frac{2000 \dfrac{\text{mol A}}{\text{hr}}}{\left(96 \dfrac{\text{liter}}{\text{hr}\cdot\text{kg cat}}\right)\left(0.1 \dfrac{\text{mol A}}{\text{liter}}\right)} \left(4 \ln \frac{1}{0.65} - 1.05\right)$$

$$= 140 \text{ kg catalyst}$$

EXAMPLE 5. Reactor size from rate concentration data

For the reaction of Example 2 suppose the following rate concentration data are available

C_A, mol/liter	0.039	0.0575	0.075	0.092
$-r'_A$, mol A/hr·kg cat	3.4	5.4	7.6	9.1

Directly from this data, and without using a rate equation, find the size of packed bed needed to treat 2000 mol/hr of pure A at 117°C (or $C_{A0} = 0.1$ mol/liter, $\varepsilon_A = 3$) to 35% conversion.

Note: Rate information such as this can be obtained from a differential reactor (see Table E2), from an integral reactor (see Table E3*b*), or from other types of experimental reactors.

SOLUTION

To find the amount of catalyst needed without using an analytic expression for the rate concentration relationship requires graphical integration of the plug flow performance equation, or

$$\frac{W}{F_{A0}} = \int_0^{0.35} \frac{dX_A}{-r'_A}$$

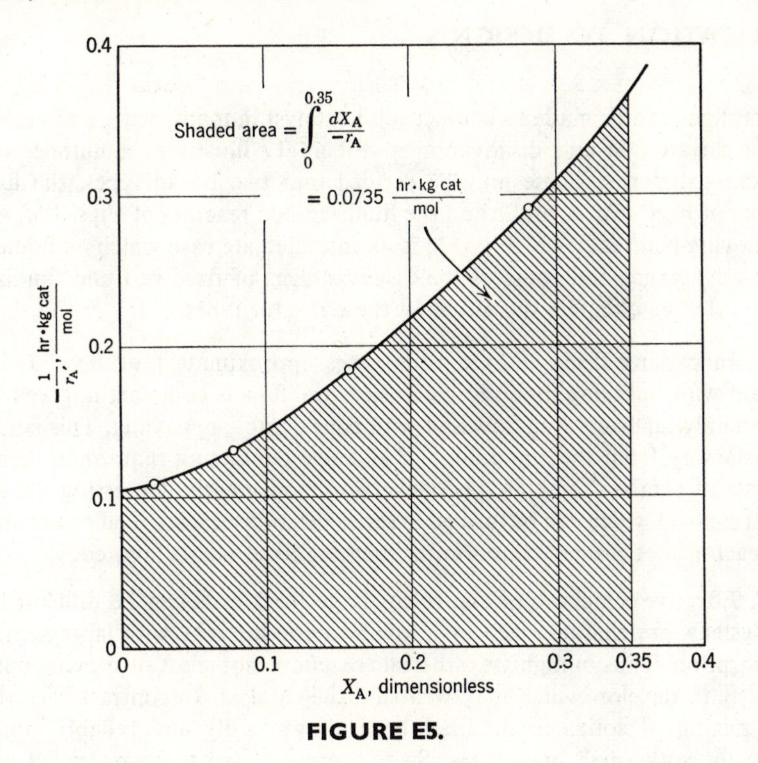

FIGURE E5.

The needed $1/-r'_A$ versus X_A data are determined in Table E5 and are plotted in Fig. E5. Integrating graphically then gives

$$\int_0^{0.35} \frac{dX_A}{-r'_A} = 0.0735$$

Hence

$$W = \left(2000\,\frac{\text{mol A}}{\text{hr}}\right)\left(0.0735\,\frac{\text{hr}\cdot\text{kg cat}}{\text{mol A}}\right) = 147 \text{ kg cat}$$

TABLE E5

$-r'_A$ (given)	$\frac{1}{-r'_A}$	C_A (given)	$X_A = \frac{1 - C_A/0.1}{1 + 3C_A/0.1}$
3.4	0.294	0.039	0.2812
5.4	0.186	0.0575	0.1563
7.6	0.1316	0.075	0.0778
9.1	0.110	0.092	0.02275

APPLICATION TO DESIGN

Reactant gas can be made to contact solid catalyst in many ways, and each has its specific advantages and disadvantages. Figure 17 illustrates a number of these contacting patterns. These may be divided into two broad types, the fixed-bed reactors of Figs. 17*a*, *b*, and *c* and the fluidized-bed reactors of Figs. 17*d*, *e*, and *f*. The moving-bed reactor of Fig. 17*g* is an intermediate case which embodies some of the advantages and some of the disadvantages of fixed-bed and fluidized-bed reactors. Let us compare the merits of these reactor types.

1. In passing through fixed beds gases approximate plug flow. It is quite different with bubbling fluidized beds. Here the flow is complex, not well known, but certainly far from plug flow and with considerable bypassing. This behavior is unsatisfactory from the standpoint of effective contacting requiring much larger amounts of catalyst for high gas conversion, and greatly depressing the amount of intermediate which can be formed in series reactions. Hence if efficient contacting in a reactor is of primary importance then the fixed bed is favored.

2. Effective temperature control of large fixed beds can be difficult because such systems are characterized by a low heat conductivity with large heat release or absorption. Thus in highly exothermic reactions hot spots or moving hot fronts are likely to develop which may well ruin the catalyst. In contrast with this, the rapid mixing of solids in fluidized beds allows easily and reliably controlled, practically isothermal, operations. So if operations are to be restricted within a narrow temperature range, either because of the explosive nature of the reaction or because of product distribution considerations, then the fluidized bed is favored.

3. Fixed beds cannot use very small sizes of catalyst because of plugging and high-pressure drop, whereas fluidized beds are well able to use small-size particles. Thus for very fast reactions in which pore and film diffusion may influence the rate, the fluidized bed with its vigorous gas–solid contacting and small particles will allow a much more effective use of the catalyst.

4. If the catalyst has to be treated (regenerated) frequently because it deactivates rapidly, then the liquid-like fluidized state allows it to be pumped easily from unit to unit. This feature of fluidized bed contacting offers overwhelming advantages over fixed bed operations for such solids.

With these points in mind, let us proceed to Fig. 17. Figure 17*a* is a typical packed bed reactor embodying all its advantages and disadvantages. Figure 17*b* shows how the problem of hot spots can be substantially reduced by increasing the cooling surface. Figure 17*c* shows how intercooling can still further control the temperature. Note that in the first stage where reaction is fastest, conversion is kept low by having less catalyst present than in the other stages. Units such as these can all be incorporated in a single shell or can be kept separate with heat exchangers between stages.

Figure 17*d* shows a fluidized reactor for a stable catalyst which need not be regenerated. The heat exchanger tubes are immersed in the bed to remove or add heat and to control the temperature. Figure 17*e* shows operations with a deactivating catalyst which must be continually removed and regenerated. Figure 17*f* shows a three-stage countercurrent unit which is designed to overcome the shortcomings of fluidized beds with regard to poor contacting. Figure 17*g* shows a moving-bed reactor. Such units share with fixed beds the advantages of plug flow and disadvantages of large particle size, but they also share with fluidized beds the advantages of low catalyst-handling costs.

Many factors must be weighed to obtain optimum design, and it may be that the best design is one that uses two different reactor types in series. For example, for high conversion and a very exothermic reaction we may well look into the use of a fluidized bed followed by a fixed bed.

The main difficulties of design of catalytic reactors reduce to the following questions: how to account for the nonisothermal behavior of packed beds, and how to account for the nonideal flow of gas in fluidized beds.

Consider a packed bed with heat exchange (Figs. 17*a* and 17*b*). For an exothermic reaction Fig. 18 shows the types of heat and mass movement that will occur when the packed tube is cooled at the walls. The centerline will be hotter than the walls, reaction will be faster, and reactants will be more rapidly consumed there, hence radial gradients of all sorts will be set up.

The detailed analysis of this situation should include the simultaneous radial dispersion of heat and matter, and maybe axial dispersion too. In setting up the mathematical model, what simplifications are reasonable, would the results properly model the real situation, would the solution indicate unstable behavior and hot spots? These questions have been considered by scores of researchers, numerous precise solutions have been claimed; however, from the point of view of prediction and design the situation today still is not as we would wish. The treatment of this problem is quite difficult, and we will not consider it here. A good review of the state of the art is given by Froment (1970).

The staged adiabatic packed bed reactor of Fig. 17*c* presents a different situation. Since there is no heat transfer in the zone of reaction the temperature and conversion are related simply, hence the methods of Chapter 8 can be applied directly. We will examine numerous variations of staging and heat transfer to show that this is a versatile setup which can closely approximate the optimum.

The design of gas fluidized beds is still unreliable today because the flow pattern which will develop in a bed cannot be predicted with assurance. This form of contacting is particularly susceptible to maldistribution of gas, partial settling of solids and other erratic behavior. Despite this difficulty, and to illustrate the approach which should be used, we conclude this chapter with a brief presentation

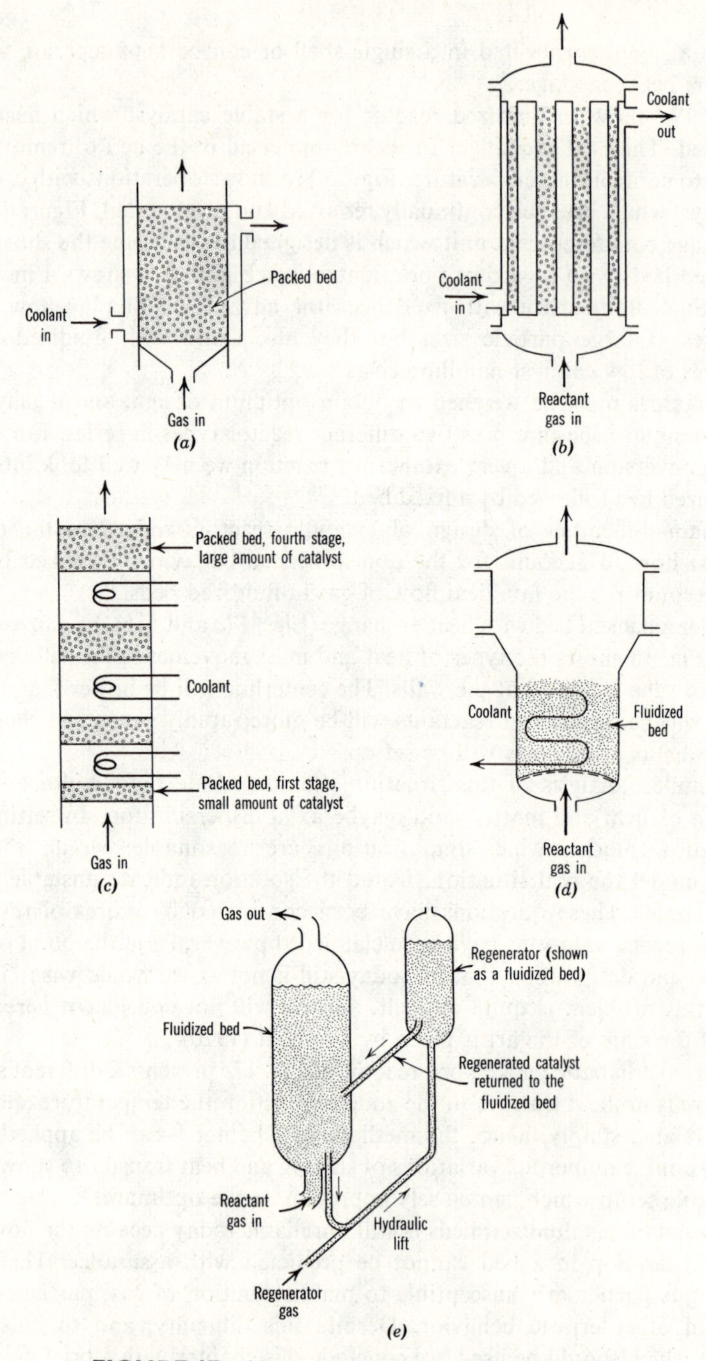

FIGURE 17. Various types of catalytic reactors.

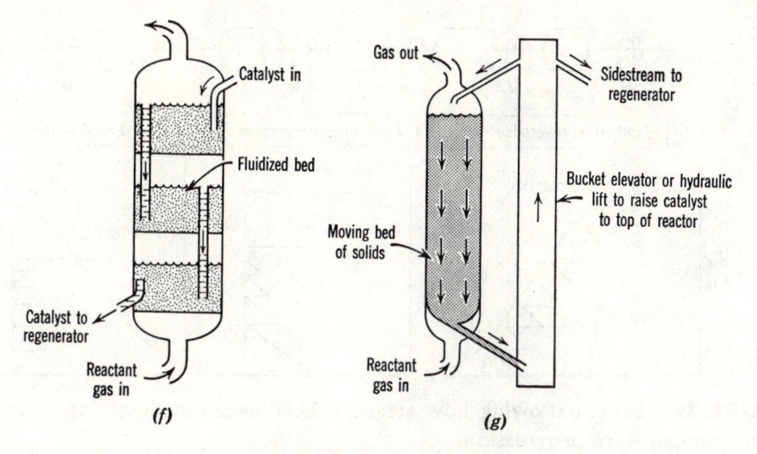

FIGURE 17. *Continued.*

of a model which accounts for the main features of the nonideal contacting in these reactors.

Staged Adiabatic Packed Bed Reactors

With proper interchange of heat and proper gas flow staged adiabatic packed beds become a versatile system which is able to approximate practically any desired temperature progression. Calculation and design of such a system is simple and we can expect that real operations will closely follow these predictions.

We illustrate the design procedure with the single reaction $A \rightarrow R$ with any kinetics. This procedure can be extended to other reaction types without difficulty. We first consider different ways of operating these reactors, we then compare these and point out when one or other is favored.

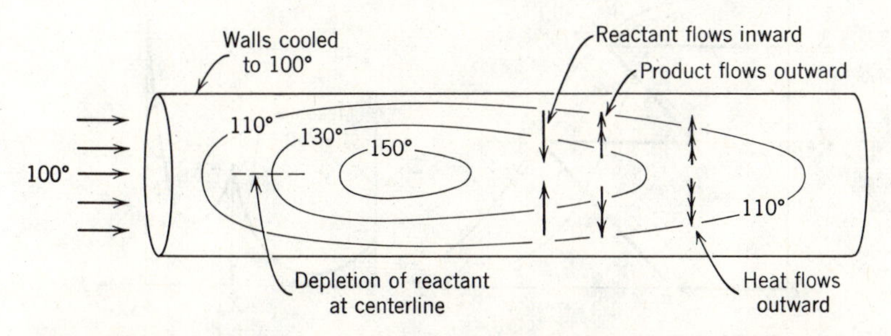

FIGURE 18. Temperature field in a packed bed reactor for an exothermic reaction showing how this field creates radial movement of heat and matter.

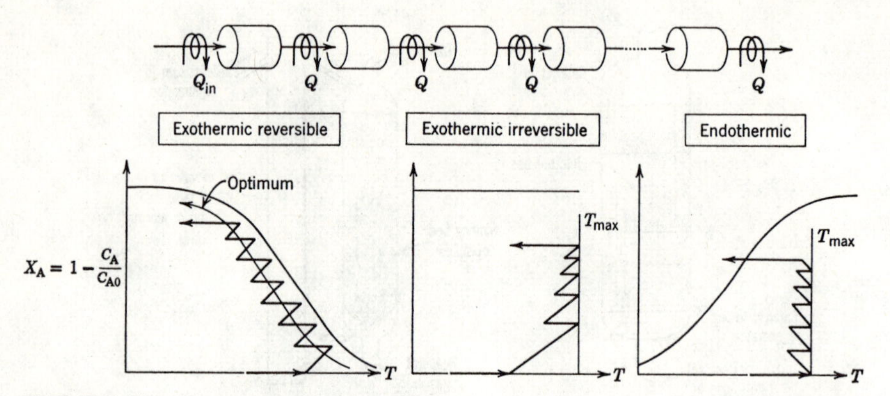

FIGURE 19. Sketch showing how staged packed beds can closely approach the optimal temperature progression.

Staged Packed Beds with Intercooling.* The reasoning in Chapter 8 shows that we would like the reacting conditions to follow the optimal temperature progression. With many stages available this can be closely approximated, as shown in Fig. 19.

For any preset number of stages the optimization of operations reduces to minimizing the total amount of catalyst needed. Let us illustrate the procedure for two-stage operations with *reversible exothermic reactions*. The method of attack is shown in Fig. 20. In this figure we wish to minimize the total area under the

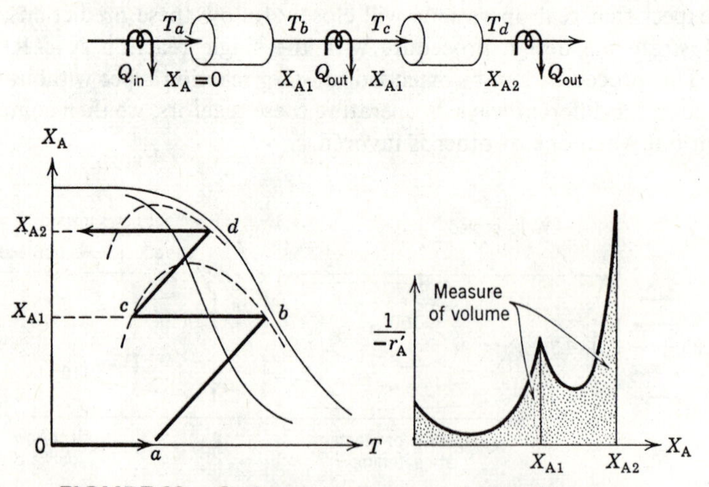

FIGURE 20. Optimum two-stage packed bed reactor.

* This section follows directly from pp. 218–226 of Chapter 8. Hence it is suggested that the reader familiarize himself with that section before proceeding here.

$1/-r'_A$ versus X_A curve. In searching for this optimum we have three variables which we can set at will: the incoming temperature (point T_a), the amount of catalyst used in the first stage (locates point b along the adiabatic), and the amount of intercooling (locates point c along the bc line). Fortunately, we are able to reduce this 3-dimensional search (5-dimensional for 3 stages, etc.) to a one-dimensional search where T_a alone is guessed. The procedure is as follows:

1. Guess T_a.

2. Move along the adiabatic line until the following condition is satisfied:

$$\int_{\text{in}}^{\text{out}} \frac{\partial}{\partial T}\left(\frac{1}{-r'_A}\right) dX_A = 0 \tag{61}$$

This gives point b in Fig. 20, thus the amount of catalyst needed in the first stage as well as the outlet temperature from that stage. Especially in preliminary design it may not be convenient to use the criterion of Eq. 61. A simple alternative is a trial-and-error search. Usually two or three carefully chosen trials keeping away from low rate conditions will yield a good design, close to the optimum.

3. Cool to point c which has the same rate of reaction as point b; thus

$$(-r'_A)_{\text{leaving a reactor}} = (-r'_A)_{\text{entering the next reactor}} \tag{62}$$

4. Move along the adiabatic from point c until the criterion of Eq. 61 is satisfied, giving point d.

5*a*. If point d is at the desired final conversion then we have guessed T_a correctly, hence the interstage heat transfer and the distribution of catalyst among stages (given by the areas under the $1/-r'_A$ versus X_A curves) are such as to minimize the total catalyst volume.

5*b*. If point d is not at the desired final conversion try a different incoming temperature T_a. Usually three trials will very closely approach the optimum.

For three or more stages the procedure is a direct extension of that presented here, and it still remains a one-dimensional search. This procedure was first developed by Konoki (1956*a*) and later, independently, by Horn (1961*a*).

Overall cost considerations will determine the number of stages to be used, so in practice we examine 1, then 2, etc., stages until a minimum cost is obtained.

Criteria similar to Eqs. 61 and 62 have been developed for most of the contacting patterns to be considered. However, these criteria are rather more time-consuming to use than Eqs. 61 and 62. Hence a few trials keeping as far as possible from regions of low rates will closely approach optimal operations, and this is the procedure recommended.

Let us next consider the two other cases of Fig. 19. For *irreversible exothermic reactions* the criterion for optimal operations has also been presented by Konoki (1956*b*). For *endothermic reactions* the optimal criterion has yet to be developed.

In all these cases a trial-and-error search keeping far from the regions of low rates is recommended.

Staged Packed Beds with Recycle. Here we have a flexible system which can approach mixed flow and as such is able to avoid regions of low rates. Figure 21 illustrates two-stage operations with a recycle ratio $\mathbf{R} = 1$, and a feed temperature T_f. Extension to three or more stages follows directly.

Konoki (1961) presents the criterion for optimal operations; however, in preliminary design a few good guesses will suffice to closely approach optimal operations.

In recycle operations the heat exchangers can be located in a number of places without affecting what goes on in the reactor. Figure 21 illustrates one of these; other alternatives are shown in Fig. 22. The best location will depend on convenience for startup, and on which location gives a higher heat transfer coefficient (note that the exchanger arrangement of Fig. 22*a* has a higher through flow of fluid than the arrangement of Fig. 22*b*).

Staged Mixed Reactors. For very high recycle the staged recycle reactors approach mixed flow. As shown in Fig. 23, in this case the reactors should operate on the line of optimum temperature progression, the best distribution of catalyst among the stages being found by the maximization of rectangles (see Fig. 6.11 and Problem 6.16). In effect we need to choose the distribution of catalyst so as to maximize area KLMN which then minimizes the shaded area in Fig. 23.

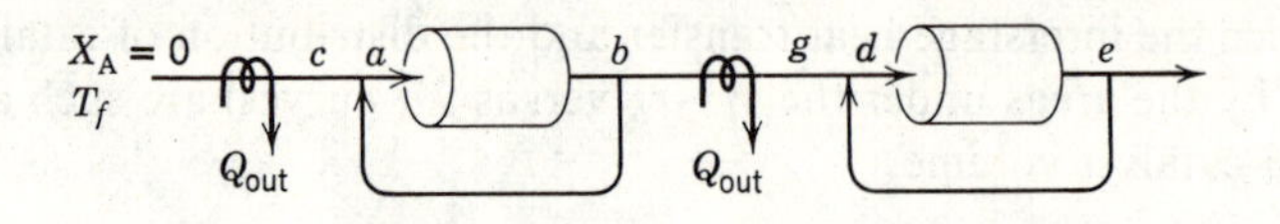

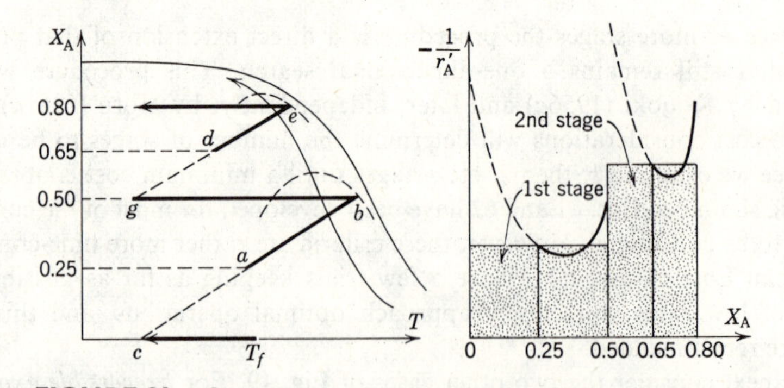

FIGURE 21. Optimum two-stage packed bed reactor with recycle. The conversions shown represent a recycle ratio $\mathbf{R} = 1$ in both stages.

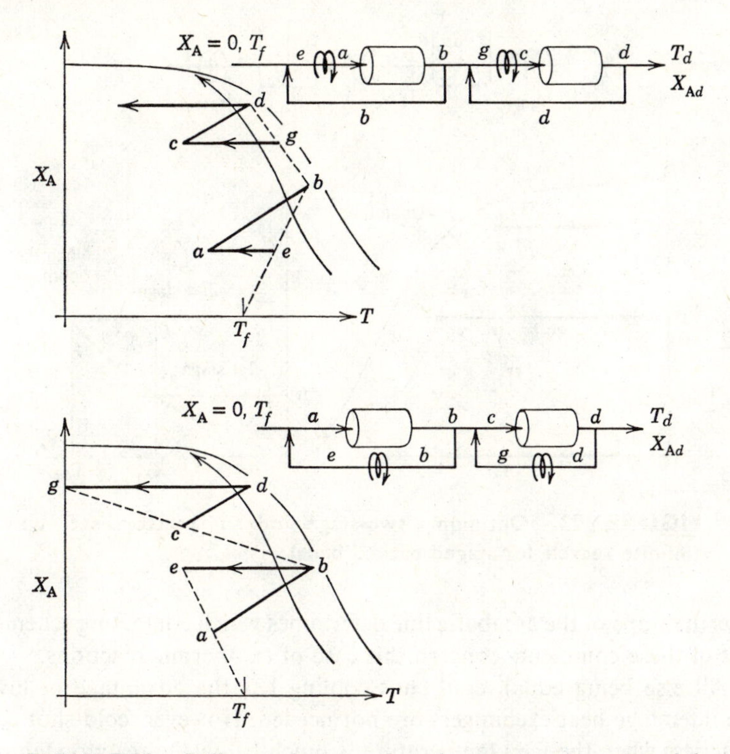

FIGURE 22. Different location for the heat exchangers while keeping the same reactor conditions as in Fig. 21.

Cold Shot Cooling. One way of eliminating the interstage heat exchangers is by properly adding cold feed directly into the second and succeeding stages of the reactor. The procedure is shown in Fig. 24. The criterion for optimal operations of such an arrangement is given by Konoki (1960), and in somewhat different form, by Horn (1961*b*). They found that the extent of interstage cooling is still given by Eq. 62, and this is shown in Fig. 24.

With cold shot cooling the calculation of reactor volumes by the $1/-r'_A$ versus X_A curve becomes more complicated because different amounts of feed are involved in each stage. We can also cold shot cool with inert fluid. This will affect both the $1/-r'_A$ versus X_A and T versus X_A curves.

Choice of Contacting System. With so many contacting alternatives let us suggest when one or other is favored.

1. For endothermic reactions the rate always decreases with conversion; hence we should always use plug flow with no recycle (see Chapter 8). For exothermic

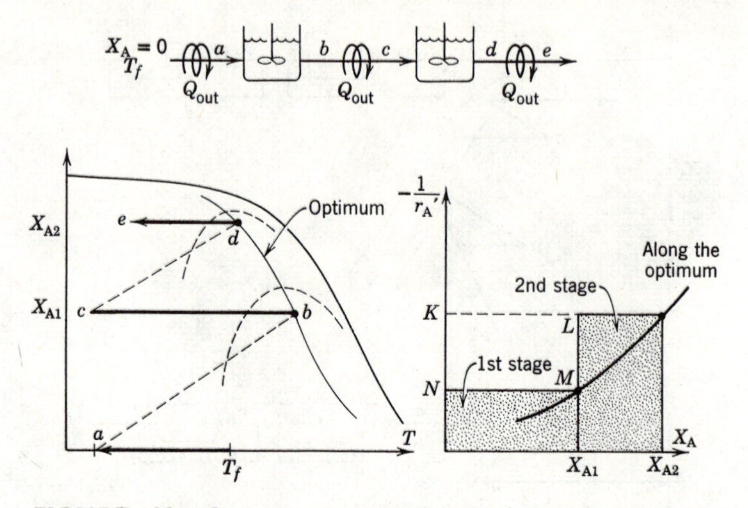

FIGURE 23. Optimum two-stage mixed reactor set up (infinite recycle for staged packed beds).

reactions the slope of the adiabatic line determines which contacting scheme is best. The rest of these comments concern this case of exothermic reactions.

2. All else being equal, cold shot cooling has the advantage of lower cost because interstage heat exchangers are not needed. However, cold shot cooling is only practical when the feed temperature is much below the reaction temperature, and, in addition, when the temperature does not change much during reaction. These conditions can be summarized as follows:

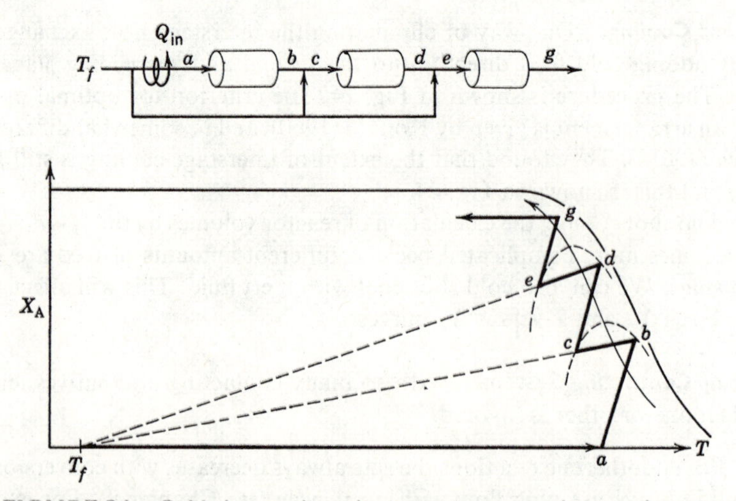

FIGURE 24. Cold shot cooling eliminates interstage heat exchangers.

Cold shot cooling is practical when

$$T_{\text{reaction}} - T_f > \frac{-\Delta \mathbf{H}_r}{\mathbf{C}_p}$$

Two situations, one when cold shot cooling is practical, the other when it is not, are shown in Fig. 25.

3. For exothermic reactions if the slope of the adiabatic line is low (large temperature rise during reaction) it is advantageous to avoid the low temperature regime where the rate is very low. Thus use high recycle approaching mixed flow. On the other hand, if the slope is high (small temperature rise during reaction) the rate decreases with conversion and plug flow is to be used. Typically, for pure gaseous reactant the slope of the adiabatic is small; for a dilute gas or for a liquid it is large. As an example, consider a reactant with $\mathbf{C}_p = 10$ cal/mol·°K and $\Delta \mathbf{H}_r = -30{,}000$ cal/mol. Then

For a pure reactant gas stream:

$$\text{slope} = \frac{\mathbf{C}_p}{-\Delta \mathbf{H}_r} = \frac{10}{30{,}000} = \frac{1}{3000}$$

For a dilute 1% reactant gas stream:

$$\text{slope} = \frac{\mathbf{C}_p}{-\Delta \mathbf{H}_r} = \frac{1000}{30{,}000} = \frac{1}{30}$$

For a one-molar liquid solution:

$$\text{slope} = \frac{\mathbf{C}_p}{-\Delta \mathbf{H}_r} = \frac{1000}{30{,}000} = \frac{1}{30}$$

The adiabatic lines for these cases are sketched in Fig. 26, and illustrate this point.

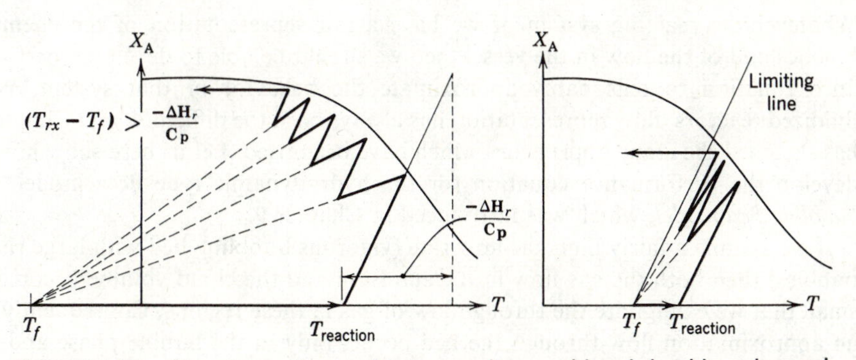

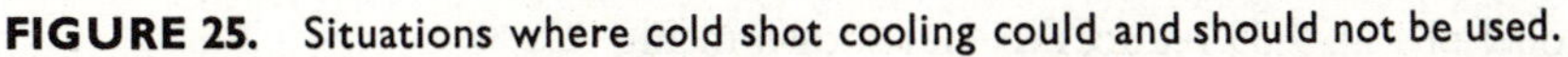

FIGURE 25. Situations where cold shot cooling could and should not be used.

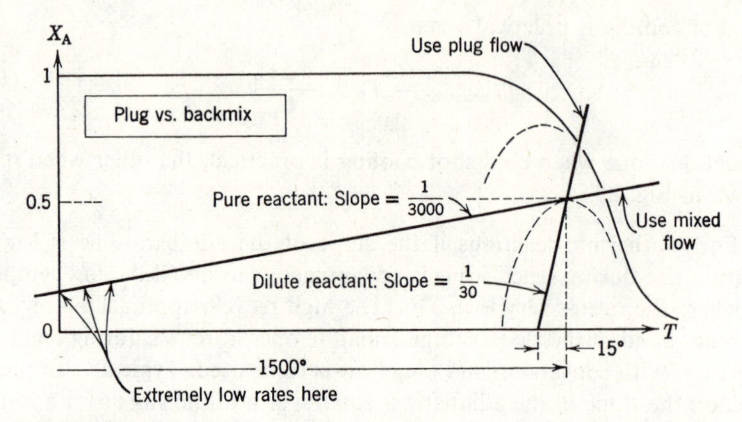

FIGURE 26. Sketch showing why plug flow is used for steep adiabatic lines, and mixed flow (packed beds with large recycle) for lines with small slope.

4. For exothermic reactions in staged reactors the above discussion can be summarized as follows:

For pure gas use high recycle approaching mixed flow.
For dilute gas (or a liquid) requiring no large preheating of feed use plug flow.
For a dilute gas (or a solution) requiring large preheating to bring the stream up to reaction temperature use cold shot operations.

Fluidized Bed Reactor

Whatever the reacting system, if we have a fair representation of the chemical kinetics and of the flow in the vessel then we should be able to develop a performance equation to reasonably approximate the behavior of that system. With fluidized reactors, flow representation has always been the difficulty, and Chapter 9 has sketched the many approaches which have been tried. Let us here show how to develop the performance equation for the hydrodynamic type flow model, the *bubbling bed model*, which was introduced in Chapter 9.

If we assume a fairly high gas flow rate (vigorous bubbling bed with large rising bubbles) then both the gas flow in the emulsion and the cloud volume become so small that we can ignore the throughflow of gas in these regions. Consequently, as an approximation flow through the bed occurs only in the bubble phase and the model of Fig. 9.39 reduces to that shown in Fig. 27.

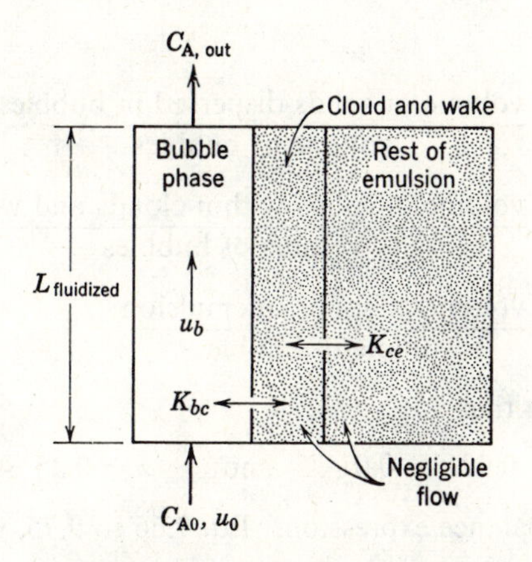

FIGURE 27. Model of the bubbling gas fluidized bed used for reaction applications.

Next, if the reaction is first-order catalytic with $\varepsilon_A = 0$, or

$$-r'''_A = -\frac{1}{V_{solids}} \frac{dN_A}{dt} = kC_A \tag{63}$$

we then have for the reaction and disappearance of A in rising bubble gas (see Fig. 27):

$$\begin{pmatrix}\text{disappearance}\\ \text{from bubble gas}\end{pmatrix} = \begin{pmatrix}\text{reaction}\\ \text{in bubble}\end{pmatrix} + \begin{pmatrix}\text{transfer to}\\ \text{cloud and wake}\end{pmatrix} \tag{64a}$$

$$\begin{pmatrix}\text{transfer to}\\ \text{cloud and wake}\end{pmatrix} = \begin{pmatrix}\text{reaction in}\\ \text{cloud and wake}\end{pmatrix} + \begin{pmatrix}\text{transfer to}\\ \text{emulsion}\end{pmatrix} \tag{64b}$$

$$\begin{pmatrix}\text{transfer to}\\ \text{emulsion}\end{pmatrix} = \begin{pmatrix}\text{reaction}\\ \text{in emulsion}\end{pmatrix} \tag{64c}$$

In symbols this expression becomes

$$-r_{A,b} = -\frac{1}{V_{bubble}} \frac{dN_A}{dt} = \gamma_b k C_{Ab} + K_{bc}(C_{Ab} - C_{Ac}) \tag{65a}$$

$$K_{bc}(C_{Ab} - C_{Ac}) = \gamma_c k C_{Ac} + K_{ce}(C_{Ac} - C_{Ae}) \tag{65b}$$

$$K_{ce}(C_{Ac} - C_{Ae}) = \gamma_e k C_{Ae} \tag{65c}$$

where by definition

$$\gamma_b = \frac{\text{volume of solids dispersed in bubbles}}{\text{volume of bubbles}} \tag{66a}$$

$$\gamma_c = \frac{\text{volume of solids within clouds and wakes}}{\text{volume of bubbles}} \tag{66b}$$

$$\gamma_e = \frac{\text{volume of solids in emulsion}}{\text{volume of bubbles}} \tag{66c}$$

Experiment has shown that

$$\gamma_b = 0.001 \sim 0.01 \qquad \text{and} \qquad \alpha = 0.25 \sim 1.0 \tag{67}$$

Also by the material balance expressions, Eq. 9.66 to 9.75, we can show that

$$\gamma_c = (1 - \epsilon_{mf})\left[\frac{3u_{mf}/\epsilon_{mf}}{u_{br} - u_{mf}/\epsilon_{mf}} + \alpha\right] \tag{68}$$

and

$$\gamma_e = \frac{(1 - \epsilon_{mf})(1 - \delta)}{\delta} - (\gamma_c + \gamma_b) \tag{69}$$

On eliminating all intermediate concentration in Eq. 65 we find that

$$-r_{A,b} = \left[\gamma_b k + \cfrac{1}{\cfrac{1}{K_{bc}} + \cfrac{1}{\gamma_c k + \cfrac{1}{\cfrac{1}{K_{ce}} + \cfrac{1}{\gamma_e k}}}}\right] C_A \tag{70}$$

Integrating for plug flow of gas through the bed yields the desired performance expression, or

---

$$\ln \frac{C_{A0}}{C_A} = \left[\gamma_b k + \cfrac{1}{\cfrac{1}{K_{bc}} + \cfrac{1}{\gamma_c k + \cfrac{1}{\cfrac{1}{K_{ce}} + \cfrac{1}{\gamma_e k}}}}\right] \frac{L_{\text{fluidized}}}{u_b} \tag{71}$$

---

where approximately

$$\frac{L_{\text{fluidized}}}{u_b} = \left(\frac{1 - \epsilon_{\text{packed}}}{1 - \epsilon_{mf}}\right) \frac{L_{\text{packed}}}{u_{br}} \tag{72}$$

and where the five terms in brackets represent the complex series-parallel resistances to mass transfer and reaction or

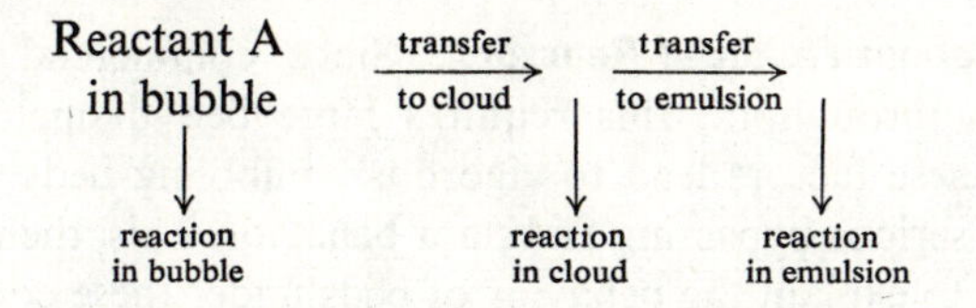

For very fast reaction (large k value) very little A gets as far as the emulsion and the first two terms in the bracket dominate. For slow reaction the latter terms in the brackets become increasingly important.

Since the bubble size is the one quantity which governs all the rate quantities with the exception of k, we can plot the performance of a fluidized bed as a function of d_b, as shown in Fig. 28. Note that large d_b gives poor performance because of extensive bypassing of bubble gas, and that the performance of the bed can drop considerably below mixed flow.

For multiple reactions the effect of this flow is much more serious still. Thus for reactions in series the lowering in amount of intermediate formed can be and usually is quite drastic.

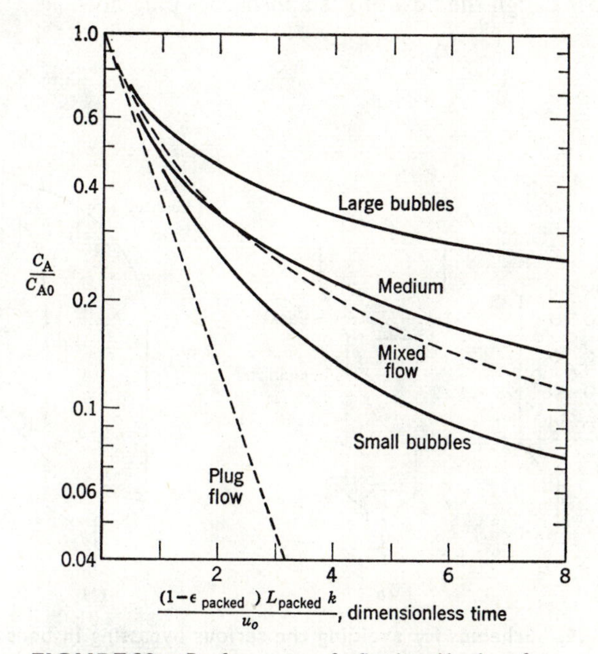

FIGURE 28. Performance of a fluidized bed as a function of bubble size, as determined by Eq. 71.

Examples illustrating these effects, and additional discussions and problems on this model, are found in Kunii and Levenspiel (1969).

Final Comments about Fluidized Reactors. Since commercial scale operations involve high gas throughput, this requires large bed diameters and high gas velocities. Both these factors lead to vigorously bubbling beds with large bubble size with all their serious bypassing and poor behavior. This, then, is the reason for the intense effort to explain the behavior of beds under these conditions.

From the design standpoint, however, it is preferable to avoid such behavior rather than explain and operate under these conditions. To this end the following ideas are used (see Fig. 29).

1. Insert internals into the bed to hinder bubble growth and cut down bubble size.
2. Use a fluidized-packed bed where the gas first passes through the fluidized section, then the packed section. Schemes such as these are attractive where very high gas conversions are required.
3. Use a high enough gas velocity to cause appreciable carryover (and return) of solids. Under these conditions the bed behaves as a lean emulsion with no gas bubbling and no serious bypassing. Much higher gas velocities can be used here, and with proper design the flow of gas approaches plug flow.

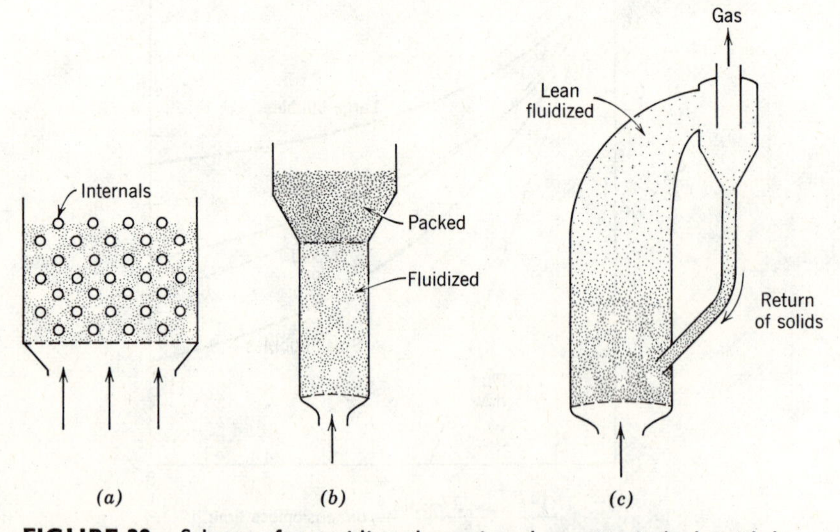

FIGURE 29. Schemes for avoiding the serious bypassing in beds with large bubbles: (*a*) bubble size limited by space between internals; (*b*) packed bed allows high conversion; (*c*) lean fluidization throughout.

EXAMPLE 6. Conversion in a fluidized bed

In a laboratory packed bed reactor ($L_m = 10$ cm and $u_o = 2$ cm/sec) conversion is 97% for the first-order reaction A → R.

(*a*) What would be the conversion in a larger fluidized bed pilot plant ($L_m = 100$ cm and $u_o = 20$ cm/sec) in which the estimated bubble size is 8 cm?

(*b*) What would be the conversion in a packed bed under these conditions?

Data. From experiment:

$$u_{mf} = 3.2 \text{ cm/sec}$$
$$\epsilon_{mf} = \epsilon_m = 0.5$$

From the literature:

$$\mathscr{D} = \mathscr{D}_e = 0.204 \text{ cm}^2\text{/sec}$$
$$\alpha = 0.34$$

Let subscript m refer to the fixed bed or settled bed condition.

SOLUTION

Assuming plug flow in the fixed bed laboratory reactor the performance equation becomes

$$\frac{V_s}{F_{A0}} = \int \frac{dX_A}{-r_A'''} = \int \frac{dX_A}{kC_{A0}(1 - X_A)}$$

Integrating gives

$$-\ln(1 - X_A) = \frac{kC_{A0}V_s}{F_{A0}} = \frac{k(1 - \epsilon_m)L_m}{u_o} \tag{i}$$

from which the rate constant for the reaction is found to be

$$k = \frac{u_o}{L_m(1 - \epsilon_m)} \ln \frac{C_{A0}}{C_A} = \frac{2 \text{ cm/sec}}{(10 \text{ cm})(1 - 0.5)} \ln \frac{100}{3} = 1.4 \text{ sec}^{-1}$$

(*a*) *Fluidized bed operations.* The performance equation for fluidized bed operations is given by Eqs. 71 and 72. Evaluating the terms in this expression we find from Eq. 9.68:

$$u_{br} = 0.711(980 \times 8)^{1/2} = 63 \text{ cm/sec}$$

from Eq. 9.69:

$$u_b = 20 - 3.2 + 63 = 79.8 \text{ cm/sec}$$

from Eq. 9.70:

$$\delta = \frac{20 - 3.2}{79.8} = 0.2105$$

from Eq. 9.76:

$$K_{bc} = 4.5\left(\frac{3.2}{8}\right) + 5.85\left(\frac{0.204^{1/2} \times 980^{1/4}}{8^{5/4}}\right) = 2.892 \text{ sec}^{-1}$$

from Eq. 9.77:

$$K_{ce} = 6.78\left(\frac{0.5 \times 0.204 \times 79.8}{8^3}\right)^{1/2} = 0.855 \text{ sec}^{-1}$$

from Eq. 68:

$$\gamma_c = (1 - 0.5)\left[\frac{3(3.2)/0.5}{63 - 3.2/0.5} + 0.34\right] = 0.34$$

from Eq. 69 (note that since $\gamma_b \leqslant 0.01$ it can be ignored)

$$\gamma_e = \frac{(1 - 0.5)(1 - 0.2105)}{0.2105} - (0.34 + 0) = 1.53$$

Replacing all these values into Eqs. 71 and 72 gives (again note that γ_b can be ignored).

$$\ln\frac{C_{A0}}{C_A} = \left[\sim 0 + \cfrac{1}{\cfrac{1}{2.892} + \cfrac{1}{(0.34)(1.4) + \cfrac{1}{\cfrac{1}{0.855} + \cfrac{1}{(1.53)(1.4)}}}}\right]\frac{100}{63}$$

$$= (0.79)\frac{100}{63} = 1.253$$

or

$$\frac{C_A}{C_{A0}} = e^{-1.253} = 0.285$$

hence the conversion

$$X_{A,\text{fluidized}} = 71.5\%$$

(*b*) *Packed bed operation.* From Eq. (i)

$$\ln\frac{C_{A0}}{C_A} = \frac{(1.4 \text{ sec}^{-1})(1 - 0.5)(100 \text{ cm})}{20 \text{ cm/sec}} = 3.5$$

hence

$$\frac{C_A}{C_{A0}} = e^{-3.5} = 0.03$$

or

$$X_{A,\text{packed}} = 97\%$$

REFERENCES

Aris, R., *Chem. Eng. Sci.*, **6**, 262 (1957).
Bischoff, K. B., *Chem. Eng. Sci.*, **22**, 525 (1967).
Boudart, M., *A.I.Ch.E. Journal*, **2**, 62 (1956).
Butt, J. B., Bliss, H., and Walker, C. A., *A.I.Ch.E. Journal*, **8**, 42 (1962).
Carberry, J. J., *A.I.Ch.E. Journal*, **7**, 350 (1961).
———, *A.I.Ch.E. Journal*, **8**, 557 (1962*a*).
———, *Chem. Eng. Sci.*, **17**, 675 (1962*b*).
———, *Ind. Eng. Chem.*, **56**, 39 (Nov. 1964).
———, *Catalysis Reviews*, **3**, 61 (1969).
Chou, C. H., *Ind. Eng. Chem.*, **50**, 799 (1958).
Corrigan, T. E., *Chem. Eng.*, **61**, 236 (Nov. 1954); **61**, 198 (Dec. 1954); **62**, 199 (Jan. 1955); **62**, 195 (Feb. 1955).
———, *Chem. Eng.*, **62**, 203 (May 1955); **62**, 227 (July 1955).
Froment, G. F., First International Symposium on Chemical Reaction Engineering, Washington D.C., June 1970.
Hamilton, C. J., Fryer, C., and Potter, O. E., Proceedings Chemeca '70 Conference, Butterworths, Sydney, 1970.
Horn, F., *Z. Elektrochemie*, **65**, 295 (1961*a*).
———, *Chem. Eng. Sci.*, **14**, 20 (1961*b*).
Hougen, O. A., and Watson, K. M., *Chemical Process Principles*, Part III, John Wiley & Sons, New York, 1947.
Hutchings, J., and Carberry, J. J., *A.I.Ch.E. Journal*, **12**, 20 (1966).
Konoki, K. K., *Chem. Eng.* (Japan), **21**, 408 (1956*a*).
———, *Chem. Eng.* (Japan), **21**, 780 (1956*b*).
———, *Chem. Eng.* (Japan), **24**, 569 (1960).
———, *Chem. Eng.* (Japan), **25**, 31 (1961).
Kunii, D., and Levenspiel, O., *Fluidization Engineering*, Chaps. 8 and 14, John Wiley & Sons, 1970.
Livbjerg, H., and Villadsen, J., *Chem. Eng. Sci.*, **26**, 1495 (1971).
McGreavy, C., and Cresswell, D. L., *Can. J. Ch.E.*, **47**, 583 (1969*a*).
——and——, *Chem. Eng. Sci.*, **24**, 608 (1969*b*).
McGreavy, C., and Thornton, J. M., *Can. J. Ch.E.*, **48**, 187 (1970*a*).
———, *Chem. Eng. Sci.*, **25**, 303 (1970*b*).
Perkins, T. K., and Rase, H. F., *A.I.Ch.E. Journal*, **4**, 351 (1958).
Prater, C. D., *Chem. Eng. Sci.*, **8**, 284 (1958).
———, and Lago, R. M., *Advances in Catalysis*, **8**, 293 (1956).
Rowe, P. N., and Partridge, B. A., *Trans. Inst. Chem. Engrs.*, **43** T157 (1965).
Tartarelli, R., *Chim. Ind. (Milan)*, **50**, 556 (1968).
Thiele, E. W., *Ind. Eng. Chem.*, **31**, 916 (1939).

Walas, S., *Reaction Kinetics for Chemical Engineers*, McGraw-Hill, New York, 1959.
Weisz, P. B., *Chem. Eng. Prog. Symp. Series*, No. 25, **55**, 29 (1959).
———, and Hicks, J. S., *Chem. Eng. Sci.*, **17**, 265 (1962).
———, and Prater, C. D., *Advances in Catalysis*, **6**, 143 (1954).
Weller, S., *A.I.Ch.E. Journal*, **2**, 59 (1956).
Wheeler, A., *Advances in Catalysis*, **3**, 250 (1951).
Yang, K. H., and Hougen, O. A., *Chem. Eng. Progr.*, **46**, 146 (1950).
Smith J.M. "Chemical Engineering kinetics", McGraw-Hill 2nd ed.

PROBLEMS

These are loosely grouped as follows:

Problems 1–10:	Direct application of performance equations. These problems should be tried first.
Problems 11–20:	Interpreting experiment to find which factors influence the rate.
Problems 21–30:	Harder, miscellaneous.
Problems 31–39:	Staged adiabatic packed bed reactors.
Problems 40–44:	Fluidized bed reactors.

1. While being shown around Lumphead Laboratories, you stop to view a reactor used to obtain kinetic data. It consists of a 5 cm ID glass column packed with 30 cm active catalyst. Is this a differential or integral reactor?

2. A solid catalyzed first-order reaction, $\varepsilon = 0$, takes place with 50% conversion in a basket type mixed reactor. What is the conversion if the reactor size is trebled and all else—temperature, amount of catalyst, feed composition, and flow rate—remains unchanged.

3. The following kinetic data on the reaction $A \rightarrow R$ are obtained in an experimental packed bed reactor using various amounts of catalyst and a fixed feed rate $F_{A0} = 10$ kg-mol/hr.

W, kg cat	1	2	3	4	5	6	7
X_A	0.12	0.20	0.27	0.33	0.37	0.41	0.44

(*a*) Find the reaction rate at 40% conversion.
(*b*) In designing a large packed bed reactor with feed rate $F_{A0} = 400$ kmol/hr how much catalyst would be needed for 40% conversion.
(*c*) How much catalyst would be needed in part (*b*) if the reactor employed a very large recycle of product stream.

4. The second order reaction $A \rightarrow R$ is studied in a recycle reactor with very large recycle ratio, and the following data are recorded:

Void volume of reactor: 1 liter
Weight of catalyst used: 3 gm
Feed to the reactor: $C_{A0} = 2$ mol/liter
$v_0 = 1$ liter/hr
Exit stream conditions: $C_{A.out} = 0.5$ mol/liter

(*a*) Find the rate constant for this reaction (give units).
(*b*) How much catalyst is needed in a packed bed reactor for 80% conversion of 1000 liter/hr of feed of concentration $C_{A0} = 1$ mol/liter.
(*c*) Repeat part (*b*) if the reactor is packed with 1 part catalyst to 4 parts inert solid. This addition of inerts helps maintain isothermal conditions and eliminate hot spots.

Note: Assume isothermal conditions throughout.

5. Determine the order of reaction and the weight of catalyst needed for 35% conversion of A to R for a feed of 2000 mol/hr of pure A at 117°C and 3.2 atm. For this reaction the stoichiometry is $A \rightarrow R$ and the kinetic data are given in Example 2; also assume plug flow in the packed bed reactor.
Note: Compare answers with Examples 2 and 4. The difference in values found is due to ignoring the expansion, ε_A, in treating this system.

6. Repeat Problem 5 using the data of Example 3 instead of Example 2. Again compare with the results of Examples 3 and 4.

7. Kinetic experiments on the solid catalyzed reaction $A \rightarrow 3R$ are conducted at 8 atm and 700°C in a basket type mixed reactor 960 cm^3 in volume and containing 1 gm of catalyst of diameter $d_p = 3$ mm. Feed consisting of pure A is introduced at various rates into the reactor and the partial pressure of A in the exit stream is measured for each feed rate. The results are as follows:

Feed rate, liters/hr	100	22	4	1	0.6
$p_{A,out}/p_{A,in}$	0.8	0.5	0.2	0.1	0.05

Find a rate equation to represent the rate of reaction on catalyst of this size.

8. At 700°C the rate of decomposition, $A \rightarrow 3R$, on a specific catalyst of given size is found to be

$$-r'_A = -\frac{1}{W}\frac{dN_A}{dt} = \left(10\,\frac{\text{liter}}{\text{hr}\cdot\text{gm cat}}\right)C_A$$

(See previous problem for the data leading to this equation.) A pilot plant is to be built. This is to be a tubular packed bed 2 cm ID using 25% of these active catalyst pellets evenly mixed with 75% inert pellets to insure isothermal operations. For 400 mol/hr feed consisting of 50% A–50% inert gas at 8 atm and 700°C what must be the length of reactor so that $p_{A,out}/p_{A,in} = 0.111$.

Data: Catalyst and inert pellets are porous, of diameter $d_p = 3$ mm, particle density $\rho_s = 2$ gm/cm³. Bulk voidage of packed bed = 50%.

9. The solid-catalyzed decomposition of gaseous A proceeds as follows:

$$A \rightarrow R \qquad -r_A = kC_A^2$$

A tubular pilot plant reactor packed with 2 liters of catalyst is fed 2 m³/hr of pure A at 300°C and 20 atm. Conversion of reactant is 65%.

In a larger plant it is desired to treat 100 m³/hr of feed gases at 40 atm and 300°C containing 60% A and 40% diluents to obtain 85% conversion of A. Find the internal volume of the reactor required.

10. Repeat the previous problem with the modified stoichiometry

$$A \rightarrow R + S$$

11. What is the most reasonable interpretation, in terms of controlling resistances, of the kinetic data of Table P11 obtained in a basket type mixed reactor if we know that the catalyst is porous. Assume isothermal behavior.

TABLE P11

Pellet Diameter	Leaving Concentration of Reactant	Spinning Rate of Baskets	Measured Reaction Rate
1	1	high	3
3	1	low	1
3	1	high	1

12. Repeat the previous problem with one change; the catalyst is known to be nonporous.

13. What can you tell about the controlling resistances and about catalyst porosity from the data of Table P13 obtained in a recycle type mixed reactor. In all runs the leaving stream has the same composition, and conditions are isothermal throughout.

TABLE P13

Quantity of Catalyst	Pellet Diameter	Flow Rate of Given Feed	Recycle Rate	Measured Reaction Rate
1	1	1	high	4
4	1	4	higher still	4
1	2	1	higher still	3
4	2	4	high	3

14. What can you tell about the controlling resistances and activation energy for the reaction from the data of Table P14 obtained in a basket type mixed reactor? Assume a uniform temperature throughout the pellet, equal to that of the surrounding gas.

TABLE P14

Flow Rate of Gas	$C_{A,in}$	$C_{A,out}$	Amount of Catalyst	Pellet Diameter	Spinning Rate of Baskets	Temperature, °C
5	5	1	10	1	low	344
2	4	1	6	2	high	344
8	2	1	4	2	high	372
9	3	1	9	2	low	372

15. In the absence of pore diffusion resistance a particular first order gas phase reaction proceeds as reported below.

$$-r'''_A = 10^{-6} \text{ mol/sec}\cdot\text{cm}^3 \text{ cat}$$

at

$$C_A = 10^{-5} \text{ mol/cm}^3 \text{ at 1 atm and } 400°\text{C}$$

What size catalyst pellets should we use ($\mathscr{D}_e = 10^{-3}$ cm²/sec) to insure that pore resistance effects do not intrude to slow the rate of reaction?

16. A liquid phase hydrogenation of an unsaturated hydrocarbon is catalyzed by porous pellets about 3 mm in diameter, the reaction is approximately first order with respect to the limiting reactant, hydrogen, and we hope to achieve useful reaction rates of the order

$$-r'''_{H_2} = 10^{-6} \text{ mol/sec}\cdot\text{cm}^3 \text{ cat}$$

From vapor-liquid equilibrium data the solubility of hydrogen in hydrocarbon fluid is estimated to be $C_{H_2} = 10^{-5}$ mol/cm³ at 1 atm. Assuming ideal behavior we also expect this solubility to be proportional to the pressure of hydrogen above the liquid phase. The effective diffusivity of reactants is estimated to be $\mathscr{D}_e = 2 \times 10^{-5}$ cm²/sec and approximately independent of concentration of the components, since this is a liquid system. From this information estimate the pressure at which we should operate to insure that pore diffusion effects are absent.

17. (*a*) What fraction of the overall resistance to mass transfer and reaction is provided by the gas film in a catalytic decomposition if the observed rate based on volume of catalyst bed is

$$k = 4/\text{time}$$

and if the gas film resistance as estimated by the dimensionless correlations of mass transfer is

$$k_g = 5 \text{ mm/time}$$

(*b*) If gas film resistance were negligible, what would be the observed rate of reaction expressed on a mass basis?

Data: Bed porosity = 0.33
Diameter of spherical catalyst pellets = 1 mm
Bulk density of bed = 2 gm/cm^3

18. It is suspected that the pilot plant of Problem 8 will be operating in the regime of strong pore diffusion, and that nonisothermal effects within pellets and across the gas film may also intrude. Check for these effects, and if present, tell how they might affect reactor performance.

Data: Effective mass diffusivity in pellet = 0.09 cm^2/hr
Effective thermal conductivity of pellet = 1 kcal/hr·m·°K
Pellet-fluid heat transfer coefficient = 100 kcal/hr·m^2·°K
Heat of reaction $(-\Delta H_r)$ = 20 kcal for the stoichiometry A → 3R.

19. Experiments in a basket type mixed reactor on the solid catalyzed decomposition A → R → S give the results of Table P19.

TABLE P19

Size of Porous Pellets	Temperature	W/F_{A0}	$C_{R,max}/C_{A0}$
6 mm	300°C	25	23%
12 mm	300°C	50	23%

Under the best possible reaction conditions (always at 300°C) what is the maximum concentration of R we may expect? How do you suggest that this be obtained?

20. Table P20 summarizes the results of three series of runs in a packed-bed reactor on the solid-catalyzed first-order decomposition A → R → S.

TABLE P20

Size of Porous Catalyst, mm	Temperature of Run, °C	W/F_{A0}	$C_{R,max}/C_{A0}$
3	300	27	0.50
6	300	54	0.50
6	320	21	0.50

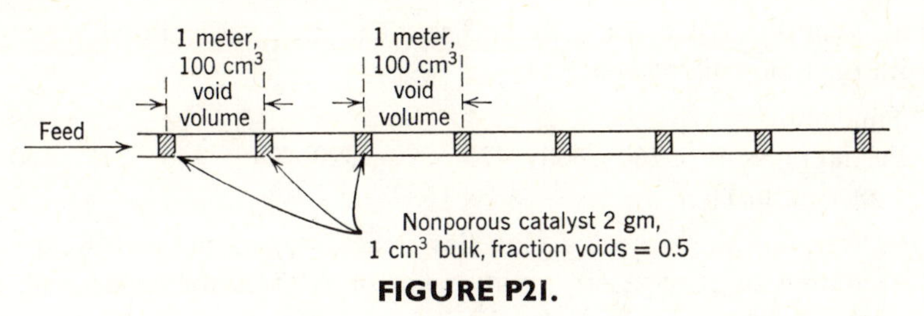

FIGURE P21.

Further experiments anywhere between 200 and 340°C are planned to search for the conditions under which production of R is maximized. What operating conditions (catalyst size and temperature) should we explore, and what fractional yield of R may we expect to find?

21. Because the catalytic reaction A → R is highly exothermic with rate highly temperature-dependent, a long tubular flow reactor immersed in a trough of water, as shown in Fig. P21, is used to obtain essentially isothermal kinetic data. Pure A at 0°C and 1 atm flows through this tube at 10 cm^3/sec, and the stream composition is analyzed at various locations.

Distance from feed input, meters	0	12	24	36	48	60	72	84	(∞)
Partial pressure of A, mm Hg	760	600	475	390	320	275	240	215	150

(*a*) Determine what size of plug flow reactor operating at 0°C and 1 atm would give 50% conversion of A to R for a feed rate of 100 kmol/hr of pure A.

(*b*) Suppose that this data had been obtained for a reaction whose stoichiometry is A → 2.5R. Find the volume of reactor required for the same feed rate and conversion as given in part (*a*).

22. A closed-loop experimental system as shown in Fig. P22 is used to study the kinetics of a catalytic reaction A → R. Pure A is introduced into the systems and

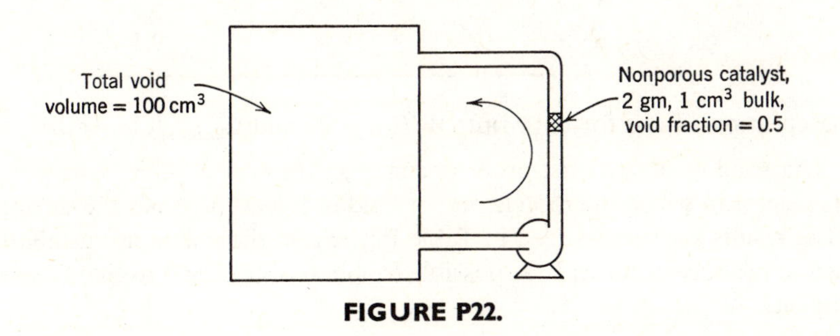

FIGURE P22.

is circulated at 0°C and 1 atm at 10 cm^3/sec. The stream is analyzed from time to time with the following results:

Time, min	0	2	4	6	8	10	12	14	(∞)
Partial pressure of A, mm Hg	760	600	475	390	320	275	240	215	150

(*a*) Determine the size of a plug flow reactor operating at 0°C and 1 atm required to effect a 50% conversion of A to R for a feed rate of 100 kmol A/hr.

(*b*) Repeat part *a* with the modification that an inert at a partial pressure of 1 atm is present in the closed loop so that the total pressure at the start is 2 atm.

23. From a feed containing reactant A at $C_{A0} = 1$ mol/liter, 100 mol/hr of R are to be produced in a catalyst-packed tubular reactor. The reaction in the packed bed can be taken as zero order, or

$$A \rightarrow R, \qquad -r'_A = 1 \frac{\text{mol}}{\text{hr} \cdot \text{kg cat}}$$

The cost of feed stream is \$1/mol A.
The cost of reactor and catalyst is \$1/hr · kg cat.
Unreacted A cannot be recovered.

(*a*) What is the optimum conversion and the cost of R at this conversion.

(*b*) Repeat part (*a*) if the packed bed reactor is operated with a large recycle stream.

24. (*a*) Table P24 records data for the solid-catalyzed reaction $A \rightarrow R \rightarrow S$ occuring isothermally in an experimental recycle reactor. Under what conditions (what controlling resistance, packed or fluidized-bed operations) will the concentration of intermediate be maximized? What is the expected value of $C_{R,max}/C_{A0}$?

TABLE P24

Size of Porous Catalyst, mm	W/F_{A0}	$C_{R,max}/C_{A0}$
3	27	0.5
6	54	0.5

(*b*) Repeat part (*a*) if for both runs $W/F_{A0} = 27$ and $C_{R,max}/C_{A0} = 0.17$.

25. The solid-catalyzed first-order decomposition $A \rightarrow R \rightarrow S$ is studied in a recycle reactor in which the recycle rate of fluid is at least 50 times the throughput rate. The results are summarized in Table P25. From these data an installation is planned to produce as much R as possible from a feed identical to that used in the experiments.

TABLE P25

Size of Porous Catalyst, mm	Temperature of Runs, °C	$\tau' = WC_{A0}/F_{A0}$	$C_{R,max}/C_{A0}$
3	300	30	0.17
6	300	60	0.17
6	320	30	0.17

(*a*) Choose between a packed-bed or fluidized-bed reactor.
(*b*) Choose between 3 mm and 6 mm catalyst.
(*c*) In the range between 280 and 320°C select an operating temperature.
(*d*) Determine the weight-time $\tau' = WC_{A0}/F_{A0}$ to be used.
(*e*) Predict the expected $C_{R,max}/C_{A0}$.

26. Diffusion is temperature-dependent. Consequently evaluate the activation energy to represent this process both at room temperature (about 300°K) and at high temperature (about 1200°K) knowing that
(*a*) $\mathscr{D} \propto T^{3/2}$ for ordinary molecular diffusion of gases
(*b*) $\mathscr{D} \propto T$ for ordinary diffusion in liquids
(*c*) $\mathscr{D} \propto T^{1/2}$ for Knudsen diffusion in gases.

27. *Homogeneous and catalytic rates.* The homogeneous decomposition $A \rightarrow R$ is studied in a stainless steel paddle type mixed flow reactor (volume of voids in the reactor, $V = 0.8$ liter, total surface in the reactor, $S = 800\ cm^2$) with the following results:

τ, sec	40	10	for $C_{A0} = 100$
$C_{A,out}$, millimol/liter	20	40	

We suspect that the stainless steel surface catalyzes the reaction. To verify this suspicion additional surface is introduced in the reactor ($S = 1500\ cm^2$, $V = 0.75$ liter) and more runs are made as follows:

τ, sec	26.7	7.5	for $C_{A0} = 100$
$C_{A,out}$	20	40	

Find the kinetics of this decomposition, and if homogeneous and catalytic reactions are occurring simultaneously give rate expressions (with units) for both processes.

28. *Active Site Mechanism.* To illustrate the method for developing rate equations from the active site theory, consider adsorption controlling for the catalytic reaction $A \rightarrow R$, and assume the following elementary mechanisms.

Step 1. A molecule of reactant A in the gas phase is captured by an active site ℓ to form the intermediate $(A \cdot \ell)^*$. Since adsorption controls an equilibrium develops between free A and adsorbed A, thus

$$A + \ell \underset{2}{\overset{1}{\rightleftarrows}} (A \cdot \ell)^*$$

Step 2. This intermediate occasionally decomposes to release product R and thereby free the active site for further use

$$(A \cdot \ell)^* \xrightarrow{3} R + \ell$$

Finally the total number of active sites available per unit surface of catalyst is ℓ_0. These sites can be either in the free state or may have A adsorbed on them, thus

$$\ell_0 = \ell + (A \cdot \ell)^*$$

Show that the rate equation obtained from this model is of the form

$$-r_A = \frac{k_1 C_A}{1 + k_2 C_A}$$

Note: In comparing with enzyme substrate kinetics (see Problem P2.19 or the discussion following Eq. 3.60) we observe a close parallel in the form of the theories and in the final rate expressions. The reason for this is because in both cases the catalytic agent (enzyme in one case, active sites in the other) is present in definite amount and is either in active or inactive form.

29. The catalytic gas-phase decomposition of A yields a variety of products which for the sake of simplicity can be designated as R (desired product) and S (undesired product). Under optimum conditions of maximum R yield, the stoichiometric relationship characterizing the overall reaction is $A \rightarrow 0.8R + 3.2S$. This occurs in the presence of cadmium-impregnated WW pumice catalyst (porosity = 0.375) at 459°C.

Experimental studies in a constant-volume bomb at 459°C using a 50% A, 50% inert mixture give the following results:

Time, sec	0	30	60	90	120	150
Total pressure, atm	4.00	5.17	6.43	7.60	8.79	9.97

600 mol R/hr are to be produced from a feed of pure A ($0.40/mol) in a packed tubular reactor. Any pressure from atmospheric up to 20 atm absolute may be used. Naturally, the cost of reactor and supporting equipment will depend not only on its size but also on the pressure selected. This cost on an hourly basis is

$$\$20.00 + (\$0.04/\text{lit})(\text{pressure in atm})^{0.6}, \qquad \pi > 1 \text{ atm}$$

and includes cost of catalyst replacement because of poisoning, etc.

(*a*) For optimum conditions assuming isothermal plug flow operations, find the operating pressure, the fractional conversion A, the size of reactor, and the unit cost of producing R.

(*b*) Feed consisting of 25% inerts instead of pure A can be purchased at \$0.32/mol A. How would a change to this new feed affect the operations?

(*c*) If the stoichiometric equation were $A \rightarrow 0.8R + 0.2S$ and the rate equation were that of part (*a*), in what way would this affect the answer to part (*a*)?

(*d*) If tracer experiments indicate that the flow pattern in the reactor can be approximated by assuming that one-sixth of the fluid bypasses the reactor completely, the rest passing through the catalyst in ordinary dispersed plug flow, how would this affect the answer to part (*a*)? Take the particle Reynolds number = 350, particle size = 6 mm, reactor diameter = 1 m.

30. *Nonideal Flow in a Packed Bed.* A gas decomposes according to the elementary reaction $2A \rightarrow R + S$ when in contact with a solid catalyst.

The kinetics of this reaction is investigated in a packed-bed reactor using 2.5 mm spherical beads in the following manner. A 4 cm layer of nonactive beads is laid down on a support. This is followed by a 1-cm layer of catalyst beads and by another 4-cm layer of nonactive beads. For a gas flow rate corresponding to a particle Reynolds number of 23, the reactant is 99% decomposed. Find the error in the calculated second-order rate constant which would result if plug flow is assumed. Assume isothermal conditions throughout.

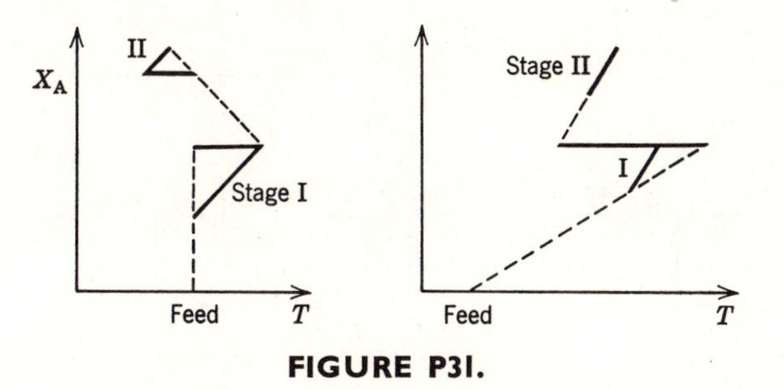

FIGURE P31.

31. Figures P31 represent different arrangements of two recycle reactors.

(*a*) Sketch flow sheets for these operations, and therein show the location of the heat exchangers, indicate whether they cool or heat, and estimate the recycle ratios for the streams.

(*b*) Is it possible to rearrange the location of heat interchange so as to avoid using external heat sources, or to eliminate either of the two heat exchangers? If so, show how; however whatever modification you suggest keep conditions within the reactors unchanged.

32. Figures P32 represent different arrangements of two reactors. Sketch the flow sheet for these operations, show the location of heat exchangers, and per 100 units of feed estimate how many units flow through the various streams.

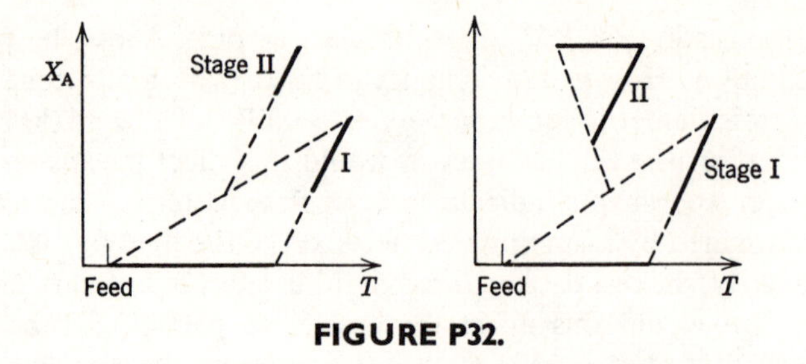

FIGURE P32.

33. For staged operations of an exothermic reaction the heat exchangers can be placed in various locations. For the arrangements of two packed beds with recycle shown in Fig. P33 sketch the corresponding X_A versus T chart locating all pertinent points.

Data: For these four cases take the same temperature for the feed, take the same temperature for the fluid leaving the first reactor, and take $X_{A1} = 0.6$, $X_{A2} = 0.9$, $\mathbf{R}_1 = 2$, $\mathbf{R}_2 = 1$.

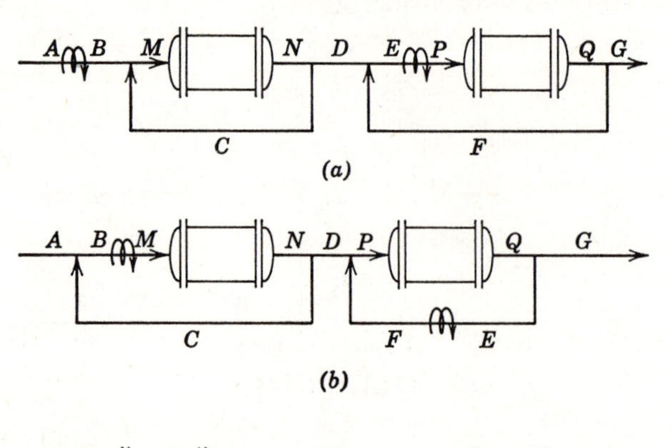

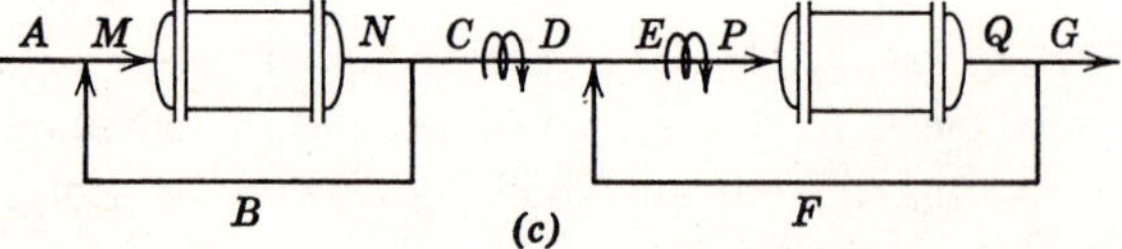

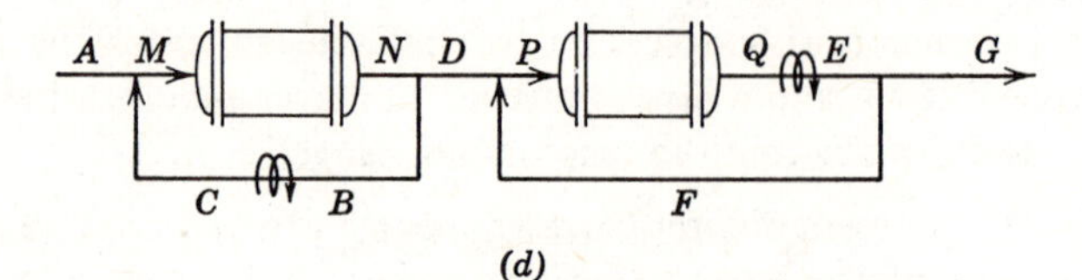

FIGURE P33.

Note for Problems 34–39. Examples 8.1 and 8.2 develop the rate-temperature-conversion chart for a specific homogeneous reaction. The result is shown in Fig. E.2 of Chapter 8, and inside the dust jacket of this book. Let us suppose that this chart represents the kinetics of a liquid phase catalytic reaction with $C_{A0} = 1$ mol/liter and rate given by $-r'_A$, mol/min·kg cat. The following 6 problems deal with this reaction.

34. For the above mentioned reaction and with a more concentrated feed, $C_{A0} = 10$ mol/liter at 25°C, and an allowable temperature range from 5°C to 95°C, present a reasonable design for treating $F_{A0} = 1000$ mol/min to 80% conversion in a two stage adiabatic packed bed system with appropriate heat exchange. Summarize your answer with a sketch showing the amount of catalyst to be used in each stage, the temperature of the streams, and the location and duty of the heat exchangers. Do not use any recycle of fluid.

35. Repeat the previous problem with one modification: the outlet conversion is to be 90%, not 80%.

36. For the reaction mentioned before Problem 34, a feed $C_{A0} = 10$ mol/liter at 25°C, and an allowable temperature range from 5°C to 95°C present a reasonable design for treating $F_{A0} = 1000$ mol/min to 80% conversion in a two-stage adiabatic packed bed reactor with very large recycle. Summarize your answer by a sketch showing the amount of catalyst to be used in each stage, the temperature of the streams, the location and duty of the heat exchangers.

37. Repeat Problem 36 with two modifications; the maximum allowable temperature is 50°C and any two-stage reactor system can be used.

38. Repeat Problem 36 with two modifications: any recycle ratio up to $\mathbf{R} = 5$ may be used, and the allowable temperature range is from 25°C and 95°C.

39. For the reaction mentioned before Problem 34 in a 3-stage packed bed reactor with very large recycle, appropriate heat exchange, and allowable temperature range from 5°C to 95°C, and 80% conversion overall,

(*a*) find the percentage distribution of catalyst among stages

(*b*) compare this requirement with that for the optimum temperature progression which uses the very minimum amount of catalyst.

40. How does the conversion in the fluidized bed of Example 6 compare with that for mixed flow of gas in the reactor?

41. A gaseous reactant A disappears by first order reaction, A → R, and in a vigorously bubbling fluidized bed conversion is 50%. If the amount of catalyst in this bed is doubled and all else remains unchanged *including the mean bubble size*, estimate the resultant conversion.

42. In the fluidized pilot plant reactor of Example 6 conversion is 71.5%. How much more catalyst is needed to raise the conversion to that obtainable in the packed bed, or 97%, assuming that the effective bubble size remains unchanged.

43. (*a*) In Example 6 what fraction of the conversion takes place within the bubble, and what fraction of the remaining resistance to reaction is due to bubble-cloud interchange.

(*b*) Repeat if the reaction rate constant is 100 times as large.

Data: Assume $\gamma_b = 0.01$.

44. To double the treatment rate of gas in the fluidized pilot plant reactor of Example 6 the gas velocity and the bed height are both doubled. Under these conditions the effective bubble size is estimated to be 17.5 cm. What conversion can we expect?

45. Early experiments by Rowe and Partridge (1965) in bubbling fluidized beds show that the wake to bubble ratio $\alpha = 0.25 \sim 0.67$; however, more recent experiments by Hamilton *et al.* (1970) show that $\alpha \cong 1$. This difference could be because different ways were used to measure the wake volume and because different solids were used.

Let us see if changing α has a serious effect on the predicted conversion of reactant gas in the fluidized bed. For this repeat Example 6 with $\alpha = 1.02$ instead of 0.34.

15

DEACTIVATING CATALYSTS

The previous chapter assumed that the effectiveness of catalysts in promoting reactions remains unchanged with time. Often this is not so, and as a rule the activity decreases as the catalyst is being used. Sometimes this drop is very rapid, in the order of seconds; sometimes it is so slow that regeneration or replacement is needed only after years of use. In any case with deactivating catalysts regeneration or replacement is necessary from time to time.

If deactivation is rapid and caused by a deposition and a physical blocking of the surface this process is often termed *fouling*. Removal of this solid is termed *regeneration*. Carbon deposition during catalytic cracking is a common example of fouling

$$C_{10}H_{22} \rightarrow C_5H_{12} + C_4H_{10} + C\downarrow_{\text{on catalyst}}$$

If the catalyst surface is slowly modified by chemisorption on the active sites by materials which are not easily removed then the process is frequently called *poisoning*. Restoration of activity, where possible, is called *reactivation*. If the adsorption is *reversible* then a change in operating conditions may be sufficient to reactivate the catalyst. If the adsorption is not reversible then we have *permanent* poisoning. This may require a chemical retreatment of the surface or a complete replacement of the spent catalyst.

Deactivation may also be *uniform* for all sites, or it may be *selective*, in which case the more active sites, those which supply most of the catalyst activity, are preferentially attacked and deactivated.

We will use the term *deactivation* for all types of catalyst decay, both fast and slow; and we will call any material which deposits on the surface to lower its activity a *poison*.

This chapter is a brief introduction to operations with deactivating catalysts. We will consider in turn:

The mechanism of catalyst decay,
The form of rate equation for catalyst decay,
How to develop a suitable rate equation from experiment,
How to discover the mechanism from experiment,
Some consequences in design.

Although this is basically a rather involved subject, still its very importance from the practical standpoint requires at least an introductory treatment.

MECHANISMS OF CATALYST DEACTIVATION

The observed deactivation of a porous catalyst pellet is dependent on a number of factors: the actual decay reactions, the presence or absence of pore diffusion for reacting species and poisons, the way poisons act on the surface, etc. We consider these in turn.

Decay Reactions. Broadly speaking, decay can occur in four ways. Firstly, the reactant may produce a side product which deposits on and deactivates the surface. This is called *parallel deactivation.* Secondly, the reaction product may decompose or react further to produce a material which then deposits on and deactivates the surface. This is called *series deactivation.* Thirdly, an impurity in the feed may deposit on and deactivate the surface. This is called *side-by-side deactivation.*

If we call P the material which deposits on and deactivates the surface, we can represent these reactions as follows:

Parallel deactivation:

$$A \rightarrow R + P\downarrow \qquad \text{or} \qquad A \begin{matrix} \nearrow R \\ \searrow P\downarrow \end{matrix} \tag{1}$$

Series deactivation:

$$A \rightarrow R \rightarrow P\downarrow \tag{2}$$

Side-by-side deactivation:

$$\left.\begin{matrix} A \rightarrow R \\ P \rightarrow P\downarrow \end{matrix}\right\} \tag{3}$$

The key difference in these three forms of decay reactions is that the deposition depends respectively on the concentration of reactant, product, and some other substance in the feed. Since the distribution of these substances will vary with position in the pellet, the location of deactivation will depend on which decay reaction is occurring.

A fourth process for catalyst decay involves the structural modification or sintering of the catalyst surface caused by exposure of the catalyst to extreme conditions. This type of decay is dependent on the time that the catalyst spends in the high temperature environment, and since it is unaffected by the materials in the gas stream we call it *independent deactivation*.

Pore Diffusion. For a pellet, pore diffusion may strongly influence the progress of catalyst decay. First consider parallel deactivation. From Chapter 14 we know that reactant may either be evenly distributed throughout the pellet ($mL < 1$ and $\mathscr{E} = 1$) or may be found close to the exterior surface ($mL > 1$ and $\mathscr{E} < 1$). Thus the poison will be deposited in a like manner; uniformly for no pore resistance, and at the exterior for strong pore resistance. In the extreme of very strong diffusional resistance a thin shell at the outside of the pellet becomes poisoned. This shell thickens with time and the front moves inward. We call this the *shell model* for poisoning.

On the other hand, consider series deactivation. In the regime of strong pore resistance the concentration of product R is higher within the pellet than at the exterior. Since R is the source of the poison, the latter deposits in higher concentration within the pellet interior. Hence we can have poisoning from the inside out for series deactivation.

Finally, consider side-by-side deactivation. Whatever the concentration of reactants and products may be, the rate at which the poison from the feed reacts with the surface determines where it deposits. For a small rate constant the poison penetrates the pellet uniformly and deactivates all elements of the catalyst surface in the same way. For a large rate constant poisoning occurs at the pellet exterior, as soon as the poison reaches the surface.

The above discussion shows that the progress of deactivation may occur in different ways depending on the type of decay reaction occurring and on the value of a pore diffusion factor. For parallel and series poisoning the Thiele modulus for the main reaction is the pertinent pore diffusion parameter, for side-by-side reactions the Thiele modulus for the deactivation is the prime parameter.

Nonisothermal effects within pellets may also cause variations in deactivation with location, especially when deactivation is caused by surface modifications due to high temperatures.

Figure 1 summarizes this discussion by showing the various ways that the surface can be deactivated.

Form of Surface Attack by Poison. Consider an element of active catalyst surface in a reactive gas environment. The active sites of this surface may be immobilized

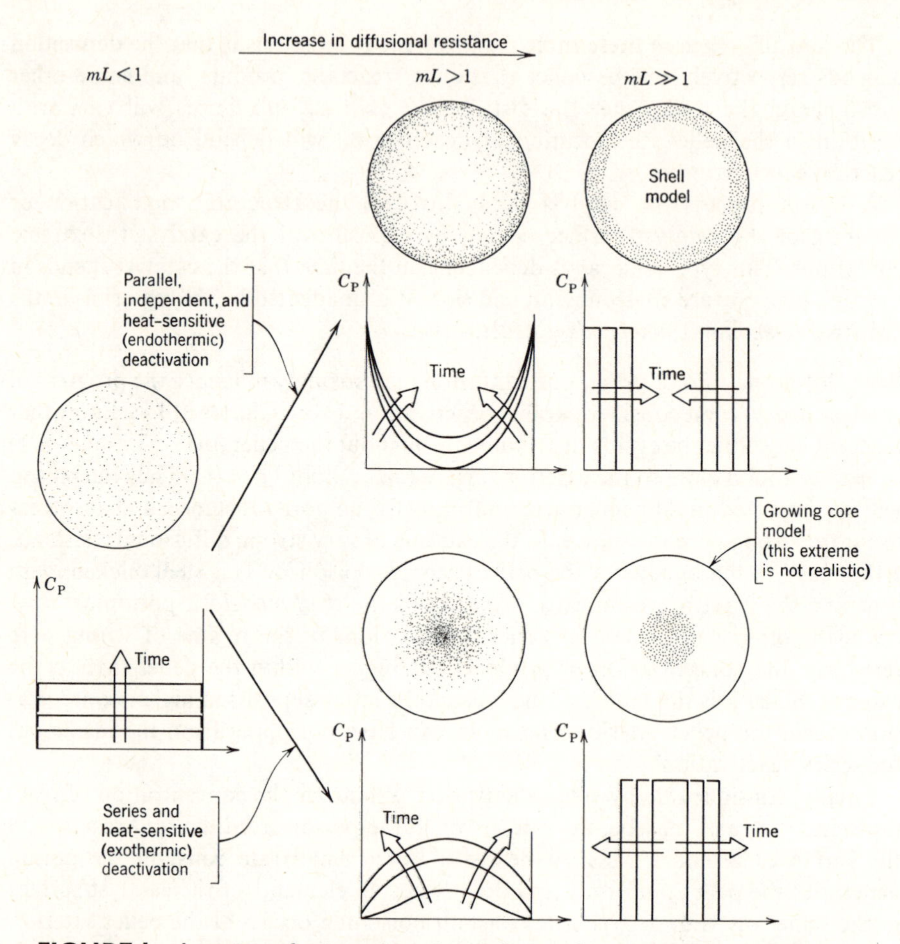

FIGURE 1. Location of poison deposit is influenced by the diffusional effect and by the type of decay reaction.

in distinctly different ways. On one hand, this may take place uniformly; in other words all sites, the very active and the slightly active, are attacked indiscriminately. On the other hand, it may be that the more active sites are preferentially attacked and immobilized. Let us call these cases *homogeneous site-attack* and *preferential site-attack*.

Homogeneous site-attack is representative of poisoning by physical deposition on the surface, such as fouling. On the other hand, preferential site-attack is likely to occur during chemisorption of small amounts of poisons.

In homogeneous site attack another factor may be at work. If the physical deposition produces a growing porous layer the activity may decrease gradually as reactant experiences increasing difficulty in diffusing through this thickening layer.

Additional Factors Influencing Decay. Numerous other factors may influence the observed change in activity of catalyst. These include pore mouth blocking by deposited solid, equilibrium or reversible poisoning where some activity always remains, the action of regeneration (this often leaves catalyst with an active exterior but inactive core).

Most important of all, the observed deactivation may result from a number of processes at work simultaneously; for example, the speedy immobilization of the most active sites by P_1 which will not attack the less active sites, and then the slower attack of the remainder of the sites by P_2.

Although the possible influence of all these factors should be examined in the real case, in this introductory treatment we will concentrate on the first two factors: the decay reaction and pore diffusion. There are enough lessons here to illustrate how to approach the more complete problem.

THE RATE EQUATION

The activity of a catalyst pellet at any time is defined as

$$\mathbf{a} = \frac{\text{rate at which the pellet converts reactant A}}{\text{rate of reaction of A with a fresh pellet}} = \frac{-r'_A}{-r'_{A0}} \tag{4}$$

and in terms of the fluid bathing the pellet the rate of reaction of A should be of the following form:

$$\begin{pmatrix}\text{reaction}\\ \text{rate}\end{pmatrix} = f_1\begin{pmatrix}\text{main stream}\\ \text{temperature}\end{pmatrix}\cdot f_3\begin{pmatrix}\text{main stream}\\ \text{concentration}\end{pmatrix}\cdot f_5\begin{pmatrix}\text{present activity of}\\ \text{the catalyst pellet}\end{pmatrix} \tag{5}$$

Similarly, the rate at which the catalyst pellet deactivates may be written as

$$\begin{pmatrix}\text{deactivation}\\ \text{rate}\end{pmatrix} = f_2\begin{pmatrix}\text{main stream}\\ \text{temperature}\end{pmatrix}\cdot f_4\begin{pmatrix}\text{main stream}\\ \text{concentration}\end{pmatrix}\cdot f_6\begin{pmatrix}\text{present state of}\\ \text{the catalyst pellet}\end{pmatrix} \tag{6}$$

In terms of nth-order kinetics, Arrhenius temperature dependency, and isothermal pellet, these word expressions become, for the main reaction:

$$-r'_A = k\cdot C_A{}^n\cdot \mathbf{a} = k_0 e^{-E/RT}\cdot C_A{}^n\cdot \mathbf{a} \tag{7}$$

and for deactivation which in general is dependent on the concentration of gas phase species i:

$$-\frac{d\mathbf{a}}{dt} = k_d \cdot C_i^{n'} \cdot \mathbf{a}^d = k_{d0} e^{-E_d/RT} \cdot C_i^{n'} \cdot \mathbf{a}^d \tag{8}$$

where d is called the *order of deactivation*, n' measures the concentration dependency and E_d is the activation energy or temperature dependency of the deactivation.

For porous pellets in the regime of negligible diffusion resistance k and n are the true rate constant and reaction order of the main reaction, respectively. However, in the regime of strong resistance to pore diffusion k and n become the observed quantities which account for these pore diffusion effects (see Equations 42 and 44 in Chapter 14).

For the deactivation we will see that the diffusional effects are incorporated into the order of deactivation. Thus, the order d is an important clue to the role of pore diffusion.

For different decay reactions we may expect different forms for the above equations. Thus

For parallel deactivation ($A \to R$; $A \to P\downarrow$)

$$\left.\begin{aligned} -r'_A &= kC_A{}^n\mathbf{a} \\ -\frac{d\mathbf{a}}{dt} &= k_d C_A{}^{n'}\mathbf{a}^d \end{aligned}\right\} \tag{9}$$

For series deactivation ($A \to R \to P\downarrow$)

$$\left.\begin{aligned} -r'_A &= kC_A{}^n\mathbf{a} \\ -\frac{d\mathbf{a}}{dt} &= k_d C_R{}^{n'}\mathbf{a}^d \end{aligned}\right\} \tag{10}$$

For side-by-side deactivation ($A \to R$; $P \to P\downarrow$)

$$\left.\begin{aligned} -r'_A &= kC_A{}^n\mathbf{a} \\ -\frac{d\mathbf{a}}{dt} &= k_d C_P{}^{n'}\mathbf{a}^d \end{aligned}\right\} \tag{11}$$

For independent deactivation (concentration independent)

$$\left.\begin{aligned} -r'_A &= kC_A{}^n\mathbf{a} \\ -\frac{d\mathbf{a}}{dt} &= k_d\mathbf{a}^d \end{aligned}\right\} \tag{12}$$

In certain reactions, such as isomerizations and cracking, deactivation may be caused both by reactant and product, or

$$\begin{matrix} A \to R \\ A \to P\downarrow \\ R \to P\downarrow \end{matrix} \quad \text{and} \quad -\frac{d\mathbf{a}}{dt} = k_d(C_A + C_R)^{n'}\mathbf{a}^d \tag{13}$$

Since $C_A + C_R$ remains constant for a specific feed this type of deactivation reduces to the simple to treat independent deactivation of Eq. 12.

Although the above nth-order expressions are quite simple they are general enough to embrace many of the decay equations used to date.

The Rate Equation from Experiment

Experimental devices for studying deactivating catalysts fall into two classes: those which use a batch of solids, and those which use a through flow of solids. Figure 2 shows some of these devices.

Because of the ease of experimentation the batch-solids devices are much preferred; however, they can only be used when deactivation is slow enough (in the order of minutes or longer) so that sufficient data on the changing fluid composition can be obtained before exhaustion of the catalyst. When deactivation is very rapid (in the order of seconds or less) then a flowing-solids system must be used. Cracking catalysts whose activity half-lives can be as short as 0.1 second fall into this class.

The method of searching for a rate equation is analogous to that of homogeneous reactions: start with the simplest kinetic form and see if it fits the data. If it doesn't, try another kinetic form and so on. The main complication here is that we have an extra factor, the activity, to contend with. Nevertheless, the strategy is the same; always start by trying to fit the simplest rate expression.

In the following sections we treat the batch-solids devices in some detail, according to Levenspiel (1972), and then briefly consider the flowing-solids system.

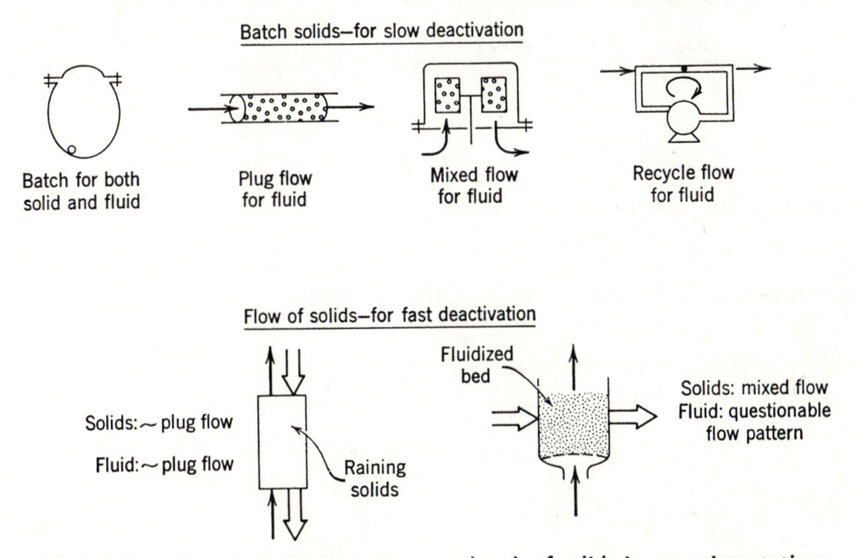

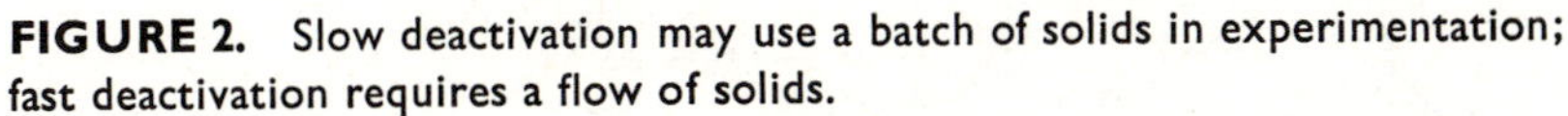

FIGURE 2. Slow deactivation may use a batch of solids in experimentation; fast deactivation requires a flow of solids.

Now the type of batch-solids reactor which we find convenient to use depends on whether the deactivation expression $d\mathbf{a}/dt$ is concentration independent or not. When it is concentration independent any type of batch-solids system may be used and can be analyzed simply, but when it is concentration dependent then unless one particular type of reactor is used, the analysis of the experimental results becomes horrendously awkward and difficult.

We treat these two classes of rate equations in turn.

Batch-Solids: Determining the Rate for Independent Deactivation

Let us illustrate how to interpret experiments from the various batch-solids reactors of Fig. 2 and how to manipulate the basic performance equations for these reactors by testing the fit for the simplest of equation forms for independent deactivation.

$$-r'_A = kC_A\mathbf{a} \qquad \text{with} \qquad \varepsilon_A = 0 \tag{14a}$$

$$-\frac{d\mathbf{a}}{dt} = k_d\mathbf{a} \tag{14b}$$

This represents first-order reaction and first-order deactivation which, in addition, is concentration independent.

Batch-Solids, Batch-Fluid. Here we need to develop an expression relating the changing gas concentration with time. Using time as the one independent variable throughout the run the kinetic expressions of Eq. 14 become

$$-\frac{dC_A}{dt} = \frac{W}{V}\left(-\frac{1}{W}\frac{dN_A}{dt}\right) = \frac{W}{V}(-r'_A) = \frac{kW}{V}C_A\mathbf{a} = k''C_A\mathbf{a} \tag{15}$$

$$-\frac{d\mathbf{a}}{dt} = k_d\mathbf{a} \tag{16}$$

Integrating Eq. 16 yields

$$\mathbf{a} = \mathbf{a}_0 e^{-k_d t}$$

and for unit initial activity, or $\mathbf{a}_0 = 1$, this becomes

$$\mathbf{a} = e^{-k_d t} \tag{17}$$

Replacing Eq. 17 in Eq. 15 we then find

$$-\frac{dC_A}{dt} = k''e^{-k_d t}C_A$$

and on separation and integration

$$\ln\frac{C_{A0}}{C_A} = \frac{k''}{k_d}(1 - e^{-k_d t}) \tag{18}$$

This expression shows that even at infinite time the concentration of reactant in an irreversible reaction does not drop to zero but is governed by the rate of reaction and of deactivation, or

$$\ln \frac{C_{A0}}{C_{A\infty}} = \frac{k''}{k_d} \tag{19}$$

Combining the above two expressions and rearranging then yields

$$\ln \ln \frac{C_A}{C_{A\infty}} = \ln \frac{k''}{k_d} - k_d t \tag{20}$$

A plot as shown in Fig. 3 provides a test for this rate form.

The batch–batch reactor becomes a practical device when the characteristic times for reaction and deactivation are of the same order of magnitude. If they are not and if deactivation is much slower, then $C_{A\infty}$ becomes very low and difficult to measure accurately. Fortunately, this ratio can be controlled by the experimenter by proper choice of W/V.

Batch-Solids, Mixed Constant Flow of Fluid. Inserting the rate of Eq. 14a into the performance expression for mixed flow gives

$$\frac{W}{F_{A0}} = \frac{X_A}{-r'_A} = \frac{X_A}{k\mathbf{a}C_A} \tag{21}$$

and on rearrangement

$$\frac{C_{A0}}{C_A} = 1 + k\mathbf{a}\left(\frac{WC_{A0}}{F_{A0}}\right) \tag{22}$$

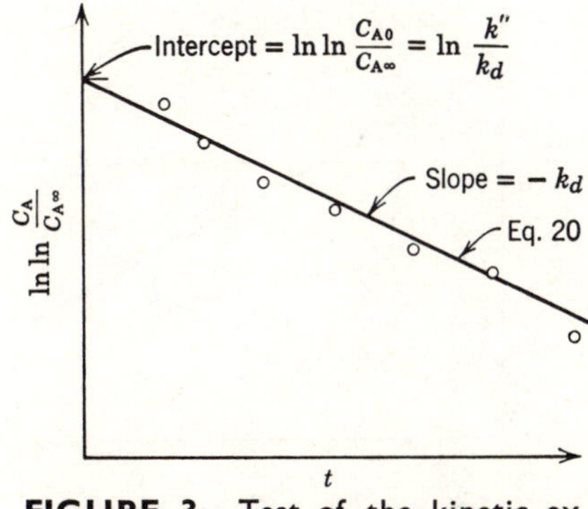

FIGURE 3. Test of the kinetic expressions of Eq. 14 using a batch-solids, batch-fluid reactor.

where the term in brackets is a capacity factor. By analogy to the space-time for homogeneous systems we may call this the *weight-time*. Thus

$$\text{Space-time:} \quad \tau = \frac{C_{A0}V}{F_{A0}}, \qquad [\text{min}]$$

$$\text{Weight-time:} \quad \tau' = \frac{C_{A0}W}{F_{A0}}, \qquad \left[\frac{\text{gm cat} \cdot \text{min}}{\text{liter}}\right] \tag{23}$$

Replacing Eq. 23 in 22 then gives

$$\frac{C_{A0}}{C_A} = 1 + k\mathbf{a}\tau' \tag{24}$$

In this expression the activity varies with the chronological time. To eliminate this quantity integrate Eq. 14*b* (see Eq. 17) and insert in Eq. 24. Thus

$$\frac{C_{A0}}{C_A} = 1 + ke^{-k_d t}\tau' \tag{25}$$

Rearranging gives, in more useful form,

$$\ln\left(\frac{C_{A0}}{C_A} - 1\right) = \ln(k\tau') - k_d t \tag{26}$$

This expression shows how the reactant concentration at the outlet rises with time, while the plot of Fig. 4 provides the test of this kinetic equation. If the data fall on a straight line then the slope and intercept yield the two rate constants of Eq. 14.

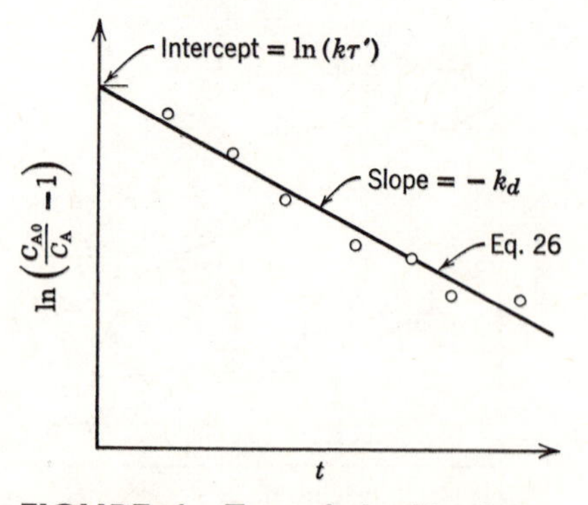

FIGURE 4. Test of the kinetic expression of Eq. 14 using a batch of solids and steady state mixed flow of fluids.

We should mention that this and the succeeding batch-solids derivations are based on the pseudo steady-state assumption. This assumes that conditions change slowly enough with time for the system to be at steady state at any instant. Since a batch of solids can only be used if deactivation is not too rapid this assumption is reasonable.

Batch-Solids, Plug Constant Flow of Fluid. For plug flow the performance equation combined with the rate of Eq. 14*a* becomes

$$\frac{W}{F_{A0}} = \int \frac{dX_A}{-r'_A} = \int \frac{dX_A}{k\mathbf{a}C_A} = \frac{1}{k\mathbf{a}} \int \frac{dX_A}{C_A} \tag{27}$$

Integrating and replacing **a** by the expression of Eq. 17 gives

$$\frac{WC_{A0}}{F_{A0}} = \tau' = \frac{1}{k\mathbf{a}} \ln \frac{C_{A0}}{C_A} = \frac{1}{ke^{-k_d t}} \ln \frac{C_{A0}}{C_A} \tag{28}$$

which becomes, on rearrangement,

$$\ln \ln \frac{C_{A0}}{C_A} = \ln (k\tau') - k_d t \tag{29}$$

Figure 5 shows how to test for the kinetics of Eq. 14, and shows how to evaluate the rate constants with data from this type of reactor.

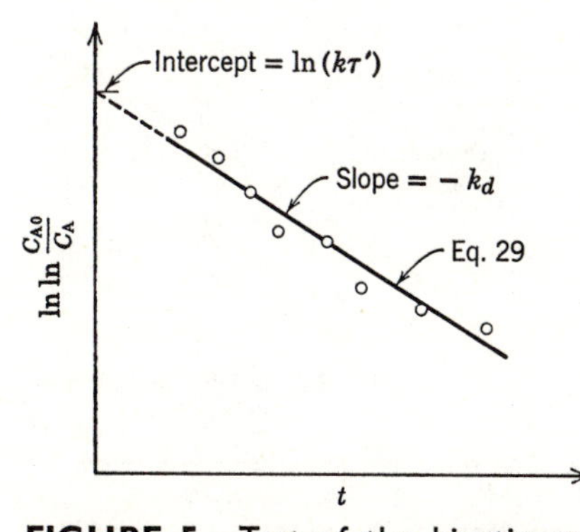

FIGURE 5. Test of the kinetic expressions of Eq. 14 using a batch of solids and steady state plug flow of fluid.

Batch-Solids, Mixed Changing Flow of Fluid (to keep C_A fixed). For steady flow in a mixed reactor we have found

$$\frac{C_{A0}}{C_A} = 1 + ke^{-k_d t}\tau' \tag{25}$$

To keep C_A constant the flow rate must be slowly changed with time. In fact, it must be lowered because the catalyst is deactivating. Hence the variables in this situation are τ' and t. So, on rearranging we have

$$\ln \tau' = k_d t + \ln\left(\frac{C_{A0} - C_A}{kC_A}\right) \tag{30}$$

Figure 6 shows how to test for the kinetic expressions of Eq. 14 by this procedure.

Actually there is no particular advantage in using varying flow over constant flow when testing for the kinetics of Eq. 14 or any other independent deactivation. However, for other deactivation kinetics this reactor system is by far the most useful because it allows us to decouple the three factors C, T, and **a** and study them one at a time.

Batch-Solids, Plug Changing Flow of Fluid (to keep $C_{A,out}$ fixed). At any instant in the plug flow reactor Eq. 28 applies. Thus noting that τ' and t are the two variables we obtain, on suitable rearrangement,

$$\ln \tau' = k_d t + \ln\left(\frac{1}{k}\ln\frac{C_{A0}}{C_A}\right) \tag{31}$$

Figure 6, with one modification (the intercept given by the last term of Eq. 31), shows how to test for the kinetic expressions of Eq. 14 with this device.

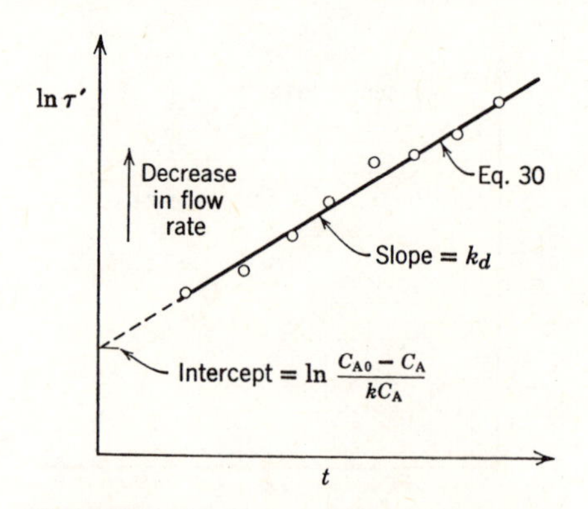

FIGURE 6. Test of the kinetic expressions of Eq. 14 using a batch of solids and changing flow of fluid in a mixed reactor so as to keep C_A constant.

So far we have illustrated how to use a batch, plug flow and mixed flow, of fluid to search for the rate constants of a particular rate form, Eq. 14. The recycle reactor of Fig. 2 can also be used; however, except for high recycle rates where it approaches the behavior of the mixed reactor, it offers no particular advantages.

As long as the deactivation is concentration independent then the activity and concentration effects can be decoupled and any of the above experimental devices will give simple easy-to-interpret results. Thus the above analyses can be extended with no difficulty to any reaction order n and any deactivation order d, as long as $n' = 0$.

On the other hand, if the deactivation is concentration dependent, or $n' \neq 0$, then the concentration and activity effects do not decouple and analysis becomes difficult unless the proper experimental device is used, one deliberately chosen so as to decouple these factors.

We consider this case next.

Batch-Solids: Determining the Rate for Parallel, Series or Side-by-Side Deactivation

To decouple the activity and concentration effects in the rate we must choose a device which allows the deactivation order to be studied without interference of concentration effects. The key then is to keep constant the concentration term in the deactivation equation of Eqs. 9, 10, or 11 while searching for the deactivation order. Later the concentration effect can be studied. As examples of pertinent concentrations:

For parallel deactivation (Eq. 9) keep C_A constant.
For series deactivation (Eq. 10) keep C_R constant.
For side-by-side deactivation (Eq. 11) keep C_P constant.

The mixed flow reactor with controlled and changing flow rate of feed can satisfy this requirement for all of these kinetic forms.

Parallel deactivation. We illustrate the method of analysis with general n, n', and d order kinetics for parallel deactivation. At constant C_A Eq. 9 becomes

$$-r'_A = (kC_A{}^n)\mathbf{a} = k'\mathbf{a} \tag{32}$$

$$-\frac{d\mathbf{a}}{dt} = (k_dC_A{}^{n'})\mathbf{a}^d = k'_d\mathbf{a}^d \tag{33}$$

So for the mixed reactor and the rate of Eq. 32

$$\frac{W}{F_{A0}} = \frac{X_A}{-r'_A} = \frac{X_A}{k'\mathbf{a}} = \frac{C_{A0} - C_A}{C_{A0}k'\mathbf{a}}$$

or

$$\tau' = \frac{C_{A0} - C_A}{k'\mathbf{a}} \tag{34}$$

Integrating Eq. 33 for different orders of deactivation and replacing in Eq. 34 gives on rearrangement:

For *zero-order deactivation*

$$\frac{1}{\tau'} = \frac{k'}{C_{A0} - C_A} - \frac{k'k'_d}{C_{A0} - C_A}t \tag{35}$$

For *first-order deactivation*

$$\ln \tau' = \ln \frac{C_{A0} - C_A}{k'} + k'_d t \tag{36}$$

For *second-order deactivation*

$$\tau' = \frac{C_{A0} - C_A}{k'} + \frac{(C_{A0} - C_A)k'_d}{k'}t \tag{37}$$

For *third-order deactivation*

$$\tau'^2 = \left(\frac{C_{A0} - C_A}{k'}\right)^2 + \left(\frac{C_{A0} - C_A}{k'}\right)^2 2k'_d t \tag{38}$$

For *dth-order deactivation*

$$\tau'^{d-1} = C_1 + C_2 t \tag{39}$$

Figure 7 shows how to test for *d*th-order deactivation. If the data fall on a straight line the guessed order is correct. The slope and intercept of the line then give the constants k' and k'_d.

After the order of deactivation is found then the concentration and temperature dependency can be determined. This is not difficult to do.

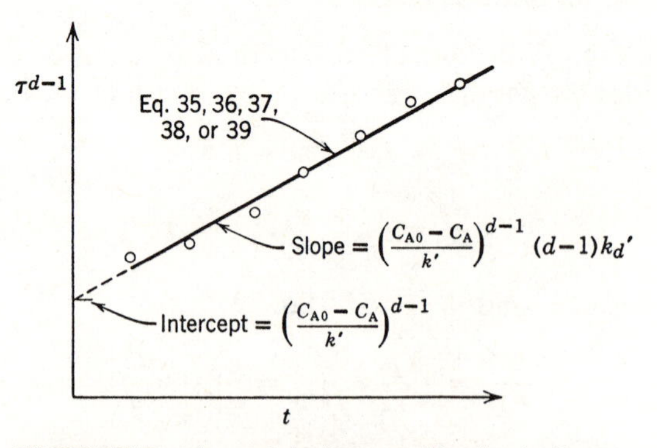

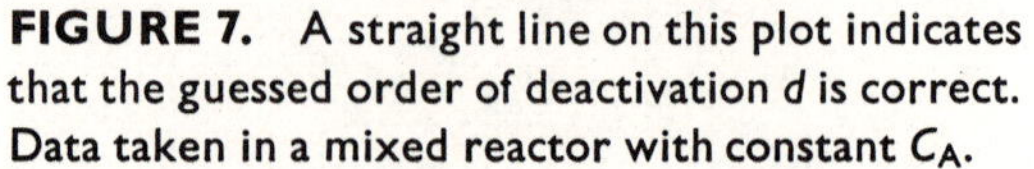

FIGURE 7. A straight line on this plot indicates that the guessed order of deactivation d is correct. Data taken in a mixed reactor with constant C_A.

Series and Side-by-side Deactivation. The above development shows how to treat parallel deactivation. The analysis is quite similar for series deactivation if C_A, hence C_R, is kept constant, and for side-by-side deactivation if C_P and if possible C_A are kept constant. In the latter case if C_A cannot be kept constant the analysis is still not particularly difficult.

Flowing Solids Experimental Reactors

For fast deactivation a reactor with throughflow of both gas and solids must be used. Under steady-state conditions and with flow pattern known (this is usually the difficulty with this type of setup) integration and testing of some of the simpler kinetic forms can be done without much difficulty. It should be noted that the flowing solid can be looked upon as containing a reactant (the activity) which is consumed or depleted in the reactor. Thus we treat this as a heterogeneous gas–solid reacting system with solid as a macrofluid and gas as a microfluid.

The examples and problems at the end of this chapter give practice in treating some of the many combinations of rate equations and reactor types, both batch-solids and flow-solids.

Finding the Mechanism of Decay from Experiment

The main source of information for the decay mechanism comes from the chemistry of the reaction, knowing what are the impurities in the feed, observing the deposition of carbon, etc. Additional clues are obtained from:

1. Physical examination of sliced pellets: this may show the distribution of poison.
2. Experiments with different sized pellets: a change in initial rate indicates pore diffusional resistance for reactant, or
3. Kinetic experiments: of particular importance here is the order of deactivation.

Briefly consider this third point. For parallel deactivation the order of deactivation is a unique function of the Thiele modulus of the main reaction, and as shown in Fig. 8 the order shifts from 1 to 3 as pore resistance becomes increasingly strong. Thus when

$$(mL)_A < 1 \cdots d = 1$$

$$(mL)_A \gg 1 \cdots d \to 3$$

From another point of view, when the Thiele modulus becomes large the shell model is approached (see Fig. 1), and for this extreme, theoretical considerations show that the deactivation order becomes 3.

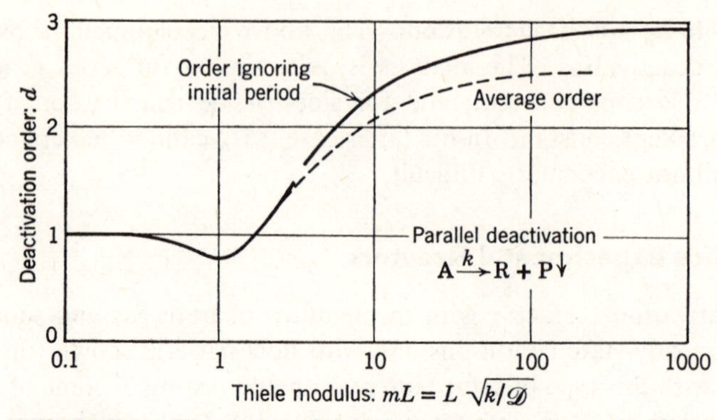

FIGURE 8. The order of deactivation depends on the Thiele modulus for the reaction; adapted from Khang (1971).

For side-by-side deactivation the Thiele modulus for the poisoning reaction is of primary importance, thus when

$$(mL)_{\text{deact}} < 1 \cdots d = 1$$

$$(mL)_{\text{deact}} > 1 \cdots d \to 3 \quad \text{whenever} \quad (mL)_{\text{deact}} \lessgtr (mL)_{\text{react}}$$

For series deactivation the order of deactivation, for all practical purposes, is insensitive to and unaffected by pore diffusional effects. Thus

$$d \cong 1 \qquad \text{for all } (mL)$$

Khang (1971) qualitatively treats these cases.

As a word of warning, this presentation is greatly simplified by the assumptions:

1. Only one mechanism at work.
2. No nonisothermal effects.
3. No diffusion through the deposited poison. If there is such an effect then even with uniform deposition the deactivation order becomes 3 instead of 1.
4. Homogeneous site attack. This assumption is not likely to be reasonable for poisoning by chemisorption. Preferential site attack should give a higher deactivation order than shown above.

Although we may be unsure of the mechanism of decay, firm kinetic equations for decay and reaction are the most important measures needed for scale-up and design.

DESIGN

When reacting fluid flows through a batch of deactivating catalyst the conversion drops progressively during the run, and steady state conditions cannot be maintained. If conditions change slowly with time, then the average conversion during a run can be found by calculating the steady-state conversion at various times and summing over time. In symbols, then,

$$\bar{X}_A = \frac{\int_0^{t_{run}} X_A(t)\, dt}{t_{run}} \tag{40}$$

When conversion drops too low the run is terminated, the catalyst is either discarded or regenerated, and the cycle is repeated. Example 2 illustrates this type of calculation.

There are two important and real problems with deactivating catalysts.

The operational problem: how to best operate a reactor during a run. Since temperature is the most important variable affecting reaction and deactivation this problem reduces to finding the best temperature progression during the run.

The regeneration problem: when to stop a run and either discard or regenerate the catalyst. This problem is easy to treat once the first problem has been solved for a range of run times and final catalyst activities. (*Note:* each pair of values for time and final activity yields the corresponding mean conversion.)

The operational problem has been solved analytically for a rather general family of kinetic equations:

$$-r'_A = kC_A{}^n\mathbf{a} = (k_0 e^{-E/RT})C_A{}^n\mathbf{a} \tag{41}$$

$$-\frac{d\mathbf{a}}{dt} = k_d\mathbf{a}^d = (k_{d0}e^{-E_d/RT})\mathbf{a}^d \tag{42}$$

Note the restriction here, that deactivation is concentration independent.

For a batch of solids and either plug or mixed flow of fluid the analysis leads to two types of temperature policies, either a rising temperature with time, or the maximum allowable temperature. The policy to use depends somewhat on the reaction order and conversion level, but primarily on the activation energies of the two processes, reaction and deactivation. Thus if deactivation is very temperature sensitive compared to reaction, or specifically if

$$E_d \geqslant E\left\{1 - \frac{n}{n-1}[1 - (1 - X_{A,out})^{n-1}]\right\} \qquad \text{for plug flow} \tag{43}$$

or if

$$E_d \geqslant E\left\{\frac{1 - 2X_A - (n-1)X_A{}^2}{[1 + (n-1)X_A]^2}\right\} \qquad \text{for mixed flow} \tag{44}$$

then the optimal policy is a rising temperature with time *so as to keep the conversion from the reactor unchanged.* This operational criterion can be written as

$$\left.\begin{aligned} X_{A,out} &= \text{constant during the run} \\ \text{or} \qquad k\mathbf{a} = (k_0 e^{-E/RT})\mathbf{a} &= \text{constant during the run} \end{aligned}\right\} \tag{45}$$

Thus as the activity drops we must raise the temperature, hence k, to just counter this drop.

Here is the procedure for operating at a particular conversion level in a flow reactor with a given feed rate. Pick T_{initial} so as to obtain $X_{A,\text{out}}$ and then continually boost the temperature so as to maintain this conversion level. When $X_{A,\text{out}}$ can no longer be maintained because the catalyst activity is exhausted, shut down and regenerate. If this run time is too short or long adjust the conversion level or amount of catalyst accordingly. In any case the conversion level should be kept constant during any run.

When deactivation is only slightly temperature sensitive compared with reaction (when the conditions of Eqs. 43 and 44 do not hold) then the optimal policy requires operating at the highest allowable temperature either part of the time or all of the time.

One point to mention about Eqs. 43 and 44: when the conversion level is high enough then the constant conversion policy is always optimal, for any and all values or the activation energies.

These conclusions and additional quantitative results are given by Szepe (1966).

Consider systems where deactivation does not follow Eq. 42, in particular where deactivation is concentration dependent,

$$-\frac{d\mathbf{a}}{dt} = k_d C_i^{n'} \mathbf{a}^d \tag{46}$$

Here different regions in the plug flow reactor will deactivate at different rates, hence for optimal operations each point in the reactor should have its own temperature history. This situation is difficult to treat.

One simplification is possible, however, where the solids are uniformly mixed such as in a batch-solids fluidized bed reactor, or a fluidized-packed bed reactor. Since each particle moves about the bed it may be viewed to contact gas of some average composition. In such a case Eq. 46 simplifies to

$$-\frac{d\mathbf{a}}{dt} = (\overline{k_d C_i^{n'}})\mathbf{a}^d = k_d' \mathbf{a}^d \tag{47}$$

and the optimal progression should be similar to that for a reaction following the kinetics of Eqs. 41 and 42.

There are numerous variations and extensions to these problems, such as to find the actual temperature progression with time, to maximize operations while staying within certain temperature bounds, to maximize product distribution, to operate a batch-solids batch-fluid reactor optimally, etc. Analytical solutions to these problems for certain restricted kinetics are reported by Szepe (1966), Chou *et al.* (1967), Ogunye and Ray (1968), and Lee and Crowe (1970). The more general problem of arbitrary kinetics is difficult to solve and requires a numerical search technique to find solutions which are close to optimal; see Ogunye and Ray (1971).

EXAMPLE 1. Kinetics from a changing flow basket reactor

The kinetics of a particular catalytic reaction $A \rightarrow R$ are studied at temperature T in a basket reactor (batch-solids and mixed flow of gas) in which the gas composition is kept unchanged, despite deactivation of the catalyst. What can you say about the rates of reaction and deactivation from the results of the following three-hour runs? Note, to keep the gas concentration in the reactor unchanged the flow rate of reactant had to be lowered to about 5% the initial value.

Data

Run 1

$C_{A0} = 1$ mol/liter
$X_A = 0.5$
$\pi = 1$ atm

t, time from start of run, hr	0	1	2	3
τ', gm cat·min/liter	1	e	e^2	e^3

Run 2

$C_{A0} = 2$
$X_A = 0.5$
$\pi = 2$

t, hr	0	1	2	3
τ', gm cat·min/liter	1	e	e^2	e^3

No impurities are present in the feed and the catalyst is not affected by temperature, hence deactivation is a result of either a series or a parallel decay mechanism, but not side-by-side or independent decay.

SOLUTION

Start by finding the order of deactivation from the progress of a single run. Then comparing runs at different concentration levels find the concentration dependency of the deactivation and of the reaction. The temperature dependency (activation energies) cannot be determined since all data were taken at one temperature. Finally the general rate form to which we hope to fit the data is

$$-r'_A = kC_A{}^n\mathbf{a}$$

$$-\frac{d\mathbf{a}}{dt} = k_d C_i{}^{n'}\mathbf{a}^d, \qquad i = \text{A or R} \tag{9) or (10}$$

For the reactor set up used here Eqs. 35 to 39 represent the progress of reaction for various orders of deactivation. Noting that

$$\ln 1 = 0, \qquad \ln e = 1, \qquad \ln e^2 = 2, \qquad \ln e^3 = 3$$

and from the data that

$$\ln \tau' \propto t$$

we see immediately from run 1 (or run 2) that first-order deactivation (Eq. 36) fits the data. Thus the rate form reduces to

$$-r'_A = kC_A{}^n\mathbf{a}$$

$$-\frac{d\mathbf{a}}{dt} = k_d C_A{}^{n'}\mathbf{a}$$

Next, compare the results obtained at the two concentration levels of A and R. Since the progress of the deactivation is identical the deactivation is independent of any concentration effect, and since the fractional conversion of reactant is concentration independent the reaction is of first order. Thus the rate form we are left with is that of Eq. 14, or

$$-r'_A = kC_A\mathbf{a}$$
$$-\frac{d\mathbf{a}}{dt} = k_d\mathbf{a} \tag{14}$$

On a ln τ' versus t plot such as Fig. 6 we find for the data

$$\text{slope} = 1 = k_d$$

$$\text{intercept} = 0 = \ln\left(\frac{C_{A0} - C_A}{kC_A}\right)$$

Evaluating k_d and k then gives, finally

$$\left.\begin{aligned} -r'_A &= C_A\mathbf{a}, \ [\text{mol/liter}\cdot\text{min}] \\ -\frac{d\mathbf{a}}{dt} &= a, \ [\text{hr}^{-1}] \end{aligned}\right\}$$

We could have reached the solution directly by guessing the simplest rate form, which happens to be Eq. 14, and verifying that it fits the data.

EXAMPLE 2. Reactor performance with a batch of decaying catalyst

At 730°K the isomerization of A to R (rearrangement of atoms in the molecule) proceeds on a slowly deactivating catalyst with a second order rate

$$-r'_A = kC_A{}^2\mathbf{a} = 200C_A{}^2\mathbf{a}, \quad [\text{mol A/hr}\cdot\text{gm cat}]$$

Since reactant and product are similar in structure deactivation is caused by both A and R. With diffusional effects absent, the rate of deactivation is found to be

$$-\frac{d\mathbf{a}}{dt} = k_d(C_A + C_R)\mathbf{a} = 10(C_A + C_R)\mathbf{a}, \quad [\text{day}^{-1}]$$

We plan to operate a packed bed reactor containing $W = 1$ metric ton of catalyst for 12 days using a steady feed of pure A, $F_{A0} = 5$ kmol/hr at 740°K and 3 atm ($C_{A0} =$ 0.05 mol/liter).

(*a*) What is the conversion at the start of the run?
(*b*) What is the conversion at the end of the run?
(*c*) What is the average conversion over the 12-day run?

SOLUTION

First consider the activity of the catalyst. Since $C_A + C_R = C_{A0} = 0.05$ mol/liter a all times the rate equation for deactivation becomes

$$-\frac{d\mathbf{a}}{dt} = \left(10\frac{\text{liters}}{\text{mol}\cdot\text{day}}\right)\left(0.05\frac{\text{mol}}{\text{liter}}\right)\mathbf{a} = \left(0.5\frac{1}{\text{day}}\right)\mathbf{a}$$

Integrating gives with t in days

$$\text{At any time } t: \qquad \mathbf{a} = e^{-0.5t} \tag{i}$$

$$\text{Initially, at } t = 0: \qquad \mathbf{a} = 1 \tag{ii}$$

$$\text{After 12 days:} \qquad \mathbf{a} = e^{-6} = \frac{1}{400} \tag{iii}$$

Next consider isomerization in the packed bed. The weight-time is

$$\tau' = \frac{WC_{A0}}{F_{A0}} = \frac{(10^6 \text{ gm})(0.05 \text{ mol/liter})}{(5 \times 10^3 \text{ mol/hr})} = 10 \frac{\text{gm cat} \cdot \text{hr}}{\text{liter fluid}}$$

Noting that deactivation is slow (steady state assumption can be made), the performance equation at any instant, by analogy to Eq. 5.19, is

$$\tau' = -\int_{C_{A0}}^{C_A} \frac{dC_A}{-r'_A} = -\int_{C_{A0}}^{C_A} \frac{dC_A}{k\mathbf{a}C_A{}^2} = \frac{-1}{k\mathbf{a}} \int_{C_{A0}}^{C_A} \frac{dC_A}{C_A{}^2}$$

Integrating and rearranging gives

$$1 - X_A = \frac{C_A}{C_{A0}} = \frac{1}{1 + k\mathbf{a}C_{A0}\tau'} \tag{iv}$$

(*a*) To find the conversion at the start of the run replace Eq. (ii) in (iv). Thus in consistent units

$$\frac{C_A}{C_{A0}} = \frac{1}{1 + (200)(1)(0.05)(10)} = \frac{1}{1 + 100}$$

or

$$X_{A,\text{initial}} = 99\%$$

(*b*) For the conversion at the end of the 12-day run replace Eq. (iii) in (iv). Thus

$$\frac{C_A}{C_{A0}} = \frac{1}{1 + (200)(1/400)(0.05)(10)} = \frac{4}{5}$$

or

$$X_{A,\text{after 12 days}} = 20\%$$

(*c*) The time average outlet composition (see Eq. 46) is given by

$$\overline{\left(\frac{C_A}{C_{A0}}\right)} = \frac{1}{t_{\text{run}}} \int_0^{t_{\text{run}}} \left(\frac{C_A}{C_{A0}}\right) dt \tag{v}$$

Replacing Eq. (i) and (iv) in (v) gives

$$\overline{\left(\frac{C_A}{C_{A0}}\right)} = \frac{1}{t_{\text{run}}} \int_0^{12} \frac{dt}{1 + k(e^{-k_d(C_A + C_R)t})C_{A0}\tau'}$$

$$= \frac{1}{12} \int_0^{12} \frac{dt}{1 + (200)(e^{-0.5t})(0.05)(10)} = \frac{1}{12} \int_0^{12} \frac{dt}{1 + 100e^{-0.5t}}$$

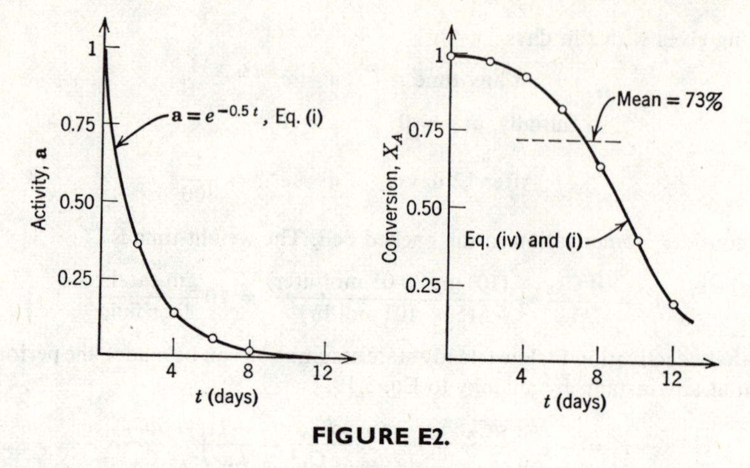

FIGURE E2.

A table of integrals gives for this integral

$$\left(\overline{\frac{C_A}{C_{A0}}}\right) = \frac{1}{12}\left[t + 2\ln\left(1 + 100e^{-0.5t}\right)\right]\Bigg|_0^{12}$$

and on evaluation we find

$$\left(\overline{\frac{C_A}{C_{A0}}}\right) = 0.27 \qquad \text{or} \qquad \bar{X}_A = 73\%$$

Figure E2 shows the activity and conversion as functions of time as calculated by these equations.

EXAMPLE 3. Reactor with fresh flowing catalyst

Consider a reactor with mixed flow of catalyst and either plug flow or mixed flow of gas. Gas reacts, catalyst deactivates, and the kinetics of these processes are represented by

$$A \rightarrow R, \qquad -r'_A = kC_A\mathbf{a} \tag{i}$$

$$-\frac{d\mathbf{a}}{dt} = k_d\mathbf{a} \tag{ii}$$

Find the fractional conversion of gas and the activity of the leaving catalyst stream in terms of the variables of the system

(*a*) for plug flow of gas

(*b*) for mixed flow of gas.

Figure E3 shows the pertinent variables of this steady state system.

SOLUTION

To solve a problem of this type first write an expression for the activity of the catalyst, then write the performance expression for the gas, then combine. In this problem we obtain simple explicit expressions. In most cases this is not the case.

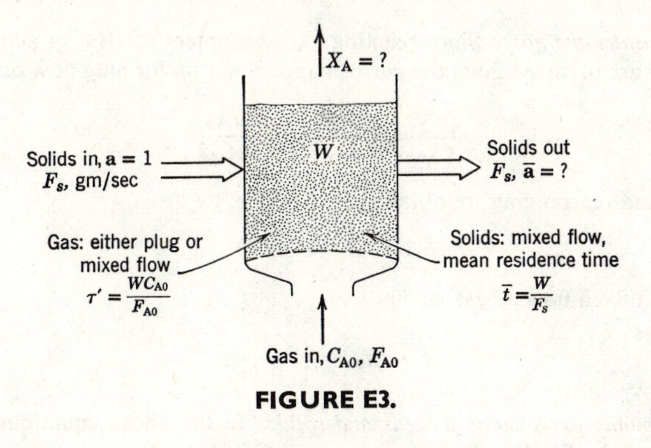

FIGURE E3.

Step 1. *Consider the solids.* Let us look at the activity as a reactant which is consumed. Then for the solid stream as a whole (treated as a macrofluid) the exit activity is given in general by

$$\bar{\mathbf{a}} = \int_0^\infty \mathbf{a} \cdot \mathbf{E}_s \, dt \qquad \text{(iii)}$$

For mixed flow of solids the exit age distribution is

$$\mathbf{E}_s = \frac{1}{\bar{t}} e^{-t/\bar{t}}$$

and for any particular particle integration of Eq. (ii) gives

$$\mathbf{a} = e^{-k_d t}$$

Introducing the above two expressions into Eq. (iii) gives

$$\bar{\mathbf{a}} = \int_0^\infty e^{-k_d t} \cdot \frac{1}{\bar{t}} e^{-t/\bar{t}} \, dt$$

Integrating then yields the expression for the activity of the solids in the reactor and in the exit stream, thus

$$\bar{\mathbf{a}} = \frac{1}{1 + k_d \bar{t}} = \frac{1}{1 + k_d W/F_s} \qquad \text{(iv)}$$

Note, since the deactivation is of first order the macrofluid could be treated as a microfluid (see Chapter 10). Thus for mixed flow the performance expression Eq. 5.13 becomes

$$\bar{t} = \frac{1 - \bar{\mathbf{a}}}{(\text{rate of deactivation})} = \frac{1 - \bar{\mathbf{a}}}{k_d \bar{\mathbf{a}}}$$

which is identical to Eq. (iv).

Step 2. Consider the gas. Since reacting gas encounters catalyst of activity $\bar{\mathbf{a}}$ everywhere (solids are in mixed flow) the performance equation for plug flow of gas becomes

$$\frac{\tau'}{C_{A0}} = \int \frac{dX_A}{-r'_A} = \int \frac{dX_A}{kC_A\bar{\mathbf{a}}}$$

Integrating and rearranging we obtain

$$1 - X_A = e^{-k\tau'\bar{\mathbf{a}}} \tag{v}$$

Similarly for mixed flow of gas we find

$$1 - X_A = \frac{1}{1 + k\tau'\bar{\mathbf{a}}} \tag{vi}$$

Step 3. Combine expressions for gas and solid. In the above equations for the gas phase $\bar{\mathbf{a}}$ is given by Eq. (iv), thus we find on combining

---

$$1 - X_{A,\text{plug}} = e^{-k\tau'/(1+k_d\bar{t})} = e^{-R/(1+R_d)}$$

$$1 - X_{A,\text{mixed}} = \frac{1}{1 + k\tau'/(1 + k_d\bar{t})} = \frac{1 + R_d}{1 + R_d + R}$$

where

$$R = k\tau' \quad \text{and} \quad R_d = k_d\bar{t}$$

---

are first-order rate groups. We conclude that the conversion of gas is dependent on the first-order rate groups for reaction and deactivation.

EXAMPLE 4. Reactor-regenerator system with deactivating catalyst

Consider a system consisting of reactor and regenerator with a steady state circulation of solids among the units. Reactant gas flows through the reactor, forms product, and at the same time causes rapid catalyst deactivation. The spent catalyst flows into the regenerator where its activity is partially restored on contact with regenerator gas, and is then returned to the reactor.

Assume that solids are in mixed flow in the two units and that the kinetics of the various processes are as follows:

For reacting gas: $\quad A \rightarrow R, \quad -r'_A = kC_A\mathbf{a}$ (i)

For deactivation: $$-\frac{d\mathbf{a}}{dt} = k_d\mathbf{a} \tag{ii}$$

For regeneration: $$+\frac{d\mathbf{a}}{dt} = k_r(1 - \mathbf{a}) \tag{iii}$$

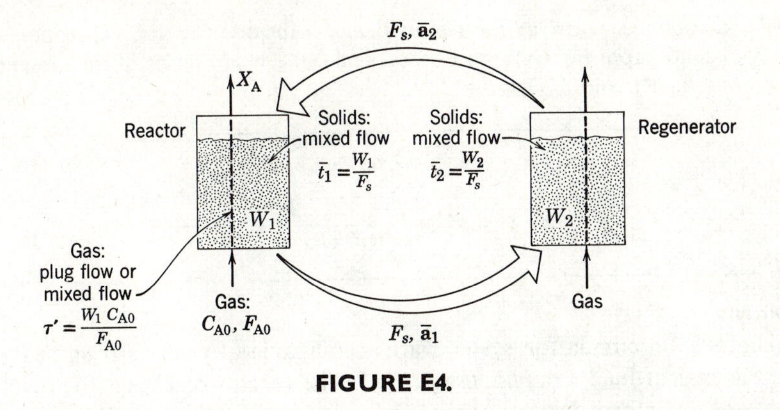

FIGURE E4.

Find the fractional conversion of reactant gas and the activity of the catalyst streams in terms of the variables of the system

(*a*) for plug flow of gas in the two units.

(*b*) for mixed flow of gas in the two units.

Figure E4 shows the pertinent variables of this steady state system.

SOLUTION

As in Example 3 write expressions for the activity of the catalyst, for conversion of gas, and then combine.

Step 1. *Consider the solids.* Treating the solids as a microfluid (allowed since deactivation and regeneration are both linear processes, see Example 3 or Chapter 10) and applying the expressions for mixed flow, Eq. 5.13, we find

For reactor:
$$\bar{t}_1 = \frac{\bar{a}_1 - \bar{a}_2}{k_d \bar{a}_1}$$

For regenerator:
$$\bar{t}_2 = \frac{\bar{a}_1 - \bar{a}_2}{k_r(1 - \bar{a}_2)}$$

Combining and rearranging we find the activity in the reactor to be

$$\bar{a}_1 = \frac{R_r}{R_d + R_d R_r + R_r} \qquad \text{where} \begin{cases} R_d = k_d \bar{t}_1 \\ R_r = k_r \bar{t}_2 \end{cases} \tag{iv}$$

Step 2. *Consider the gas.* As in Example 3 apply the performance equation for the reactant gas. This gives

For plug flow:
$$1 - X_{A,\text{plug}} = e^{-k\tau' \bar{a}_1} \tag{v}$$

For mixed flow:
$$1 - X_{A,\text{mixed}} = \frac{1}{1 + k\tau' \bar{a}_1} \tag{vi}$$

Step 3. Combine expressions for gas and solid. Introducing the expression for the activity of solids from Eq. (iv) into Eqs. (v) and (vi) gives in terms of the dimensionless rate groups R_r, R_d, and $R = k\tau'$

$$1 - X_{A,\text{plug}} = e^{-RR_r/(R_d + R_d R_r + R_r)}$$

$$1 - X_{A,\text{mixed}} = \frac{1}{1 + RR_r/(R_d + R_d R_r + R_r)}$$

Comment

Frequently solid circulation systems are accompanied by very strong heat effects. Take the case of fluid catalytic cracking. In the reactor cracking (the reaction by which long chain hydrocarbons are transformed into molecules of lower molecular weight) is highly endothermic, carbon deposits on the surfaces and the particles cool. These particles flow into the regenerator where carbon is burned off, activity is restored and the pellets heated. They then return to the reactor. The circulation of solids, then, is thus seen as the means for supplying heat to the reactor.

As may be suspected in these situations thermal considerations often control the overall kinetics. Such systems are more fully treated by Kunii and Levenspiel (1969).

REFERENCES

Chou, A., Ray, W. H., and Aris, R., *Trans. I. Ch.E.*, **45**, T153 (1967).

Khang, S. J., M.S. Thesis, Oregon State University, 1971.

Kunii, D., and Levenspiel, O., *Fluidization Engineering*, John Wiley & Sons, New York, 1969.

Lee, S., and Crowe, C. M., *Can. J. Chem. Eng.*, **48**, 192 (1970).

Levenspiel, O., *J. Catal.*, **25**, 265 (1972).

Ogunye A. F. and Ray, W. H., *Trans. I. Ch.E.*, **46**, T225 (1968).

———, *A.I.Ch.E. Journal*, **17**, 43 (1971).

Szepe, S., Ph.D. Thesis, Illinois Institute of Technology, 1966; also see Szepe, S., and Levenspiel, O., *Chem. Eng. Sci.*, **23**, 881 (1968); "Catalyst Deactivation" Fourth European Symposium on Chemical Reaction Engineering, Brussels, Sept. 1968, Pergamon, London, 1971.

PROBLEMS

The problems are grouped as follows:

Problems 1–9: Batch of solids, or slow deactivation
Problems 10–18: Flowing solids, or fast deactivation
Problem 19: Different.

1. A recycle reactor with very high recycle ratio is used to study the kinetics of a particular irreversible catalytic reaction, A → R. For a constant flow rate of feed ($\tau' = 2\,\text{kg}\cdot\text{sec/lit}$) the following data are obtained:

Time after start of operation, hr	1	2	4
X_A	0.889	0.865	0.804

The progressive drop in conversion suggests that the catalyst deactivates with use. Find rate equations for the reaction and for the deactivation which fit these data.

2. The following two kinetic runs are made in a basket reactor operated as in Example 1. What information does this give you of the rate equation for the reaction and for the deactivation?

Run 1

$C_{A0} = 1$ mol/liter	t, hr	0	1	2	3
$X_A = 0.5$	τ', gm·min/liter	1	e	e^2	e^3

Run 2

$C_{A0} = 2$ mol/liter	t, hr	0	1	1	3
$X_A = 0.667$	τ', gm·min/liter	2	$2e$	$2e^2$	$2e^3$

3. The following data on an irreversible reaction are obtained with decaying catalyst in a batch reactor (batch-solids, batch-fluid). What can you say about the kinetics?

C_A	1.000	0.802	0.675	0.532	0.422	0.368
t, hr	0	0.25	0.5	1	2	∞

4. The basket reactor operated as in Example 1 gives the following data. What does this tell of the rates of deactivation and of reaction?

Run 1

$C_{A0} = 1$ mol/liter	t, hr	0	1	2	3
$X_A = 0.5$	τ', gm·min/liter	1	e	e^2	e^3

Run 2

$C_{A0} = 3$ mol/liter	t, hr	0	0.5	1	1.5
$X_A = 0.667$	τ', gm·min/liter	2	$2e$	$2e^2$	$2e^3$

5. The reversible catalytic reaction

$$A \rightleftarrows R, \qquad X_{Ae} = 0.5$$

proceeds with decaying catalyst in a batch reactor (batch-solids, batch-fluid). What can you say of the kinetics of reaction and deactivation from the following data:

t, hr	0	0.25	0.5	1	2	(∞)
C_A, mol/liter	1.000	0.901	0.830	0.766	0.711	0.684

6. What can you say of the kinetics of reaction and deactivation from the following kinetic runs taken in a basket reactor operated as in Example 1:

Run 1

$C_{A0} = 2$ mol/liter	t, hr	0	1	2	3
$X_A = 0.5$	τ', gm·min/liter	1	e	e^2	e^3

Run 2

$C_{A0} = 20$ mol/liter	t, hr	0	0.5	1	1.5
$X_A = 0.8$	τ', gm·min/liter	1	e	e^2	e^3

7. With fresh catalyst the packed bed reactor is run at 600°K. Four weeks later when the temperature reaches 800°K the reactor is shut down to reactivate the catalyst. In addition, at any instant the reactor is isothermal. Assuming optimal operations what is the activity of the catalyst at the time of reactivation.
Data: The rate of reaction with fresh catalyst is

$$-r_A = kC_A^2, \qquad k = k_0 e^{-7200/T}$$

The rate of deactivation is unknown.

8. Repeat Example 2 with one modification as follows: the rate of deactivation is given by

$$-\frac{d\mathbf{a}}{dt} = 200(C_A + C_R)\mathbf{a}^2, \ [1/\text{hr}]$$

9. Under conditions of strong pore diffusion the reaction A → R proceeds at 700°C on a slowly deactivating catalyst by a first-order rate

$$-r'_A = 0.03C_A\mathbf{a}, \ [\text{mol/gm}\cdot\text{min}]$$

Deactivation is caused by strong absorption of unavoidable and irremovable trace impurities in the feed, giving third-order deactivation kinetics, or

$$-\frac{d\mathbf{a}}{dt} = 3\mathbf{a}^3, \ [1/\text{day}]$$

We plan to feed a packed bed reactor ($W = 10$ kg) with $v = 100$ liters/min of fresh A at 8 atm and 700°C until the catalyst activity drops to 10% of the fresh catalyst, then regenerate the catalyst and repeat the cycle.

(*a*) What is the run time for this operation?

(*b*) What is the mean conversion for the run?

Note the difference in shape of the conversion-time curve with that of Example 2, caused simply by a difference in the order of deactivation.

10. A catalyst with an activity half-life of 1 sec is to be used in a solid circulation cracker-regenerator system treating 960 tons/day of feed oil. Conversion of oil feed is satisfactory as long as the mean activity of catalyst in the reactor is no lower than 1% that of fresh catalyst. If the fluidized reactor contains 50 tons of catalyst which is completely regenerated find the necessary circulation rate of solids between reactor and regenerator to maintain this catalyst activity in the bed.

11. We are planning an FCC (fluid catalytic cracking) system to crack petroleum vapors and obtain useful reaction products. This system consists of two fluidized beds (reactor and regenerator) with fine catalyst in a suspended state circulating continuously from unit to unit. Oil vapor is fed to the reactor where it contacts the catalyst and cracks. Carbon is deposited on the catalyst during reaction and the catalyst thereby deactivates. In the regenerator the carbon deposit is burned off and the activity of the catalyst is completely restored. At present the conversion of feed oil A is $X_A = 0.50$ and the average activity of the catalyst in the reactor is $\bar{\mathbf{a}} = 0.01$. Let us see how the following changes in operating conditions affect the conversion:

(*a*) Double the feed rate of reactant gas.
(*b*) Double the circulation rate of catalyst.
(*c*) Double the reactor size.

Note: Any factor (feed rate, reactor size, circulation rate) not mentioned is kept unchanged. Also, assume that gas is in plug flow, solids in mixed flow, and that the kinetics can be described as follows:

$$\text{For reaction:} \qquad -r'_A = kC_A\mathbf{a}$$

$$\text{For deactivation:} \qquad -\frac{d\mathbf{a}}{dt} = k_d\mathbf{a}$$

12. Repeat the previous problem with one modification: assume that the gas is in mixed flow, not plug flow.

13. Consider the reactor-regenerator system and the kinetics of Example 4. For a given treatment rate and conversion level of gas, and for a given circulation rate for the solids there must be some size ratio for reactor and regenerator which will minimize the inventory of solids, hence total volume of the system. Find this size ratio.

14. In Example 4 if the rate of deactivation were concentration-dependent, as follows

$$-\frac{d\mathbf{a}}{dt} = k_dC_A\mathbf{a}$$

and all the rest remained unchanged how would this change the final expressions for gas phase conversion.

15. Flowing solids are used primarily when decay is rapid, and if this is fast enough, deactivation proceeds according to the shell model—with third-order or close to third-order deactivation, not first-order which we have assumed in all problems so far. Let us treat the more realistic cases here. Develop equations for the conversion of gas and activity of leaving catalyst for the setup of Example 3 and the following kinetics:

$$\text{For reaction:} \quad A \rightarrow R, \quad -r'_A = kC_A\mathbf{a}$$

$$\text{For deactivation:} \quad -\frac{d\mathbf{a}}{dt} = k_d\mathbf{a}^3$$

Show how to solve these expressions to obtain the kinetic constants.

16. Repeat the previous problem with one change, the kinetic expression for deactivation is given by

$$-\frac{d\mathbf{a}}{dt} = k_d C_A \mathbf{a}^3$$

17. A first-order gas phase reaction is taking place in the reactor of a reactor-regenerator system with circulating catalyst as in Fig. E4. In this system we have:

Catalyst entering reactor: $\bar{\mathbf{a}}_2 = \frac{2}{3}$

Catalyst leaving reactor: $\bar{\mathbf{a}}_1 = \frac{1}{3}$

Conversion of gaseous A: 63.2%

Without changing the circulation rate and flow pattern of gas (assume it to be plug flow) we could introduce vertical baffles in the bed so that the solids approximate plug flow instead of mixed flow in the two units. How would this affect the conversion of gas?

Data: Assume the following rate expressions:

$$\text{For deactivation:} \quad -\frac{d\mathbf{a}}{dt} = k_d\mathbf{a}$$

$$\text{For regeneration:} \quad -\frac{d\mathbf{a}}{dt} = k_r(1 - \mathbf{a})$$

18. Repeat the previous problem with the modification that gas is in mixed flow in both units.

19. Consider the chemical basis of muscle activity and fatigue as follows: When a muscle is rested and fresh the rate at which it can do work (for example, pedal an experimental bicycle) is W_0. Exertion, however, causes fatigue. In chemical terms exertion produces lactic acid in the muscle cells and this lowers the rate at which the muscle is able to do work. This acid is then destroyed in the cells by oxidation to CO_2 and H_2O.

Develop a simple kinetic model which you think may be a fair first approximation to this process. Then show how to evaluate by experiment the pertinent parameters of this model.

Note: This process is part of the Krebs cycle.

NAME INDEX

SUBJECT INDEX